T0182188

ON THE ROAD
TO WORLDWIDE SCIENCE

ON THE ROAD TO WORLDWIDE SCIENCE

CONTRIBUTIONS TO SCIENCE DEVELOPMENT

MICHAEL J. MORAVCSIK

World Scientific
Singapore • New Jersey • Hong Kong

Published by

World Scientific Publishing Co. Pte. Ltd.
P O Box 128, Farrer Road, Singapore 9128

USA office: World Scientific Publishing Co., Inc.
687 Hartwell Street, Teaneck, NJ 07666, USA

UK office: World Scientific Publishing Co. Pte. Ltd.
73 Lynton Mead, Totteridge, London N20 8DH, England

ISBN 9971-50-617-3
     9971-50-620-3 (pbk)

Printed in Singapore by General Printing Services Pte. Ltd.

PREFACE

The vagaries of history have cleft the world in twain. With many shades and nuances, there are developed countries (DCs), where the goods of this earth are distributed in reasonably generous and equal amounts, and the underdeveloped, or, to use an euphemism, less developed countries (LDCs), who suffer from many lacks, are heavily dependent and economically wanting. The fact that DCs possess a strong scientific and technological infrastructure and employ science and technology for their progress, and LDCs do not, or do so to a feeble extent, is certainly no random coincidence, since science, aided and abetted by social, economic and political forces, has been the most powerful instrument in catalyzing the process of development. The latter, incidentally, should include not only economic, but also social and cultural elements.

It is no wonder, then, that LDCs have wanted for some decades to "jump on the bandwagon" and to harness their progress onto science and, of course, technology. The results have not been terribly encouraging, and it is therefore opportune to analyze what are the elements in the subtle alchemy of the growth of science and technology which are lacking in some countries and abound in others. A corollary of such diagnosis is to establish a treatment for the disease. Michael Moravcsik undertakes this analysis and outlines therapy in the thirty-odd essays published here on the subject covering a twenty-five year span.

Mike Moravcsik's tall frame and his *simpático* demeanor have been frequently seen in developing countries, where he has acquired first-hand knowledge on science there. His understanding of, and in some mysterious way (he is a DC theoretical physicist) his affinity with, the problems of science and development have made him specially apt for the job. His articles, published at various times and in many journals, are at times repetitive, understandably, but there are many conducting threads which I shall try to uncover, adding my own comments, resulting from first-hand experience in a developing country, Venezuela, and from many contacts abroad.

At this point, some remarks may be made about the differences between basic science, applied science and technological research. Admittedly, they constitute nowadays a continuum of processes, with many grey areas between them, but the distinction is still useful as a basis for policy decisions. The fundamental difference lies in the *motivation* of the doers — to know nature disinterestedly *or* to act in a direct or indirect way upon it. Also in the *products* of each kind of research. Another difference, which I did not find mentioned by Moravcsik, is the *audience* at whom the research is aimed. This is constituted by scientific *peers* for basic science, with a paper as the chief formal means of communication, and the public in general for the technologist, when his work emerges, to be sold and judged on the market. This difference leads to marked divergences between the *ethos* of both activities and the behavior of both groups. The first, as Derek de Solla Price humorously puts it, is papyrophilic (writes papers and books), and the second is papyrophobic (conceals his results until they find their way in the market).

While basic and technological research stand at both extremes of a spectrum, applied research, which shares some of the characteristics of both, is a special case. Like technology, it must be performed in economically or socially "relevant" areas, while it shares with basic research a degree of academic freedom and is usually promptly published in specialized journals. Actually, applied research in DCs is for the most part sterile. It is often "second class" science and seldom has any tie with production. And there is nothing sadder than "applied" science which is not applied. Basic science is a good *per se,* and technology gives rise to marketable goods, but applied science which remains fallow is a non-entity!

Science for whom? For *all* people and for *all* nations, of course, since I cannot conceive of a science whose dividends are not widely shared. *Science for what?* Economists, politicians and the public alike all agree that it should be for the material welfare of human beings, attained by the development and application of technology. With this we have no quarrel, but, in order to reach this desired end, one condition and one consequence must be accepted. The condition is that science (the *understanding* of nature) becomes part of each nation's culture and the consequence is that, as this develops, the people's view of the world and way of life will be modified.

What science? For both DCs and LDCs, not all science, or application of science if you so wish, is good, and the "perverse effects" of certain science and technology have often been commented upon (1). One of these "perverse effects" is research of new and more "perfected" weapons — and often their introduction into the Third World — which constitute a grotesque distortion of what development is meant to be.

Science in the Western sense — as contrasted with empiricism or with the so-called "ethnosciences", no matter what Thomas Kuhn's view on the subject — is a communal, cumulative, experimental and objective activity which, with Greek, Babylonian and other antecedents, appeared on the European scene some 400 years ago. It eroded mythologies and undermined religions, but gave man an instrument for controlling nature, for good or for bad. It demanded objectivity in its method, and humility in its gait. At the same time, technology — how to do — became dependent on science — how to know — specially in the last 100 years. It can be argued, and Moravcsik does it eloquently, that science, *in a cultural sense,* and technology go together and that, therefore, the first thing DCs must do, even at an early stage of their development, is to establish groups doing quality *basic* (so-called *pure, disinterested* or *academic*) science. This marriage works in a four-fold way: by satisfying the intellectual hunger of gifted members of the community — always an *elite,* let us not conceal it; by elevating the *morale* of nations; by serving as a basis for rigorous training of technologists and by helping to make informed choice of technology to be transferred from other latitudes. It is Moravcsik's opinion, and my own, that, without such an indigenous base of quality science, independent technological research cannot grow. By "quality science", I mean obtaining reasonable results, starting from existing knowledge (the so-called "literature"), by means of observation, experimentation and logical thinking, subject always to facts, and publishing the findings in well-refereed journals, from where the generalized consensus of other scientists can be reached. At its best, "quality science" is moved by inspired theory. At an early — and also in later stages of growing science — the primary problem is then to identify talented individuals and give them the means to do research at the frontier of their specialties, rather than to build up grandiose and detailed plans that subsequently gather dust on the shelves of National Science Councils and the like. The choice of proper individuals as researchers can only be done by other researchers, who know best how to identify budding talent. Of course, the establishment of basic science in a country does not guarantee the growth of technology — other measures of social, economic and political nature, must be implemented — but it is a necessary background for the solid implantation of endogenous technological research.

Not that planning, in a general way, is superfluous. While science must be left free, technology, and even applied science, should be planned from the top, with scientists heavily involved in the process, in order to pinpoint areas where application is most likely to occur in a given country, and in order to promote production. Administratively speaking, this is a much more complex process than for basic science, since it involves complicated and widespread social and economic considerations.

From early schooling to University and, of course, post graduate education, an active and participatory knowledge, rather than rote memory and early specialization, as seems to be the case in Moravcsik's experience in other regions, and in our own Latin America, is to be emphasized. Science must be learned through a living experience; it is, as we say in Spanish, a *vivencia*. I have been asked many times by younger people to recommend to them books on the methodology of science, and have always refused to do so. While some ancillary methods — such as the handling of statistics — can be learnt in textbooks, the hard core of scientific behavior and *ethos* can still only be learned through contact, for many months or years, with experienced researchers, in a socratic, personal manner. Hence the need to send personnel abroad in the first stages of development. Specially in universities, the teaching of science should be done by people who perform active scientific investigation, so that their lessons are imbued with living, constantly changing theories, rather than by obsolescent textbook material. At the level of mass media, an aggressive information campaign should be encouraged in order to promote the population's "scientific literacy" which creates a proper receptive social *milieu* wherein both science and scientists can grow and prosper.

There is in developing countries a psychological obstacle to endogenous science progress, what I have termed elsewhere (2) "peripheral complex". This consists of the general belief that scientists in LDCs cannot contribute effectively to the growth of knowledge, because of innate, or possibly cultural, "defects". Such a fundamentally erroneous belief can be dispelled by scientists in LDCs contributing significantly to the growth of knowledge. This is another product of science, hygienic so to speak, and serving to give confidence to the public at large, and to developing scientists themselves.

Since science is a communal undertaking, isolation from other scientists is a common plight of LDCs scientists. Perhaps the chief task of a science administrator in such countries is to try to alleviate it. In the short term, well-stocked science libraries are indispensable, and frequent (at least once a year) trips abroad plus sabbatical leaves are useful. In the longer term the next step should be to form groups of scientists within the country who can talk to each other. It is well to bear in mind that this latter process may give results in no less than 20 years.

The practically unavoidable result of the mutation constituted by the growth in science in LDCs is the modification of their view of the world — the so-called *Weltanschauung* — with the destruction or change in their aspirations and habits. The late Argentinian scientist Jorge Sabato used to say: "Our aim in Latin America is to get the benefits of science and technology without giving up the *siesta*". Probably a chimerical undertaking, since *siesta* is fast disappearing in our latitudes, as science and technology are making inroads, requiring new ways of living. Similarly, we Latin Americans must learn to abandon our traditional unpunctuality and to better value the importance of time. Unpunctuality in merely social events can be accepted as a form of agreed-upon behavioral convention, but it is deadly in science, constitutes a sympton of disorganization and can interfere markedly with the conduct of research. While we demand the fruits of science and technology, we continue at the same time to waste time, to act under no pressure, in a relaxed way. In brief, we want to have our cake and eat it too.

There are other illusions LDCs must dispel before they can make headways on the road to endogenous science. One is the belief that they must somehow invent a "new type" of science radically different from Western science. Yet, in the spread of science in the United States, Canada, and others, when they were LDCs, no such phenomenon emerged. And Western science has amply shown to be logically satisfactory, predictive, and practically efficacious.

Another illusion is that present "dependence" upon DCs is the only or at least the main cause of our backwardness. The existence of such dependence is real enough and I do not dispute it, but it is due mostly to the lack of science and technology in LDCs, and the root way to remove it is to do science, active and endogenous, of the highest quality.

Another notion is that the machiavellian action of multinational corporations prevents us from doing research in our own countries. There is some truth in this, at least for technological research, since multinational companies tend to do their most relevant research in their country of origin. Thus, for example, there was no research worth the name on oil in Venezuela (where 92% of foreign income comes from this commodity) before 1976, the year the oil industry was nationalized. In that same year, INTEVEP, the Venezuelan oil research institute, was founded and it has thrived ever since. But it must be added that the rapid success of INTEVEP was due largely to the previous existence in the country of basic chemists and physicists, who were able rapidly to take over the mostly applied and technological research carried out in the Institute. As Moravcsik aptly puts it: "It is much more comforting to believe that one's own failures are due to the conspiracy of external forces".

An indispensable item in science and technology policy is evaluation of the results of research. We have mentioned that from the administrative point of view planning was a simpler task for basic than for applied and technological research. In evaluating the results, the contrary is true. The aims of technological research are usually clear-cut and palpable. In the last resort, they consist of the marketability of a product and its benefits, which can be measured, although it may not be possible to unweave the respective contributions of research and technology and other factors, such as marketing. Not so for basic science: the goal here is, first, a published article and, two, the effect of this article on the state-of-the-art through its influence on the thinking of other scientists working in the area. One first step — highly unsatisfactory but often practiced — is simply to count the number of articles appearing in journals. The second step is to consider only well-refereed, usually international journals. The third step is to count in *Science Citations Index (SCI)* the number of references made to each article (its "impact"). This is not very satisfactory for LDCs, since only some 4000 journals, out of a possible 70000 existing in the world, are listed in SCI, most of them published in DC countries — the so-called (by DCs) *mainstream* of science. It remains that, for basic science, the old-fashion empirical approach is still the best, consisting as it does of evaluation made by knowledgeable directors of research, themselves scientists, of course, helped by committees of peers and also by the tools (SCI and the like) mentioned above.

There are many more detailed items, of course, which can be found in Moravcsik's useful collection of essays. The whole phenomenon of development through science if of extreme complexity — and, as the author states: "... our knowledge of how to formulate a systematic scientific and technological development program in any country, advanced or less developed, is extremely rudimentary". But we do know a good many things about the process, thanks to the efforts of people like Moravcsik. In addition, his book will come as a stimulus to other people to deal with the subject. In particular, scientists and writers from developing countries should address themselves to the problem, with which they have a living experience, since "the literature dealing with problems of science in developing countries is being written primarily by people from the developed countries" (Moravcsik). Such bias must be corrected somehow.

References

(1)   ROCHE, M.: *¿Ha Contribuido la Ciencia al Desarrollo?*; *Interciencia*, **11**: 216—220, 1987.

(2)   ROCHE, M.: *Early History of Science in Spanish America*; *Science*, **194**: 806—810, 1975.

July, 1987   Marcel Roche
Instituto Venezolano
de Investigaciones
Científicas, IVIC

PREFACE

To the subject of scientific development in the Third World, Professor Michael Moravcsik brings the precision of a scientist, and the zeal of a missionary convinced of the virtues of modern science and modernity. There can be no question as to the author's qualifications as a physicist, or of his active involvement in numerous attempts to stimulate scientific growth in developing countries. The set of essays comprising this book are the clear and concise expression of a mind capable of choosing between facts, discerning those which conceal something and recognizing that which is concealed. To be sure, not every argument or conclusion reached herein will be popular with all readers. But even when the reader disagrees completely, the issues raised are so important and clearly posed that they shall surely force his gratitude.

"Scientific development" has come to be virtually an article of faith, and paying homage to it is a familiar ritual in Third World countries. The incense, however, is frequently burned at the wrong altar because it is only certain particularized applications of science—not the absorption and internalization of its other myriad aspects—which are really meant. It is, of course, perfectly natural that developing societies should seek modern agricultural techniques, factories for producing cement and steel, and so on, and identify these as "science". But, the author reminds us, there is far more to science than this. Indeed, science has as much — if not more — to do with philosophy and a world outlook as with industrial and technical production. As a methodological procedure, it combines observation, experiment, classification, and measurement to test various laws, hypotheses, and theories. Built on a great stream of thinkers and workers lies the vast cumulative tradition of scientific knowledge. And, in its philosophical dimensions, science is perhaps the most powerful influence moulding man's beliefs and attitudes towards the universe. Science lives in the minds of men and women, not inside equipment and machines.

The point, therefore, is that scientific knowledge cannot be simply bartered for material goods and services. In one of his essays, co-authored with John Ziman, the author attacks the notion that the advanced countries should somehow export the great surpluses of their knowledge factories in exchange for cocoa, bananas, oil, and copper. Quite apart from enhancing the relations of dependence, he argues, this transfer simply cannot be effected as a commodity exchange. A substantial indigenous scientific infrastructure is vital for any meaningful communication to occur between developing and developed countries in this sphere. This may well be a truism, but it is one which gets here the forceful emphasis it deserves.

In identifying the principal hurdles faced by Third World science, Professor Moravcsik articulates what is often felt but seldom stated. The problems are far more complex than one might infer from the remedies that are usually suggested. In considering the issue of increased spending for science in the Third World, this immediately becomes apparent. There is no question that Third World countries, as a whole, spend much too little on scientific development. Considering the material and intellectual needs of such countries, one could easily argue for, say, a tenfold increase in the amount. There are definitely countries which would benefit immensely from some such change, and one could confidently predict a speedy transformation of their entire techno-socio-economic structure. On the other hand, there are also examples where, if all other factors are left constant, this change would probably do little. This is to be expected wherever the principal bottlenecks involve a rigid and irrational bureaucracy, or vital decision making powers are concentrated in the hands of an unsympathetic and non-understanding leadership, or when corruption pervades the entire fabric of society. In such countries, various externalities have a tendency for soaking up arbitrary amounts of funds. The creation of new scientific institutions on mere administrative impulse affords one such example. Bearing pretentious titles such as "Centre of Excellence" in this or that field, a veritable population explosion of new institutes and organizations has occurred. Formally, one is obliged to count this as progress. But, in real terms, notable success never occurs unless the new institution is endowed with a crucial, valuable, and scarce resource — the leadership of competent, imaginative, and dedicated scientists belonging to the country itself.

Living in the post-colonial era, but one in which the unhappy division between the metropolis and the periphery persists, the political, economic, cultural, and sometimes military, domination by the West inspires a reaction against the perceived instruments of domination. Modern science, sadly enough, has become increasingly an object of attack as well. One consequence is that, from time to time, there appears the notion of a science different from "Western science". There is no question that technologies differ from place to place. But is there such a thing as "Third World science" or "Islamic science" or "Marxist science"? The author takes an explicit and unapologetic stand: his answer is a simple no. One cannot agree more. The idea of any special science in this sense negates the concept of science as a universal, independent, rational human endeavour which can transcend divisive national and ideological boundaries. Upon examination, one is struck by the fact that such constructs derive from passionately held beliefs of how matters ought to be, rather than what is required by the imperatives of logical and empirical enquiry. Neither the premises, or the conclusions, of any "special science" are the least bit in doubt. It seeks to reaffirm what is already known, not search into the unknown. No new mathematical principles are sought, no experiments will be designed for its verification, and no new machines will be built on account of it. Like creationism in the West, the numerous varieties of "special science" are a reaction against modern science, and not a new direction of science. Their pursuit is nought but a fruitless chase.

And now to move on to more delicate, controversial, and perhaps irresolvable matters. The need for such a discussion is forced by the value structure implied by science and modernity, and by the explicit positions taken on this in the book.

Science has evidently prospered in some Third World environments. In others it has not. Why? Access to financial and material resources certainly cannot be the whole story, even though these are universally emphasized and are undoubtedly essential. This question forces upon us the importance of considering the overall idea system of society, and how it influences the acquisition of positive, rational knowledge. Overall idea systems — by which shall be meant beliefs, attitudes, social mores, general assumptions, and specific religious and ideological positions — are of the profoundest importance in human history. Julian Huxley compared them to skeletons in biological evolution: they provide the framework for the life that animates and clothes them, and in large measure determine the way it shall be lived. Embodied within an idea system, to an extent which may be greater or lesser, is rationality — by which is meant that matrix of connections assigning cause to effect. Rational thought and science have their origin in man's drive to have power over events in the external world. Without this impelling Nietzchean obsession, humans become mere buoys that float on the waves. A society oriented towards fatalism, or one in which an interventionist Deity forms part of the matrix of causal connections, is bound to produce less individuals inclined to probe the unknown with the tools of science.

So let us accept, at least as a provisory hypothesis, that the idea system of a society has more to do with its scientific advancement than any other factor. Combine this with the fact that a great revolution took place in Europe some 400 years ago as a consequence of which experimentation, quantification, prediction, and control became the paradigm of a new culture. It was only after this that a mysterious and capricious universe could be understood as mechanical, orderly, and in which "number holds sway over the flux". Add to all this the continuing scientific and technological pre-eminence of the West. What emerges? That the Western idea system, based on a Greco-Roman legacy and Christian ethic, is superior to any of its contemporaries?

The author's position on the merits of European civilization is perfectly explicit. His essay, authored jointly with John Ziman, begins thus:

"With European industrial civilization comes European science. It is a package deal. The question whether a culture thus superseded or repressed had its own form of science has become purely academic: the process of economic growth and social development is entirely predicated on the "rational materialism" of post-Renaissance Europe and its North American colonies ... In the present discussion, it is taken for granted that European science should become a dominant cultural force throughout the world."

Whether or not our reaction to this is typical of a person from a country with a colonial past, we cannot tell. But, inspite of having sworn allegiance to both the techniques and philosophy of scientific rationality, and though sadly aware of the shortcomings and defects present in our environment, we must nevertheless confess that such explicit Eurocentrism makes us deeply uncomfortable. Presuming that the book seeks to spread science rather than "civilization", such a position does harm to its mission by introducing an extraneous issue. The temptation for retort becomes irresistible: Do the contributions of the Chinese, Islamic and Hindu civilizations warrant such a peremptory dismissal? Were not Auschwitz and Hiroshima consequences of a supposedly rational and scientific European culture? How should we assess a civilization which has created the concept of "megadeaths" and "mutual assured destruction"?

Personally, we feel that such divisive issues should, and could, have been avoided. Instead of tracing the roots of science to narrow geographical or cultural domains, it would be infinitely more preferable, and also more accurate, to regard the development of modern science in a universal context — as a collective effort of mankind in which different civilizations have contributed their share. One can persuasively argue that science is the natural outcome of human intelligence. Rational man has emerged from the realms of biological evolution endowed with innate mental structures capable of abstract thought. Epistemological and linguistic research bring evidence that humans are intricate "pre-wired" computers needing only external stimuli to set cognitive and creative processes into operation. Regarded in this manner, science pretty much had to develop sooner or later in the course of man's progress. That it should have developed over the last few thousand years is probably quite accidental. There were countless ages before that during which there was no knowledge, and there may be countless ages without knowledge in the future. On a cosmic scale, the history of knowledge and science is profoundly irrelevant, reminding us of the futility of parochial pride in the historical cultures which we accidentally happen to be associated with.

Differences aside, one must acknowledge this book as a valuable manual on science development. It sets out the role of science in technology transfer with unusual clarity, identifies both the immediate and long term needs of science institutions, discusses the role of scientists as science managers, brings to attention the importance and feasibility of scientometric studies, emphasizes the need for communication among scientists, analyzes the problems and potentialities of overseas university education, and much more. One hopes that it will be widely read, and with care.

Pervez Hoodbhoy
Physics Department
Quaid-e-Azam University
Islamabad, Pakistan

Abdus Salam
Director
International Centre for Theoretical Physics
Trieste, Italy

ACKNOWLEDGEMENT

The Publisher and Editor wish to thank the authors and the following publishers/institutions for permission to reproduce published papers in this volume:

American Association for the Advancement of Science
American Association of Physics Teacher
American Institute of Physics
Association of Geoscientists for International Development
Council on Foreign Relations, Inc.
Department of Biochemistry, Mahidol University, Thailand
Educational Foundation for Nuclear Science
Elsevier Scientific Publishers B. V.
Indian Science News Association
Interciencia
International Council on the Future of the University
International Commission on Physics Education
International Science Policy Foundation
IOP Publishing Ltd
Macmillan Journals Ltd
Ministry of Education, State of Bahrain
Ministry of Science, Technology and Energy, Bangkok
National University of Malaysia
North-Holland
Organisation for Economic Co-operation and Development
Pergamon Press
Rijksuniversiteit Gent
Society for Social Studies of Science, Illinois Institute of Technology
The University of Michigan
United States Advisory Commission on Public Diplomacy
Universite Libre de Bruxelles, Institut de Sociologie

While every effort has been made to contact the publishers of reprinted papers prior to publication, we have not been successful in a few cases. Where we could not contact the publishers, we have acknowledged the source of the material. Proper credit will be given to these publishers in future editions of this work after permission is granted.

CONTENTS

ON THE ROAD
TO WORLDWIDE SCIENCE

I. THE TASK AND ITS FRAMEWORK

A. The Task in a Context

PARADISIA AND DOMINATIA:
SCIENCE AND THE DEVELOPING WORLD

By Michael J. Moravcsik and J. M. Ziman

WITH European industrial civilization comes European
science. It is a package deal. The question whether a culture
thus superseded or repressed had its own form of science has
become purely academic: the process of economic growth and social
development is entirely predicated on the "rational materialism" of
post-Renaissance Europe and its North American colonies. This fact
may well be deplored, but can scarcely be denied. The very word
technology, denoting a practical technique that has been studied and
transformed in the light of scientific rationality, betrays our values
and intentions as it displaces the crafts from town and village, from
workshop and field, throughout the world.

Nevertheless, European science, both intellectually and practically,
diffuses very slowly and unevenly into the culture of a developing
country. This is not because it is firmly resisted by alternative metaphysical systems to which it seems antagonistic, but because the actual
agents of diffusion are weak and uncoordinated. In Western Europe
and the United States, scientific knowledge is a natural product as
well as a fuel of the advanced industrialized society; in a country such
as *Paradisia*—the pseudonym conceals no particular country, but
refers perhaps to a whole class of medium-sized states such as Ghana
or Iran, Korea or Peru—it is a foreign import, an exotic plant that
has not yet established and seeded itself in new soil.

Modern science comes into Paradisia, from distant Europe or the
United States, along three distinct channels. In the first instance, historically speaking, it came through the *academic* channel, into schools,
colleges, and universities. Whether these institutions are hundreds of
years old or are quite modern is irrelevant; the natural sciences were
not introduced into the academic curriculum in Western countries
until the nineteenth century, and did not play a very large part in
formal education until quite recently. Paradisia took its time in following the lead of the powerful metropolitan country—shall we call
it *Dominatia?*—in whose political or at least cultural sphere of influence it then lay captive, so that the teaching of physics and chemistry,
botany and zoology, in a few secondary schools and colleges, was
perfunctory and of little practical significance. Until the Second
World War, the only genuine academic centers of scientific knowledge and practice would have been in some small schools of engineer-

ing, medicine and agriculture, where external standards of technical expertise could be imposed by expatriate professional practitioners.

The expansion of secondary and higher education in Paradisia in the past 30 years is immensely impressive, and much has been changed. In particular, the creation or expansion of fully fledged, independent universities, offering bachelor's and master's degrees in all the major academic disciplines, has transformed the opportunities for advanced instruction in the basic sciences. The student of a sophisticated subject such as nuclear physics or genetics need no longer go abroad as he once did for even the most elementary training. On the campus of his national university he will find Paradisians with doctorates from MIT or Cambridge, Paris or Moscow, in precisely these subjects. Yet the number of students who actually receive such instruction is often pitifully small, perhaps no more than a few dozen graduates in science in the whole country each year—a fraction that can only be expressed, like the concentration of a trace impurity in a chemical compound, in parts per million of the total population.[1]

Some aspects of scientific technology enter Paradisia along the *technical* channel, carried by a flood of imported or locally manufactured industrial products—automobiles, transistor radios, pharmaceutical and agricultural chemicals, electric power plants, etc. Starting with elementary practical skills, such as driving a bulldozer or installing plumbing, there is a natural progression through the expertise of the motor mechanic, machine-tool operator, or electronic technician, to the more "intellectual" tasks of the civil and electrical engineers or factory managers. Daily life in Paradisia (especially in the cities) has become dependent on such people, who are in very short supply. At the lowest levels the need is met by self-taught practice or crude apprenticeship, but the facilities for rational technical training at the middle level are almost nonexistent. Deploring this defect—but not regarding it as any of their business to cure—the multinational corporations that dominate trade in sophisticated industrial products bring in their own technical experts for after-sales service, or set up complete "turn-key" factories where almost every operation is automated or prescribed in a manual for the benefit of unskilled local operatives. Indeed, motivated by prestige, the rulers of many developing countries demand sophisticated and even fully automated production methods, despite their capital intensity in the midst of unemployment. The "science" that infects Paradisia along this channel is often quite

[1] For figures of researchers per million population, see M. J. Moravcsik, *Science Development,* Bloomington (Indiana): International Development Research Center, 1975, Chapter 3B. For the leading scientific countries the number is 1,000–3,000, while for Paradisia it is perhaps 20–100.

out of touch with a coherent, rational, applicable body of knowledge.

Finally, science is injected into Paradisia through a third channel—government organizations dedicated to particular technological applications and deliberately charged with specific *"missions"* to serve as sources of modern scientific ideas for economic, agricultural, medical and industrial development in particular technological areas. Typical of such organizations would be an Atomic Energy Commission with the task of bringing all the benefits of nuclear energy to Paradisia. With its own little research reactor (usually a gift from a benevolent great power), reasonable technical facilities, a fair degree of autonomy within the government machine—and a substantial fraction of the total science budget—this is often the only real center of modern scientific technique in the whole country. By training its research workers abroad, and through its affiliations to the International Atomic Energy Agency, it maintains adequate links with the scientific world at large.

Even though the research that is done in such an establishment may be of very low quality by international standards, it may serve as a vehicle for a broader scientific effort than its nominal mission would suggest. But for all the political goodwill, high hopes, and considerable expenditures that have gone into such organizations, they have made very little impact on the general social development of countries such as Paradisia. Indeed, the over-ambitious promise—which naturally could not be fulfilled—to transform the economic and cultural situation by the magic wand of a highly sophisticated scientific technique has done a great deal of harm, discrediting "science" in the eyes of politicians, intellectuals, and practical men.

In all that follows it is important to keep in mind that these three channels carry little traffic, and are almost unconnected with one another. Of course, if one were content to stay in the superficial world of words and titles, one might get the impression that science development has a prominent role in Paradisia. Like almost all developing countries it has at least one organization proudly proclaiming the importance of science. To reinforce this impression, speeches and pronouncements on development policy, both in Paradisia and in the many organizations for international aid, abound in the reaffirmation of the importance of "science-and-technology" or "R and D."

Behind the facade, however, one often finds no more than the fragments of a scientific community, disorganized, disunited, of limited professional competence, poverty stricken, intellectually isolated, and directed toward largely romantic goals—or no goals at all. This is a harsh judgment, especially when applied to the activities of a small group of exceptionally hard-working, sincere, and dedicated people, and to the results of high-minded efforts over several decades by essentially well-meaning organizations such as international agencies and government bureaus. Indeed, the truth about the state of science in many developing countries is so hard to bear that it does not easily pass the lips of the proud Paradisian himself nor of the courteous foreign expert enjoying the hospitality of a short visit. The printed word on development carries scarcely a hint of what is so often said in private among close friends and colleagues.

And yet the task to be undertaken is not hopeless. The example of Japan, whose industrial power is now matched by the intellectual resources of a first-class scientific community, beckons on such great nations as India and Brazil. Since the Second World War, Australia and Canada have "taken off" scientifically, and are no longer mere intellectual provinces of more powerful empires. In countries as politically diverse as Argentina and Romania, centers of scientific excellence are beginning to make their mark on the international scene. Just as the category "developing country" includes a great variety of economic, environmental, cultural and political circumstances, so also it may cover a very wide range in the conditions of science, technology and higher education.

In the present discussion, it is taken for granted that European science should become a dominant cultural force throughout the world. But there are many thoughtful and good men who assert that development of this kind is by no means beneficial in ultimate human terms and who oppose science itself as the very source of the malady. This point of view is not untenable, although it makes too little of the irrationality and possible inhumanity of alternative metaphysical systems, and does not allow for the decoupling of the cognitive, cultural and attitudinal content of science from many of its current applications. If we ignore this attitude here, it is not because it is absurd or despicable but because it is very far from what is demanded by enlightened Paradisians who see very clearly that without the help of science the road leads only to disaster. And it is not our business, here in comfortable Dominatia, to tell them that they cannot have what we so manifestly enjoy.

II

In the effort to discover the reasons for the disparity between aspiration and achievement in science in Paradisia, one becomes aware of a combination of factors, involving, on the one hand, a lack of understanding of the nature of science and its evolution, and, on the other hand, a confusion in the goals and processes of development. To these must be added indifference to some of the crucial potential elements in scientific progress, and organizational and administrative practices which are quite mismatched to real needs.

The attitude of politicians, economists, and even of many scientists toward the role of science in development is clouded by fundamental misconceptions. After an initial period of euphoria, in which it seemed that all the practical problems of poverty, hunger and economic backwardness would be solved at a stroke by the new understanding that would flow from the establishment of modern science in an old country, a reaction set in. The rhetoric was displaced by skepticism and cynicism: despite lip service to the contrary, there is now widespread disbelief that science can contribute significantly to development, except at the level of immediate practice.

This skepticism is largely due to the extended time scale and subtlety of both the development process in general, and the contribution of science to it. In all walks of human life, short-term considerations tend to win out over potential long-term benefits. This tendency is epitomized in the use of the word "relevance," which has come to mean direct connection with an immediate material problem. In this sense, scientific research is often felt to be "irrelevant" for the less-developed countries by comparison with problems like providing food for today and tomorrow.

Taking a broader view, however, national development must be seen as the process whereby the material, spiritual, cultural and psychological well-being of the country and its citizenry is improved and assured into the future. To carry out this process, one must tend both to the needs of this year and to the foreseeable requirements of the next few decades. Furthermore, one must cater for both material and non-material needs, of which the former are but instruments for the fulfillment of the latter. No society, whatever race, religion, ideology or century it belonged to, ever considered the only purpose in life to be to provide its members with food and creature comforts. The nature of non-material aspirations changes from society to society, and has its own evolutionary dynamic, becoming what we call the history of mankind. In considering the role of science in our own epoch, in relation to many different countries with very varied cultures, we must always think of development in this broad sense. This will become very clear when we discuss the factor of morale.

What then is the justification of science in Paradisia in this enlarged concept of development? As is often the case with important and successful human activities, a number of different arguments combine into a broadly based and solid rationale. One might start with the long-term character of science and education. Most of those who doubt the relevance of science to present conditions in the developing countries will agree that these countries are likely to need indigenous scientific activity three or four decades from now. That is to say, Paradisia ought by then to be able to share in the give-and-take of research in the pure and applied sciences, contributing to and drawing knowledge equitably from the international pool. If such a level of respectability and solvency in the world marketplace of knowledge cannot be achieved overnight, it is a realistic goal for half a century hence.

The development of a viable national infrastructure in the sciences is a lengthy process, even where it is favored by material, cultural and social factors. In both the United States and Japan, for example, it took several decades to progress from an embryonic, imitative and subordinate scientific infrastructure to significant power in world science. In the less favorable initial circumstances of many developing countries, the same process can scarcely be more rapid. Thus the process must be started immediately. As the elderly French marshal told his gardener, when informed that an ornamental tree would take 100 years to grow to full stature: "Then we have no time to lose; plant it today!"

Yet there are still those who seem unconvinced of the urgency of the need for an indigenous scientific community in a country such as Paradisia. With scientific knowledge and technological know-how available for all on the world market, would it not be easier to import and use whatever is needed, without building "local production facilities" in these commodities? This commercial metaphor is appropriate, since the essence of this argument is an appeal to the principles of classical liberal economics—the open market, free trade, economies of scale, and the division of labor. Let the great knowledge factories of the advanced countries export their great surpluses of information, fact and theory, in exchange for cocoa, bananas, oil and copper, and eventually, in exchange for industrial products manufactured in the newly developed nations.

At the most elementary level, this analogy is entirely fallacious. Scientific knowledge lacks many of the necessary attributes of a commercial commodity, and cannot be fairly bartered for material goods and services.[2] It is not possible simply to "import" science and technology in the absence of an indigenous scientific and technical com-

[2] For a more direct attack on the attempt to represent knowledge as a quantifiable economic category, see J. M. Ziman, Book Review, *Minerva*, July 1974, p. 384.

munity. And even if this were possible, it would be unsatisfactory and undesirable for economic, political and psychological reasons.

In importing scientific and technical information (for example, through patents), one first needs to identify the purpose for which importation is necessary; then one must select appropriate items from a wide and varied catalogue; finally one must acquire, absorb and adapt existing science and technology to the task at hand. These processes cannot be carried out effectively without a substantial indigenous scientific infrastructure. This is evident, for example, from the history of Japan, where the large-scale importation of scientific techniques in recent times was preceded by decades of development of an indigenous scientific and technological community.

Particular stress must be laid on the adaptation of science and technique to meet *local* needs. New research, applied and basic, may have to be performed locally or regionally before the major problems faced by a given country on account of its particular geographical location or social history can be dealt with. Rice cultivation in Southeast Asia, soil desalination in Pakistan, ranching in East Africa, deep-sea fishing in Peru are examples of important human activities which cannot be understood and improved merely by the application of general biological and ecological principles taken from the international scientific literature.

Beyond the solution of specific material problems, the existence of an indigenous scientific community is even more essential for science to achieve a broad social impact. For example, the natural sciences are an essential component in education for the medical, engineering and teaching professions. In principle, such education could be carried out by persons whose only contact with science was a one-time university degree. Practice shows, however, that without the benefit of continued personal involvement in scientific research, such people not only quickly fall behind the fast-growing content of the sciences but also become divorced from the spirit of science as the art of problem-solving. Evidence of this is unfortunately all too obvious on the Indian subcontinent, where the stifling system of rote learning and regurgitation for examinations is closely connected with these ubiquitous shortcomings in the science instructors. It is not surprising that the doctors, engineers and teachers educated in such a system lack initiative and imagination in their professional work.

The educational role of the natural sciences goes far beyond the training of particular professionals. The development process as a whole is based on the assumption that change in time is a regular and desirable feature of social life, that human fate can, at least to some extent, be shaped by the application of knowledge and effort, and that the experience of others before us can be helpful to us in making further progress. By contrast, the traditional milieu of a less-developed country often is imbued with a static view of life, a fatalism in which events are determined by incomprehensible extra-human factors, and a compulsion to repeat previous practices rather than improve them. Thus, contact with modern science, with its expanding and cumulative character, can have a very broad cultural and attitudinal impact. To put it another way: those who oppose development on the material plane as a challenge to traditional culture are quite right to see science as a fifth column sapping this culture at its spiritual roots.

There is, however, yet another facet of science which takes us back to the earlier discussion of non-material aspirations. In post-Renaissance European culture, the pursuit of scientific knowledge is considered to be more than a means to a better material life: it is an important medium for higher activities made possible by the partial satisfaction of our immediate needs. The immense efforts made to answer "irrelevant" questions concerning the nature of the Universe, the origins of life, the laws of sub-atomic physics or the topology of n-dimensional algebraic varieties are not directed toward greener grass, more lavishly buttered parsnips, or better mousetraps; they are actuated by religious and aesthetic impulses of harmony, coherence, comprehension and intellectual order by which all who partake of our civilization are bound to be moved.

In this context, it is essential to realize that the huge gap between Dominatia and Paradisia is as much psychological as material. When we talk about the continued dependence of the less-developed countries on the advanced countries, we do not mean only that formal political authority has become economic dominance; we mean that the advanced countries continue to exert an overwhelming influence on the rest of the world in intellectual, cultural, social and spiritual ways. This non-material dependence is quite as demeaning, frustrating—and ultimately alienating—as economic and political exploitation. For this reason, continued scientific dependence on the advanced countries is an unacceptable future for any self-conscious developing country. We would be wise to encourage genuine competition (and cooperation) in our own game, rather than risk the very real possibility of repudiation of the whole scientific enterprise by the majority of mankind!

Finally, we come to the factor of morale. In the enormous effort to close the gap, a crucial element is a belief that the task can be accomplished and that it is a purpose of human life worth sacrificing for. This inner strength and spiritual fortitude is a potent force in all human activities. While high morale may rest primarily on philosophical and other non-material factors, it can be strengthened by instances of success. In the matter of social development, scientific achievement can well serve as an indicator of such success. Thus, the Nobel Prize of Raman or Houssay is tangible evidence that India or Argentina can compete favorably with the United States in some areas of scientific research, even if it is not likely to catch up in the near future in steel production or in the number of TV sets. It must be admitted that the recent Indian detonation of a nuclear device, diplomatically deplorable and militarily irrelevant as it may be, has had the same effect on the self-image of the Indian scientific and technological community. Such spectacular successes are not necessarily expensive and yet help bolster the morale of the country across the whole broad front of development.

In summary, the conventional wisdom that fails to see an effective role for science in many countries at their present stage of development misses the point in several respects. It conceives development only in immediate and narrowly material terms. It also thinks of science in narrow terms, failing to appreciate the broad impact of its problem-solving attitude and its cumulative, optimistic, self-reliant spirit on the development of a country. This same conventional wisdom underestimates the time and effort needed to plant science firmly on new soil, and thus neglects the long-term task of establishing an indigenous scientific infrastructure in the less-developed countries—an objective that should, in reality, have the highest priority.

III

What are the major defects of science in Paradisia, and what are the obstacles to curing them? Here again, it is necessary to have a good grasp of the meaning of science as a multifaceted human activity, and to dig deeply beneath the surface, to uncover the real weaknesses and to devise genuine remedies.

The greatest obstacles to science development are the obstacles to social and economic development in general—poverty, ignorance, and maladministration. It scarcely needs to be said that the material facilities—laboratory apparatus, workshops, repair services, technical staff, buildings, secretarial assistance, power supply, transport, chemical supplies, electronic spare parts, climatic protection, computers, audio-visual aids, photographic equipment, books, experimental animals, electron microscopes etc., etc., etc.—are grossly inferior, in almost all scientific institutions in Paradisia, to what would be regarded as appropriate in Dominatia. Yet this does require continued emphasis, for there are many scientists in the advanced countries who lack the personal experience, or the testimony of their foreign colleagues, or the small quantity of imagination required to grasp this point. Modern science in Dominatia is affluent and competitive; the profligate style of its operations, geared to the anxiety of the professional scientist to "get there fustest with the mostest," is a luxury that cannot be extended over the whole scientific community in a poor country with a low standard of living and a very modest tax base. It is very difficult indeed to do good science, at the international standard, on one-tenth of the real resources available to one's contemporaries elsewhere. There is little consolation in the thought that the Cavendish Laboratory, in the great days of Lord Rutherford, spent only a few thousand pounds a year on material facilities; the comparison is with the physical sciences in the United States today, where the overhead costs, per Ph.D., run to something like $50,000 per year.[3]

It is important, also, to realize how unevenly such facilities may be distributed among scientific institutions in the same country. As we have already observed, money for equipment and other technical resources tends to flow more freely into the "mission" channels within the government machine than it does to equally qualified scientists in academic circles. In India—a land notable for princely wealth alongside utter destitution—there are several laboratories as lavishly equipped as any in Europe or the United States, in the same cities as universities where even sealing wax and string put a burden on the departmental budget. Such inequities, and the lack of corporate identity among those who allow them to exist, are further obstacles to science development.

A constant topic of complaint among scientists from all developing countries is the administrative climate in which they must do their work. The standard pejorative term is "bureaucracy"—by which is meant a whole complex of time-wasting, irrational and rigid procedures, accentuated by a hierarchical and authoritarian decision-making apparatus, that hinder the simple spontaneous initiatives demanded in active research.[4] Because the results of an experiment cannot be foreseen, the needs of future research cannot be anticipated over the long time scales that become customary in more routine ad-

[3] For figures on the cost of research per researcher in various countries, see M. J. Moravcsik, *op. cit.*, Chapter 6B. For India, for example, the comparable figure is $8,000 per year.

[4] Edward Shils has described the problem well. "The organization of laboratories . . . will have to make provisions to avoid the frustration of this scientific disposition by the dead hand of a desiccated and embittered older generation or by an unsympathetic and non-understanding bureaucracy." "Scientific Development in the New States," *Bulletin of the Atomic Scientists*, Feb. 1961, p. 48.

ministrative tasks. But the very absence in the general culture of the attitude of mind characteristic of science makes these needs incomprehensible or deeply suspect in the political sphere, so the unfavorable climate persists against all protests.

And behind the administrative morass of the scientific and academic institutions, the governmental system stretches away into regions of more opaque motive and deeper ignorance, headed often as not by politicians, high civil servants, businessmen, or soldiers who do not even understand the difference between fundamental research and technological development, or the connection between the two.

Then we must also look at the scientists themselves and their response to the difficult circumstances in which they find themselves. After all, the fundamental deficiency is the lack of adequately trained people at all levels—technicians, lecturers, research workers, professors, directors of laboratories, and even ministers of science and education. As we have already remarked, secondary school, college and university teaching of science is not very inspired in Paradisia. The teachers themselves are so ill-trained that only the best pupils feel the excitement and grasp the meaning of their subjects. Up to the level of the bachelor's or master's degree, teaching is often pedantic, and learning may be little better than memorizing standard bookwork. Practical laboratory and field work is hampered by lack of equipment and other resources. The student emerges from such a course with a few technical tricks and only the vaguest notion of what it is all about. It is not surprising, therefore, that those lucky few that then go abroad for graduate study find the research training leading to a doctorate very heavy going. Personal problems of living in another country and having to grapple with a foreign language do not make it any easier. The experience of Ph.D. supervisors in advanced countries is clear: the graduate student from a developing country *can* reach the same scientific level as his contemporaries in all branches of sophisticated modern science, provided that he is given plenty of time to break away from the habits of rote learning and become a self-reliant and independent research worker.

Generally speaking, the young scientist who returns to Paradisia with a Ph.D. from a reputable institution in an advanced country is competent to do research *in the sort of institution where he was trained*. But science in Dominatia is quite different from what is possible, at the present time, in Paradisia. Although he may be perfectly familiar with the general economic and social conditions of his country, the returning scientist is seldom prepared for the extremely hard task that lies ahead of him. Disillusionment, defeat, flight abroad, or cynical engagement in academic politics are familiar symptoms of this deficiency.

The real problem with his training is that it is much too narrowly technical. With a Ph.D. in, say, algebraic botany, the newly trained scientist sees himself as a professional algebraic botanist (specializing in cubic roots!) and can imagine no other goal than to set up a research laboratory in this topic in the University of Paradisia or in some appropriate division of the Paradisian Council for Science Research. Although he may be reasonably self-winding within his own field, he still lacks self-confidence and self-reliance in both the practical and intellectual spheres.

Thus, on the one hand, he may not be accustomed to improvising experimental equipment with his own hands: his early education, perhaps as a relatively well-to-do member of Paradisian society, is unlikely to have encouraged the development of simple practical skills, while research in Dominatia is performed with sophisticated, expensive apparatus purchased ready-made. One of the most depressing sights to be seen on a brief courtesy visit to a university or government laboratory in a developing country is a large and complicated piece of apparatus, such as a spectrophotometer, or a liquid helium plant, that was bought at considerable sacrifice to support the research of some bright young man returned from abroad, and is now lying under dust sheets, out of repair for lack of maintenance or discarded when that same bright young man took a job back in the United States. By the same token, the best evidence of a sound scientific basis is an experiment going ahead competently, with home-made equipment— not at the primitive level of sealing wax and string, but simply constructed in the departmental workshop, incorporating pieces of technical bric-a-brac, such as meters, pumps, radio sets etc., that can be easily picked up in any contemporary city.

On the other hand, the conventional graduate school program is seldom seen by the student as training in the general art of research. He is not encouraged to regard himself as competent to undertake investigations and to solve problems of all kinds within the widest scope of his discipline. The remarkable experiences of the Second World War, when nuclear physicists demonstrated their skills in the interpretation of aerial photographs, and zoologists made important contributions to radar technique, are quite forgotten. The cobbler, he would say, must stick to his last: on with the determination of cubic roots by the method of algebraic botany.

This is an extremely important point, for it brings us to one of the gravest weaknesses in the development of science in countries such as

Paradisia—the extraordinary incoherence and inconsequence of the research that is actually attempted, especially in academic circles. There is seldom any effort to link a research topic with any local industrial, agricultural, geographic or otherwise practical issue, or to direct it toward significant ends within the discipline itself.

There is no tradition of purposeful research, mainly because there are no external users around to convey their needs and wants to the scientists, and there is no established culture of such links between the academic and the practical. Thus the doctrine of academic freedom, according to which every professor should be quite at liberty to work at any scientific topic that pleases his fancy, is carried to an extreme. As Stevan Dedijer has said, "Even 'applied' research becomes pure research in developing countries because there is no application."[5]

The student who wins an overseas scholarship is not directed into a particular line of research that might be strongly favored in the future, but is encouraged to apply for a place at the most prestigious university in Dominatia, to take a Ph.D. in whatever happens to be the current fad in advanced science. After a few years, the young men return to staff the department, and begin to compete for facilities for a miscellany of unconnected research topics. The very obvious human difficulties of arranging cooperation under such circumstances are heavily compounded by the attitude of technical specialization and expertise that is now characteristic of the scientific life.

One can go further, to uncover a most unpleasant psychological phenomenon—the justification of the esoteric by its own evident uselessness. Like a member of Thorstein Veblen's Leisure Class exhibiting his power by his extravagant "conspicuous consumption," the Paradisian professor, completely out of contact with local needs and local culture, persuades himself that he must be doing a very important job by the very fact that, on the face of it, it is entirely irrelevant. By the transcendence of its claims to give some ultimate benefit to humanity, he protects his research, which is incomprehensible even to his colleagues, from assessment in terms of common sense.

As we have already emphasized, the criterion of immediate relevance should not be applied too strictly to the program of scientific research in a developing country. What may seem a grave practical problem today may look quite insignificant in 20 years' time. But in the effort to establish science in a new cultural environment a choice must often be made between many possible lines of research. In the circumstances such a choice ought not to be made on the basis of ill-

[5] Quoted by Dilip Mukerjee, in A. B. Shah (ed.), *Education, Science Policy, and Developing Societies,* Bombay: Manaktalas, 1967, p. 376.

informed and wildly uncertain guesses as to what will prove to be scientifically important in the future, but with concern for its immediate or potential applicability to material problems or connection with local cultural or environmental features of the country. But this obvious principle comes into conflict with the debased metaphysic and value system of the conventional Dominatian graduate school, where technical specialties and highly fashionable "breakthroughs" are exalted, at the expense of a general philosophy of science as a means of satisfying our insatiable curiosity concerning nature in all its aspects and the solution to practical human problems. The most pathetic request to be heard on a visit to Paradisia comes from scientists asking: "What research should we be doing?" The only proper answer to this is reminiscent of Cornelius Vanderbilt's reply to the man who asked about the cost of running a yacht: "If you have to ask that question, then you are not competent to undertake it!" But this is only a symptom of the failure to internalize the scientific attitude and the scientific method, which are equally applicable to, and obtain equal satisfaction from, the solution of all apparently difficult problems and unanswered questions.

IV

The weaknesses of science in the less-developed countries are not due entirely to poverty, an unfavorable cultural environment and deficiencies of education; the individual scientist also is isolated from his fellows by a lack of solidarity within the local scientific community and weak links with the outside world.

Science is a collective undertaking. There is much truth in the cliché that the progress of science is like building in brick; the contributions of individual scientists throughout the ages are laid one on another, the bricks in each course being created to rest upon those previously built into the structure; without constant cooperation among the bricklayers, progress is slow and the architecture goes awry. Such is, indeed, the fundamental source of the creative power and objectivity of the "scientific method."[6]

This objectivity underlines another characteristic of scientific activity, namely *universalism* or (in this wicked world of sovereign states) *internationalism*. To the earth-bound layman, there is something almost indecent and treasonable in the ease with which scientists from different countries, political systems and philosophical outlooks can communicate and interact with one another on scientific matters.

---

[6] See Robert K. Merton, *The Sociology of Science,* Chicago: University of Chicago Press, 1973, and J. M. Ziman, *Public Knowledge,* New York: Cambridge University Press, 1968.

SCIENCE AND THE DEVELOPING WORLD   713

Scientists have an international language (a jargon compounded of mathematical symbols and Broken English), commute tirelessly to countless international conferences, communicate through truly international journals, and migrate from country to country with very slight disruption of their scientific output. That is not to deny that they are men of our times, subject to all the imperatives of nationalism and patriotism and to the terrors and wrongs of war, tyranny and exile; but the work about which they care and their true ties of community are directed outward to the "invisible college" of their scientific discipline. Without such ties and the recognition of their research by significant other scientists, they are afflicted with anomie, and wither on the vine. In this whole discussion of science in developing countries, the key word is *isolation*. On account of the collective, cumulative and relatively objective character of the natural sciences, constant contact with fellow scientists is a necessity for fruitful scientific work. Such contact includes ready and quick access to journals, reports and reprints, interaction with visitors, attendance at conferences, opportunities for short visits to other institutions, periodic sabbatical leave to be spent at important centers for basic and applied research, opportunities for collaboration in research with scientists in other institutions, etc. In all such activities, scientists in the less-developed countries are handicapped—by poverty, by geographical distance, by political and administrative barriers.

It is hard to grasp the significance of this isolation without feeling it for oneself. The tough-minded experimental physicist from Dominatia, on his first visit to Paradisia, is not unprepared for the lack of technical facilities, and can imagine himself rolling up his shirt sleeves and improvising his own apparatus. But how can research be possible in a country where the *Physical Review* comes a year late, where the best library has no more than tattered copies of the undergraduate textbooks of a generation back, where the last visit of a foreign physicist was 18 months ago, where there is nobody within 500 miles who can understand or criticize the work one is doing, where there has never been a conference or a seminar where research results are discussed, and where the next opportunity for foreign travel seems at least two years in the future? It is easy to predict that the scientific productivity even of our Dominatian colleague would decay rapidly to zero under those conditions!

As we shall be arguing below, the international scientific community could do a great deal to open up channels of communication with scientists in the less-developed countries. But the improvement of *local* communications and cooperation within the small scientific community in each country cannot be catalyzed from outside. Whatever the social and psychological causes, science in Paradisia is gravely weakened by the lack of a corporate spirit among the few serious scientists in the country. Pretentiously titled research councils and academies do not, in fact, succeed in welding the research of individually weak groups into a rational program. The academic, technical and mission channels by which science diffuses into Paradisia have little connection with one another. Government laboratories do not collaborate with universities, and neither party seems to concern itself about imported technology. Personal rivalries and institutional conflicts that almost seem to spur creativity in the much larger scientific community of an advanced country are crippling within a small group with very limited resources. This is a subject on which it is difficult to make general remarks, for the historical and cultural circumstances differ markedly from country to country, and the outsider rarely has the opportunity to discover the facts and comment fairly. To recall an old scandal, however, there is no doubt that great harm was done to Indian science in the 1930s by the schism that tore apart the Indian National Academy of Sciences under the dictatorial presidency of Sir C. V. Raman—an episode that exemplifies the dangers of authoritarianism in a small group that lacks a corporate identity and yet does not interact strongly with society at large.

V

The establishment of science in the less-developed countries is a heavy burden on the scientists of those countries. So heavy is it, indeed, that many of them cast the load from their own shoulders, and emigrate to the major centers of research in the advanced countries. Given their personal circumstances, who is to blame them? Being accustomed to job mobility ourselves, and not scrupling to buy their services as students, technical assistants or colleagues, we are in no position to preach to them on patriotism and their duty to their own people.

But we certainly owe more sympathy and understanding to those who stay in, or return to, Paradisia to work at science. When it comes to the plight of scientific colleagues, former students, the authors of scientific papers that they cite, participants in their conferences, Dominatian scientists seem to close their minds. The technical scientific press, for example, is almost completely silent on the issues discussed in the present article.

The first thing to be done, then, is to bring this whole matter into the public consciousness of the scientific communities of the advanced

SCIENCE AND THE DEVELOPING WORLD

countries. It is not our job to tell the scientists of Paradisia how they should go about their work, but we should certainly create a platform for a wide debate on these controversial issues. In the last 30 years, many actions have been taken on the basis of a very poor understanding of the nature of science and of development. Many of these misconceptions spring from romantic notions about research, or grossly optimistic assumptions about its potentialities, that can be traced back to the ideology current among scientists of a previous generation. Perhaps, with the rise of movements for social responsibility in science, we are not now ruled by that ideology in quite such narrow terms. But it lives on unreformed, in many distant countries, among the influential elders of government and academic science, who are now under strong attack from younger scientists who resonate to the radical critical call.

Although the critics are often no wiser than those whom they attack, this controversy is, nevertheless, the first sign of genuine vigor and spiritual health in the scientific communities of many developing countries. But the matter is, as we have tried to point out, not at all simple, and first-order, short-range slogans like "relevance" and "academic freedom" need to be qualified, compromised and brought into a coherent synthesis if there is to be real progress. Although each instance of the debate has its local reference, it is really a matter of worldwide concern, in developing and advanced countries alike. A well-informed public discussion in the international scientific press, at international conferences, in the international agencies, and within the scientific communities of the major advanced countries is called for, to build up an appropriate doctrine—a body of principles, examples, priorities and ideals—to replace the stale and shallow clichés of "the conventional wisdom."

It must be emphasized again that science is not an industrial organization or bureaucratic institution; it grows and lives by close interpersonal relationships—collaboration and competition, teaching and learning, mastery and apprenticeship. To the extent that science in the developing countries is a natural extension of the international scientific community, it must depend on just those person-to-person links that make this community a reality. Scientists in the advanced countries have a responsibility to inform themselves of the circumstances of their foreign graduate students and to direct them into fields of research that will not be futile and sterile when they return home. They have a duty to respond favorably to requests for advice, and to put themselves to trouble to lecture and teach abroad, *in partibus infidelis,* as they would to their Dominatian students. A small number of academic scientists in Europe and North America do indeed have high standing in this respect, as sympathetic and understanding friends of science in the developing countries, but this cannot be said of the scientific community as a whole.

In this context, it is perhaps worth remarking that the United States is at a disadvantage compared with Britain and other former colonial powers where there is a long tradition of expatriate teaching, technical and medical work, and research in Third World countries. In almost every department of a British university one may find academic staff who have taught in Nigeria, or practiced a profession in Kenya, or spent some years in an administrative post in India or Pakistan. This tradition is now decaying, as each developing country fills its professional and governmental positions with its own citizens. But there still remain valuable channels of personal contact, such as the custom of sending professors from Britain to many Commonwealth universities as external examiners to maintain academic standards and to advise on curricula.

Although a small but growing number of American scientists have had opportunities to observe science in developing countries at first hand, it is more difficult for them to appreciate the problems and the needs. And one must not suppose that a prestige visit by a Nobel laureate, giving two popular lectures, dining with the Minister of Education and the Rector of the University and then flying on, is an adequate substitute for months or years of patient teaching by a younger, less-celebrated scientist, who may finally come to understand how his temporarily adopted country really works. No single act of deliberate policy could do more for science in the less-developed countries than a political, administrative and fiscal device that opened up a wide range of temporary scientific appointments, by fellowship or exchange visit, in such countries, and made them attractive to young scientists in the advanced countries as a normal stage in a research career.

VI

A particular responsibility for the state of science in Paradisia rests upon those Paradisian scientists who have emigrated and who now work in the advanced research laboratories and universities of Dominatia. The phrase "brain drain" is an unkind metaphor which misrepresents a very complex political, economic, and psychological phenomenon.[7] The case of an academic research scientist, for example, may be quite different from that of a physician or engineer, in that,

---

[7] See, generally, Committee on the International Migration of Talent, *The International Migration of High-Level Manpower*, New York: Praeger, 1970.

in spite of evident need, there may be no opportunity at all for him to employ his talent in his own country, and little to be gained for his fellow-countrymen if he were to attempt to do so. Furthermore, the intrusion of politics into scientific activities also can result in the up-rooting of talented researchers into migration or exile, as recent examples in Argentina, Chile, and Cuba demonstrate. Apart from general factors of standard of living and personal freedom which may make this form of exile preferable to life in one's native land, one should perhaps see this as a symptom rather than a cause of the weakness of science in Paradisia.

Nevertheless, patriotic duty cannot be entirely shrugged off, and the time may come when return seems a credible personal policy. It seems best if this can be arranged in parallel with others of like mind, so that a group of experienced scientists come back to Paradisia as a research team, forming the nucleus of a viable new institution, as has happened recently in Brazil and in the Republic of Korea.

But what happens next? The sociology of this characteristic human situation has not been carefully studied, but one can observe two significant features. On the one hand, jealousy and tension can arise between the returned prodigal and the "good boys" who have stayed at home (or returned after taking a Ph.D. abroad). Since the latter know the local political ropes and have been digging themselves well into the establishment, the return of a distinguished scientist from long exile may lead to nothing but the frustration of high hopes and the stifling of his own creative talents. On the other hand, such a person brings with him an invaluable commodity—personal knowledge of the way in which science in an advanced country really works. We have spoken of the "diffusion" of science into the less-developed countries; it must be recognized that what is transferred is a sensibility, a style of action, a system of interpersonal relations which cannot be learnt from books but which is only acquired by many years of immediate experience. One of the dangers of trying to spread science too quickly is that these characteristics of the scientific method will not be adequately internalized by, shall we say, the Ph.D. with no more than three or four years in a graduate school, and are not, therefore, carried safely back home. If he is modest and thoughtful, wise and strong (poor fellow—it is a hard task!), the Paradisian who comes back to a senior appointment after many years at Bell Labs or Berkeley, at Oxford or Saclay, can bring into his university department or research laboratory the high standards of scientific achievement and the rationales of technique, education and administration that one takes for granted in a modern scientific institution.

VII

The scientific community is loosely organized and governs itself through a variety of institutions such as professional institutes, learned societies, academies, etc., from the local and national level up to such international federations as the International Council of Scientific Unions (ICSU). It is through such bodies that cooperative efforts on behalf of science development in the Third World should properly be directed.

Unfortunately these organizations take almost no interest in such matters. The multidisciplinary American Association for the Advancement of Science, other than occasionally holding a meeting in Mexico City, has yet to show results of activity in science development in spite of a sizable grant from the Rockefeller Foundation to explore the possibilities in this direction. The American and British Institutes of Physics are not directly involved in international scientific cooperation, while the American Association of Physics Teachers has half a dozen activists working on volunteer overtime on an annual budget of a few hundred dollars. A similar situation prevails in national organizations of chemists, biologists and other scientists. The National Academy of Sciences in Washington and the Royal Society in London are somewhat more active, but mainly because they are the official channels through which some governmental aid funds are distributed.

This state of affairs is not altogether surprising. Naturally enough the dominant concern of scientific communities in advanced countries is to further the progress of science (i.e., the acquisition and application of scientific knowledge) in the short run at a maximal rate. Consequently, they operate on the "rich get richer and the poor get poorer" principle: those individual groups, institutions, and countries which are already proficient and successful in doing science are made the primary foci of the benefits of the worldwide scientific infrastructure, and those which are just starting and have little achievement to show are neglected. The proposition that science could advance much faster if the "other" three-quarters of humanity made a proportionate contribution to the scientific community cuts little ice. And although scientists are often involved in political and social matters entirely outside their range of competence, they generally show little interest in the suggestion that the internal practices of the international scientific community could be altered without significant detriment to their own work but with major benefit to the eight percent of the community that happen to reside in less-developed countries. Through their lack of imagination and their utter self-interest, the

SCIENCE AND THE DEVELOPING WORLD   719

leaders of the scientific community miss the opportunity to make momentous social and political contributions to humanity.

An enormous contribution to science in the developing countries could be made simply by enforcing equity within the world scientific community—that is, by treating the scientist in Paradisia on an equal footing with his colleague in Dominatia. To illustrate the present lack of equity, consider scientific communications. Access to journals in Paradisia is limited because the most important journals are published in the scientifically advanced countries, because they are expensive and must be paid for in hard currency, and because they are heavy and quick access by airmail costs more. Reports and reprints reach Paradisian research centers only sporadically, because most of these documents originate in scientifically advanced countries, are expensive to ship abroad, and are produced only in limited numbers which usually are sent to the most prestigious research groups, i.e., those in advanced countries. Scientific conferences are usually held near the major world centers of scientific activity, so that scientists from developing countries have difficulty in finding sufficient travel funds to attend them.

Moreover, at conferences with attendance by invitation only, scientists from the less-developed countries are under-represented since they have not had the opportunity to accumulate the visible credentials (i.e., published papers) that would seem to justify such invitations. Scientific centers in the developing countries are infrequently visited by other scientists because they are geographically remote from the standard interconference travel itineraries of the majority of scientists.

The same strictures apply to science education, the tools of scientific research, science organization, scientific manpower development, and to all other aspects of the scientific enterprise: the less-developed countries are not only far behind, but are handicapped in their access to the communal tools which are needed to make further progress. It is not sufficient, within the world community of science, to give the rich and powerful precisely equal opportunities with the poor and weak to spend their money and cast their votes. Only a deliberate policy of internal social justice can keep this community together in what may be a very hard and divisive political era. What would be the annual cost, for example, of sending free copies of the major scientific journals to all accredited scientific institutions in the less-developed countries? Perhaps ten million dollars (who knows?), which is only a tiny fraction of yearly expenditure on scientific information in the advanced countries. The problem is not whether it *ought* to be done, but in deciding which of many appropriate social institutions—industry, government, the professional organizations, etc.—should actually bear the cost.

Unfortunately, at the international level where such matters should really be discussed, the professional organizations are weak, poorly financed and politically divided. Scientists from the less-developed countries themselves should be much more vocal and aggressive in shaping policies and plans within the world scientific community. It is regrettable that the existing literature and activity concerned with the problems of science in development are dominated by contributions from scientists outside these countries, and that representatives of Third World science seldom plead their case before meetings of the international scientific community.

VIII

Since science development is a worldwide task, whose characteristic problems are not confined to a single country or region, a large effort surely ought to be made by the international agencies established under the aegis of the United Nations. The Office of Science and Technology at the U.N. headquarters is concerned with international agreements having scientific ingredients, and the other specialized agencies such as the International Atomic Energy Agency (IAEA), the World Health Organization and the Food and Agriculture Organization are charged with the progress and propagation of science-related technologies and practices. These agencies certainly contribute a great deal to the diffusion of scientific knowledge along "mission" channels to the less-developed countries, but their activities do not impinge directly on the growth of indigenous science in those countries. In a complete survey one might also mention regional organizations, such as the Organization for Economic Cooperation and Development and the Organization of American States, which are active in some aspects of science policy, scientific exchanges and communications.

In principle, however, the U.N. Educational, Scientific and Cultural Organization (UNESCO) is supposed to be the primary organ for scientific cooperation and development. It has programs in science education, including some efforts at curriculum development, and in science policy, which generate poorly distributed documents (occasionally of high quality), and a few formal and repetitive conferences. In fact many of its activities are grandiose in conception, technically ill-informed, inadequately financed and largely irrelevant to the immediate if mundane needs of scientists and technologists in the less-

SCIENCE AND THE DEVELOPING WORLD 721

developed countries. As with most of the U.N. agencies, UNESCO programs are hampered by the need to satisfy political conditions—"neutrality" and parity of national prestige—and by a heterogeneous international civil service of very uneven quality, so that the projects that finally see the light are often noncontroversial, unimaginative and a poor value for the money that is spent on them. Very few of the U.N. science activities directly involve rank-and-file scientists from the advanced countries, or even from the developing countries, upon whom the main burdens and responsibilities for their own science development must finally rest.

An exception, which might serve as a model for other institutions, is the International Center for Theoretical Physics (ICTP) in Trieste, Italy, which is maintained jointly by UNESCO, the IAEA and the Italian government, (with important contributions from the U.N. Development Program [UNDP], the Swedish International Development Agency, the Ford Foundation and other agencies). The ICTP provides a place where working scientists from the developing countries can stay for several months at a time, carry out research, participate in advanced courses and seminars, and make personal contact with the leading members of the international scientific community.[8] Despite a modest budget and limited administrative resources, the ICTP has already performed a valuable service as the means by which many of the younger physicists in some of the less-developed countries have been kept alive professionally in their own countries. It is also unique as a high-level scientific institution where scientists from the developing countries gather together in their own right; it has been interesting to observe the growth of a corporate awareness of their common problems in this group over the past decade, and the extent to which the romantic view of science for its own sake has been modified by the demand that it should be directed to more immediate human goals.

Unfortunately, the ICTP stands practically alone as an international institution dedicated to the creation of strong indigenous science in the less-developed countries. Similar institutions oriented toward the experimental sciences would require larger funds and more sophisticated administrative and technical bases. The existing international research institutes connected with the Third World—the International Centre for Insect Physiology and Ecology in Kenya, and the International Rice Research Institute in the Philippines—are not perhaps on a sufficiently large scale to show their full possibilities.

It is interesting to note, however, that the idea of creating very large

[8] See J. M. Ziman, "The Winter College Format," *Science*, Jan. 29, 1971, p. 352.

and sophisticated international research institutes to tackle particular technical problems is now very fashionable, and budgets to the tune of hundreds of millions of dollars are glibly sketched out by high-level committees. Whether many of these splendid dreams will come to fruition remains to be seen; what we must ask is whether these will be, once more, mere offshoots of the scientific establishment of the advanced countries, or whether they will make a contribution to science development in the broadest sense.

National governmental development agencies often have much larger financial resources than parallel international organizations, and hence could be a vigorous source of assistance to less-developed countries in building their science. Yet in general such agencies are dominated by narrow, short-term thinking, focused primarily on technology, and usually enjoy a monopolistic position in their country which encourages a monolithic structure reluctant to experiment. Although science development is within the mandate of many of them, it remains very much on the periphery of their concerns. These agencies have failed to involve a significant number of scientists in their activities, either from the "donor" or from the "recipient" countries. As a result, many of the programs that do exist fail to be of real help to the working scientist in Paradisia.

IX

The gist of our argument is simply stated: insufficient thought and effort is being given to creating and maintaining indigenous scientific activity in the less-developed countries. The actions of the countries concerned are uncoordinated, and are often based on very poor grasp of the real issues. The world scientific community is negligent of the plight of this small fraction of its members. The international agencies lack the resources, and the big aid programs, such as those of the U.S. government, lack the understanding, to act effectively. A highly significant fraction of general social and economic development, with enormous long-term leverage, is being left to the mercy of a variety of haphazard forces—political whim, crude economic theory, intellectual fashion, administrative convenience.

It is evident, however, that this is not a problem that can be solved simply by the expenditure of very large sums of money under the guidance of a committee of science policy experts. When we talk of a "scientific community" in Paradisia, we refer to a few hundred highly individual, highly trained and intellectually independent people, whose attack on the problems of nature cannot be commanded like the advance of a battalion of infantry. Only careful personal attention to their aspirations, needs and capabilities can really help them or direct their energies in the right directions. The cost of such attention is much higher in professional scientific grasp and human understanding than it is in dollars and cents. How much easier it is for a leading scientist in Dominatia to advise his government to give a $50,000 instrument to the University of Paradisia than it is to spend a few days there, explaining tactfully that the experiments proposed for this instrument are now obsolete and suggesting worthwhile alternatives! How much simpler to use the Paradisian graduate student as another hand in a Big Science research team than to devise a Ph.D. topic that will train him to build his own apparatus and to think for himself!

The prevailing style of social and political action is thus quite unsuitable. It involves large institutions, organizational charts, massive budgets, and professional administrators. This style can be highly successful in managing a project which is purely technical, which has primarily in-house or directly contracted extramural components, and which operates on a large budget. But science development has no reliable technology, cannot be done in-house, cannot rely on large contractors, and does not, and need not, operate on huge budgets. It is a program aimed at improving social and human conditions, for which the administrative machinery for "putting-a-man-on-the-moon" is entirely inappropriate. The proper aim is to encourage and stimulate the latent potential at the grass roots of the world scientific community. If only ten percent of the world's scientists could be effectively involved in cooperation with the scientific communities in the less-developed countries, a great stride would have been taken across the gap.

But in addition to the mobilization at the grass roots of the world-wide scientific community itself, there is also a need to reorient the existing developmental organizations toward the actual need of science building in Paradisia. Such a reorientation would involve both fresh action and fresh research on which such action can be based. We need a better quantitative assessment of scholarships and fellowships for scientists from the less-developed countries to study in the advanced countries, or of the cost to distribute the major learned journals to all the institutions that ought to have them. We must have a better understanding of the ways multinational corporations could contribute to science development, or of the ways the U.N. agencies could bolster science in the less-developed countries by farming out their research activities to universities and research institutions in those countries. We would want to have a realistic estimate of the cost of research facilities needed to reattract to their home countries Paradisians who now are working as trained scientists in Dominatia. Generalities about these questions are available, but hard information to serve as a basis for specific action is scarce.

It is insufficient to relegate these investigations entirely to developmental organizations. Here also, the involvement of individual scholars is needed, whether scientists themselves, or economists, historians, sociologists, or political theorists. At the moment there are very few of these who regard the worldwide social phenomenon of science development as a worthy subject for research. As a result, whatever decisions are taken by developmental organizations are, by default, often based on a combination of preconceived academic notions, wishful thinking, and administrative convenience.

It is not the purpose of this article to suggest specific organizational frameworks for the reorientation advocated here, even though the perceptive choice of such frameworks is essential to success. Preceding such a choice, we must have a widespread awareness of the problems of science development and their great urgency. That is the aim of this discussion.

THE CONTEXT OF CREATIVE SCIENCE*

MICHAEL J. MORAVCSIK

It is almost a truism to say that virtually every human activity aims at some appreciable and gratifying result. Tillers of the soil, for millennia, have recognized this strive for creativity in their festivals and religious ceremonies. Craftsmen in the Middle Ages built much of their social organization around visible products of their labor. In more recent times, countries and communities characterize their economic efforts in terms of the products created thereby, and workers, even in mass production plants, where the results of individual efforts can be ascertained only as an ingredient in some collective undertaking, take some pride in the output of their factory as a whole. Teachers find gratification in their outstanding students who bring to fruition the persistent process of education. Even the psychologically most deadly and dehumanizing occupation, that of a lower level government bureaucrat, tries to find gratification in some small and often formalistic "achievement" of the office or the division.

This desire to be creative within the framework of one's work is, of course, very much present in science also. Yet, one certainly cannot say with certainty that every scientific group, or every country undertaking scientific work, is in fact successful in such creativity. The reason for this is partly organic and partly semantic. It is organic in that sometimes success is not attained inspite of definite goals and this failure can be definitely established. But often it is semantic in that the goals of creativity may be ill defined, ambiguous, or might differ from viewer to viewer (or participant to participant), and that the measures of success themselves may be vague or illusive. In this essay, therefore, I want to analyze the context of creative science, that is, the meaning of creativity in this area and the way one can gauge creativity. It would be ludicrous to claim that the essay will come anywhere near to a definite and final clarification of these issues. And yet, it is very worthwhile to ask questions and contribute to some answers. Scientific activity is very much in the forefront these days, because of public resources needed to pursue it, because of economic, technological, moral and social issues with which science is, rightly

* Written during a year-long visit to the University of Sussex, Falmer, Brighton BN1 9QH, Sussex, England. The visit was in part supported by the U.S. National Science Foundation.

SUMMARY Any kind of human activity seeks some visible and gratifying result. This is also true for science: those practicing or supporting it like to see creativity which produces tangible results. Sometimes this goal cannot be attained because the goals and purpose of science are not clarified and because, even with set goals, the means for evaluating the results are lacking. In this essay, therefore, the goals, motivations, and purpose in science are analyzed and a survey is then given of the various measures for the output of science. The essay deals only with science, and does not discuss analogous questions in connection with technology.

In discussing the goals, it is first suggested that, like many human activities with a substantial and lasting appeal, science has a broad spectrum of goals, some of which might even contradict others. On the whole, however, this multitude of motivations can be summarized in three broad headings: (a) Science as a basis for material development, (b) Science as an outlet for cultural, intellectual, and spiritual development, and (c) Science as an influence on human world view.

The first of these is often emphasized, even though the intricacies of the process whereby scientific discoveries lead to material improvements are very often misunderstood. The second, science as a "higher" human aspiration, is sometimes condemned but at the same time glorified, perhaps because it is difficult to admit that throughout human history all societies have found it necessary to be also concerned with "useless" aspirations, even when the material improvement of their lot was far from complete. It is also stressed that the materially "useful" and "useless" aspects of science can hardly be separated from each other (which, however, should not prevent countries from trying to obtain a certain balance between "fundamental" and "applied" scientific activities in the short run).

The third goal of science, that of contributing to the human world view, is of particular importance in the context of development, because the anti-authoritarian, objective, optimistic, and anti-fatalistic character of the scientific method is a necessary prerequisite for evolution in the modern sense of the word.

These goals suggest three questions that should be asked about every piece of scientific research to be undertaken: a) Is the researcher genuinely interested in and excited about the problem? b) Will the results of the research contribute to our understanding of nature so that other researchers will become interested in the results and will build further research on them? c) Will the results of the research help toward an application to a material or social problem so that technologists or others might welcome the results and utilize them? For creative science the reply to the first of these questions and to at least one of the other two should be "yes". These criteria are emphasized because at the present time a certain piece of research is not infrequently justified by the sole criterion that it has never been performed before. Such work often turns out to be sterile both in its effect on our understanding of nature and in its influence on applications.

To conclude the first part of the discussion, it is suggested that each scientific community must take an active role in formulating and maintaining its own goals, motivations, and purpose, and not let bureaucrats, politicians, businessmen, or party functionaries impose them from the outside.

In the second part, dealing with a survey of the evaluative means of science, first some systematic and fairly quantitative methods, many of them very recent, are discussed. They all pertain to the scientific literature, and deal with the number of scientific authors, with the number of scientific papers, and with the number of citations scientific papers receive in the scientific literature. Some interesting results have emerged from the study of these measures of scientific output, but at the same time there are also serious shortcomings in the use of them.

The other type of evaluative tool used in gauging the output of science is the peer judgement and review, in which scientists are asked to judge their colleagues in the same field. This has been used for a long time by scientific journals, by grant-giving agencies, and in other situations

Dr. Michael J. Moravcsik *Professor of Theoretical Physics, University of Oregon, USA. Author of some 140 articles on elementary particle and nuclear physics and on science policy and the "science of science", with particular emphasis on less developed countries. Author of "Science Development", a book published in 1975, Institute of Theoretical Science University of Oregon, Eugene, Oregon, USA., 97403.*

or wrongly, connected in the public mind and because science plays a prominent role in the international culture of the 20th century even apart from its material connections and consequences. And yet, in dealing with science, communities, countries, or organizations, can exhibit remarkably simpleminded attitudes which result in being satisfied with, or even aiming at, science which is that only in name, and which could hardly be called creative by any definition. The magic of quantification lures people into neglecting the fundamental discussion about goals and purpose which seldom can be done numerically, and impatient frustation with the difficulty of quantifying scientific output tempts people into substituting input for output. Thus, for example, we find the assertion in a national development plan that the aim for science is to produce, say, 3,560 more scientists in the next three years (as if the purpose of science were the production of scientists).

This carelessnes in talking about (and taking action in connection with) science is a disease of "undeveloped", "less developed", and "advanced" countries. Thus a discussion of this topic is perhaps a most appropriate subject for a journal that aims to appeal to the scientific communities of all the Americas, including some "very developed" and some "very undeveloped" countries. Furthermore, large but generally less developed countries with a quite sizeable scientific manpower, such as India or the People's Republic of China, might particularly feel the relevance of such a discussion since in the development of such a sizable manpower over a relatively short period quantity invariably gets the upper hand at the cost of quality.

The discussion will be in two parts. In one, I will analyze the goals of science and thus establish a framework for talking about creativity in science. In the other, I will turn to methods of evaluating scientific output in the light of the creative goals outlined. Finally, I will give a brief summary of the main conclusions and relate them to a cardinal element of science policy. It should be understood that this essay deals only with science and does not discuss the analogous problems pertaining to technology.

Goals, Motivation and Purpose of Science

Is is a sometimes confusing, or even disturbing, but at the same time extremely salutary and enriching fact that goals, purposes, values, and "philosophies", vary greatly within the human community from group to group, country to country, and age to age. It is confusing because the analysis and direction of human behavior and activity would be infinitely simpler and

also. Among the problems connected with this method, there is the difficulty that, when used in an international context, nationalistic feelings and politics might make peer review suspect. It is suggested, however, that international peer reviewing, arranged through scientific and informal channels, and without much publicity, has not been given an adequate opportunity to prove itself.

As a concluding remark, it is predicted that concern with science policy is likely to become even more important in the near future, so much so that every young scientist migh be required to show some acquaintance with it before obtaining and advanced degree. Hence scientific communities should increase their participation in the consideration of problems like those discussed in this article.

....

more logical if such diversity did not exist. Accordingly, individuals or cliques who have managed to amass great power have tried, from time to time throughout history, to eliminate this variety and to mold, by force, the world or part of the world, according to their particular set of tastes and values. They have invariably failed and are also failing today.

In the light of this overall kaleidoscopic state of affairs, it is not surprising if a particular activity, such as science, finds no single goal, no single purpose, no single motivation, and no single justification either. In fact, in order to appeal to and have support from a broad segment of humanity, it must have a variety of goals, a multitude of rationales, and those participating in it must find a broad spectrum of justifications for doing so.

While this statement appears rather trivial, it must precede our discussion because to some people having many justifications, some even contrary to others, is is automatically a sign of a sinister, Macchiavellian undertaking. Thus, to take a simple example, the fact that science can, simultaneously, please the sick by offering cure and please the general by offering new weapons appears to some as an *a priori* strike against science. I would like to suggest that perhaps the contrary is true, and that activities which have meaning in only one single philosophy and value system are, in fact, likely to be less interesting, less significant, and also less dynamic, in the long run, than activities with a multidimensional appeal. Thus to my mind, there is nothing inconsistent or hypocritical in a situation where say, Bell Telephone Laboratory supports some research in the hope that perhaps some day something marketable by the company would emerge from it, while at the same time the scientist performing the research does so primarily because of the personal excitement and esthetic satisfaction he obtains from it. This is a point well worth remembering in a discussion dealing with "pure" vs. "applied" research, a discussion that is ubiquitous in science policy bodies.

While accepting the inevitability, and, in fact, desirability of a variety of goals, motivations, and purposes surrounding science, one can nevertheless try to order and classify them. While this can be done in a number of ways, I would like to suggest, for our discussion, a tripartite classification which could be headed by

the titles of, a) Science as a source of material improvement, b) Science as a manifestation of human curiosity, estheticism, and "higher" undertaking, and c) Science as an influence on the human world view.

The first of these is perhaps paramount in national or international science policy circles, even though the process by which, or the extent to which, science brings about material improvement is little understood or, more precisely, predominantly misunderstood. It pertains to the relationship between science and technology (the difference between the two in itself being often misunderstood or ignored), and the ways whereby science can be "applied".

Science as a manifestation of human curiosity and esthetic striving, in other words, science for its own sake, has a curious position. On the one hand, science policy makers often assume that the population as a whole has no interest in this aspect of science at all. On the other hand, actual contact with the same population quickly yields a different conclusion. Far more often than not, the "average man" queries the scientist first about some "fascinating" but esoteric scientific discovery, gleaned from a newspaper column or from radio reports. That science also makes food or colored TV sets appears to occupy a distinctly secondary place in the layman's mind. As a result of this, but again paradoxically, national spokesmen (self-appointed or otherwise), who on the one hand heap scorn on "science for its own sake", will quickly turn around and take great pride in the "national" accomplishments in "pure" science.[1]

What is, of course, forgotten and ignored in such ambiguous attitudes is the fact, thousands of years old, that all societies find it necessary to turn toward "useless" aspirations and goals in addition to taking care of the necessary tasks of providing themselves with food and shelter. In fact, societies which lose sight of goals other than the immediate material ones tend to die, being swept away by other dynamic societies, fuelled by a strong concern with their own "useless" aspirations.[2]

This is a point to be kept in mind particularly because it has been claimed that less developed countries, with material standard of living far below those of the "advanced" countries, have no business at this time to be concerned with anything but their material improvement. According to this view, there is science for the "rich" and science for the "poor", the former being pure luxury, and the latter having the exclusive aim of improving the material means of the country.[3] I always found this view not only absurd in a practical sense (since scientific research cannot be cleanly classified in terms of which of the three main aims it serves), but also utterly insulting to the less developed countries themselves.[4]

In fact, the entire argument about "useful" and "useless" science is, to a large extent, a red herring. Nuclear physics in 1939 was a most esoteric science, undoubtedly viewed by some as science "for the rich", and its practice an activity only for its own sake. Yet fifteen years later, many less developed countries would have immensely benefited from domestic know-how in nuclear physics as used in power generation, in agricultural research, in medical diagnosis, and many other areas. Needless to say, such a know-how was not present in those countries in 1954, and in fact is hardly present even today.

Countless other examples could be brought up to illustrate the inseparability of "useless" and "useful" science, and the utilitarian and selfdirected aspects of science. It is one of the marvelous features of the world that such "idealistic" or "self-less" efforts, in the exploration of the secrets of nature for their own sake, are rewarded, in the long run, inevitably, with by-products for the improvement of man's material life.[5]

Finally, besides science as a source of material improvement and science as a manifestation of human curiosity and estheticism, we have science as an influence on man's world view. This aspect of science, seldom if ever discussed in science policy documents, has an impact which is hard to overestimate. There are several aspects of this influence.

Perhaps foremost, science tells us that very powerful knowledge comes to us not from what grandfather says, or what the government decrees, or what the chairman of the board of our company believes, or what we read in the Little Red Book, but instead, from turning to nature itself and asking questions and receiving answers through experimentation. As a corollary, science also tells us that knowledge is by no means a closed system, and new questions, new efforts, yield new answers and new knowledge.[6].

Second, science also tells us that the answer to these questions can be obtained quite unambiguously, by a methodology that scientists around the world subscribe to and accept, regardless of their nationality, race, religion, social status, political ideology, or moral values.

Thirdly, science is based on strong optimism, in that scientists always implicitly assume that given the patience, talent, and resources, every question will yield an answer, every scientific problem a solution.

And, finally, science tells us that through knowledge one also gains power and control over nature, so that one's fate is not determined entirely by external circumstances but that it is possible for man to navigate purposefully amidst the dangers and accidents of natural phenomena.

It would of course be a gross exaggeration to claim that the above traits of science will help man with all the problems he faces. But since in a non-scientific civilization the above considerations generally do not carry weight at all, their introduction into the local world view can make an enormous difference. In fact, one can confidently claim that without these elements in the outlook the development of a community or country in the contemporary sense of the word is almost impossible.

For the propagation of the scientific attitude among the population as a whole, it is necessary that indigenous creative science be present in the country itself, so that its practitioners can interact with the rest of the country and transmit the proper spirit. It hardly helps to have scientists only in the formal sense, people with degrees who have not practiced science for 30 years, and for whom science consists of a dry recitation of facts. From such persons one could learn only the wrong attitude and spirit, and miss precisely those elements of scientific thinking that are most essential in a dynamic and self-confident outlook. Considering these general goals of science, can we now formulate guidelines for the scientists in his choice of scientific problems and in his structuring of his research work? In a general way this can indeed be done. In particular, I would like to suggest the following checklist when making such scientific choices:

A) Are you genuinely intrigued by, interested in, and enthusiastic about the problem you attack?

Personal involvement in this sense is an absolute prerequisite for creative scientific work. For one reason or another, the researcher must feel a personal drive towards solving the problem. If the motivation is purely external, the work is likely to be at best routine, listless, and void of significant results.

B) Does the solution of the problem represent a positive step within the broader context of science? Would any other scientist in the world be interested in the solution, would he use it to build further research on it? Does the solution contribute noticeably to a better understanding of the laws of nature?

C) Does the solution of the problem represent a positive step in terms of the application of the results toward an approach to a material or social problem? Will some technologist be interested in the result and be stimulated by it to develop, build, or design.

I would like to suggest that within the context of creative science virtually any research problem chosen by scientists should evoke a "yes" on question A, and a "yes" on at least one of the two questions B and C.

Outlined in such blunt terms the above criteria appear to be self-evident. Yet they needed to be stressed, because a sizeable fraction of research work around the world, some in "advanced" and some in "less developed" countries, is motivated not by answers to any of the above questions but rather by the single question: "Has this piece of research been performed previously?". If the answer is "no", it is taken as sufficient justification for undertaking the investigation. The result of such work is hardly creative science. Instead, it produces results which, though never attained before, are of interest to practically nobody besides the author, contributes virtually nothing to our overall understanding of nature, are highly unlikely to be of any use in application and simply add to the bulk of piecemeal data on certain narrow areas of natural phenomena, or to the collection of theoretical calculations either of a purely formalistic bend or of phenomena which are of no conceptual or practical importance. Being esoteric in this sense is then often claimed to be a sign of depth, and the work is claimed to be a piece of "pure" research, thus giving a bad name to basic research in an environment where its value is often misunderstood even without such misrepresentations.[7]

As to the criterion in terms of the third goal of science, that of propagating a scientific outlook, most pieces of research, "pure" or applied, which have high merit under the above three questions, will help towards this third goal also, provided that the scientist in question has the appropriate personality and makes the necessary effort to become a "scientific missionary". Admittedly, one cannot expect every scientist to excel in such communication with non-scientists. But some will and must. Contacts in everyday life, with government officials, through radio and newspapers, through popular lectures, through exhibits and demonstrations, offer a huge variety of outlets for the creative scientist to whom science is a way of life, a method of asking questions and receiving answers, a key to problem solving, and who wishes to give the public a testimony about the enrichment that such a way of life can offer.

Finally let me mention that a discussion of the goals of science has to be an explicit and ongoing concern of the scientific community in any country and, particularly, in the scientifically less advanced countries where such questions received less exposure in the past. The absence of such a substantial, frank, and detailed dialogue in most countries appears to me to be one of the most regrettable faults of scientific communities and a demonstration of their weakness or incoherence. In the absence of such a discussion there will either be a vacuum, or the imposition of "goals" by government bureaucrats, ideologists, or politicians, usually with disastrous results for the development of indigenous creative science.

The Evaluation of Scientific Output

Let us now assume, perhaps somewhat utopically, that in a given scientific community the problem of setting goals, defining a purpose, and establishing motivations for scientific activity has been solved. We then still need ways of ascertaining whether the actual work of this community indeed conforms with the criteria for creativity thus established.

It is well to admit right from the start that there are, at least at the present, no such ways which would not be open to valid criticism as to their soundness, self-consistency, and comprehensibility. The reason for this is rather clear. The product of scientific work is ideas, observations, understanding and knowledge, and it is extremely difficult to quantify these. In contrast, the output of technology is patents or actual prototypes of gadgets or procedures. These are tangible and can be numerically dealt with much easier. Similarly, the products of agriculture are crops which can be measured easily, the output of factories are goods which can be counted and appraised.

It is for this reason that economists, who usually play a dominant part in national planning, are often confused and insensitive about science as an element of national development. As mentioned earlier, they may, consequently, use input for output ("the quota for scientific output is 3,560 new scientists in the next three years"), or they fail to differentiate between science and technology and thus look for signs of scientific output in the statistics of agricultural or industrial production. When they fail to find them (as they are bound to), they assert that science has no role in economic development, or that the particular science in that country does not do its job.

Fortunately, however, the picture is not completely bleak. There are some methods of gauging the creativity and productivity of science, even though these methods need considerable improvement, generalization, and supplementation. Some of them are traditional, others fairly recent, so that a brief review of the overall status of the field might be helpful.

Methods of evaluation can be roughly divided into two types: a) Systematic, impersonal and fairly

quantitative methods, b) Personal procedures involving peer judgement and review.

The first type is generally centered around scientific publications. The assumption here is that since scientific research is a collective undertaking among scientists of all countries and all times, scientific achievements can be monitored best at the point where they enter the scientific public domain that is, when they are communicated by the author to the rest of the community. This assumption is probably fairly sound. It is then also assumed, [8] now standing on much shakier grounds, that this communication is virtually always reflected in a journal publication, and hence the quantitative assessment of the scientific literature is equivalent to a quantification of scientific output. This assumption ignores all informal written and spoken interaction among scientists, such as preprints, conferences, or person-to-person discussions in offices and laboratories, channels which, as every scientist knows from personal experience, play an immensely important role in doing science.

Even apart from its questionable basic assumption, the quantification of scientific literature runs into difficulties also in its practical realization. This is evident from a quick survey of what is actually done in practice. Perhaps the simplest approach is to count the number of scientific authors, that is, the number of scientists who have more than a certain minimal number of scientific papers per year to their credit in recognized scientific journals. It is interesting that even this very simpleminded approach, highly vulnerable to obvious objections, can supply some apparently intriguing results, [9] at least when applied to national scientific output. I doubt, however, that anybody would want to make use of this indicator to compare, say, two departments at a university, or two research groups in a laboratory.

The next step is to count publications rather than scientific authors, thus giving greater weight to the more prolific scientist. Such an emphasis on the length of publication lists is said to be common practice among deans at universities who judge promotions, but is used in other contexts also. This measure obviously ignores the possible differences in standards among various journals, and the importance (or quality) of the article itself, let alone the even more mechanical aspects such as the length of the article, or the degree of difficulty of the experiment or calculation.

The obvious next step, therefore, is to try to weight publications by their scientific "importance" or "quality". A way to do this which recently became quite popular is to count the number of citations [10] an article receives in the scientific literature. The assumption here is that a more meritorious article will be more often cited than a routine one.

Tracing citations has in fact yielded some interesting results. This is perhaps less so in the context of science policy where one is interested in evaluating the quality of scientific output. Instead, citations proved fruitful in the "science of science", that is, in the study of the structure of the scientific method and of the scientific community. For example, citations (and the more refined concept of co-citations[11]) helped to "map out" scientific specialties, and to write a history of certain discoveries and of the group of researchers (often geographically widely dispersed) who were connected with it.

Yet citations can also be attacked on many grounds. [12] An article can be cited for many reasons, some of which might not be scientific, while others might not shed favorable light on the cited article. Thus the next step is to classify citations according to their context [13] in the citing articles. While this can most probably be done fairly unambiguously, it requires (at least at the moment) human attention (and possibly attention by a trained scientist), and so it cannot be done on a large enough scale for all purposes.

With all these uncertainties, whether one would want to use any of these measures of science policy considerations depends on the accuracy required. For example, if one can show that the number of papers published per year per scientist does not vary by more than a factor of 5 among averages for various countries, and then one compares two countries where the number of papers published per year differs by a factor of 10,000, it makes little difference whether one uses the number of scientific authors or the number of published articles. What is important is to constantly keep in mind, in a quantitative way, the limitations of each of these measures, and thus avoid using them beyond their capacity. With this warning, however, I would strongly recommend that those active in science policy, organization and management, become familiar with developments in the study of these indicators.

The second type of evaluative tool is the personal assessment through peer judgment and review. In this the performance of a scientist or a group of scientists is gauged by other scientists familiar with the work in question. This is perhaps the oldest and most prevalent evaluative method used in science. Most journal editors use scientists for refereeing papers submitted to them. [14] Virtually all governmental agencies in the United States that distribute research support use a broad segment of the scientific community itself to evaluate research grant applications from scientists. On the whole, this method of evaluation can claim much success. It also has, however, many weaknesses.

For one thing, when a whole field of science is under scrutiny, the method becomes controversial, because, on the one hand, it is claimed that those within the field cannot objectively evaluate themselves while, on the other hand, it is also claimed that those in other fields do not have the knowledge and perspective to assess the field under scrutiny. Thus, for example, if an overall assessment of the effectiveness of research in high energy physics were proposed (as, I think, it should), it would take some time to find an investigating team that would be even moderately acceptable to everybody involved.

Second, even if only a particular proposal is being assessed and therefore appropriate experts are available, they must be selected with a full knowledge of the positive and negative interpersonal relationhips within the scientific community which might distort the assessment. These factors cannot always be taken into account completely.

Among the other blemishes of the peer review system, I want to mention one of particular importance in the framework of this article written for an international audience, namely the problem of national sensitivities. It is often thought that because international assessing teams are supposed to have many members from the scientifically advanced countries with a large scientific manpower, their conclusions would be tinged with political controversy and, hence, would not be heeded.

I would like to suggest that since hardly any definite precedents are on record of such international teams of assessment, [15] the above pessimistic predictions are at the moment mere theorizing and hence one should undertake a much more extensive program of experimentation with these methods before sentencing them to oblivion. In particular, I want to suggest that if such evaluations are done, not through formal international organizations but through the informal internal channels of the scientific community, and if the assessments are carried out without fanfare, newspaper coverage, and ceremonies, and if the evaluation teams are composed of competent scientists from many countries who are under no political or ideological pressure by their own governments, then the method can work without arousing undue nationalistic feelings.

The utilization of the entire international scientific community for evaluation would be an immense help to many countries with a small scientific manpower where the number of practitioners in a given field is too small for a domestic peer review system to work, and where, because of the small size of the community and other reasons, personal animosities among scientists are often much more pronounced than in the scientifically large countries. The system would also be a boom to the local or regional scientific journals in refereeing articles.

An efficient creation of such an international pool of peer reviewers would probably require some organizing within the scientific community, but, as mentioned before, without the involvement of those national and international organizations which are highly visible and politically constituted, such as (speaking now about the pan-American context) the United States Agency for International Development (US AID), the Organization of American States (OAS), UNESCO, or others. Instead, one could perhaps utilize bodies like Centro Latinoamericano de Fisica (CLAF), the American Association for the Advancement of Science (AAAS), the American Physical Society (APS), etc. Alternatively, one could build an even more informal network of connections that could arrange such assessments. There is no reason why science in the United States itself could not also be subjected to assessments by such interamerican teams.

The existence of such an international evaluating standard of recognized quality and manifest impartiality (at least within the context of local skirmishes) would clearly be very helpful to scientists of excellence in the smaller countries, and probably would also be appreciated in those countries by those science policy makers who are cognizant of the problems of building a scientific infrastructure but who have been hampered in the past by the difficulty of making reliable judgements. Thus, the method certainly deserves much more active attention than it has had in the past.

Conclusions

In the previous two sections I attempted to analyze the need for purposefulness and goal definition in science if it is to be creative and then outlined some ways, however imperfect, of assessing the success of science in view of the goals set. In particular, I listed three types of goals and purposes for science: Science as a basis for material development, science as an outlet for cultural, intelectual, and spiritual development, and science as a component in the world view. I indicated that different groups, different countries, and different ages may emphasize different aspects within these three categories, and may assign different weights to them, but generally all three types must be present to a substantial degree for science to achieve true creativity. In view of these classes of goals, I listed three questions that a scientist (or a supporter of scientific work) should ask himself in connection with a piece of research to be undertaken. The three questions have to do with personal involvement, with the impact of the research on the rest of science, and the impact of the research on possible applications. I then asserted that virtually any research must appeal to the researcher and must have appreciable merit either in terms of contributing to the whole of science, or to some application, or to both. Research which can be justified only because it has never been performed before does not deserve support or effort. [16]

Turning to the evaluation of science, I briefly outlined recent developments in finding impersonal indicators of scientific output, particularly in terms of various quantifications of scientific literature, and urged those involved in science decision making to become acquainted at least with what has been done in this area. I also outlined the personal peer review system and urged a greater utilization of it on an international scale.

I want to conclude this article by an observation that follows naturally from the above discussion. I emphasized the necessity for each scientific community setting its own goals doing that for itself lest such a definition of purpose and motivation be imposed from the outside. Accordingly, the check list on criteria involved scientists querying themselves about the estimated impact of the research to be done. Then, in the section on evaluation, not only did the impersonal methods involve the literature of the scientists themselves and the classification of citation types by scientists, but the personal method depended almost entirely on scientists participating in the assessment process.

It is therefore difficult to avoid the firm conclusion that the health of creative science depends crucially on scientists playing a significant role in science policy. Clearly science policy has aspects which connect the scientific community with society as a whole, and such points of contact must be manned by a combination of scientists and others. But many aspects of science policy are internal, and these can and should be taken care of successfully by scientists themselves. Four decades ago, when science was small and inconspicuous, there might not have been a need for such an involvement by scientists. Today it is unavoidable. Inspite of this, the prevalent attitude in the scientific circles of most countries is to ignore the acquisition of knowledge about the workings of science and how it is to be provided for and hope that somebody somehow will take care of it competently so that scientific work can continue smoothly.

I predict that the time is not far when every young scientist, at the time when he receives his doctoral degree, will be required to show at least some acquaintance also with "the science of science" and with science policy and management, just as engineers are now often given some training in business and economics, and doctors are increasingly oriented, during their training period, to the social and psychological context in which they will have to operate. So, as a concluding point, I would like to urge the scientific communities to rise to the ocassion early and greatly enhance their participation in the consideration of the kind of problems discussed in this article.

NOTES

1. One of the numerous examples is the People's Republic of China, where there are constant ideological statements condemning, as a cardinal sin, the indulging in Science for its own sake. At the same time, a delegation from that country recently visited major accelerator centers in the United States to decide what type of accelerator the People's Republic should build for its own high energy physics, a scientific prestige field which, at the moment, offers practically no application whatever. A similar episode in technology is the concrete interest of the People's Republic in the Anglo-French prestige supersonic Concorde, even after most other airlines dropped out. One might, however, debate whether excessive nationalism is or is not "utilitarian".

2. For an easy and most rewarding introduction to the rise and fall of civilizations, I would recommend Kenneth Clark's "Civilization: A personal View". See Clark, 1970.

3. Two examples of this view in the Latin American context can be found in Varsavsky, 1967 and Lopes, 1966.

4. An unusually eloquent statement of this point can be found in an article by Marcel Roche, (1966), which I wish to quote: "One often hears the opinion expressed that only research which is immediately useful should be publicly supported. This is understandable but unfortunate. Latin America will start to contribute significantly to humanity's scientific progress —and to its own material well-being at the same time— when it loses its complex about the need for practical results, and simply develops a passion for knowledge rather than a simple desire for material progress. The day our community, and our scientists, discover the sense of purpose in science —whether pure or applied— we shall be able to utilize to the full, without social distortions, our real scientific potential, whatever it may be, in both the pure and applied areas."

5. This part of the discussion should not be construed as an advocacy of no balance whatever between "pure" and "applied" research in a country. On the contrary, a balance is necessary for indigenous science to be creative. But attaining such a quantitative balance is, contrary to popular belief, not really a problem in real life. In fact, virtually all countries around the world, "advanced" or "less developed", spend about 10-20% of their total research and development expenditure on basic research. That this does not appear to be so to a casual observer is mainly due to the fact that applied research in many countries is of such low quality so as not to be applicable, at which point it is either completely invisible, or, even worse, is classified as "pure" research (because it has nu applications). The main problem, here is also quality and not quantity. For a more thorough discussion of this and other points in this article, see Moravcsik, 1975.

6. For an unusually interesting study of the view of nature in a non-Western society, see some research on Nepal in Dart and Pradhan, (1967).

7. The purposelessness of scientific activity in some less developed countries was discussed quite explicitely in a recent article by Ziman and Moravcsik, (1975).

8. The role of the formal scientific literature as a tool for the study of the scientific community has been expounded extensively in the numerous articles and books of the Yale historian of science Derek J. De Solla Price.

EL CONTEXTO DE LA CIENCIA CREATIVA

MICHAEL J. MORAVCSIK

RESUMEN

Cualquier actividad humana busca inevitablemente algún tipo de gratificación o resultado visible. Esta tendencia puede aplicarse por igual a la ciencia: aquellos que trabajan con ella, o la respaldan, buscan la creatividad que produce resultados tangibles. Esto no siempre puede lograrse debido a que la finalidad y los propósitos de la ciencia, por lo general, no están bien definidos y porque, aún teniendo metas concretas, los medios que se utilizan para evaluar los resultados de la producción científica dejan mucho que desear. Por ello, en este ensayo se analizan las metas, las motivaciones y los propósitos de la ciencia y se hace una encuesta de los diferentes medios que pueden usarse para evaluar la producción científica. El ensayo se ocupa únicamente de la ciencia y no considera problemas análogos relacionados con la tecnología.

Al discutirse las metas, se sugiere, primeramente, que la ciencia, al igual que muchas otras actividades humanas con atractivo profundo y duradero, tiene un amplio espectro de metas, algunas de las cuales pueden ser contradictorias. En general, sin embargo, esta multiplicidad de motivación puede resumirse bajo tres amplios renglones: a) La ciencia como una de las bases para el desarrollo material, b) La ciencia como una actividad que contribuye al desarrollo cultural, intelectual y espiritual y c) La ciencia como una influencia sobre la visión del mundo.

Con mucha frecuencia se destaca el primero de estos renglones a pesar de que, muy a menudo, se conoce mal el proceso complejo por medio del cual los descubrimientos científicos conducen al progreso material. La ciencia como aspiración humana "elevada" algunas veces es censurada, aunque al mismo tiempo sea exaltada posiblemente porque no es fácil admitir que, en todas las sociedades, a través de toda la historia humana, se ha considerado indispensable el interés por las aspiraciones "sin importancia" aun cuando el bienestar material del grupo no fuese completo. Igualmente se hace hincapié sobre el hecho de que los aspectos materialmente "útiles" e "inútiles" de la ciencia difícilmente pueden desligarse los unos de los otros (lo cual, sin embargo, no debe impedir que los países traten de lograr, a corto plazo, cierto balance entre las actividades científicas "fundamentales" y "aplicadas").

La tercera meta de la ciencia, la que contribuye a la visión del mundo que tiene el hombre, es de particular importancia en el contexto de desarrollo, pues el carácter anti-autoritario, objetivo, optimista y anti-fatalista del método científico es un pre-requisito indispensable para la evolución, entendida en el sentido moderno de la palabra.

Estas metas sugieren que hay que hacerse tres preguntas cada vez que se vaya a emprender una investigación científica: a) ¿Está el investigador realmente interesado o estimulado por el problema? b) ¿Contribuirán los resultados de la investigación a nuestra comprensión de la naturaleza para que otros investigadores puedan a su vez interesarse en los resultados y elaborar nueva investigación basándose en ellas? c) ¿Podrán los resultados de la investigación ser aplicados para resolver algún problema material o social para que los tecnólogos, u otros, puedan aprovechar los resultados y utilizarlos? Para la ciencia creativa la respuesta a la primera y al menos a una de las otras dos preguntas debería ser afirmativa. Se insiste sobre estos criterios porque, en la actualidad, con frecuencia se justifican algunos tipos de investigación por el solo hecho de que nunca han sido realizados. Tal investigación suele ser estéril, tanto en su efectividad para mejorar nuestra comprensión de la naturaleza como en su posible aplicación.

Para concluir la primera parte de la discusión, se sugiere que todas las comunidades científicas deben jugar un papel activo en la formulación y definición de sus propias metas, motivación y propósitos y no deben permitir que éstos sean impuestos, desde afuera, por burócratas, políticos, hombres de negocio o funcionarios de partido. En la segunda parte, que examina los medios para evaluar los resultados de la ciencia, se mencionan algunos métodos sistemáticos, bastante cuantitativos, muchos de ellos de reciente formulación. Todos han sido extraídos de la literatura científica y tratan del número de autores, el número de artículos científicos y el número de veces que éstos aparecen citados en la literatura científica. El estudio de estas medidas de la producción científica arroja algunos resultados interesantes pero, al mismo tiempo, muestra los graves inconvenientes que pueden acarrear.

El otro tipo de instrumento de evaluación que se utiliza para calcular la producción de la ciencia es el de crítica y revisión realizada por colegas. En este caso, se les pide a los científicos que juzguen y revisen el trabajo de sus colegas en el mismo campo. Este medio ha sido utilizado desde hace mucho tiempo por revistas científicas, organismos que otorgan fondos y en otro tipo de situaciones. Pero, en el plano internacional, tiene el inconveniente de que sentimientos nacionalistas y políticos pueden hacer sospechoso el juicio. Se sugiere, sin embargo, que este mismo método, a través de canales científicos informales y sin mucha publicidad, podría dar mejores resultados.

Para concluir el artículo, se predice con bastante certeza que el interés por la política científica aumentará en el futuro próximo, tanto así que a cada científico joven se le exigirán conocimientos sobre la materia antes de recibirse en cualquier postgrado. Por tanto, las comunidades científicas deberían aumentar su participación en la consideración de problemas como los que se discuten en este artículo.

O CONTEXTO DA CIÊNCIA CRIATIVA

MICHAEL J. MORAVCSIK

RESUMO

Qualquer atividade humana procura inevitavelmente algum tipo de gratificação ou resultado visível. Esta tendência pode ser aplicada igualmente à ciência: aqueles que nela trabalham, ou que a respaldam, buscam a criatividade que produz resultados palpáveis. Isto nem sempre pode ser logrado devido a que a finalidade e os propósitos da ciência, em geral, não estão bem definidos e ainda porque, existindo metas concretas, os meios que se utilizam para avaliar os resultados da produção científica deixam muito a desejar. Por isto, nesse ensaio se analisam as metas, as motivações e os propósitos da ciência e se faz uma averiguação dos diferentes meios que podem ser usados para avaliar a produção científica. O ensaio se ocupa unicamente da ciência e não considera problemas análogos relacionados com a tecnologia.

Ao discutir-se as metas, sugere, primeiramente, que a ciência, assim como muitas outras atividades humanas dotadas de atrativo profundo e duradouro, tem um amplo espectro de metas, algumas das quais podem ser contraditórias. Geralmente, porem, esta multiplicidade de motivações pode se apresentar sob três amplos aspectos: a) A ciência como uma das bases para o desenvolvimento material; b) A ciência como atividade que contribui para o desenvolvimento cultural, intelectual e espiritual; c) A ciência como uma influência sobre a visão do mundo.

Com muita frequência se destaca o primeiro destes aspectos apesar de que, muito raramente se conhece bem o processo complexo pelo qual os descobrimentos científicos conduzem ao progresso material. O segundo ponto de vista —a ciência como aspiração humana "elevada" —algumas vezes é censurado, ainda que, ao mesmo tempo, também seja exaltado, possivelmente porque não é fácil admitir que, em todas as sociedades, através de toda a história humana, se considerou indispensável o interesse pelas aspirações "sem importância", ainda quando o bem estar material do grupo não fosse completo. Igualmente, se faz finca-pé sobre o fato de que os aspectos materialmente "úteis" e "inúteis" da ciência dificilmente podem se desligar uns dos outros (o que, entretanto, não deve impedir que os países tratem de lograr, a curto prazo, certo equilíbrio entre as atividades científicas "fundamentais" e "aplicadas").

A terceira meta da ciência, a que contribui para a visão do mundo que tem o homem, é de particular importância no contexto do desenvolvimento, pois o caráter anti-autoritário, objetivo, otimista e anti-fatalista do método científico é um pré-requisito indispensável para a evolução entendida no sentido moderno da palavra. Estas metas sugerem que devem ser feitas três perguntas cada vez que se vai empreender uma investigação científica: a) Está o investigador realmente interessado ou estimulado pelo problema? b) Contribuirão os resultados da investigação para nossa compreensão da natureza para que outros investigadores possam, por sua vez se interessar pelos resultados e elaborar nova investigação baseando-se neles? c) Poderão os resultados da investigação ser aplicados para resolver algum problema material ou social para que os tecnólogos, e outros, possam aproveitar os resultados e utilizá-los? Para a ciência criativa a resposta à primeira pergunta, e pelo menos a uma das outras duas perguntas, deveria ser afirmativa. Insiste-se sobre estes critérios porque, na atualidade, muitas vezes se justificam alguns tipos de investigação unicamente pelo fato de que nunca foram realizadas anteriormente. Tal investigação se apresenta estéril, tanto em sua efetividade para melhorar nossa compreensão da natureza como em sua possível aplicação.

9. Perhaps the most interesting example for such a result is the correlation by De Solla Price (1969, for example) between the total annual gross national product (GNP) and the number of scientific authors. While both quantities vary, from country to country around the world, by many factors of ten, a plot of the logarithms of these two quantities against each other shows all countries of the world (regardless of race, ideology, political system, economic structure, geographical location or anything else) to lie around a straight line, and the scatter is at the most a factor of 3 on each side of the line.

10. The study of citations and studies using the tool of citations have grown into a rather large field. For some typical sources with many additional references, see Cole and Cole, 1973, 1967; Weinstock, 1971; Line and Sandison, 1974.

11. An introduction to co-citations can be found for example, in an article by two of the originators of this approach, Small and Griffith, 1974.

12. For an example of such a critique, see Moravcsik, 1973.

13. Some aspects of the typing and content analysis of citations can be found in the following two articles: Moravcsik and Murugesan, 1975; and Murugesan and Moravcsik, 1976. In a third article submitted for publication in Science and Culture, the Indian journal dealing with matters surrounding science, Murugesan and myself (joined by Evelyn Shearer) used the typing of citations to compare Indian scientific articles with articles from other countries.

14. For a thorough study of some aspects of the peer review system in journals, see the article by Merton and Zuckerman, 1971; in which they analyze the operation of the refereeing system in Physical Review, the largest physics journal in the world.

15. To be sure, there are some examples of situations which could perhaps be called international peer review, such as the external examination system in the former British colonies, visiting study groups in Korea sent there by US AID, or some OAS reviews in Latin America. Yet all of these have certain special features which make them different from what I am suggesting here.

16. It would be inappropiate to end this article without pointing out that most of the ideas contained in it, while "subjectively original", that is, expressing my own experience and conclusions, are hardly "objectively original", that is, they have certainly been expressed by others before. Science policy and the "science of science" are just beginning to become formal, "academic" disciplines. At the moment, only very few educational centers offer any curriculum in science policy, and there are practically no textbooks, reference books, or even systematic review articles dealing with it. Formalization, while carrying with it the danger of ossification and dogmatization, has the advantage of allowing the field to become cumulative in that a certain body of knowledge becomes accepted as generally known and hence new literature can be built on top. Science policy has not quite reached this stage yet, and, hence, literature is to some extent repetitive as those more recently entering the field arrive at conclusions rather similar to those of their predecessors. While such consensus is heartening, it also indicates that the wiews of the predecessors have not been paid enough attention in practice, their advice not needed sufficiently, so that repetition is needed. In any case, for those interested in exploring more deeply what has been said on science policy and the science of science, I would recommend the writings of Joseph Ben-David, Harvey Brooks, Vannevar Bush, the two

Cole brothers Jonathan and Stephen, Warren Hagstrom, Don K. Price, Derek De Solla Price, Steven Toulmin, Alvin Weinberg, and John Ziman, just to name an extremely incomplete and rather kaleidoscopic list which, however, will lead the reader to others through the references mentioned by these authors.

REFERENCES

Clark, K. (1970): *Civilisation: A Personal View.* (Harper and Row, New York).
Cole, S. and Cole, J. (1967): Scientific output and recognition: A study of the operation of the reward system in Science. *American Sociological Review 32:* 377-399.
Cole, J. and Cole S. (1973): *Social Stratification in Science.* (University of Chicago Press).
Dart, F. and Pradhan, P. L. (1967): Cross-cultural teaching in Science. *Science, 155:* 649.
De Solla Price, D. (1969): Measuring the size of Science. *Proc. Israel Acad. of Science and Humanities, 4:* (6) 98-111.
Line, M. B. and Sandison, A. (1974): Progress in documentation. *Journal of Documentation, 30:* 283.
Lopes, L. J. (1966): Science for development: A view from Latin America. *Bull. Atomic Scient. 22:* (Sept) 7.
Merton, R. K. and Zuckerman, H. (1971): Institutionalized patterns of evaluations in Science, *Minerva IX* (1): 66-100.
Moravcsik, M. (1973): Measures of scientific growth. *Research Policy (3):* 266-275.
Moravcsik, M. (1975): *Science Development. The Building of Science in Less Developed Countries.* (International Development Research Center, Bloomington, Indiana).
Moravcsik, M. and Murugesan, P. (1975): Some results on the function and quality of citations. *Social Studies of Science 5:* (1) 86-92.
Murugesan, P. and Moravcsik, M. (1976): Variations of the nature of citation measures

Para concluir a primeira parte da discussão, sugere-se que todas as comunidades científicas devem desenvolver um papel ativo na formulação e definição de suas próprias metas, motivação e propósitos e não devem permitir que estes sejam impostos, de fora, por burocratas, políticos, homens de negócios ou funcionários de partido.

Na segunda parte, que examina os meios para avaliar os resultados da ciência, mencionam-se alguns métodos sistemáticos, muitos deles de recente formulação. Todos foram extraídos da literatura científica e tratam do número de autores, do número de artigos científicos e do número de vezes que este aparecem citados na literatura científica. O estudo destas medidas de produção científica apresenta alguns resultados interessantes mas, ao mesmo tempo, mostra os graves inconvenientes que podem ocasionar.

O outro tipo de instrumento de avaliação que se utiliza para calcular a produção da ciência é o de crítica e revisão realizada por colegas. Neste caso, se pede aos cientistas que julguem e revisem o trabalho de seus colegas no mesmo campo. Este meio vem sendo utilizado desde muito tempo por revistas científicas, em organismos que outorgam fundos e noutro tipo de situações. Mas, no plano internacional, esse processo tem o inconveniente de que sentimentos nacionalistas e políticos podem tornar o juizo suspeito. Sugere-se, contudo, que este mesmo método, através dos canais científicos informais e sem muita publicidade, poderia dar melhores resultados.

Para concluir o artigo, prediz-se com bastante certeza que o interesse pela política científica aumentará no futuro próximo, tanto assim que a cada cientista jovem exigir-se-á conhecimentos sobre a matéria antes de receberem qualquer título de pós-graduação. As comunidades científicas deveríam, pois, aumentar sua participação na consideração de problemas como os que se discutem neste artigo.

with journals and Scientific specialties. *Journal of Amer. Soc. Information Science* (in press).
Roche, M. (1966): Social aspects of Science in a developing country. *Impact, 16:* 51-60.
Small, H. and Griffith, B. (1974): The structure of scientific literature I: Identifying and Graphing Specialties. *Science Studies, 4:* 17-40.

Varsavsky, O. (1967): Scientific colonialism in the hard sciences. *The American Behavioral Scientist, 10:* 22.
Weinstock, M. (1971): Citation Indexes. *Encyclopedia of Library and Information Science 5:* 16-40.
Ziman, J. and Moravcsik, M. (1975): Paradisia and Dominatia: Science and the developing world. *Foreign Affairs, 53:* 669-724.

Reprinted with permission from *CASME* **2** (1981) 16–27. (This Journal is now published as the CASTME JOURNAL, the organ of the Commonwealth Association of Science, Technology and Mathematics Educators (CASTME), a professional body under the auspices of the Commonwealth Foundation.)

What is a developing country?

Michael J Moravcsik, Institute of Theoretical Science, University of Oregon, Eugene, Oregon

Abstract

Since the term "developing country" is inevitably with us, an analysis is attempted to give it a realistic definition in the context of science and technology. A number of "one-dimensional" possibilities for such a definition are discarded. It is then claimed that a composite, multi-dimensional set of indicators is needed, corresponding to the many aspects and the interacting parts of the science and technology system. Some of the needed indicators can be constructed and quantified fairly well, but others, at least at the moment, cannot. An appropriately selected group of scientists and technologists, in conjunction with some "objective" impersonal indicators, remains the best present method of evaluating and assessing the functioning of science and technology in a country, hence of determining the degree of "developedness" of the country in science and technology.

1. Does the question make sense?

It is our habit to view the world by classifying it into categories, and then juxtaposing them, or even pitting one against the other. One of these very commonly heard categorizations is that of a "developing country", also referred to as an "underdeveloped country", or an "emerging country", or the "South" in the "North - South" confrontation. This terminology arises in the context of issues which can become highly emotional, political, and conspicuous, therefore it would be advisable to ascribe a precise meaning to this terminology.

And yet, in practice these terms appear to be used in a multitude of different ways, with real or perceived "hidden" meanings and connotations which are sometimes felt to be condescending, demeaning, ethnocentric, descriminatory, and tendencious. The ambiguity of the concept is at least one reason why, from time to time, new terms, thought to be more euphemistic, are substituted for the old ones to describe the concept, and each time the new term appears to focus on a different aspect of the concept. ("Underdeveloped", "developing", "emerging", "South" etc.) It is, therefore, the aim of this article to try to clarify the meaning of "developing country" in order to avoid such misunderstandings and ambiguities.

But does the question make sense in the first place? Can one talk about the category of developing countries? There might be some doubt about this on at least two counts.

CASME Journal Vol 2 No 1 Nov/Dec 1981   *17*

First, there is the question of the continuum *versus* descrete characterizations. It is easier for us, logically, to make discrete categories, but in reality the phenomena in the world form continua in which the differences and distinctions are not qualitative, but quantitative. In the case of developing countries, to place, say, Upper Volta and Mexico in the same qualitative category is, to say the least, an oversimplification of the situation.

Yet this problem does not make our original question senseless and unrealistic. One must simply realize that although there are a number of borderline cases of countries making a transition between categories, and although there are large quantitative differences even within the mainstream population in each category, the classification into the dichotomy of developing and the other countries is a useful one as a first approximation. As we will see, the indicators of being a developing country will apply to various countries to a varying extent, but the fact that such indicators can be formulated at all suggests that the concept is a useful one.

The second doubt that might arise about the utility of the category of developing country pertains to what aspect of the country we consider. For example, in terms of GNP per capita, there are many leading countries in the world, such as Kuwait, which are nevertheless "developing" in many other respects. To take a different type of example, some countries, "developing" in some sense, have a cultural tradition that is many times more ancient than that of some of the countries which are regarded as the leading countries in the world in other respects.

In this discussion we will concentrate entirely on science and technology (S & T), and will use the concept of development in that framework. In other words, we want to discuss what it means that a country is a "developing" one as far as science and technology are concerned. Once thus narrowed, the question in the title is fairly unambiguous, as we will see after we have carefully analyzed its meaning.

With this preliminary assurance that our discussion has at least some chance of being meaningful, I want to turn next to an explicit list of some criteria which are false and inappropriate when it comes to defining what a developing country is.

II. *Wrong criteria*

In this section I want to list eight criteria, each of which may seem to delineate the distinction, as far as science and technology are concerned, between developing countries and the others, but which in fact I believe to be very weak and, in some respects, even spurious in themselves, especially each by itself.

1/ Nomenclature

The language used in talking about developing countries is highly unfortunate and sloppy, hence the criteria used to distinguish such countries from others also tend to be faulty. As mentioned above there are various synonyms used, developing = less developed = emerging = South, but all of these are misleading. To demonstrate this, we might ask what the "other" countries should be called. The antithesis of "developing" is either "non-developing" or "developed",

CASME Journal Vol'2 No 1 Nov/Dec 1981

neither of which fits the "other" countries. The term "advanced" is also used for the other countries, which in my view is better since it denotes being ahead in a certain line of chronological progression without connoting that the progression has stopped.

On the other hand, most recently there are also examples for countries which appear to "underdevelop" [1], that is, decline in their fostering and utilization of science and technology. If this is so, the various countries should really be placed somewhere along an arch, with the "developing" countries low on the ascendant leg, some of the "advanced" countries on the descendant leg. All this presupposes that the general line of development in science and technology for all countries of the world is, has been, and will be approximately the same., something I believe is the case, but which nevertheless cannot be proven, and certainly does not go unchallenged nowadays.

To summarize, the criteria based on the common meanings of the terminology are false simply because the terminology itself is ambiguous and misleading.

2/ Barbarians?

One of the connotations of "developing" which is much resented is that the developing countries are in the process of emerging from a "savage", cultureless state into a "civilized" state. Clearly, "developing" in the sense we want to use it in the framework of science and technology should not have even a trace of such a connotation. As briefly mentioned earlier, many of the developing countries have a cultural tradition which goes back many centuries and which is often very sophisticated, rich, and successful. The existence of such a tradition in no way conflicts with a "developing" status in science and technology. For one thing, science and technology do not form a culture, although they may be ingredients of some cultures, and there have been very rich cultures which placed little, if any, emphasis on science and technology. Thus, we must not feel that being called a developing country is in any way a condescending label for the country's cultural status.

3/ Science and Modern Science

Being labelled "developing" in the area of science and technology does not mean that the country's history is entirely void of concern with natural phenomena and their utilization. [3] To clarify this point, let me make a distinction between "old" science (for want of a better name) and "modern" science. The latter is what arose in Europe some 400 years ago, creating its own methodology, philosophy, and infrastructure, whereas the former has existed almost everywhere, and almost always. The "old" science is most exclusively what in modern science would be called phenomenology, that is, the cataloguing of facts and observations without making an attempt at a conceptual and theoretical explanation, which in turn can give rise to predictive power.

The difference between the two sciences is in their repective

CASME Journal Vol 2 No 1 Nov/Dec 1981   19

successes: virtually all of what we now know about the universe was
acquired through modern science in the last 400 years, and mainly in
a tiny geographical part of the world, while the old science, practised universally and from time immemorial, has contributed only a
tiny fraction of our present scientific knowledge, and some of even
that has become widely known through being rediscovered, independently, by modern science.

Thus, the "developing" label for a country must be used in terms
of modern science. It may very well be that a country had flourishing
"science" 1,500 years ago, or that it has an extensive tradition in
which doctors, native medicine, calendar-making, or other manifestations of "ethnoscience". All this amounts to little when it comes
to assessing the impact of science on the country along the three-
pronged dimensions of technological capability, human aspirations,
and attitude toward the world.

4/ Self-sufficiency?

One might think that the antithesis of a "developing" country,
which is thus dependent on other countries for what it has not been
able to develop yet, is an "advanced" country, which through its
accomplishment has attained self-sufficiency. Taking the word
"dependence" in its simplest and more literal sense, the above conclusion is quite erroneous, hence should not be used as a criterion
for defining a developing country. Even the scientifically most
advanced country, the United States, creates only about 25% of the
world's scientific knowledge [4], that is, it depends on 3/4 of its
science from other countries around the world. Similar statements
hold also for technology. Smaller advanced countries are of course
in an even more lopsided position. As I discussed in a recent
article,[5], if the concept of dependence is employed in a much more
sophisticated way, including material and psychological elements,
it may have more relevance to the concept of development.

5/ Trade balance

Since economists play a vocal role in the discussion of issues
pertaining to development, it is worthwhile, specifically, to debunk
some of the economic indicators which are sometimes used to suggest
that a country is "developing" or not. One of these is the question
of trade balance, a constituent in the concept of economic dependence.
The train of thought is that a developing country which lacks the
skills and the capacity of producing goods of substantial value is
likely to export more in value than they import.

There are, however, too many exceptions to this rule. The
United States, hardly a developing country by the usual definition,
has had a negative trade balance for some years now. On the other
hand, countries exporting large amounts of agricultural products,
(i.e. rice in Thailand, rubber in Malaysia) can have a positive
trade balance if the standard of living in the country is sufficiently low. These examples have not even touched on the oil-rich
developing countries which also constitute exceptions. All in all,

CASME Journal Vol 2 No 1 Nov/Dec 1981   20

it is evident that the size of the trade balance has only a very tenuous connection with the developedness of the country in science and technology.

6/ GNP vs. scientific output curve

We have known for some time that there is an approximate correlation between the GNP of a country and its scientific authors [6]. Furthermore, it has been shown that this correlation has less scatter if we stipulate two different curves [7] along which the points lie, which represent countries with a given GNP and a given scientific output: one curve for the "developed" countries, one for the "developing" ones. Since there are certainly numerous instances of two countries, one developed and one developing, having the same GNP (though very different GNP/capita), the claim of the existence of two curves has substance.

Nevertheless, it is not practically feasible to use this criterion by itself to decide whether a country is developed or developing. The dispersion along each curve is considerable, enough to make the distributions overlap considerably, and there is also a certain amount of tautology involved in such a criterion, since the two curves were originally hypothesized on the basis of some definition of what a developing country is.

7/ Bureaucratic infrastructure for S & T

It is sometimes claimed [8] that a sign of becoming a developed country is that an elaborate bureaucratic infrastructure is in place for science and technology: there are national S & T plans, there is a ministry or a national research council for S & T, etc. Such claims have none but the most superficial validity. For one thing, the United States never had a national S&T plan, and yet is certainly among the developed countries. On the other hand, many countries with practically no functional scientific and technological output are burdened with elaborate bureaucratic paraphernalia for science and technology, so much so that the few scientifically trained people in the country are quickly absorbed in such bureaucratic operations, and are therefore prevented from doing scientific work.

8/ Input-output

Since input S & T is so much easier to measure than output, it is also claimed sometimes that the "developness" of the country in S & T can be gauged by the quantitative input figures the country devotes to science and technology. Such indicators ignore quality altogether, thus assume that from the quantity of the input one can tell the quality also, something that is hardly so. Furthermore, such indicators assume that output in science and technology is proportional to input in terms of money and manpower, which is demonstrably untrue, see point 6 above. For example, it is known that the productivity per scientist can vary from country to country by as much as a factor of ten [9]. Thus any purely quantitative input indicators are suspect, and cannot be used to define developing countries.

44

III. *Making a composite indicator*

In surveying the various wrong indicators in the previous
section, one realizes that while some of them contain some grain of
truth (in fact, might even overlap with the individual indicators
developed later in this section), they have two main deficiencies.
First, each of them pertains only to one small aspect of a complex
picture, that is, even if pertinent, each of them is one-dimensional.
Second, even taken together, they constitute only a part of the
whole picture, and describe only elements of that picture and not
their interactions.

To view real-life problems in only one dimension is a frequent
defect of analyses. Such a one-dimensional view has many superficial
attractions. It is simple: in it, the situation can be described in
a highly polarized way, with a clearly defined ranking possible along
the single dimension, and suggestions for improvements can accordingly
be easily formulated. In all these respects, however, the one-dimensional picture differs radically from the real-life situations, which
are invariably multi-dimensional[10].

We will attempt, therefore, to construct a multi-dimensional
concept of a developing country. In such a picture, indicators of
developedness will be viewed as a set, and we will be concerned with
whether the system which is described by such a set of indicators
functions or not interactively. As an analogue, one may want to
think of a set of indicators describing a car. These may pertain to
its build, its structure, its funancial value, its component parts,
the sound it makes,etc., and then, very importantly, with whether
the contraption actually fulfills the function of a car.

The system we want to view in this regard is illustrated in
Figure 1. Our aim is to say whether a country is advanced or developed in terms of its indigenous science and technology, which
are represented by parts of the two circles. The motivations or justifications[11] for science and technology are indicated in the form
of the four quadrangles, and then lines with arrows mark the connections and interactions between these elements. All lines should be
interpreted as indicating output, that is, being indicative of the
component of the picture from which the line emanates. Only those
elements and connections are indicated which are important for the
present discussion.

The picture is certainly a multi-dimensional one. For one thing,
the motivations themselves extend over three main dimensions: material production, aspirations and the world view. Within each of
these, there are many different aspects, sufficiently distinct to
merit being called a dimension.

It may be worth noting that the connections between science and
material production are not direct, but run through technology. I
consider this an important point. In the absence of a suitable technological infrastructure, there is very little local scientists can
do in the short run for the material productive sector of the country,

45

CASME Journal Vol 2 No 1 Nov/Dec 1981   22

hence they should not be expected to do so, and blamed for it if
they do not. The confusion of science with technology [12], and
of the role of a scientist with that of a technologist, is a common
problem in analyses of developing countries.

It would now be reassuring and convenient to be able to claim
that we have a reliable quantitative indicator for each of the elements of Figure 1. In that case, the next step would be to agree
on some kind of quantitative weighting of the importance of these
indicators. Having done so, it is easy and unambiguous to define an
overall "distance" or "magnitude" in this multi-dimensional space,
which then can serve as a quantitative measure of the developedness
of the country in science and technology.

Unfortunately, we are very far from being able to do all that[13].
Some of the elements in the picture can be ascribed some kind of
quantitative measure, but many of them have defied such an attempt so
far. The fact remains, therefore, that the most functional method
of assessment of the state of this system is still the personal
evaluation by one or several scientists and technologists, active and
knowledgeable in their technical fields, unencumbered by organizational,
political and personal ties, experienced in the specific problems
of the country, and familiar with whatever "objective" indicators for
the various elements have been suggested by recent research in the
science of science.

To illustrate the complexity of the task, it might be worthwhile
mentioning, in brief, telegraphic style, some of the considerations
that go into the various elements of the picture.

Line A measures the output of the direct scientific research
activity. This is by now reasonably amenable to quantitative evaluation by various scientometric means [14], such as author-, public-
ation- and citation indices, though controversy remains on the extent
to which such compilations include scientific work in developing
countries which is primarily of local interest.

Line B measures the analogous quantity in technology, for which
there are also some quantitative measures, such as patent counts,
though these are also not free of problems when applied to developing
countries.

Lines C and D measure the extent to which the indigenous scientific and technological community is integrated into the worldwide
community, and can take advantage of developments elsewhere to build
itself, something that can be gauged by fairly objective indicators.

The extent to which local technology can utilize science originating from outside and inside the country, respectively, is described by lines E and F. This is much more difficult to ascertain,
since the scientific input in a given piece of technological development work is both direct and indirect [15], making use of both "old"
and "new" science. Yet this link is crucial, since the age of purely
trial-and-error based technology has passed.

CASME Journal Vol 2 No 1 Nov/Dec 1981 23

The dialogue and influence in the opposite direction, namely from local technology to local science, is indicated by G, and includes the conveying of technological problems that await scientific input, the technological support in terms of equipment and supplies of local scientific research work, etc.

Finally, the two self-interaction links T and U, denoting the cooperation within the local scientific community and within the local technological community, represent important indicators of the health of the respective communities.

Although far from easily measurable, the connections A-G discussed thus far constitute a more manageable task compared to connections H-O, since these link science and technology with the much more intangible entities of human aspirations and Man's view of the world. Yet, unless science and technology are internalized in a country by a substantial fraction of the people in terms of these aspirations and of their prevailing world view, the attempts of science and technology to affect material production are likely to falter.

But to measure links H-O, one needs to penetrate the thinking of the population as a whole, that of the political leaders, of the educated segment, of the village people [16], of the public media, etc. In-depth studies in psychology, in public opinion, in the way people do things (as distinct from what they say), in the personal philosophies of individuals are needed to assess this dimension of science and technology. Such information is very rarely available in the developing countries, especially from unbiased sources not intent on grinding an axe.

A particularly important psychological aspect to consider is the feeling of dependence [17] which enters especially markedly into the connections with the non-material aspirations of science, that is, into H,I,L and M.

When assessing links P-S, we encounter difficulties of a different kind. It would be tempting to evaluate the effect of science on material production by measuring the output of the productive sector of the economy of the country, in the industrial or agricultural areas. This, however, is quite unrealistic. The point is that there are so many stages between science and technology on the one hand, and the output of the productive sector of the economy on the other, stages that have little if anything to do with science and technology, that to blame the weaknesses of this long chain of stages of deficiencies on science and technology is unjustified. Land policies, banking practices [18], agricultural extension systems, organizational and managerial questions, family traditions and many other considerations could very well share or carry the blame. To my knowledge, very few if any studies exist in the context of developing countries which succeeded in isolating the direct effect of science and technology from these other factors.

In listing all these factors and in constructing the system

CASME Journal Vol 2 No 1 Nov/Dec 1981

described by Figure 1 we have actually slipped over from the seem-ingly simple question of a definition ("what is a developing country?" to the much more basic question of assessing and evaluating the functioning of science and technology in a country [6]. This transit-ion was not unintentional. The two problems are indeed closely related, and the inadequacy and frailty of our efforts to create a methodology for assessment are directly reflected in our inability to offer a good and functional definition.

So where does all this leave us? Although we are still in the midst of resolving the problems discussed here, a few rather definite conclusions can be listed:

1. The terminology of developing and advanced countries is widely in use and would be difficult to erase from our dictionaries. Hence, we must attempt to clarify these terms as much as possible.

2. In this discussion we concerned ourselves exclusively with developing countries in the context of science and technology.

3. One-dimensional definitions of what a developing country is, of which many have been offered, are inadequate and misleading.

4. The suitable concept of development in science and tech-nology involves a whole large set of indicators, and the interactive aspects of a system. In other words, a developing country is one in which many elements of this set and/or many of the connecting links are missing or very weak.

5. While some parts of this system can be measured through fairly objective and quantitative indicators, others, equally im-portant, cannot.

6. On the whole, apprpriately selected members of the worldwide scientific and technological community, in combination with objective indicators wherever they exist, remain the best way to evaluate the development of a country in science and technology, hence the best way to decide to what extent a country is "developing" and to what extent it is "advanced".

ACKNOWLEDGEMENT

I am indebted to Edith Moravcsik for a constructive critique on the first draft of the manuscript.

48

Figure 1

*Schematic representation of the system of science and technology
in a country, with some of the interconnections among the elements.
The figure is discussed in detail in the text. Only those elements
and connections are indicated which are important for the present
discussion.*

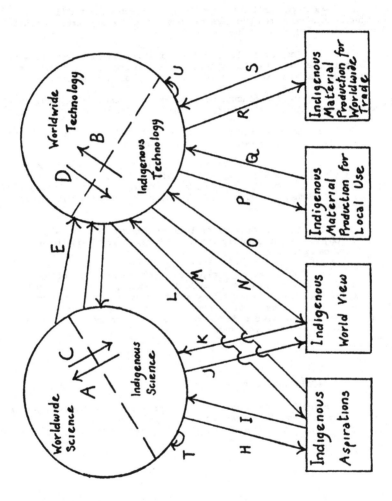

CASME Journal Vol 2 No 1 Nov/Dec 1981   26

REFERENCES

I first discussed this phenomenon about a decade ago. See
for example M J Moravcsik, "The Transmission of a Scientific
Civilization", Science and Public Affairs 29:3, 25 (1973)
and M J Moravcsik, "A Chance to Close the Gap?" Science
and Culture 39:3, 204 (1973)

I argued this point in M J Moravcsik, "Do Less Developed
Countries have a Special Science of their Own?" Interciencia
3:1,8 (1978).

A very illuminating debate on this point was published
recently in Interciencia, under the title of "Science and
National Identity", containing comments from a number of
people. See Interciencia 6:3, 165 (1981).

The literature on the distribution of science in the world
is by now sizeable. See in particular writings by Derek de
Solla Price, Davidson Frame, Francis Nairn, Herbert Inhaber,
just to mention a few, as well as the various compilations
of the Institute of Scientific Information in Philadelphia.

M J Moravcsik, "Dependence", Human Systems and Management,
(in press, 1981).

The pioneering paper on this is that of Derek de Solla Price
"Measuring the Size of Science", Proceedings of the Israeli
Academy of Sciences and Humanities 4, 98 (1969).

J D Frame, "National Economic Resources and the Production
of Research in Lesser Developed Countries", Soc. Stud. of
Sci. 9, 233 (1979).

For details on this as well as a survey of assessment
efforts, see M J Moravcsik, "Assessment of Science in Developing Countries", in M Srinivasa (ed.) Technology Assessment and Development, Praeger Special Studies Series, (in
press, 1981).

For an example see B V Rangarao, "Scientific Research in
India: An Analysis of Publications", J Scient. Ind. Res.
26, 166 (1967), in which quanitative data is given to demonstrate that it takes about 10-12 scientist years to produce one publication in India.

For a discussion of one- vs. multi-dimensioanl ways of looking at problems, see M J Moravcsik, How to Grow Science,
Universe Books, New York, 1980. Appendix 1.

For a discussion of the motivations for and the justific-
ations of science, see M J Moravcsik, "The Context of
Creative Science", Interciencia 1, 71 (1976),reprinted in
Everyman's Science (India) 11,3 97 (1976).

CASME Journal Vol 2 No 1 Nov/Dec 1981   27

12.   This confusion shows up particularly clearly when it
      comes to undertaking applied scientific research. See
      M J Moravcsik, "Linking Science with Technology in Dev-
      eloping Countries", Approtech 2:2, 1 (1979) and
      W P Pardee and M J Moravcsik, "Creating an Effective
      Applied Scientific Research Program in a Developing Country"
      submitted for publication (1980).

13.   A graphic way of illustrating this is to look at the excell-
      ent series called "Science Indicators", published period-
      ically by the U.S. National Science Foundation. They un-
      doubtedly represent the latest and ultimate in the state of
      the art of science indicators. It is enough to peruse the
      table of contents of these volumes to see how limited their
      scope is compared to the task of assessment in terms of the
      motivations for science as discussed in this paper.

14.   For a brief overview of some of these scientometrics means,
      see M J Moravcsik, "A progress Report on the Quantification
      of Science", J. Scient. and Industr. Res. (India) 36, 195
      (1977). See also the comprehensive review by Francis Nairn,
      "Evaluative Bibliometrics", PB 252 339, Report to the Natio-
      nal Science Foundation, reproduced by the National Technical
      Information Service, Washington D.C. (1976). Another
      review from a different perspective can be found in
      G N Gilbert, "Measuring the Growth of Science - a Review of
      Indicators of Scientific Growth", Scientometrics 1, 9 (1978)

15.   For some remarks on the science-technology link along these
      lines,see M J Moravcsik, "Reflections on the Science-Tech-
      nology Link", submitted for publication (1981).

16.   Pioneering and unfortunately little emulated work in this
      area was done by the late Francis Dart, see for example
      F Dart, "The Rub of Cultures", Foreign Affairs 41, 360
      (1963); F Dart and P L Pradhan, "Cross Cultural Teaching
      of Science", Science 155, 649 (1967); F Dart, "Science
      and the Worldview", Physics Today, 25, 48 (1972); F Dart,
      "The Cultural Context of Science Teaching", Search 4, 322
      (1973).

17.   This is discussed in greater detail in M J Moravcsik,
      "Dependence", Human Resources and Management(in press,1981).

18.   For some examples, see some of the studies by the World Bank
      such as J Ramesh, "National Financial Institutions and Tech-
      nological Development", (Report to the OECD, Sept.1979);
      P Shapiro, "Technology and Science in the Acitivities of
      the World Bank", Science and Technology Series 30 (1978);
      and the book "Science and Technology in World Bank Oper-
      ations", World Bank, Washington (1980).

Leonardo, Vol. 11, pp. 214–216.
© Pergamon Press Ltd. 1978. Printed in Great Britain.

0024—094X/78/0701—0214$02.00/0

DEVELOPING COUNTRIES AND THE FRUITS OF SCIENCE

Michael J. Moravcsik*

1.

Once upon a time there was a man who lived in a little house. He was a poor man, and had to work hard to provide for himself the bare necessities for survival. His hut was on a flat piece of land, surrounded by weeds and scrubby brush, just as it was when he came to settle there.

When, at the end of a long day's work, he sat down to rest after his evening meal and looked out the window at the jumble of greenery surrounding him, he felt dissatisfied. It surprised him that this should be so, even with a full belly and a chance to find some leisure. But then he realized that he was unhappy because he remembered a trip to the mountains and recalled the miraculous feeling of beholding tall trees overhead, the awe inspired by their reaching to the sky, the pleasure of listening to the rustling of their leaves. That wonderful experience was in sharp contrast with the stark reality of the ugly weeds around his house.

He decided that he must have some trees. It was not easy to turn this resolution into reality. He spent days and weeks observing the trees in the forest and trying to transplant them into his yard. And finally he succeeded. Trees began to grow where only weeds had been, and he felt happy and contented, even though he was still struggling hard just to survive.

Years passed and the trees matured, and one year, to his amazement and surprise, they began to yield fruit. The fruit was large and juicy and most desirable to look at and to eat. Year after year the amount of fruit multiplied, and he became wealthy and well nourished, eating some of the fruit and selling the rest. As he grew older, he enjoyed an easier and more pleasant life.

About that time a young man moved to his area and built a house next to his. He was a poor fellow and, like his neighbor in his younger days, worked hard to survive. As he sat, after a long day's work, on the porch of his house, his eyes kept gazing at the wonderful fruit growing across the fence in his neighbor's yard. He envied the neighbor for being so rich. From time to time, he was also taken by the feeling of how nice it would be to be shaded by those beautiful green leaves and waving branches, but he quickly dismissed this feeling as not appropriate for a poor man.

Finally, he resolved that he must also have some fruit like that, and he must have it quickly. He reasoned that, as he was poor and getting older, he could not wait much longer to become more affluent. He decided, therefore, that he would produce the fruit and the fruit only. His reasoning was simple. First he decided that most of the

roots, branches and leaves on the tree were useless, so he could reduce their number. Let the rich bother with waste. Second, he came to the conclusion that with the neighbor's trees already there, he could avoid the arduous task of transplanting trees from the forest.

He asked his neighbor for some twigs from the fruit trees and stuck them into the ground in his back yard. But no fruit was produced, in fact, the twigs wilted and died. He tried again; only this time he obtained small saplings from his neighbor. They began to grow, and so he proceeded to tend them in his own way. He trimmed the roots from time to time, so they should not reach too far into the ground and use up too much nourishment and water. As small branches and some leaves began to grow, he cut off all branches but one (the one that looked similar to the branch carrying fruit on his neighbor's tree), and on that branch he carefully trimmed the leaves so that nothing useless would be produced by his trees.

But the result was a failure. His trees did grow, but they had weird shapes and yielded no fruit, in spite of his spending much time and money tending them. Year after year, whenever he wanted to eat fruit, he had to squeeze some money out of his meager income to buy it from his rich neighbor.

This failure was a heavy burden on the young man's mind. It was not only a matter of having to pay for the fruit he wanted. He increasingly realized that the burden arose mainly from a feeling of inferiority, from a sense of competitive impotence relative to his neighbor.

He finally became weary of all this and decided to have a long talk with his rich neighbor to find out the secrets of growing trees that richly bear fruit. His neighbor, however, did not tell him any deep secrets. He said nothing about magic nutrients, clever tricks and sophisticated procedures that assure fruit on the trees. All he told him was his own life story, his desires and motivations in growing trees in the first place, and his surprise at eventually finding the fruit on the trees. Then, in a reflective mood, he added: 'I think my real fortune was in not having had a neighbor whose tempting fruit trees I could behold. Had I had one, I should easily have been led astray, just as you were by my trees. When one wants to reproduce the results of others, one is often blinded by them and fails to see the complexity of the essence that brought about those results. Not having such results to replicate, I was lucky enough to have been carried away with an aspiration to catch the essence and thus was rewarded also with the results.'

2.

What is commonly called the modern scientific revolution began in Western civilization about 400 years

*Physicist, Institute of Theoretical Science, University of Oregon, Eugene, OR 97403, U.S.A. (Received 29 Nov. 1977)

ago, when some pioneering minds became dissatisfied with the static and fatalistic outlook of the Middle Ages. During the Middle Ages what was believed to be true about the world was transmitted by a few specialists who consulted texts by accepted authorities, including the belief that the world was governed by the will of the Christian God and that humans were only His passive pawns. Through the efforts of these pioneers the strong aspiration arose to reshape the prevailing mental picture of the world on the basis of direct observation and experimentation.

It is important to realize that the motivation underlying the modern scientific revolution was not the material improvement of the life of humans. Little did these pioneers realize what a stupendous impact the new way of learning to understand the world would have in this respect. In fact, this impact was not felt until about 300 years after Copernicus (1473–1543). I believe the reason for this is easy to understand.

In the early days of modern science, investigation was focused on natural phenomena that are directly available to the senses for observation. Thus the branches of physics dealing with mechanics (the effect of forces on solids, liquids and gases), light and sound were first pursued. These branches do have, of course, a direct bearing on many areas of technology, but, for example, tools, pumps, steam engines, mechanical weapons, etc. could be and were developed empirically, that is by trial-and-error methods, rather than by those of new science. Until about 100 years ago, since the invention of simple hand tools in the distant past, empirical methods were used in technology. Even today, some technological innovations are made by clever, resourceful people without a scientific background, provided the innovations involve phenomena directly available to the senses.

Until about 100 years ago, the motivation of scientists was mainly of a philosophical, intellectual and cultural character, and this motivation, as an inquisitive methodology, is considered to underlie basic or fundamental research. Thus science in Western civilization had two to three centuries to implant in societies a new world-view in terms of this motivation. Since then, science has acquired a broader scope. Scientists began to give attention to aspects of nature that are not directly accessible to the senses (electric phenomena that cannot be either seen or heard; atomic physics and then particle physics, chemical processes, microbiology, etc. which involve phenomena that cannot be perceived directly). Knowledge of these phenomena cannot be obtained by solely empirical methods, and inventors without a scientific background are not likely to contribute to a technological exploitation of these domains.

In principle, if an imaginative person without a scientific background was provided with a box containing pieces of copper, germanium, plastics, steel, wood, glass, etc., the person could, by trial and error eventually invent a transistor radio, but it is likely to require an effort of many millions of years! It became apparent that what is now called advanced technology (which is the most influential part of present-day technology) is dependent on persons who grasp the understandings of nature provided by science, and this has led to a new motivation for doing science, namely applied scientific research. By the time this happened, however, the subtleties of how to pursue science was learned and fairly widely absorbed as a general kind of human activity. This new motivation could be incorporated into the already existing structure of basic science and 'science policies' of governments and individuals now include technological motivations as one of the justifications for doing science.

Recently, however, Western civilization, or at least some aspects of it, has tended to spread throughout the world. Sporadic signs of this trend began about 150 years ago, and they became clearly evident during the past 50 to 75 years. The notable example is Japan, where the trend started more than 100 years ago, and where a number of favorable circumstances supported it. However, in other so-called 'developing countries', with or without a Western colonial experience, one can speak about a substantial influence of and the beginning of the growth of modern science and advanced technology during a shorter period of time. For the purposes of my discussion, the important factor to note is that this influence began when science and advanced technology already were visibly related to each other, that is, the technological motivation for pursuing science was broadly accepted.

One should note that the first influence on developing countries came from advanced technology in the form of automobiles, radios and of other kinds of consumer products. Manufacturing methods were transferred so that the production of such goods, with relatively low labor costs, would improve the foreign trade balance of a country. Thus the impression in these countries was strengthened that through what seemed to them the mysterious thing called 'science and technology' they could obtain political power and wealth commesurate with that of advanced-technology countries.

To be sure, science as an intellectual and cultural force also began to seep into these countries, but very slowly. It has up to now affected only a very small number of those who were educated in science and advanced technology abroad. These had the talent, inclination and nonconformist attitude that permitted them to absorb the scientific spirit and to make it an integral part of their lives. In most developing countries, however, these are not the ones who initiate governmental policies for science and technology, just as the early pioneers of science in the West played a small role in the social and political development of their societies.

In contrast, those who do make the decisions on science and technology were exposed to an education that shielded them from 'unnecessary and useless' ideas and views and that avoided an introduction to the historical, philosophical and sociological aspects of science. Their attention was directed to the fruit in the neighbor's garden of my introductory story. They are, thus, obsessed with the view that, since they are responsible for improving the condition of the people in their poverty-striken countries, they should not indulge in anything in science that does not promise immediate material results. They formulate 'science policies' that trim the roots, cut the 'spurious' branches and eliminate the 'useless leaves' in the hope that fruit will be produced more quickly and cheaply. Such an approach cannot but lead to failure.

I believe that this approach is myopic and is not a consequence of the poverty that prevails in most developing countries. Several of them have rich traditions where 'useless' products continue to be made, for example temple wall paintings in Thailand, wood carvings in Nigeria, masks in Zaire and batiks in Indonesia. The contrast between such manifestations and the blindness to the philosophical and aesthetic aspects of science is striking but not surprising, for art for magical, religious and decorative purposes has been practiced in these societies for centuries, but science was introduced only few decades ago.

Not only do we see, however, blindness and incomprehension as regards the essence of science but often seemingly also with respect to its results. The functional evaluation of the results of scientific research is almost completely missing in the developing countries. What their governments require, instead, is vacuous monthly 'progress reports' along formalistic lines, which can be conveniently placed into filing cabinets. In view of the expressed desire to produce fruit like the neighbor does, such indifference to the evaluation of the results is at first astonishing.

On second thought, however, it is not, since, as in the case of the young man of my story, the desire in the developing countries for growing fruit like the neighbor is *not* mainly a result of the attempt to have fruit to eat but a craving for not being inferior to the neighbor in fruit growing. That is, the main motivation for 'development' is not to acquire material wealth but to eliminate a sense of inferiority in a world containing advanced-technology countries. As, over the years, I visit scientific and technological institutions in many developing countries and am given guided tours complete with commentary, it is virtually impossible for me not to come to the conclusion that the 'game' is to exhibit the existence of 'science-and-technology' in these countries to demonstrate that they are equal partners in the world community of countries.

Such a motivation is likely to play a decisive role in determining the direction of development in these countries. However, it should not be allowed to become the dominant one. If it does, fruit that *looks* like the neighbor's may result, but it will be sour and wormy, inedible.

I am convinced that for the scientific revolution to pervade a country what is needed is long-term, broadly based education for future generations, the energetic support of the few individuals of the country who have already been initiated into the scientific spirit, and a steady interaction by these countries with persons from advanced-technology societies where the scientific revolution has, at least to a considerable extent, been absorbed in the general culture. Stress should be placed on the benefits to be gained from science in molding a new outlook on the world, rather than only on its contribution to making physical survival easier. In this respect, one might encounter resistance even from some Western 'development experts' (especially political economists) who exhibit an astonishing ignorance about what it takes to introduce productive science in developing countries.

Intellectual and philosophical transformations of societies, drastic change in a people's outlook on the world and new cultural trends cannot be produced quickly. There is no known way to grow the tree of science so that it produces fruit in a week after planting, but there are many ways of tending the tree so that no fruit will be collected 20 years later. If the unavoidable features of introducing science and advanced technology could be convincingly communicated to the eager people in developing countries, these people would avoid many disappointments in the future and gain material as well as many other benefits.

Reprinted by permission of *Physics Bulletin*
29 (1978) 249.

OPINION

THE TWO WAYS OF MOUNTAINEERING

By Professor Michael J Moravcsik

Suppose you live in a small town near a mountain range and want to get proficient in mountaineering. One way is as follows. To start, you put on some old tennis shoes and walk up the nearest hill. Perhaps you puff by the time you make it but the next time, and the time after, it is easier and you get up faster.

Next, you get a small knapsack and a pair of stronger shoes and attempt the closest mountain of moderate height. Again after a few tries this goes well and you are then ready to get hold of a puptent and a sleeping bag and venture for an overnighter. In this you may want to get advice from some mountaineer friends who have done it before. Soon enough you are quite skilful and confident in such hikes, and decide to embark on some rope climbing and glacier crossing, in the company of your more experienced friends who are willing to lend you ropes and crampons to begin with.

Eventually you graduate to even more challenging technical achievements, and there you have it. All through this process, when anybody asks, you glowingly relate how high you have gotten, how many miles you have covered, how demanding a rock wall you have conquered. Never during this process do you lose sight of the main purpose, that of reaching the mountain tops and becoming an accomplished mountaineer.

There is, however, another way. You start by buying handbooks for mountaineers describing in great detail the techniques of organising expeditions. This is followed by your contacting and joining the Universal Expedition, Safari and Climbing Organisation (UNESCO), receive its magazine and obtain its equipment guides.

You then begin investing in arctic tents, Himalayan packs, ice-axes, space-age freeze-dry food and expeditionary climbing boots. You also draw up elaborate plans for your future expeditions, including coordinating managers, guides, supply organisations and time schedules, and even establish an office for such activities with the prominent banner of Centre for Reaching Alpine Peaks (CRAP). During all this you find no time to visit even the nearest hilltop but when queried you proudly boast about your shiny equipment and busy Centre, give stirring speeches about the importance of mountaineering and sometimes even parade up and down the street in your new climbing outfit.

Ridiculous? Obviously, in this context, but apparently not enough so when these two approaches are projected into a different, perhaps less familiar, area.

Suppose you want to build up and utilise science and technology in a country where there is practically none. Again there are two ways.

You can start by providing the few scientists and engineers you have with an environment where they can exert their talents and energies. After a while this manpower expands, attains variety, its direction and orientation solidifies and it has a greater impact on the rest of the country.

As already existing equipment becomes clearly inadequate for the work in progress, new acquisitions are made. Eventually the community becomes large enough to need some firmer administrative structure, which you provide on a minimal level as the necessity for it arises.

You strengthen your community by assisting interaction with colleagues locally and abroad. All this time, you assess this development by marking the rising *output* of your manpower in terms of research papers, reports, patents, contributions to production methods and alike.

There is, however, the other way. In that case you start with a 'grand plan for science and technology for development', involving organisational charts, targets and chains of command. You also erect splendid buildings and acquire shiny equipment and in these national centres of research you place administrators.

You join international organisations and send your bureaucrats to international meetings on science policy where they learn about the use of the Delphi method in policy formation and agree to create yet another international organisation.

You distribute some research funds according to whether the title of the project appears to harmonise with the 'national plan', and whenever somebody asks you about the development of science and technology in the country, you are careful to cite only *input* quantities: the number of buildings, the number of administrative bodies, the amount of equipment and perhaps the number of people with degrees in science and technology (regardless of quality).

Whether, in the midst of all these formalistic manipulations, costly and time-consuming as they are, any science or technology is produced or utilised is not asked and not assessed.

At the present, the second approach is all too common, and hence the goal of the development of science and technology and their utilisation suffer in many countries. Will the 1979 United Nations conference on science and technology for development serve as a focus to recast the perspectives and reassess the approaches? □

Two perceptions of science development

Michael J. MORAVCSIK

Institute of Theoretical Science, University of Oregon, Eugene, OR 97403, U.S.A.

Two very different perceptions of the nature of science in the Third World are summarized. The areas considered are aims, the structure of science and technology, planning, education, manpower, communication and information, research, equipment, administration, funding, evaluation, the link between science and technology, the impact of science on Man's thinking, and international cooperation and assistance. The first of these perceptions ("A") is said to be quite close to the view prevailing among administrators concerned with science and technology in the Third World, while the other ("B") is claimed to be much closer to the realities of science as a human activity and hence to be much more functional. The aim of this juxtaposition is to encourage discussion and to catalyze experimentation with science policy measures constructed on the basis of perception B.

1. Introduction

When I first became involved in the problems of building science in the Third World, over 20 years ago, I was struck, as most other observers have been also, by the disproportionately small returns on the large amount of effort and resources allegedly devoted to this task. For example, taking a very modest $300 for per capita GNP for the Third World, and assuming an equally modest 0.1 percent of GNP devoted to science, the 3 billion inhabitants of the Third World spend yearly about a billion dollars for their scientific activities. This does not include the budgets of the innumerable international, regional, and national organizations involved in scientific cooperation and assistance with the Third World. Similarly impressive quantitative indicators can be obtained for the number of people involved in this activity, for the list of organizations and programs set up for this purpose, etc. Yet the results of all this effort appear to be very modest indeed.

It is easy and natural to try to find the reason for this apparent inefficiency in the multitude of particular elements that make up the overall effort of science development. Indeed, there is truth in this way of looking at the problem. One can find fault with some detail in every link of science development, starting from the initial education in science, all the way up to the way the science of the country is utilized. Correspondingly, programs can be devised to try to correct this or that element. I have done so myself, and I do believe that such specific programs are beneficial.

Further on in the last two decades, however, as I had the good fortune of visiting more and more countries and institutions in the Third World and had the opportunity to discuss science development with an increasing number of scientists, science managers, governmental officials, international civil servants, and others, I became more and more convinced that wrong details do not represent the whole story, and, most likely, not even the most important part of the story. Instead, I have come to the conclusion that the whole perception of the nature of science development, as it prevails among a substantial part of the administrators, officials, agency personnel, statesmen, and others who control the resources and methods used in science development, is fundamentally false in that it does not correspond to the realities surrounding science as a human activity.

The statement in the last sentence is a quite drastic one, and hence convincing people of its validity is not an easy task. In order to attempt to do so, I could write a voluminous book with statistics, case studies, lengthy arguments, and other supporting material. In a sense, my numerous articles on this subject written in the last 20 years can be regarded as an effort in that direction. Perhaps one of these days I will indeed express those ideas and other material in the form of a book. In the present article, however, I want to simply express the "perception gap." Indeed, this article is the juxtaposition of two perceptions of science development. In section 2, I will, to the

Research Policy 15 (1986) 1–11
North-Holland

best of my effort, present a discussion of that view of science development which is quite common among those involved today in science development who control the resources and trends. This view will be referred to as "perception A." The description of this view may be somewhat over accentuated in that not *all* science developers would necessarily share *all* of the elements of the view as presented here. Yet I do not believe that the picture I project is a caricature.

The presentation of this view is then followed, in section 3, by the discussion of another view of science development which I believe to be very close to the "right," that is, realistic, one. This view will be referred to as "perception B." This second view will be seen in almost every aspect to be diametrically opposite to the first view, hence my claim that the primary cause for the poor showing of science development in the last four decades is the wrong perception of the nature of science development, and that the faulty details are to a great extent the consequences of this misperception, and hence not easily correctable by individual programs aiming at the details only.

As mentioned earlier, in this article I will not offer "proofs" of my contention. It is my hope that once the two views are starkly juxtaposed, a discussion will ensue which leads to a fundamental reevaluation of the way science development has been pursued. If indeed such a discussion results in the reaffirmation of the prevailing view, such a view will have been placed on firmer foundations. If, on the other hand, the prevailing view will be found erroneous in major ways, as I believe it will be, then a broad effort to build new conceptual foundations for science development can take place.

2. Perception A

2.1. Aims

The only, or at least overwhelmingly dominant overt, aim of science in a developing country is to help technological activities related to the country's short-term economic growth. An additional covert aim in the minds of officials in the developing countries themselves is to offer a visible demonstration that the country is being "modernized."

2.2. The structure of science and technology

Science and technology are very similar. Both apply knowledge to the material activities of Man to make them better. This process is called research.

There are two kinds of research. The first, "basic" research, is defined as research that cannot be used for anything, except perhaps for the satisfaction of curiosity. The second type is called "applied" research (or, equivalently, technological development work), which is useful and produces gadgets, processes, prototypes of industrial products, in other words, something tangible.

The terms "theoretical research" and "experimental research" are also used sometimes. They mean the same as "basic" and "applied," respectively.

2.3. Planning

Scientific activities in a country must begin with planning. The more detailed the planning is (including blueprints for scientific activities in all particulars), the better it is.

Once a science plan has been formulated, the lion's share of the work has been done. Decision making and implementation are relatively easy and hence occupy a secondary position in science policy.

Since it is reasonable to assume that output is proportional to input, science plans should concentrate on input quantities, such as money, number of people, the formulation of organizations, etc.

It is reasonable to assume that the actual development of science will proceed as planned. Hence the measurement of the output of science in the previous planning period is of little importance when preparing the next plan.

Since a national plan is an economic document, and since the aim of science is to further economic growth, such plans, including plans for science, should be prepared by economists. It is unreasonable to pay much attention to what the scientists would like to put into the plan, since there is no limit to what scientists will ask for themselves.

2.4. Education

The most important demand for science education is that it be relevant, that is, it should be evidently connected, in the eyes of the science administrators, with one of the country's present economic problems. In order to guard the student from straying afield, education should stress the memorization of the relevant facts. Furthermore, in order to prepare the student for applied research, it is advisable to specialize his interest to a relevant and well-confined topic.

When sending a student abroad for education, two primary considerations should be followed. First, the education abroad should be as brief as possible, even if all the student's financial needs are covered from funds abroad, because the student is needed at home, and because there is no need for the student to learn unnecessary things abroad. Second, the student's education must be strictly confined to a prechosen project, decided by science administrators even before the student is sent abroad. In order to prevent the student from straying into another field (and also from remaining abroad after his education), he should be bonded before his departure for his education.

The most important motivation for a person is money. This explains the phenomenon of the brain drain, since the absolute level of salaries paid to scientists in Third World countries seldom can approach that prevalent in the developed countries.

Education is schooling up to, at most, the Ph.D. Although there might be a few new developments in science after the student receives his Ph.D. to which he may have to be exposed in short training courses, basically the student has acquired most of his background and tools when he receives his Ph.D. Postdoctoral training is a luxury of the rich which the Third World cannot afford.

2.5. Manpower

For the purposes of science planning, it is reasonable to assume that almost all of those who have received an advanced degree in the sciences in the past will be participating in scientific activities for some decades after obtaining their degrees.

Since the quality of Ph.D. scientists does not vary very much, it is reasonable to deal only with *quantitative* indicators regarding manpower. As a result, the sole aim of manpower development should be to attain as large numbers as possible in the shortest possible time.

In planning for manpower, it is sufficient to project the requisite number of positions for scientists and technologists together, since they do not differ from each other too much.

2.6. Communication and information

The informational and communicational needs of a scientist are primarily a matter of acquisition of facts. Hence it is desirable to place most of the stress on computerized information systems providing such facts. Once such a system is initiated on an international level, one can assume that the information indeed reaches the individual scientists.

Information and communication is required only within the confines of the scientific and technological research areas. Material on the broader context of science is an optional luxury, access to which should be postponed for some time in the future. The one exception is science policy documents, emanating from international agencies, international conferences, and similar sources, which can be obtained free of charge and should be displayed.

2.7. Research

The only, or at least overwhelmingly dominant, function of a scientist in a developing country is to participate in research on a problem which, in the eyes of a science administrator, is evidently related to one of the country's short term economic problems. The selection of these problems is made by economists in the national plan. Governmental support for research therefore should be decided on the basis of the subject matter of the proposed research, which is usually indicated in the title of the description of proposed research. It can be assumed that most research carried out by Ph.D. scientists will be competent and successful.

From seeing much research in developing countries that has no use, we can conclude that too much basic research is undertaken in those countries, compared to applied research.

A main problem with scientific research in the Third World is that those countries do Western science instead of science suited to the Third World.

2.8. Equipment

The most important ingredients of scientific research is equipment. Hence such equipment should be donated by the advanced countries and acquired by the Third World countries as a first step. For example, a newly built institute should acquire equipment even before it acquires its staff.

Once the equipment has been purchased, it can be relied on to operate trouble-free for some time to come. Hence provisions for spare parts, supplied, and repair facilities are not an important aspect of instrumentation.

2.9. Administration

The administration of science should be done by trained administrators, civil servants, and other personnel from outside the sciences. Such personnel can be adequately trained abroad by a two-year course in scientific administration. Inasmuch as scientists are also included among the administrators, they should be older scientists, no longer active in reach. Using active scientists for a limited period of time and concurrently with their scientific research activities is inadvisable.

The most important aspect of science in a country is the set of formal organizations administering it. Hence the scientific development of a country should begin by forming National Science Councils, Ministries of Science and Technology, etc., and by making sure that nationals of the country are adequately represented among the personnel of the international scientific organizations.

When establishing a new research institution, the first and most important steps are the acquisition of a building for the institution and the formation of an administrative procedure for it. It can be assumed that once these two steps have been taken, it should not be too difficult to populate the institution with productive scientists.

One of the foremost principles of the administration of science is that of centralization. Science policy in the country must be concentrated in the hands of one single agency, in order to avoid duplication and waste. Correspondingly, international scientific cooperative agencies should deal exclusively with that agency when becoming involved in any cooperation with scientific activities in the country.

2.10. Funding

The most equitable and easiest method of funding science is that of institutional funding, which

should be done as equitably as possible within the existing domestic and international political constraints. This is in accordance with the fact that most Ph.D. scientists roughly equally suitable for conducting research. As mentioned earlier, however, only relevant research should be funded.

In accordance with the principle of centralization discussed before, funding for science should be done through one single agency, so that the funds reaching the working scientists will have passed through many layers of administrative decisions, thus assuring the soundness of the project and also avoiding the possibility of graft and corruption.

2.11. Evaluation

With the help of planning for science, the outcome of scientific activities is sufficiently predictable that an *a posteriori* evaluation of scientific activities is not very important. One needs to check, nevertheless, whether scientists used the funds for the purpose they indicated originally. Evaluation therefore can be done with the help of a report written by the researcher at the conclusion of the grant period, and this report can then be checked against the original plan by science administrators of agencies. It is usually possible to tell whether there is agreement or disagreement between the two from scanning the important key words in the two reports.

The overall evaluation of science in a country can be made by looking at the country's economic development. If this is unsatisfactory, scientists can be blamed for not doing the relevant kind of science.

The evaluation of research proposals for research to be done in the future can be done by science administrators and senior scientific figures from the country itself. The most important element to concentrate on is the topic of the proposal from the point of view of whether it is relevant to the country's short-term economic goals. It is quite unnecessary to involve in the evaluation process scientists and others from outside the country, since those people do not understand the particular needs of the country and, in any case, resorting to opinions from abroad reflects negatively on the country's independence and self-sufficiency.

It is important to orient scientific research to what is relevant for short-term economic development. Thus research topics should be chosen on the basis of an overall analysis of obviously relevant needs. The choice can be best made by the administrators in charge of applied scientific institutions working in collaboration with government economists.

2.12. Contact with technology

The relationship between science and technology is simple: The results of scientific research are directly applied by technologists in producing gadgets and procedures which then are directly utilized in the productive sector of the country. The process is like a linear chain. It is, therefore, possible to evaluate the success of science through the success of technology or even through the success of the productive sector of the country. It is part of the responsibility of scientists, therefore, to be concerned with the creation of technological gadgets and procedures and with the efficiency of the productive sector.

2.13. The impact of science on people's thinking

For the most part, the influence of science on the thinking of people is restricted to the effect science has on the thinking of scientists. To a small extent, however, it is useful also for people outside science to know something about science. By far the most important from this point of view is to teach people facts, that is, some of the established results of scientific research, so that they can obey the instructions and recipes that go with the use of technological gadgets.

2.14. International cooperation and assistance

The primary purpose of international scientific assistance to a developing country is to provide whatever scientific action is needed for fulfilling the short-term economic objectives of the country. One of the best ways to do this is to perform research in the scientifically advanced countries, by researchers from those countries, and then provide these research results to the developing countries for their use. If there are some scientists in the developing countries already, some collaborative research scheme focusing on a specific and short-term research goal could also be instituted, though such cooperations are inefficient compared

to getting the research done in a scientifically well-developed country. The long-term building up of the developing countries' own scientific infrastructure is of much lower priority, and is in fact a luxury item which does not promote the short-term economic objectives of the country.

Programs to help the developing countries are best handled by large and centralized agencies, which can create huge projects. The best way to organize these projects is country by country, and within countries discipline by discipline.

The international organizations handling such assistance and cooperative projects must have a massive administration structure involving a geographically well balanced personnel, in order to assure balance in the projects and in order to give weight to the operations. It is particularly important to involve many people from developing countries in the international bureaucracy managing such projects.

Since the developing countries cannot wait very long, science development must show conspicuous and relevant results within each five-year period, and programs and decisions should be geared to this goal.

3. Perception B

3.1. Aims

The aims of science in the context of a developing country are a multitude. They can be grouped under three main headings: (1) Science is a necessary condition for technology; (2) Science is a prevalent human aspiration in the 20th century; (3) Science is a powerful influence on Man's view of the world and of his place in it. The three aims are in many ways interrelated and it is very difficult to create scientific development by just considering one of these three equally essential aims without the other two.

3.2. The structure of science and technology

Science and technology are quite different human activities, which can be told apart by their respective products: Scientific research results in knowledge, while technological development work results in prototype gadgets or procedures to do something. The two different activities attract people with different talents and aspirations, and using the two groups interchangeably usually results in work of poor quality and a discouraged working force.

The difference between "basic" and "applied" research is not in the topic of the research or in the product of the research activity (which in both cases is knowledge), but in the motivation of the researcher or of the sponsor. If the motivation is to enlarge the body of knowledge because knowing is a good and enjoyable thing in itself, the motivation is "basic," while if it is to acquire knowledge so that it can be utilized in some other activity, it is called "applied." The distinction between the two motivations is very difficult to make, since the same piece of research may be one in the short run and the other in the long run, or can be simultaneously both in the mind of the same person, or might be one in the mind of the researcher and the other in the mind of the sponsor. In order to achieve some balance between the various types of research activities, one may define, quite arbitrarily, applied research as an activity which is likely to produce results applicable in a particular field of technology within ten years, as judged by the researcher.

In *theoretical* research the *tools* are concepts, theories, models, and mathematical calculations, and the aim is to explain already known natural phenomena and to make predictions for yet unknown phenomena. In contrast, *experimental* research uses as tools equipment to create and isolate natural phenomena and measure their properties, in other words, to generate data on judiciously selected situations. The dichotomy of "theoretical" vs. "experimental" and the dichotomy of "basic" and "applied" have nothing to do with each other. In particular, a crucial and large part of applied scientific research is theoretical.

3.3. Planning

Centralized planning plays a negligible role in the scientific development of a country, especially at the initial stages. Science begins in a country through a tiny group of exceptional people with an internally motivated desire to do science, who, using ad hoc methods, create around them the infrastructure they need. Even at later stages, decision making and implementation play a much greater role than making plans. Indeed, in most

developing countries, the bloated bureaucracy associated with S&T planning is inordinately huge compared to the extent of S&T activities in the country, and it makes doing science difficult and laborious and often draws working scientists into unnecessary administrative manipulations.

Indeed, none of the countries presently leading in science began their scientific development with a centralized plan, and some of the countries at the very top still have no centralized science plans.

All this does not mean that some advance thinking about providing for science may not be helpful. Planning can be done qualitatively, semi-quantitatively, and quantitatively; it might mean some contingency planning, planning of input, or planning of output; it may be done on a short time scale or far into the future; it may be interpreted as a strict guideline or as a probability prediction; it may be aimed at quantitative targets or it may aim to improve quality; and it may be done by appropriate or inappropriate personnel. The discussion of what kind of and how much planning is helpful requires extended specifications. In any case, having done the planning only a tiny part of the task has been accomplished, and the most difficult part is still to come. The details of planning are different depending on whether one considers manpower, finances, equipment, buildings, administrative structures, or any of the many other ingredients of science. In any case, such planning should be done by a group consisting predominately (though not exclusively) of scientists personally experienced in scientific activities. The function of economists in constructing national development plans is only to combine and coordinate the needs and aspirations voiced by the many sectors of the country.

It must be recognized that planning of the output of science is practically impossible and is fraught with uncertainties. In particular, output is, for all practical purposes, not proportional to input, and two quantitatively equal inputs can easily produce outputs that differ from each other by an order of magnitude.

3.4. Education

The most important demands for science education are that it be broad and that it prepare the student for continued self-education throughout the student's entire lifetime. The main orientation of such education should be to exhibit science as a skill of asking interesting new questions and then finding approximate answers for the solution of problems. Thus memorization in science education should be absolutely minimized and instead problem solving must be stressed.

The student's advanced education abroad should be extensive enough for him to acquire a broad and functional background, including some acquaintance with the contextual aspects of science. The student's field of interest is likely to change once he has widened his horizons, and this change of interest should be encouraged in order to produce a most motivated and creative scientific manpower. Bonding a student prior to sending him abroad is an ineffectual and counterproductive process which accelerates the brain drain.

The best remedy against the brain drain is the creation of a congenial atmosphere in the home country for the scientist to work in. Salaries in this respect are only one and by no means the most important factor. An absence of bureaucracy, an availability of auxiliary services for research (spare parts, technicians, modes of scientific communication, etc.), and recognition are much more important elements.

The education of a scientist continues over his whole lifetime, and hence opportunities toward such a lifelong education must be provided. At the time of his obtaining the Ph.D., the scientist's total background and capability in science is considerably less than half of what he will need 20 years later in his career. The difference must be made up by education on the job.

3.5. Manpower

Productivity among scientists has an extremely skewed distribution. Empirical studies have shown that, for example, in terms of lifetime production of scientific research articles, for every 10,000 scientists publishing only one paper each in their lifetimes, there are only 100 publishing ten each, and only one publishing 100. Thus a relatively small segment of the overall scientific research manpower produces a substantial fraction of the overall research results. One of the foremost objectives, therefore, must be to pinpoint and encourage the few exceptional individuals in the scientific manpower pool who are unusually creative. Manpower development goals therefore have to be

specified in terms of quality much more than in terms of quantity. Manpower projections in plans must also take into account this overridingly important quality factor. Fast increasing quantitative manpower targets are dangerous, as the case histories of several countries have shown, since they neglect quality and create a large pool of functionally useless manpower that is a deadweight on further development.

3.6. Communication and information

Facts are the least important elements in the information and communication needs of a scientist. As a result, computerized information systems have a relatively unimportant role, even apart from their unreliable operation in the context of developing countries. Instead, the crucial ingredients in the communication system between scientists are person-to-person contact and written modes of communication propagating ideas, overviews, critiques, conceptual advances, speculations, and other forms which go far beyond simple pieces of facts.

Information and communication is needed by scientists in the Third World not only with respect to the strict technical aspects of science but also in the area of the context of science, that is, in the area that encompasses scientific methodology, the management of science, the sociology, psychology, history, and philosophy of science, the communication of science to the public, etc. These contextual problems are especially important in the Third World where the infrastructure of science has not been laid down yet.

3.7. Research

The most important function of a scientist in the developing country is to be a funnel of scientific ideas and information into the country, since more than 99 percent of new (and old) science is created outside the country, and hence an overwhelming fraction of science needed for activities within the country will have to come from abroad. The only way a scientist can serve as such a funnel is to maintain his own personal research on some problem in a broad area, because through such research he will be a member of the worldwide scientific community where science is created. From the point of view of being such a funnel, the exact topic on which the scientist performs his research does not matter much, though it is crucial that his research be considered good by the worldwide scientific community. To assure the latter and to monitor the quality of the local scientific activities, a vigorous evaluation procedure must be maintained continuously.

In most countries in the world, basic scientific research represents only 10–15 percent of the financial resources for research and development. In developing countries, most of the research with no use is poor applied research, and not basic research.

There is no special "Western science" or "Third World science," and no "communist science" or "capitalist science," and no "Muslim science" or "Christian science," and no "white science" or "black science." There is only one science, which all of us work on. There may be differences in detail in the practice of science in various institutions around the world, just as there are also such differences among institutions within the scientifically advanced world. These are, however, quite minor compared to the unifying features that tie together all science around the world.

3.8. Equipment

Research equipment comes second in importance behind scientific manpower. Thus equipment should be adjusted to the available people and not the reverse. It follows from this that apart from some instrumentation of completely general use, equipment should not be purchased by a Third World research institution in the absence of the person who will use it. Furthermore, a careful assessment of the potential user must be made to ascertain whether he will in fact use it. Third World laboratories are graveyards for expensive unused equipment, ordered or received as a gift without a potential user present, or for a declared potential user who was unable or unwilling to engage in research. The equipment may be unused also because of lack of spare parts, supplies, or repairing technicians. Thus no equipment should be acquired without simultaneous and adequate provisions for spare parts, supplies, and repairs. These arrangements should be kept in mind also by potential donors of such equipment to institutions in the Third World.

Much equipment can be created from locally available material and from surplus. An ability to do so must be a strong part of the education of a scientist who wants to perform research in the Third World.

3.9. Administration

It is extremely seldom that one encounters a successful administrator of science who himself is not a scientist with a set of scientific accomplishments satisfactory to himself. Such a background seems to be needed for self-confidence, for interaction with scientists on an equal footing, and for the ability to make intuitive decisions and judgments. Such administering scientists for a temporary period of time, after which they should return to direct scientific activities.

Such administering scientists should work within a minimal formal structure with very light bureaucracy. The creation of formal organizations should be postponed as much as possible, and then kept to an absolute minimum.

The flocking of scientists to international organizations with massive headquarters in faraway developed countries and with a horrendous bureaucracy is a drain on the country's useful manpower and offers little if any educational experience.

In creating new research units in the country, first and foremost is the formation of a compatible group of motivated, active, and competent scientific manpower. Showy buildings, fancy equipment, and other externalities have low priority.

To the largest extent the scientific activities of the country should be decentralized, giving a large amount of freedom of decision and action to the individual research groups. A pluralistic source of funding for research is highly desirable and conducive to creative science. Correspondingly, international organizations should, as much as possible, deal with individual research groups and not with a centralized bureaucracy.

3.10. Funding

The most productive method of funding scientific research is by individual research grants given directly to small groups of research scientists. Although some institutional funding is also necessary, that should be more the exception than the rule. Funding should be awarded on a merit basis, recognizing the lopsided distribution of excellence within the scientific community. This system entails a strict and competent assessment and evaluation system for scientific research, both prior to awarding research grants and at the conclusion of a piece of research.

While some distribution of research topics between "basic" and "applied" should be maintained, the first consideration should go to scientific excellence and not to the short-term relevance of the research topics.

3.11. Evaluation

Since planning for science (and particularly for its output) is a highly uncertain process, evaluation and assessment of scientific work should be a continual process. Evaluation cannot be done by people outside the science. Instead, it must involve scientists from both inside the country and outside of it. International peer review is a common practice around the world among the developed countries, hence aversion to it because of feelings of national pride is misplaced.

The output of science cannot be measured by the state of economic progress in the country, since many intermediate stages exist between science and production, and because science has other objectives besides aiding economic development.

Topics for research should be decided on by scientists in cooperation with the potential technological users of the applied part of the research. In considering applied research, the country's technological needs should be considered far into the future, since creating a productive tradition of research in a given area involves a long period of time.

3.12. Contact with technology

The relationship between science and technology is a complex and subtle one. Most of today's technology necessarily requires scientific input, but the science needed may be fairly old science, might be quite indirect (through instrumentation, for example), and in any case will involve science researched outside the country. Thus the primary role of the scientists is to be a funnel of scientific information into the country. This function, however, requires that the scientist be continually active in research, although the exact topic of the

scientist's research is of secondary importance.

Scientists should be used for scientific research and not technological development work, even if the temptation to do the latter is high because of the dismal state the engineering profession is in in the country.

3.13. The impact of science on people's thinking

A role of science second to none in a developing country is to make an impact on the thinking of people at large, far beyond the small group of professional scientists. The impact must be, above all, attitudinal, and propagating facts established by science is secondary in importance. In bringing about such a cultural impact of science, scientists must work closely with the public media of various sorts. Without such a cultural scientific revolution it is very difficult for the country to establish an indigenous scientific infrastructure, to create a broadly based scientific tradition, and to use science for a corresponding technological transformation.

3.14. International cooperation and management

The primary purpose of international scientific assistance to a developing country is to enable the country to evolve a strong local scientific infrastructure as soon as possible, so that the country can tend to its own scientific and technological goals using its own resources. Projects in which an advanced country does research "for" a developing country are almost entirely useless in this respect.

The above described international cooperation is best achieved in a decentralized mode, through person-to-person interaction between scientists in the country and their counterparts abroad. In addition, much can be accomplished through the worldwide scientific community by making sure that the distribution of resources needed for scientific work is equitable. Many of the measures needed do not easily divide into geographical regions or disciplinary lines, but instead attack problems across the board.

International organizations and binational assistance agencies must be light in bureaucracy and be staffed by personnel chosen on the basis of merit. Geographical balance is secondary.

The characteristic time needed for a country to

evolve a strong productive scientific infrastructure is a half century or more. Impatience resulting in forcing such a development by sacrificing quality and by a narrow interpretation of relevance may do major damage to this development process and may delay its goals.

4. Conclusion

It is not difficult to see, for example through a point-by-point comparison, that the two perceptions of the nature of science development are sharply different. The difference is so drastic that decisions and actions generated by these two perceptions will, in most cases, be almost diametrically opposite from each other.

In a sense both perceptions, as described above, are extremes. The current practice, nevertheless, is very much closer to perception A, while the view optimal for science development would, in my view, be quite close to perception B.

Each sentence in the above brief and incomplete summary of the two perceptions could very well form a basis of extensive discussion. I hope, in fact, that it will, and that, in addition to abstract discussions, some science managers will also attempt to implement some of the consequences of perception B, to see if indeed it bears richer fruits. I strongly feel that we need such a marked paradigm change if the scientific development of the Third World is to proceed at an accelerated pace and with more substantial and lasting results.

Appendix

When I first submitted this paper to *Research Policy*, the referee's report stated that perception A represents strawmanism, and that there is nobody in the Third World who "could remotely accept the first 'perception' alluded to by the author or indeed anything approximating to it. It is crude to the point of caricature and *anything* could be justified in juxtaposition to it…. If the author insists that 'perception A' is indeed true than he would need to provide some authority for its claim."

I took up his challange, and conducted a small survey of opinions regarding "perception A". In particular, I sent out "perception A" to a sample of relevant people with the following covering note: "The enclosed sheet contains a statement of science policy in the context of the developing countries including various aspects of such policy. Since I am interested in the extent to which such a statement reflects the personal views on this subject of various prominent and experienced people in the field, I am gathering some *purely statistical* information on this. I would like to ask you,

Table 1
Responses of the six respondents to the survey as described in the Appendix. The letters and number labeling the rows refer to the sections and paragraphs of "perception A" given in the text.

Section	Paragraph	Code number of the respondent					
		1	2	3	4	5	6
2.1	1	−	− −	+	+	− −	+
2.2	1	−	+		+ +	− −	− −
	2		− −	+	+		−
	3	−	− −	+	−	+ +	− −
2.3	1	+	+	−	+ +	+ +	+
	2	− −	−		−	− −	+ +
	3	− −	+	+	+	+	+ +
	4	− −	−		− −	− −	+
	5	− −	− −	−	−	− −	−
2.4	1	−	− −	+	+		− −
	2	+	−	+	+ +	−	− −
	3	−	+	−	+	− −	+
	4	−	−	− −	+	−	−
2.5	1	− −	+		+ +	+	+ +
	2	− −	−	+	+	−	
	3	− −	−	+	+	−	− −
2.6	1	− −	−	−	+	−	+ +
	2	−	−	+	+ +	+	+
2.7	1	−	+	+	+ +	−	+
	2	−	− −	+	+	− −	+
	3	−	+ +	+	+ +	−	−
2.8	1	−	− −	−	+	− −	− −
	2	−	− −	−	−		− −
2.9	1		−	+	+	− −	+ +
	2	−	−	−	+ +	+ +	+
	3		−	−	+	−	− −
	4		−	+	+ +	+ +	
2.10	1		−	+	+ +	+	
	2		− −	+	+ +	+ +	
2.11	1		+	−	+	− −	+ +
	2		−	− −	+	− −	
	3		−	+	+ +	−	+ +
	4		−	+	+ +	−	+ +
2.12	1	−	−	−	+	+	+ +
2.13	1	−	− −	+	+ +	+	+ +
2.14	1		− −	− −	−	− −	− −
	2	−	− −	−	−	− −	− −
	3		−	−	+	−	+
	4	+	+	+ +	+ +	−	

therefore, to read the rather brief two pages which are attached, and indicate its overlap with your own personal views by placing, next to each paragraph, + + if there is a very considerable amount of overlap, + if there is somewhat more overlap than disagreement, − if there is somewhat more disagreement than overlap, and − − if there is a very considerable amount of disagreement. Then please mail the sheet to me. The survey is totally anonymous, and no opinions of individuals will ever be quoted individually, nor will the list of respondents be released."

The mailing went to 14 upper-level science administrators in a variety of developing countries. I was personally acquainted with all of them, at least superficially, and that provided a good return: 6 out of 14. I did not, however, know the views of the respondents on these issues when I sent out the mailing, so there was no tendentious selection of the sample.

The responses are tabulated in table 1. As I see it, they confirm the claim made in the paper, namely that while the statement of "perception A" (as well as of "perception B") is somewhat extreme, there is considerable agreement with it among the sample. Thus the opinion of the referee quoted above appears not to be borne out by the survey.

This article was meant to arouse controversy and a reinvestigation of some of the foundations on which science development rests. It seems to have done so already with the referees. I hope it will continue to do so also with the readers.

B. Research in the Third World

Reprinted by permission of *Minerva* **2** (1964) 197.

TECHNICAL ASSISTANCE AND FUNDAMENTAL RESEARCH IN UNDERDEVELOPED COUNTRIES

MICHAEL J. MORAVCSIK

I

TECHNICAL assistance programmes to the developing countries have been in existence for some years now and they are carried out on a considerable scale. A large majority of such programmes consist of applied research and development in connection with a specific problem, such as the survey and the proposal for the solution of the salinity and waterlogging problems in Pakistan. In such programmes, " experts " on the particular problem are selected to evaluate the situation and design the solution, and then financial assistance is given to the country involved to carry out the recommendations in cooperation with or under the supervision of the " experts ". Engineers, applied scientists, legal experts, administrators, expert farmers, industrial managers, etc., are the chief participants in such undertakings. These programmes are usually of fairly short range (meaning that they affect only the next two or three five-year plans) and can produce strikingly tangible results, which contribute to the well-being of the country involved.

Another kind of assistance is concerned with education. Considerable American foreign aid funds are spent in various countries for elementary school building, on demonstration equipment in various colleges, or even on school milk programmes (which, by promising a cup of milk to the student, reduce absenteeism in schools in some cases to a surprisingly large extent). Furthermore, the Fulbright and Smith-Mundt programmes send college and university teachers to various countries to raise the level of instruction and to supplement teaching staffs which are often deficient in number as well as in quality.

There is, however, another aspect of technical assistance which so far has been neglected, which is of major importance to the developing countries and which could be undertaken with a minimum expenditure of funds and a certain amount of cooperation by the scientific communities of the United States or other countries with technical aid programmes. This is fundamental scientific research in the underdeveloped countries. By fundamental research I mean investigations undertaken in the natural sciences for the purpose of enlarging the general body of knowledge and not because the results to be obtained appear to be helpful in solving a specific practical problem. (Fundamental research in this sense is also sometimes called pure science, as opposed to applied science.)

II

It is not at all obvious that the underdeveloped countries are badly in need of fundamental scientific research today, at this very rudimentary stage of their development. Yet, the need for this kind of research is basic to the argument proposed in this article. If one travels in any of these countries and observes the high rate of illiteracy, the daily struggle to reach the bare subsistence level and the snail's pace at which the modernisation of the social structure moves, one might gain the impression that what these countries need is intense assistance to solve the most urgent economic and administrative problems and that pure research is an unjustifiable luxury. A somewhat more searching investigation will however support the contention that the underdeveloped countries must begin without delay to develop their scientific resources in the direction of fundamental research.

It is of course seldom questioned that applied research is of great importance to the underdeveloped countries even today. There is a great deal of applied research which is done in the advanced countries and which produces results which are relevant to the problems of the less developed countries. This would appear to reduce the urgency of doing even applied research in the underdeveloped countries; after all, what they must do, it could be argued, is to import and apply the results of the research done in the advanced countries. Yet, by its very nature, applied research often tends to be specialised, and conversely, particular problems require separate efforts by applied research workers. It inevitably happens, therefore, that each country finds itself faced with certain problems which must be dealt with scientifically but which are peculiar to it, perhaps because of the climate, the fauna, the natural resources, the economy, the cultural heritage or the geographical location. No advanced country can then be expected to deal with these problems and only a well trained and competent local research team will have the time and inclination to do this.

The education of such a well trained and strongly motivated body of applied scientists can be carried out only by people who are intimately acquainted with the frontiers of science in the fundamental fields and who are active research workers themselves. This is so because the fundamental science of today is the applied science of tomorrow and because the educational process in science is a long process. One or two examples should suffice to illustrate this point. It would have been proper, for instance, in the late 1930s, to include in the education of an engineer a rather thorough course in the then somewhat esoteric subject of nuclear physics, since this would have prepared these engineers to face, 10 or 15 years later, at the peak of their careers, the major developments in nuclear engineering connected with reactors. Experience has shown that it is very difficult for an already established scientist or technologist to interrupt his career to train himself in a fundamental branch of science which he missed when he was a student. Similarly, an electronic engineer receiving his training in the 1940s would have been well advised to learn a considerable amount of solid state physics, to prepare for the development of transistor technology in the 1950s.

That this point is not at all an academic one can be concretely illustrated. In most underdeveloped countries the few institutions of higher learning which exist are staffed with a large proportion of the very few local scientists or engineers who are available. Many of the more senior of these received their training in the thirties or forties at some foreign university, perhaps did some routine piece of work for a thesis, and then returned to the homeland, to be appointed to a post at a university or college. With no research being pursued there, with no competition from colleagues and in the general atmosphere in which time and change were of no great concern, they settled down to teaching the science of 1938 throughout the 1940s and 1950s, and now the 1960s. The concentration on the teaching of undergraduates accentuates this tendency to persist in the scientific ways of their own student days. By now, they have a vested interest in opposing any change in the curriculum, in the teaching methods or even in the administrative practices of the university, and since they are in senior positions, they act as a very strong barrier indeed to any modernisation of the institutions of higher learning. It is a very vivid example of what might be called scientific "featherbedding", and it is strongly self-perpetuating inasmuch as the science students of today, who thus learn at best the sciences of yesterday, will get little of the excitement of being in the forefront of the sciences, will be badly prepared for a productive scientific career and hence will be mostly concerned with trying to secure for themselves the few comfortable old-fashioned academic positions which become available at the universities as the older generations disappear. To obtain these positions, they do not have to show great brilliance or present lists of research publications, but instead have to exhibit a willingness not to rock the boat and to fit in with the old-fashioned ways which these institutions have followed for the past decades. Almost all the universities on the Indian subcontinent suffer from this malady. The situation is somewhat different in the newer universities of Africa and South-East Asia but, for closely related reasons, very similar conditions obtain. With

200   MICHAEL J. MORAVCSIK

active research being pursued at these institutions, such a process of intellectual calcification could not possibly have come about and where it has not yet occurred it can still be avoided. If active and fundamental research is not carried on in these universities, this calcification will become or remain as firm as it already is in the older universities of the " third world ".

III

The second argument in favour of basic scientific research in underdeveloped countries is also linked with the applied work done there; it asserts that applied work can be carried out successfully only if the research workers have constant access to persons working in fundamental fields. There are several reasons for this. The training of a person working in fundamental research tends to be more thorough and broader than that of persons who committed themselves early in their careers to applied work. This is particularly true for scientific personnel in underdeveloped countries, where applied research is often defined in the narrowest possible sense of the word. For instance, a basic research worker usually has a better background in general research methods, in the understanding of the interconnection of various applied problems through their common scientific basis and in the general techniques applicable in various branches of a science. Thus, a person trained in fundamental research can often serve as an " ideas man " to his colleague trained in applied research, suggesting new approaches to a problem, or carrying over analogous methods or solutions from fields unfamiliar to the workers in applied research.

Such an interplay between fundamental and applied research is an essential ingredient in the advanced countries, so much so that perhaps we take it for granted that such opportunities exist in all countries. A very large fraction of research workers in the fundamental fields in the United States spend part of their time as consultants to applied projects for the government or private industry. They serve either as members of review committees, which survey progress made in a given applied field, and suggest new directions, new approaches and techniques, or, they might be called upon to help in connection with a specific applied problem which has run into difficulties and where a broader scientific base is helpful in getting it back on the right track. The results of such cooperation between applied and pure research workers are obviously very valuable and consultants of this kind are, therefore, in great demand and consultant's fees very high.

My point is simply that this assistance in carrying out applied research should also be available to the underdeveloped countries. In fact the need there is even greater since, once an applied team runs into difficulties, the closest colleagues who might come to the rescue may be thousands of miles and large amounts of foreign exchange away. Hence day-to-day cooperation is out of the question and even a single trip for the purpose of consultation might run into financial and political obstacles. The only remedy is to establish highly competent and active local research teams in the fundamental sciences. They can then be used as consultants in times of such " crises ".

IV

The third argument supporting pure research is not related to applied work but stands on its own merit. Granting that eventually all countries should have their own basic scientific life it will still take a very long time—perhaps several scientific generations—to establish a strong tradition of pure research in a country which has previously been without science. It should therefore be part of the planning of a new country to make, in addition to five-year economic plans, a 25-year scientific plan. This would allow for the training of young scientists today in the fundamental fields, who then, years from now, with considerable experience and some notable achievements behind them, will originate and direct research carried out by a new generation of local scientists. These, having been brought up in an indigenous research atmosphere, will be able to expand their own research efforts and in turn lead another scientific generation which will then be large and strong enough to carry out, on an extensive scale, research which will be competitive with work in the advanced countries.

There are several examples demonstrating that even in relatively advanced countries, the development of a scientific tradition of excellence is a slow process. It took even communist Russia, building on an already outstanding and established scientific tradition and on the scientific institutions of Czarist Russia, about 40 years to become a major scientific nation. As for the United States, it took a tremendous influx of European scientists, coupled with the scientifically extremely stimulating atmosphere of the war years, plus 20 years, to transform American physics from a minor and sporadic effort into world leadership. When we are faced with the meagre resources and considerable cultural handicaps of an underdeveloped country, such a process should be expected to take even longer. It is therefore of utmost importance to start it as soon as possible, even if on a modest scale.

V

The fourth point in connection with research in the less developed countries concerns the training of administrators. In the new states there is a tremendous need for competent and well-trained administrators to decide on the right policy for development and to carry out those decisions according to plan. This lack of local administrators has been cited as the largest single handicap of the entire foreign aid programme of the United States. Among these administrators, many have to make decisions which deal with scientific and technological matters such as health projects, new industrial processes, improved agricultural methods, development of power resources and many others. The need is very great, therefore, for persons sufficiently well versed in science and technology to be able to make technically sound, realistic and prompt decisions. The best training for such posts is research experience and a thorough education in the fundamental sciences. This contention is borne out by the example of the United States where the administrative heads of the large government and private research and development projects are increasingly scientists with research background in some fundamental branch of science. It should be kept in mind that especially in the case of a new country, where everything starts from a low level, where decisions affecting the country for several decades have to be made, and where furthermore the number of persons making the decisions is very small, the stakes are extremely high and a few mistakes might make the difference between success and failure. It should also be remembered that because of the shortage of scientific manpower these administrators in the young countries have no advisory committees to rely on for technical advice, as is the practice in the advanced countries, and they have no one to turn to for aid in making these decisions. It is of the highest importance, therefore, that these people get as broad an education in the various fundamental sciences as is possible and that they come to understand the ethos of science through their own experience in research. The respective Chairmen of the Atomic Energy Commissions of India and Pakistan, Dr. Bhabha and Dr. Usmani, might serve as individual examples of this contention.

VI

The fifth argument for fundamental research is a psychological one; achievements in the fundamental sciences would serve as a source of great encouragement and high morale in the newly developing countries. Justly or not, pure science is generally considered one of the most sublime proving grounds for the human mind and a country, which economically, socially and politically might still have to consider itself inferior to its Western counterparts, might take special pride in the outstanding achievement of one of its sons in the natural sciences. In a small country like Denmark, the late Niels Bohr is a national hero and one could encounter a taxi driver in Lahore, perhaps illiterate, who speaks with reverence of Professor

Abdus Salam, the most eminent Pakistani physicist. More importantly, such outstanding individuals also serve as models for young local scientists, who are trained under less than ideal conditions, or who, after having been abroad, have to face the agonising decision of whether to return to their backward homeland or to succumb to the many tempting offers of employment in the advanced countries. In making this decision and in gathering strength for hard work once they decide to return home, the example of outstanding local achievements in the fundamental sciences plays a crucial role. One cannot overemphasise the importance of high morale when talking about the development of backward countries. It may be the most important single factor in deciding between success and failure.

VII

In the foregoing arguments on behalf of fundamental research in under-developed countries, I have assumed at several points that the advanced education of scientists in a country should be carried out to a large extent on their home soil. The reasons for this are as follows.

Firstly, most of these developing countries have chronic shortages of foreign exchange and education abroad uses a lot of foreign currency. Reliance on fellowships can be only on a small scale; when hundreds of scientists have to be trained each year, it is impossible to find enough outside financial help. It is much more economical, even apart from its other benefits, to establish an institution of higher learning on home soil.

Secondly, Western institutions of higher learning are becoming more and more overtaxed by demands for admission of their own nationals, so that even if a candidate from an underdeveloped country is treated on an equal footing with his Western colleague (which, for a number of reasons, might not be the case), he has a good chance of being turned down. This situation will only be aggravated in the coming years with the rising demand for higher education in the West.

Thirdly, it is very difficult indeed to establish functioning research groups when people get their education in different university systems abroad, when the assimilation of new staff workers cannot be done at an early stage and when continuity of personnel is lacking. It has been a long-standing experience in all advanced countries that the best and cheapest hands in the gruelling work of scientific research is the postgraduate student who is responsible for a very large proportion of the many man-hours involved in scientific experiments. At the same time, the student himself learns, at an early stage, about the problems peculiar to the equipment he uses, about the sources of supply for spare parts, about the peculiarities of the supporting technicians, etc. All in all, a " school " thus develops around a senior scientific person and this school can turn out to be extremely productive indeed. It should be the goal of an underdeveloped country to establish such schools on home soil in the various branches of the sciences and such schools cannot prosper without young students, especially postgraduate students.

Fourthly, living abroad during the advanced training and then returning home raises a number of problems of adjustment. For one thing, some trainees have been known to be unable to produce anything like their best in the Western countries because the problems they faced when trying to make an adjustment to the different economic, social and cultural practices were so overwhelming that their scientific studies were neglected. Others managed to adjust but in the process lost some of their sense of identity with their homeland and when the time came to return, they let themselves be lured away by the attractions of the research institutions of the countries in which they received their advanced training. In fact, the problem of how to encourage young Ph.D.s to return to their country of origin is one of the most serious in trying to develop science in a new state. The " mortality rate " is very high indeed and affects adversely the whole programme of advanced scientific training and research. Few governments would fail to grumble about spending foreign exchange to send their sons abroad only to be snatched away by the richer countries. These problems of adjustment and readjustment are really extraneous to science and hence should be eliminated if at all possible by establishing advanced institutions on home soil.

VIII

Let us now see in what way a technical assistance programme can be used to promote basic scientific activities in these countries. The three important ingredients in such activities are buildings, equipment and manpower.

The first of these, buildings, is the least difficult. Most of the young countries are well prepared, eager and willing to put up sumptuous edifices to house their research activities. There are several reasons for this. Firstly, buildings usually do not involve spending foreign exchange. To be sure, foreign architects are often engaged to make the designs but even some of these are willing to accept local currency and the architect's fee is a small fraction of the total cost anyway. The construction itself is done by a local firm, often not without faults or even corruption, but at least for the first few years the new building will serve as an imposing exhibit of the country's will to become " up to date ". And this is the second reason for the willingness to provide buildings: it is an obvious and eye-catching sign of " progress ", which can be used for impressing foreign visitors and for domestic propaganda. Only an expert, and only after a thorough investigation, can tell whether the work carried out in the building is high grade or not; the majority of the onlookers will judge the enterprise by its external manifestations, of which the building is the most evident one. The attractive buildings of universities in Mexico, Pakistan, or some of the West African countries may serve as examples.

The third reason why it is relatively easy to get buildings is that the financing of a building is a matter for a single decision, which might be made by the government in a burst of enthusiasm. Once built, it requires little further expenditure for maintenance. This is also one of the reasons why experimental apparatus, to be listed under the second ingredient, equipment, is relatively easy to get. Although this nearly always requires foreign exchange there are many advanced countries very willing and well prepared to donate it as part of foreign aid. This again has the advantage of looking impressive and in addition helping the industry of the donor country which produces the equipment. It also appears to have the advantage of involving only a single decision, although in reality this is not so. In fact, expert maintenance of such equipment is a major problem in the underdeveloped countries since the shortage of trained mechanics and technicians is even greater than the shortage of scientists and spare parts must also come from abroad. For this reason impressive equipment often stands idle for want of a very inexpensive replacement item or because nobody in the institution knows enough about the apparatus to find out what is wrong with it.

There are two immediate and easy remedies for this malady. The first would be a ready supply of spare parts in the country where the equipment is located. For instance, a storeroom could be set up which would carry all the spare parts needed for the equipment donated to all institutions in that country. Secondly, a system of roving Western mechanics could be organised until local ones have been trained. For example, one electronic technician would be assigned to Pakistan; he would constantly tour the universities, institutions, hospitals, etc. where equipment donated by the foreign government is located and could repair faulty equipment and give advice for the proper use of the apparatus. Considering the high cost of some of the electronic equipment, the cost of maintenance of this one technician would produce handsome returns in the increased efficiency of the apparatus.

The development of library facilities is another element to include in the category of equipment. This has at present a much less favoured position than experimental equipment, although it is hardly less important. Buying scientific books and periodicals requires foreign exchange and in fact, in large amounts, over a long period of time. Scientific libraries are generally

206   MICHAEL J. MORAVCSIK

not objects of national pride, being buried in one room of the upper floor of the building, and a layman generally has little basis for distinguishing a good library from a bad one. There have been several attempts made to facilitate the building up of libraries in these countries. Some books published in the United States appear in cheap "Asian editions ". There are some organisations, particularly the British Council, the Asia Foundation, and the United States Information Service, which occasionally donate subscriptions to periodicals, at least for a limited amount of time. But this is just a piecemeal solution on a small scale. The United States government should consider publishing all American scientific journals in cheap editions, which, aided by government subsidy, would be available to the underdeveloped countries either free or in exchange for local currency. At least one library in each city where considerable scientific activity is evident, should be well stocked with books, journals and research papers in temporary form, such as duplicated reports, so-called " preprints ", conference proceedings, etc. One of the greatest handicaps of a scientist working in an underdeveloped country is the feeling of isolation and this could be relieved by a prompt and ample flow of scientific information. (Other countries with technical assistance programmes could also help by supplying scientific information along the same lines.)

The most serious problems of science in an underdeveloped country are, however, those connected with the supply and maintenance of manpower of high quality. I have already discussed some of the problems in connection with giving advanced scientific education to promising students and touched on the difficulty of luring back trained personnel from abroad. Once promising young scientists return to their homeland, they are faced with further problems, such as delays in obtaining experimental equipment and problems in maintaining it, inadequate library and other information facilities, lack of contact with their colleagues in the more advanced countries, shortage of stimulating local colleagues working in the same field with whom to discuss problems, low salaries, too many routine duties, etc. These are problems which arise from national poverty, small numbers of trained personnel, faulty traditions and uncongenial administrative practices.

IX

Clearly, many of these problems have to be tackled one by one and in the context of the local conditions. In addition, however, there is also a general remedy, which might be supplied within technical assistance programmes, and this I wish to discuss in greater detail. Life in the fundamental branches of science would be greatly stimulated if active research workers from advanced countries would be willing to visit research institutes in the developing countries for a stay of, say, one year, to cooperate with the local workers in their research and organisational activities.

What would such scientists bring to an underdeveloped institution that is unique and valuable?

First, they would bring up-to-date scientific methods, information on the latest developments and up-to-date ideas. The isolation of institutes in the young countries could be greatly reduced by the wealth of information infused by the presence of persons who have lived in the most stimulating environment of one of the great Western centres of research. Such visitors' programmes are essential even for those Western research establishments which are slightly outside the mainstream of activities. They are thus even more vital in those institutions whose only contact with the latest developments of contemporary science would otherwise be through the printed page.

Secondly, one of the striking characteristics of the scientific life of the underdeveloped countries is the fact that the personnel which counts, as far as modern research is concerned, consists mostly of quite young persons. This is to be expected because higher education in most of these countries, particularly in the natural sciences, is very recent. The consequence of this is, however, that there is a crying need for originators of research, for " ideas men ", for those who can stimulate and guide the young Ph.D.s until they have acquired enough experience themselves to carry out independent research. Only a very small fraction of new Ph.D.s are sufficiently brilliant and original to be able, immediately after receiving their degree, to originate research problems or lead a research team. In advanced countries this does not pose a problem, since these young people in effect get a long post-Ph.D. training as junior members of some research team or academic staff. In an underdeveloped country, where at present such teams and staff do not exist, a research worker from a leading university or research institute in a scientifically advanced country with some experience in generating research problems and in leading research, could do wonders in making effective use of local talent which otherwise might lapse into aimless stagnation.

Thirdly, a Western research worker, even if he has no special interest or experience in scientific organisation, is likely to have acquired unwittingly the knowledge of how an advanced research institution functions. Such tacit knowledge is badly needed in the underdeveloped countries. To be sure, some of the local scientists themselves have received their advanced scientific education at a Western research institution. Their adjustment problems, however, as well as their position as foreign guests, usually prevent them from learning much about the organisation of the institute in which they are being trained. Their task is to do well in their professional studies and they consider their mission a success if this specific aim is fulfilled. There is, hence, a great need in the young research establishments in the underdeveloped countries for advice on matters of curriculum, library practices, training, organisation of seminars, store and workshop rules, etc. A particularly important aspect of this in which a Western scientist can be of great help is the attraction of short-term visitors, persons who happen to pass through the area and are willing to visit for a few days, conduct a seminar or two, and talk to staff members about their problems. Travel for scientists in the Western countries is recognised as imperative and an increasing number of scientists are going beyond the California-Moscow circuit and visiting countries in Africa, South Asia and Japan. Usually only a very small additional amount of money is needed, in local currency, to persuade the visitor to make a small detour and visit a given institution. But the contact must be made by somebody who knows (through the scientific grapevine) that the visitor is coming and who knows him well enough to invite him. A Western scientist, particularly if he has very extensive personal contacts, is invaluable in this respect.

In addition to these rather tangible benefits, the Western visitor represents, much beyond his role as a scientist, a note of encouragement, a stimulus to morale, a symbol of recognition, which is of extreme importance. In luring back young trainees from abroad, the assurance that visitors from the advanced countries will be joining them might be a decisive factor. A Western visitor can also enlarge the research group which otherwise would be too small to function properly. Finally, the visiting scientist from an eminent institution will be a symbol that the institution in question has " arrived ", that it is, in some sense, on an equal footing with its foreign counterparts as a member of the world scientific community, that it has established a strong link with the international brotherhood of scientists. The feeling of elation over this development and the high morale generated by it will compensate for many of the physical shortcomings of the research establishment and will give strength to the local scientists for the hard work facing them.

X

Thus far, this form of technical assistance, namely, the sending of research scientists for prolonged sojourns to an underdeveloped country has hardly been explored at all. The problem as such has not received adequate recognition. But even when the problem is recognised, the task will still remain more difficult to organise than the sending of equipment or the building of a dam, since it involves a large group of scientists and technologists, who will have to be convinced that the programme is valuable and they will also have to be convinced that they can assist in it without undue personal sacrifice. It requires convincing the research scientist in the United States and in other countries with technical assistance programmes, working at a national laboratory, university, or private firm, that he should transfer his own work to a remote country, at least for a year, where he might have to spend some of his time getting adjusted, or serving as an adviser to the local scientists, and where consequently his efficiency as a research worker might suffer. With some advanced planning, however, and by choosing the appropriate time for the visit in relation to the career of the visitors, such a temporary adventure might be not only not detrimental but in fact beneficial to the visitor, even from the purely scientific point of view. His year abroad might be spent catching up with broader developments in his field, with devising plans for future work, with writing a book, or just with getting a new perspective on the problems with which he has been confronted in his daily routine at home.

Channels of financial help for such an undertaking are at the present very meagre, at least compared to those available for educational projects (Fulbright, Smith-Mundt, etc.) or for technical cooperation (AID, Point Four, Colombo Plan). It would be a great help if special fellowships could be set up for this purpose, perhaps by the National Science Foundation in the United States and corresponding bodies in other countries, so that an interested scientist would not have to take on himself the responsibility for seeking financial support. Two hundred such fellowships annually, costing say, $25,000 each, would only amount to $5,000,000 a year, or a fraction of 1 per cent. of what is now spent on technical aid. Yet, 200 scientists abroad distributed over 50 underdeveloped countries which are ready to benefit from the programme outlined, would make a tremendous impact indeed on fundamental research all over the world and would in the long run pay a very great return in the form of the scientific results obtained from the institutions and the countries to which they go.

J. Sci. Soc. Thailand **1** (1975), 89–95

SCIENTISTS AND DEVELOPMENT

MICHAEL J. MORAVCSIK

Institute of Theoretical Science, University of Oregon Eugene, Oregon 97403, USA.

(*Received 10 January 1975*)

It is a great pleasure, privilege, and opportunity for me to be able to contribute to this new journal of the Thai scientific community. I am particularly pleased that at the very outset of this journal. discussion is devoted to the relationship between scientists and development, an area that I have been intensely interested and substantially involved in for the past dozen years.

The building of science in the developing countries, or science development for short, has assumed an identity of its own only in the last twenty years or so. Although it is still far from forming a well-researched, coherent discipline with ready answers to all problems, it has by now evolved a substantial body of experience and a correspondingly sizeable literature. In fact, I recently completed what I believe is the first book dealing specifically with science development[1]. Thus, it would not be possible to offer a comprehensive summary of science development within the framework of a short article in a journal. Since I hope that this new journal will continue to lend its pages to articles dealing with specific conceptual contributions to science development and with particular proposals for action programs in this area, I will devote this present article to a few fundamental remarks concerning the relationship of scientists and development. It is my hope that such general comments might serve as a framework for more detailed discussion in this journal at future times.

My train of thought will be a simple one. I will first attempt to analyze the characteristics of a scientist. This will be followed by a discussion of what we mean by development. Finally, the connection between scientists and development will be defined, and thus conclusions will be able to be drawn about how to achieve this connection.

Prof. Moravcsik received his training in mathematics and physics in Hungary and U.S.A. He is presently at the Institute of Theoretical Science, University of Oregon, Eugene, where he served a term as its Director. He is the author of two books and some 130 articles dealing with nuclear and elementary particle physics, the "science of science", and with science organization and policy, particularly in the context of less developed countries.

J. Sci. Soc. Thailand, 1 (1975)

So let me first turn to a discussion of scientists. Rather than attempting to describe scientists in terms of some formal requirements they have to satisfy, I want to enumerate the motivations and attitudes of scientists, since this will be much more important in our later discussion.

First of all, a scientist is curious about nature, and finds an esthetic satisfaction in discovering the laws of nature. This has been the primary motivation of those who engaged in scientific studies over the centuries, and it remains the most important incentive and reward for present day scientists. It is not the only motivation: An urge to convert talent into accomplishment, a sense of competition, a search for external recognition, a desire to serve humanity, and other justifications might also play an important role in fueling scientists' interests in the investigation of nature. Yet, satisfying curiosity and acquiring new knowledge has been one of the most basic aspirations of civilizations over the ages. Such an interest has not necessarily been directed toward nature as defined by the modern natural sciences. We are also intrigued by social problems, by religious and philosophical considerations, and by psychological and spiritual matters. In all these areas, however, we seek to understand, and understanding brings us satisfaction and pride. Scientists, specifically, direct their attention to the laws of nature, and devote their full energies to the satisfaction of this aspiration for knowledge. As such, they differ from many other professions, the practitioners of which are not in the position to be able to be preoccupied most of their time with such aspirations.

A second characteristic of a successful scientist is his relentless drive and virtually limitless devotion to his task of discovering new aspects of nature. This is not unrelated to his first characteristic mentioned above since he can combine his vocation with his avocation, and hence his internally propelled motivation dictates a pace not primarily determined by external working rules. In addition, the work before him is not of finite extent. Scientific knowledge is infinite, or at least it appears to be infinite to us at this time, and so only his personal skill, endurance, and energy limits the amount of work a scientist can do. Thus successful scientists often give the impression of being "possessed" by the zeal to do science. They frequently exhibit very high morale and a vitality directed toward their work which can hardly fail to produce results.

The third characteristic of a scientist is his acceptance of the criteria of truth and measures of success that govern scientific investigations. The most essential criterion by which a scientific theory is judged is whether it agrees with the results of experiments, that is, whether it conforms with scientific reality. I quickly hasten to add that in practice the application of this criterion is sometimes not as easy as it appears on paper. Experiments are often difficult to perform and occasionally incorrect results are obtained. Frequently several theories agree with particular experimental results, in which case other criteria must come into play also. But on the whole, and particularly in the long run, a scientific theory is judged by its power to predict reality as described by scientific measurements.

In other words, what the scientist produces must work. Another way to describe this characteristic is to say that the essence of a scientist is his ability as a problem

J. Sci. Soc. Thailand, 1 (1975)

solver. He faces a problem, and then has to provide an answer whose effectiveness is judged by whether it works or not. This is an attitudinal characteristic which colors the scientist's outlook far beyond his particular investigation of the laws of nature, because in other areas also, he will seek a course of action with ascertainable results. Once a scientist is known to have this attitude, his effectiveness goes beyond purely scientific problems, and he will thus be able to apply his attitude toward the resolution of other difficulties also.

The fourth and last characteristic of a scientist that I want to emphasize in this analysis is his attitude toward novelty, new ideas, and change. As I mentioned, one of the basic tenets of the scientific community is that a scientific idea, no matter how beautiful, is of value only as long as it is in agreement with reality. Let me offer a very dramatic example. The idea that nature is governed by laws which are reflection-invariant, or, in other words, that one cannot tell "real" nature from the mirror reflection of nature, grew into a very fundamental belief of scientists during the two centruies preceding the 1950's. It was an esthetically very pleasing idea, and it was confirmed by the structure of Newtonian mechanics which was so immensely successful in predicting the motion of celestial objects and in serving as a scientific basis of much of engineering throughout the centuries. Yet, this idea turned out to be incorrect, as demonstrated by experiments carried out in the 1950's, and as soon as these experiments were established, scientists had no other course than to abandon this highly cherished, traditional, and beautiful idea. They did so within a year or two from the initial discovery, and by now there is no scientist in the world who would continue to cling to the belief in reflection invariance.

In fact, change is the most fundamental hallmark of scientific work. There is a constant striving in science to formulate new ideas, to uncover new phenomena, and to modify previous concepts so that they become more powerful, with a broader validity and a greater simplicity. A person who does not believe in change and novelty as a way of life could hardly be a scientist.

Although I have hardly scratched the surface in analyzing the traits that characterize scientists, enough has been accumulated for the purposes of this discussion, and so let me now turn to the discussion of development.

I want to offer the following sentence as a definition for the concept of development: Development is the course of action taken by an individual or a group of individuals in order to achieve a greater realization of his aspirations.

This appears to be a rather general and vague definition, and yet, there is much value in starting with such a broad conceptual statement. Let me first explain its basis. Virtually all generations throughout human history shared the common experience of finding a basic incompatibility between the extent and variety of human aspirations on the one hand, and the brevity of human life on the other. Exactly what these aspirations are depends on the time period, on the particular civilization, on the particular individuals, on their cultural, religious, and social context, and many other factors. Yet, whatever our aspirations may be, we have found that they always exceed the potential of our own lives. There is, therefore, a strong urge to expand this potential as much as possible, so that we can more fully attain our aspirations. This process of expansion is called development.

92   *J. Sci. Soc. Thailand*, 1 (1975)

Development thus has two aspects: The improvement of the tools that further the realization of the aspirations, and the broadening and realization of the aspirations themselves. One of the important virtues of approaching the concept of development in the way I outlined is that it includes *both* of these aspects. More commonly, in the frequent discussions in national and international bodies, governmental organs, and even academic circles, development is thought of exclusively in terms of the material tools to improve the physical living conditions of people. While this is an important part of the first of the two aspects of development mentioned above, it is not even the totality of the first aspect, and has little direct connection with the second aspect.

Let me elaborate on this point in more detail. There is no question that a life expectancy of 70 as compared to 40 promotes the realization of the aspirations of most people, and that a person who has to devote an overwhelming fraction of his time to providing food and shelter for his bodily survival will have less of an opportunity to fulfill his more fundamental, non-material aspirations than a person who derives comfortable living from a 40 hour working week. Yet, efforts to bring about such conditions simply provide tools toward development in a more general sense. Very few of us truly believe that the purpose of our lives is to eat well and have a decent shelter over our heads. We need something beyond that, and development must make provisions for fulfilling those needs also.

It is sometimes claimed that the fulfillment of these other needs can be postponed until material conditions are sufficiently favorable. I do not believe that this is possible. All through history, people were simultaneously occupied with improving their material lot and fulfilling their non-material needs. No single individual wishes to forego the latter in favor of a greater preoccupation with the former. Thus development must address itself simultaneously to both aspects.

Let me now list four basic aspects of development as I defined it.

First, development is based on the compelling human drive toward self-realization. All of us, no matter when we were born and where we live, have some personal philosophy about the purpose and aim of our lives, and we feel an obligation to live up to this purpose and aim. The development thus defined need not be along scientific or technological lines. It might involve purely spiritual factors, or can be along the strengthening of interpersonal ties. Yet, in all cases the source of this development is a desire to realize our aspirations.

Second, development involves a willingness to devote effort and energy, and to make sacrifices toward this self-realization. One need not have a very sophisticated insight into the workings of life to comprehend that a lethargic, passive, and listless person will not achieve self-realization. Even in cultures where the direction of self-realization is believed to be determined by God or the gods whose service represents the aspiration of the people, there is a drive toward performing the acts and rituals to an ever increasing extent so as to bring about the action of the gods. Thus, development is basically an action-oriented undertaking, in which hard work, devotion, perseverence is believed to bring about the desired results of self-fulfillment.

J. Sci. Soc. Thailand, 1 (1975)

Thirdly, development is directed toward the attainment of a situation that can then be compared with the original aspirations which sparked the effort. In this sense development has a built-in yardstick. Virtually every individual finds himself, from time to time, assessing his own life in terms of whether his achievements match his aspirations. In fact, groups of individuals, that is, communities, countries, civilizations, or even humanity as a whole, tend to do the same. Much of the moral teachings in all ages stress the disparity between our aspirations on the one hand, and our capabilities and achievements on the other, and hence urge us toward greater efforts. Admittedly the yardsticks used in such comparisons are bound to be subjective in that they use the particular aspirations of the individual or group in question which might or might not be shared by other individuals or groups. Yet, the element of an internal comparison and evaluation is strongly present in terms of the "subjective reality" applicable in the particular case.

Finally, the idea of development contains the intent to bring about changes in the direction of a greater self-fulfillment of aspirations. As mentioned earlier, our aspirations define the end of a road (perhaps at an infinite distance), a road which we must make progress on in order to come closer to our goals. Such a progress need not pertain to material changes in our environment. For example, the attainment of an enlightened state of mind and spirit in some Eastern religions and philosophies is thought of as the end of a long and very difficult road which however has little to do with material prosperity or a modification of our physical environment. In other examples, however, where aspirations pertain to different goals, the results can be very tangible. The pyramid of Khufu in ancient Egypt, the creation of the Taj Mahal, or the landing of man on the Moon are examples of aspirations which are reflected in something very physical indeed.

So much for the characteristics of development. We can now turn to the relationship between this development and the scientists whose traits we described earlier. In comparing the two discussions, we can conclude immediately that a connection exists in three respects.

First and most obviously, because of the technological consequences of science, scientists can play a significant role in the improvement of the material aspects of the tools of development. This particular connection is so frequently and extensively debated that it really needs little reemphasis here. This is not to say that the debates on this subject have resolved all differences in views. On the contrary, arguments continue about the extent scientists influence technology, about the amount of science needed for a given amount of technology. about the right "type" of science and the right "type" of technology for a less developed country, etc. I would, however, like to concentrate in this analysis on the other two connections between scientists and development, because those connections are not frequently discussed and hence are neglected in development planning.

One of these is the attitudinal kinship between scientists and development. A glance at the charactristics of scientists and of development as outlined earlier reveals a striking similarity. Internally directed motivation is essential for a scientist and is crucial in the development process also. A willingness to spend energy, effort, devotion, and skill toward the desired goal is also common between the two. Both scientists and develop-

94   *J. Sci. Soc. Thailand*, 1 (1975)

ment have an agreed-on, internal yardstick against which success is measured, and which serves as a guide to define the direction of activities. Finally, a directive toward improvement and therefore change is inherent in development as well as in what scientists do.

Thus, the presence of successful indigenous science in a country should be a very powerful attitudinal influence on development along more general lines.

The final link between scientists and development is the fact that science itself serves as one of the aspirations toward which development is oriented. Achievement in scientific exploration has been the pride of many civilizations before us, and will likely continue to be so. It never ceases to impress me how people in every country and from all walks of life, when they learn in a conversation with me that I am a scientist, relate to me not as a person who can help in making material progress, but instead as a person involved in a fascinating though perhaps esoteric activity that they find attractive to participate in at least through the indirect way of conversing with me. Their questions are not directed toward technology, but instead toward the latest discoveries of astrophysics or elementary particles they may have read about in the newspaper.

In fact, science not only serves as an aspiration in itself, but science and the derivative technology create a new wealth of aspirations which previously were altogether outside the horizon of the people. In Thailand, a few centuries ago, the aspiration of intense human and interpersonal interaction with other cultures could hardly have existed, because communication with drastically different cultures by a significant segment of the population was out of the question, and hence was not even thought of. Today, with jet planes, communication satellites, radios, and mass-produced magazines, this is not a ludicrous aspiration any longer. Other examples for new aspirations created by science and technology are numerous. Printing created wide-spread literacy, and literacy opened up enormous horizons and with its novel aspirations. Space travel offers humanity a completely new view of the universe, which arouses countless new aspirations. Thus the body of aspirations itself develops, and science plays a very significant role in this evolution.

It terms of this triple link between scientists and development, can we say something about the "ideal" scientist in a less developed country? Surely so.

First, such a scientist must fully carry and maintain the characteristics of curiosity and high morale, perseverence and hard work, functional and critical problem solving, and experimentation and striving for change that I outlined earlier. A university degree does not make us a scientist, no matter where that degree came from, and a scientist can cease to remain a scientist if he loses these characteristics, even if he retains his degree title or professorship.

Second, a scientist must not only contain but radiate these characteristics, so they are evident to everybody and so that they become contagious in his environment. A scientist, beside carrying out his particular special research work, should function as a problem solver far beyond the particular areas he carries out detailed research in. This is not to say that scientists are omniscient. Not at all. But they can contribute to a broad area of development if they act in cooperation with others perhaps more knowledgeable about

J. Sci. Soc. Thailand, 1 (1975)

the details of certain problems. Perhaps most importantly, a scientist must participate in the drive to make every inhabitant of his country a "scientist" of sorts, that is, to implant into people the characteristics of a scientist that are so close to the characteristics required by development in general.

This discussion of scientists and development, as indicated at the beginning, remained on a general plane. It was my intention to suggest that if one takes a broad view at both a scientist and the process of development, one perceives a striking similarity between them and so one can also define a very fundamental and comprehensive context in which scientists are not only useful but in fact indispensible to development.

Does all this leave you still unsatisfied and unoriented? It should, because the general framework I outlined needs to be filled with particular values, standards, goals, aims, and details before it can give you specific directives for action. I feel, however, that as an outsider to the Thai scientific community (however sympathetic I may be, and in spite of having had a first hand opportunity to get acquainted with Thai science), I cannot and should not dictate or influence too much the filling in of this general framework. How a Thai scientist can best transfer his attitudes to the Thai citizen, how the Thai scientist can participate most effectively in developmental problem solving, how a Thai scientist can most successfully fulfill his own aspirations toward scientific discoveries, are questions that must be resolved internally within your own community. The experience of other countries and suggestions from knowledgeable outsiders can be helpful in generating a fruitful discussion and in considering the various options, but in the final analysis the decision must be an indigenous one. In this sense the contribution of this article was hoped to be to provide a sufficiently extensive, versatile, broad, and pershaps somewhat unconventional framework so that it can include all of the aspects of Thai sceince development in the years to come. I hope that your new journal will play a crucial role in the filling in of this frame-work and hence in the creation of a science policy for Thailand in the broadest sense of the word.

Reference

1. Moravcsik, M.J. (1975) *Science Development,* Indiana University Press, A survey of about 100,000 words in length, with a bibliography of about 500 sources. Special efforts are being made to make this book available to potential readers throughout the world in exchange for a modest amount of local currency. The usual author's royalties have been plowed back into the distribution of the book. For further information about how to acquire the book, write to John Gallman, Indiana University Press, Bloomington, Indiana 47401, USA.

DO LESS DEVELOPED COUNTRIES HAVE A SPECIAL SCIENCE OF THEIR OWN?

M. J. MORAVCSIK

SUMMARY   The belief that the developing countries should evolve a science of their own which is radically different from "Western" science can seriously hamper efforts in international scientific cooperation. Hence it is important to inquire if such a belief is tenable or not. This is the aim of this paper which deals with the natural sciences only.

The argument is divided into two parts. In the first part, science is looked at along six perspectives to see if there can be a special science for developing countries.

The first dimension is that of motivation. It is concluded that the motivation for doing science and the justification for such activities are always multidimensional, and the mix of motivations is not significantly different for the various countries around the world.

The second dimension is subject matter which, because of science dealing with "objective reality", also tends to be very similar for various countries. There are minor variations in the choice of research areas, for which five criteria are enumerated: Intrinsic interest, ripeness for exploitation, feasibility in terms of the means available, critical mass of manpower, and connections with technology. Taking into account both short and long range considerations, the balance of various research areas turns out to be not radically different for the various geographical areas of the world.

The third dimension is ideology, which is claimed by some to have an effect on the type of science practiced. It is, however, concluded that science which is to be successful in terms of bringing understanding and serving as a springboard for technology is likely to be universal and quite independent of ideological trends.

The fourth perspective pertains to methodology, and it is argued that the universal consensus of the scientific community concerning the methodology of scientific research assures that science in different locales will be largely the same in this respect also.

The fifth dimension covers applications. In this respect also, scientific activities are fairly universal, even though technological research and development can differ greatly from country to country.

Finally, the sixth dimension is the scientific infrastructure, which indeed appears to be rather different in various countries. It is shown, however, that these differences reflect mainly different stages of the same development rather than distinctly different sciences.

The second part of the paper assumes that the conclusions of the first part were in error, and somehow some new type of science can in fact be evolved in the developing countries. It is then argued, however, that inasmuch as this new phenomenon can be called science at all, it will be strengthened by most of the present day efforts in international scientific cooperation.

In particular, it is argued that such a different science, in order to have a claim to be called science, must have four characteristics: Cumulativeness, collectivity, comparison with empirical, experimental perceptions and observations, and predictive power with respect to future events. It follows from these characteristics that there is a need for an equitable propagation of present scientific knowledge to all parts of the world, a strong system of communication among all scientists of the world, and an enhancement of science education which enables the recipient to consider science as a problem solving activity guided by comparison with experiment. Since the international scientific cooperative efforts mainly aim at manpower development, education, and communication, they remain pertinent even to such different types of science. They also strengthen the morale of scientists in developing countries, without which any kind of science development is impossible.

Finally, it is argued that speculations toward the development of a "science" which lacks even the above four basic characteristics are so vague that one cannot take them into account in practical science policy decisions in which non-cooperation with the developing countries for very speculative reasons is likely to be construed to have sinister motives. It is thus concluded that international scientific cooperation should proceed on the assumption that we all strive to build basically the same science.

Throughout this discussion "science" will denote the natural sciences only. In discussions pertaining to the building of science in less developed countries, one hears, from time to time, the argument that these developing countries ought to have a special kind of science of their own, quite different from "Western" science, and therefore we should be very careful in helping these countries in the building of their science lest they be "corrupted" to adopt our science instead of acquiring their own. Some of the references for such a view have been summarized in Moravcsik (1976a).

These arguments can be heard on two differents levels. One is a purely theoretical one, and as such, can be taken to be prophetic, amusing, profound or irrelevant, according to one's own feelings, but in any case, they do not interfere with the difficult task of collaborating with less developed countries in science development. The other level, however, is more significant, in that it serves as a basis to oppose certain science development projects. If these arguments are indeed irrelevant or fallacious, therefore, they can, in this way, be in fact quite damaging to the cause of science building in developing countries. It is, therefore, worthwhile to analyze these arguments in detail and ascertain to what extent they are indeed

Michael J. Moravcsik is Professor in the Department of Physics and the Institute of Theoretical Science of the University of Oregon. His research in theoretical physics has been concerned with elementary particles and nuclear problems. For the past 15 years he has also been involved in problems of science building in the developing countries, and is the author of the book "Science Development", as well as 160 articles in theoretical physics, science policy, and the science of science. Address: University of Oregon, Eugene, Oregon 97403, USA.

pertinent to the kind of programs that exists today in the scientific collaboration with the third world. This is the aim of this article.

To start with, let us agree that we will *not* be concerned about whether developing countries should have their own kind of *technology*, distinct from that of the more advanced countries. That question, which already has a huge literature under the headings of technology transfer, appropriate technology, intermediate technology, labor-intensive vs. capital-intensive economics, etc., is quite distinct and separable from the question of whether there are two kinds of sciences: one for the advanced countries and one for the developing ones.

The Analysis

In this section we will look at science along six perspectives, six different aspects, to see if along those dimensions there are any special requirements for science in developing countries. The six aspects are: motivation, subject matter, ideology, methodology, applications and infrastructure.

Motivation

Motivations both for practicing science and for supporting scientific activities are most complex and multidimensional. There is virtually no example for a practicing scientist or of a sponsor of science who would participate in this activity for one single reason. On the contrary, studies show (see e.g. Moravcsik 1975a) that the motivation of scientists consists of many very different elements, mixed together in strengths that are not very different from each other. As to the reasons for sponsoring science, one needs only listen to politicians in any country to hear, in a sometimes bewildering melee, arguments for science involving technological applications, national pride, cultural heritage, military competition, development of a modern world view, and others.

So the question is not whether the motivations in advanced and developing countries are different from each other, but rather whether the mix of these many different motivations is (or should be) the same or not in the two environments.

While a categorical answer to this question cannot be given, there are strong indications, historical and otherwise, that the answer is in the negative. The basic aspirations of people throughout the centuries, and among various cultural and geographical groupings, are very little different. Curiosity, esthetic satisfaction, material well being, collective accomplishment ("national glory") seem to be always among the human aspirations.

A similar answer can be deduced by simply looking around the world today to assess the existing motivations for doing science in various countries. There are certainly a wide variety of cultural, religious, political, linguistic, racial, and other subdivisions among humanity but when one surveys the motivations among these countries for engaging in the practicing or the support of science, one hardly finds any differences.

There is one aspect of motivation, however, that needs to be quantitatively different in developing countries from what it is in the scientifically more advanced ones. This has to do with the impact of the scientist on his surroundings in terms of his radiating his motivation and enthusiasm for science. In a developing country, where science is not yet a traditional activity, where the practitioners are few in number, where scientific matters are not in the public consciousness, the scientists must act like a beacon, not only practicing science, but also acting as a missionary for the cause of science. One can therefore say that the motivation of a scientist in a developing country (while not neccesarily externally imposed) must be more extroverted in terms of its effect on the rest of the country. I elaborated on this point in Moravcsik (1975b).

Unfortunately, in this respect the developing countries quite often fail at the present. Scientists tend to be even more shy than their colleagues in the scientifically advanced countries, and a shining articulation of motivation and spirit is in fact rarer in the third world than it is in more advanced countries.

Subject Matter

The statement that the natural sciences deal with an "objective reality" is a very good approximation of the truth, and as a result, the subject matter of science is also preponderantly universal. This is certainly true for the general principles and the broad outlines of the sciences, and it holds the more the "harder" and "general" science we deal with. Thus, this universality is very evident in mathematics, physics, and chemistry. It is somewhat modified in biology or geology, in that the specific examples to which general principles apply may vary, for example, with geographical location.

If we try to gauge this universality, for example, in terms of the college and university type curriculum, we can say with some assurance that on the whole, virtually all "undergraduate" science (i.e. science taught in American colleges up to a Bachelor's degree) is a prerequisite for fruitful scientific activity in any part of the world. Even the beginning of graduate education falls

predominantly into this category, and it is about halfway through graduate school that specialization sets in to a sufficient degree that one could talk about a meaningful division between what might be more appropriate in advanced countries and what in developing ones.

In fact, it is mainly in connection with the specialized research for a thesis that the question arises. But even then, one must always remember that a doctoral thesis is *not* equivalent to a commitment for life to a particular research direction. A dissertation is simply a proof that the person is capable of completing a worthwhile piece of scientific research on his own. Even in the advanced countries, many, or even most, scientists nowadays drift quite far during their career from the original topic of their dissertation. A scientist who expects to work in a developing country should be prepared for even larger drifts, for even greater versality.

In selecting a research field to pursue in a developing country, perhaps five criteria must be kept in mind, some of which were also discussed in Moravcsik (1976b).

First, the problem should be interesting to the research and interesting to at least some fraction of the worldwide scientific community. This is a criterion both for so-called basic and so-called applied research. A problem that bores the researcher and which leaves cold the other scientists working in the same field is likely to be one that is not worth spending money and effort on, and one which will have no technological by-products either.

Second, the problem should be ripe for exploitation (Weinberg, 1963) and yet unexplored. These are rather self-evident criteria, valid in all scientific environments, and hence need not be discussed further.

Third, the problem should be one that can be worked on with the means that can reasonably be expected to be available in the particular scientific environment. Thus building a large radio telescope in Upper Volta (unless some large scale international cooperative venture can be organized) is not a prudent undertaking, and for, say, Bolivia to engage on its own in experimental research on thermonuclear fusion would be unreasonable, even if one or two Bolivian scientists showed great promise in this area. But even in these areas, international or regional collaboration can do wonders. Some examples are the Cerro Tololo astronomical observatory in Chile, the observatory outside Bandung, ICIFE in Nairobi, and many others.

On the positive side, certain types of research might be particularly easy or appropriate for a certain country. ICIFE capitalized on this idea, and Cerro Tololo and Bandung also profited from it. Certain physiological problems and diseases, known in Europe and North America only from textbooks, can be studied *in situ* in, say, Africa. Selecting research topics which are particularly appropriate in a given locale might also be advisable because it is in this area that the indigenous scientific community can make the most unique contributions to the overall body of science. All countries in the world (even the scientifically leading ones) import more scientific knowledge than they export, and hence an ability to acquire this internationally generated knowledge, together with the capability to make at least a modest contribution to this knowledge, is what is needed for productive "national" science.

The fourth criterion is also dictated more by necessity than by conceptual considerations. For successful scientific research it is necessary to have a minimum number of regularly interacting scientists, that is, a so-called critical mass of researchers. This well known fact reflects the collective nature of scientific research which demands collaboration between individuals in terms of exchange and critique of ideas, complementary skills and talents in a complex experimental project, etc. Indeed, the sum of the parts in such cases is much larger than one would expect from arithmetic.

What the size of such a critical mass is depends considerably on the scientific subfield, on the nature of the particular research, on the personality of the scientists, etc. As a rough estimate, 3-5 active researchers working in very closely related or identical fields and in daily contact with each other can form such a critical mass.

There are ways to attempt to create such a critical mass, but it is more likely, especially at the beginning of the scientific development of a country, that it comes about by a combination of the unplanned availability of a few exceptional scientists, combined with the understanding and skillful recognition and assistance by some local science policy decision makers.

The fifth criterion is a somewhat subtle one: it has to do with the applications to local technology. The question is subtle because, a) as everywhere, the majority of scientific knowledge used in technology comes from abroad, and hence the ability to be aware of this knowledge is of crucial importance (instead of buying technological patents from abroad), b) applied scientific research (Moravcsik, 1977), especially in a developing country with only an infant technological infrastructure, might find it difficult to make close contact with tecnological needs. If it researches a certain area and then tries to find applications for it, the appropriate technological opportunities might not be present yet. If, on the other hand, one starts with a particular technological problem and tries to perform applied

scientific research "for it", one might find that for any degree of success a large spectrum of applied scientific areas have to be mobilized simultaneously, which is often not very feasible. In practice, what one sees most often in developing countries is the focus on a particular technological problem, and a narrowly conceived, trial-and-error-based search for its solution. In this process, applied scientific research is altogether neglected, and the chances of success in any case are quite slim.

Thus, the criterion of applicability is best used in a broad sense. For example, ICIPE operates on the assumption that in Africa, insect physiology (a huge branch of biology) is likely to be of applicability somewhere and sometime, and this expectation is a realistic one. The criterion of applicability in the sense discussed above need not be applied to a small fraction (10 - 20%) of the research and development budget, which can beneficially be allocated to unrestricted ("basic") research. The short term "applications" of this research to the country's well being will be in areas other than material.

There is, however, another aspect of the science-technology interaction which deserves mention. In the context of the application of technology, one encounters frequently accompanying scientific analysis, often of a rather elementary, "back-of-the-envelope" type. Examples can be found, for instance in agricultural and health technologies, such as plant genetics, isotope tracers, dosimetry, medical biochemistry, etc., as well as in many other fields. The need for such analyses arises in so many contexts and areas, that it would be most difficult (as well as unnecessary) to cover all of them by "specialists". Instead, the need is for scientists of broad training who, regardless of the particular area of their personal research, are able and interested to supply such scientific backing. This point is worthy of emphasis, because scientific education in developing countries very often is extremely narrow, and ironically this narrowness is "justified" by evoking presumed requirements of applied research — a reasoning exactly opposite to what is required by the realities.

Incidentally, the same issue of breadth of background arises also in the requirements imposed by the country's educational needs. Such needs dictate the presence and involvement of scientists who can offer education to the younger generation which is broad and at the same time strong in the down-to-earth aspects of science also. The emphasis here is not so much on the selection of subject matter but instead on the functional problem-oriented discussion of whatever material is taught. Science education is more of an inculcation of attitudes and a general development of science-doing ability than

a transfer of a particular set of facts and material.

In choosing areas of research, a country with a small scientific infrastructure also faces the problem of having to make a small selection out of a huge variety of subfields. In doing so, some areas of research, no matter how desirable by our previous criteria, must fall by the wayside. The optimally wise choice under such circumstances combines some areas of short term impact with others with major longer range importance. As an example for the latter, some activity in nuclear physics in the developing countries in the 1930's, 40's, and early 50's would have enabled these countries to take advantage of nuclear technology very much more effectively in the 1950's, 60's, and 70's in medicine, agriculture, energy technology, and many other areas.

Thus in this respect also, the *ensemble* of developing countries covers a set of scientific areas not very different from that in which the scientifically advanced countries are involved.

All in all, we see that as far as the subject matter of science goes, there might be minor differences between advanced and developing countries, but the dominant part of science education and even much of the foundations of scientific research are universal all around the world.

Ideology

A number of claims to the effect that developing countries do or should have a "different science" are based on ideological arguments. For one thing, there have been claims in recent decades that there is Aryan and Jewish science, bourgeois and marxist science, science according to the Little Red Book, and science contrary to it, just as in previous centuries there was a claim for Christian science and heretic science. The history of such claims (at least in cases when the last word appears to have been said on the subject) tends to indicate that they were false. On more subtle levels, certain developments of science have been attempted to be linked with certain social groups or periods, such as discussed in Forman (1971). In a similar vein, the claim is being made that somehow the different cultural, social and traditional milieu of the developing countries should give rise to a new and different science.

It is difficult to refute conclusively these more subtle historical studies, since in most cases neither the history of science nor general history repeats itself completely. One can, however, argue on the strength of multiple discoveries in the sciences which are becoming increasingly more frequent. The fact that such multiple discoveries arise simultaneously in very different social and ideological environments appears to be a rather strong evidence against the existence of many different sciences. Correspondingly, one can also be quite skeptical about the possibility of the developing countries evolving a radically "new science".

Methodology

The methodology of modern science is the main basis of universality and cumulativeness. The consensus of the scientific community that comparison with experiment should be the ultimate arbitrator offers an "objective" standard that appears to be a crucial element in the progress of natural sciences. Temporary deviations from this standard, whether by carelessness, or by ideological interference from the outside, soon correct themselves simply because during the deviation science stops making progress.

It is therefore highly unlikely that a "new science" could be developed anywhere in which the methodology is strikingly different from what we use today. One can at least say with confidence that during the past 300 years, there has been no trace of such a different science, even though science has appeared in a variety of contexts and in many different philosophical and intellectual frameworks.

Applications

This point was already discussed in the context of subject matters, but some additional comments might be in order. That the specific technological applications of a given area of science will be different in different countries is not at all doubted, and it is already the case today among the scientifically advanced countries. This difference in technological applications, however, does not create a similar difference in the scientific activities of these countries. The divergence occurs mainly in the *technological* research areas and methods, and not in the *applied scientific* research activities (and of course even less in the basic scientific research). In this respect then, again, the requirements for the scientific community for developing countries are not much different from those for the advanced countries.

Infrastructure

In the infrastructure that surrounds scientists in advanced and developing countries, we finally do encounter a significant difference, at least at the present time. In most of the scientifically advanced countries, the scientific infrastructure is well developed, and, on the whole, the circumstances under which scientific work can be carried out are present. There is, of course, further development, some new institu-

tion building, some organizational modifications, and, naturally, a constant change of personnel as scientific generations enter and exit, but the basic framework is there, a tradition exists, and doing science is an accepted way of life.

In contrast, in many developing countries scientists must be concerned simultaneously with doing scientific work and creating the circumstances under which this can be done. The country has no articulate attitude toward science, institutions have not been formed, the scientific community has not gelled, auxiliary services are non-existent, etc. It is not really a matter of having a new or different science in the developing countries in this respect, but rather having the same science at a quite different stage of development. And yet, this situation imposes on scientists in less developed countries an extra responsibility, demands an extra dimension of talent, and drains an extra channel of energy. This state of affairs must be constantly kept in mind, whether one tries to evaluate research in those countries, or tries to formulate international cooperative undertaking in science.

Let me summarize what we have found so far. We studied the possible differences between advanced and developing countries in the structure and nature of their respective sciences. We found very little difference in motivations, except that the scientist in a developing country needs to be more explicit about his motivations and about the nature of his activities. On the whole, we also found little difference in much of the subject matter needed to be covered by science in the two environments, although on the level of specific research topics there might be a different weighting of the various criteria. We concluded that ideology is likely to have a negligible effect on the development of science in various areas, except that ideological interference with science might retard the rate of development. Similarly, we found that there are unlikely to be significant methodological differences between science in the advanced and the developing countries. We also found that, although technological applications might differ from place to place, this is not expected to have a significant effect in bringing about a similar differentiation of the respective sciences.

Finally, we did find some (temporary) differences between science in the advanced and in the developing countries as far as the infrastructure is concerned, that is, with respect to the circumstances and conditions under which scientific work can be carried out. But these differences are a matter of being at a different stage of evolution of the same science, rather than having a different and new science altogether.

The conclusion so far is, therefore, that there appears to be very little substance to the hypothesis that the developing countries do or should have their own separate and different "brand" of science. For a further such affirmation of the universality of science, see also Ben-David (1971), Eisemon (1976), and Raman (1976). Rahman (1975) presents an example where the argument becomes imprecise and indeterminate partly because of the failure to make a distinction between applied scientific research and technological research.

To conclude this half of our discussion, I want to emphasize that the claim so far has *not* been that social, traditional, cultural, and political factors cannot influence science in a country at all. On the contrary, these factors can, and in fact do, have a major effect on the speed by which science evolves in that country. What I do claim is that even though the speed is different, the science being developed is the same everywhere. Thus various countries differ from each other not in the type of science they practice but in the efficiency by which this universal kind of science makes progress.

Universally Necessary Components

In this section we will take a 180 degree turn from the reasoning of the previous section, and in fact we will assume that much of the conclusion of that section was in error. Thus we will assume that somehow a quite different science exists for the developing countries. For example, we can claim that the historical arguments were not convincing, or that history can always change, and that therefore the possibility of such a "new science" should be taken seriously. If so, one must ask the question of how we can cooperate best with the developing countries in the creation and strengthening of this "new science". The answer for this question will result in criteria for cooperative programs which should be necessary no matter whether "new science" exists or not, and no matter what its nature is if it exists. We will see that the set of these universally necessary components of international collaboration include many if not most of the programs existing today. In other words, in this section I wish to show that even if the "new science" exists, much of what we do today in collaboration with the developing countries is not only not damaging to this "new science", but in fact is absolutely necessary for its creation.

In arguing this point, we will have to make four assumptions about the nature of the "new science". These assumptions are very broad ones, and encapsulate those features of a scientific undertaking which appear to be indispensable for any kind of science that is to be successful in terms of the acquisition of knowledge about the universe and in terms of the utilization of this knowledge for the transmutation and control of the adverse factors facing humanity.

The first of these features is cumulativeness. By this we mean the property which allows us to claim that an ever increasing set of natural phenomena become (one way or another) understood and predictable. Without this characteristic it would be hard to talk about "progress" at all, and it would also be difficult to constantly enlarge the realm of activities aimed at utilizing our knowledge of nature for our benefit. The practical aspects of such a cumulation are discussed in Price (1965).

The second feature of any scientific activity which appears to be indispensable is collectivity based on a universal methodological consensus within the scientific community. This is most eloquently analyzed in Ziman (1968). The fact that progress made by one individual can be and is made use of by another individual for more progress, and, in fact, that progress made by one scientific generation can be used by the following generations to build on, is such a powerful factor in science that one cannot imagine how one could do without it. Indeed, human activities in which such collectivity and methodological and evaluative consensus are absent appear, from an historical perspective, spasmodic, fragmented, weak, and directionless.

The third characteristic of any science is in the requirement that it be constantly compared with and hence guided by experimental, empirical perceptions. We know from the history of science that periods in which science was divorced from an actual look at what the world is really like turned out to be stale and sterile, both in terms of insight into the universe and in terms of applications of our understanding of the universe.

The fourth feature is the ability to predict specified future occurrences. Such predictiveness is essential because all of the technological applications of science hinge on this characteristic, and hence our ability to be at least in part masters of our environment crucially depends on it also. A "scientific" theory that can only order already known occurrences but has nothing to say about phenomena not yet explored or observed has little value. Predictions are more valuable and decisive if they are highly quantitative, but because of the rudimentary state of development of some of the life sciences today, such an additional requirement of quantitativeness would be inappropriate in the present general context.

The above four criteria not only give us ways of telling what is science and what is not, but they also

help us defining the degree of success of science. The purpose of science is to present an ordered view of at least some of the phenomena around us in which future events can be foretold, and this is done partly for the purpose of intellectual and esthetic satisfaction, and partly to yield a tool with which we can influence our future. Whether these goals are accomplished by something that claims to be science can serve as a test of the veracity of this claim.

These somewhat elaborate preliminaries are needed to prevent us from lapsing into purely semantic squabbles as to what is science and what is not. Thus Lysenko's genetics was non-science not only because it substituted political pressure for scientific consensus, but also because it had no success in predicting new phenomena. Similarly, the creationists who oppose evolution theory do non-science not only because they substitute religious dogma for the scientific method, but also because they offer no predictive capability for the future. Likewise, people who replace scientific criteria by their interpretation of the Little Red Book engage in non-science as defined by our specifications. It is of course legitimate to argue about the philosophical, political, religious, spiritual, and social merits of such activities, but that does not change the fact that such undertakings are not science and will not fulfill the expectations we attach to science.

It might very well be possible to list some more features of science that are likely to be *sine qua non's* also in the future, but actually it turns out that even restricting oneself to these three, one can show that an overwhelming fraction of the programs in the scientific collaboration between the more advanced and the developing countries which exist today or which have been suggested are of the type that they will promote the evolution of science in the developing countries even if there is some yet unknown "new science".

Let me elaborate on this point. Since science is cumulative, being well aware of scientific knowledge already acquired is an absolute prerequisite for making further progress, no matter in what direction this progress lies. As a result, the' programs and proposals which are directed toward the transmission of scientific knowledge from the presently leading institutions and universities to students and researchers in the developing countries are indispensable no matter what.

Second, since science is collective and possesses a consensus, a highly developed opportunity for international communication among scientists is also a must, without which science in the developing countries will develop only very slowly, if at all.

Thirdly, since successful science must be based on empirical and experimental evidence, collaborative pro-

grams aiming at improving the capability in developing countries to undertake such a comparison with experimental information are indispensable. Similarly, efforts made to heighten the awareness and orientation of scientists toward science embedded in observations are also absolutely necessary. This includes many educational reforms as well as elements of research collaboration.

But education, manpower development, communications, and research based on observation encompass most of the main aspects of science development. It leaves out scientific organization and administration, and indeed, in these areas greater variations might be possible between various scientific communities. But this latter component of the building of science is relatively subordinate to the main and primordial elements mentioned above.

One can therefore say that no matter what science might evolve in the less developed countries, there are certain universally necessary components of science development, and that much of the present science development efforts fall within these necessary categories.

In addition to these methodological components, there is also a crucial psychological ingredient in the building of indigenous science: self confidence and high morale, a factor discussed in detail in Moravcsik (1976a), Chapter 8. While international cooperation and assistance are very important, in the final analysis, science in a country must be created by indigenous manpower. To do this, scientists in the country itself must believe that the task is attainable, and they must have the fortitude to work in the face of difficulties and temporary setbacks. It is this self confidence and high morale that will also enable scientists to absorb external information critically and without subservience, and to strike out in novel and creative directions instead of being content with an imitation of imported efforts. If these characteristics are present in a scientific community, one need not fear that it will fail in developing its own brand of science if such a special brand exists. In creating such a spirit, one very important element is the explicit recognition from and cooperation with the worldwide scientific community, which bestows equitability on the local scientific community in the hearts of the local scientists as well as in the eyes of the country as a whole which supports such scientific efforts.

Finally in order to build science in a country (the "old" kind or the "new"), at least some members of the local scientific community should have the ability, interest, and background to explore the history, philosophy, methodology and general context of science. In this respect the present situation in developing countries is very alarming. If science itself is underdeveloped there, the "science of science" is even more so. According to Price (1969), 97% of all historians of science are in the advanced countries, as compared with "only" about 92% of all scientists. Furthermore, very little attention is directed toward changing this state of affairs, apart from sterile and formalistic "science policy conferences" and fictitious exercises in writing "science development plans". Thus, a much deeper understanding of what science is and how it works seems to be another of the universally necessary ingredients for any kind of science building anywhere.

Conclusion

So far I tried to show that in the past and at the present, there is no indication that the developing countries could and do evolve a type of science radically different from that of the Western civilization, and that even if there were such a "new science", as long as it conforms to a few of the most basic and general requirements of science that contributed to its success in the past, most of the present day efforts in international scientific cooperation and assistance would be beneficial for the evolution of such a "new science".

It would, however, be possible to object to this train of thought even at this stage, and claim that the "new science" which will develop will not even have those fundamental features that I assumed in the previous section, and that we therefore at the present prevent the developing countries from evolving in the direction of this radically new science. For a strong view in favor of the cultural and social relativism of science, and expression of doubts about the cumulative nature of science, see e.g. Elkana (1971).

I would like to submit, however, that such an objection carries us clearly into the domain of science fiction on science itself, and hence, in practical decision making, it would not be able to carry its weight. We are, at the present time, confronted with a dichotomous choice: should the developing countries be given assistance to develop science broadly along the lines indicated in Sections II and III, or should we

refuse this collaboration on the extremely speculative grounds that thereby we will assist these countries toward finding the radically new type of science that none of us could pinpoint even in the vaguest terms at the present.

The choice in this dichotomy is clearly in favor of the first alternative, on at least two grounds.

First, one cannot reasonably gamble on something as hypothetical, speculative and unassessable as the existence and productivity of such a radically new type of science. This might be called the rational or systematic policy making ground.

Secondly, and in practice perhaps more importantly, the choice must go in favor of the first alternative for political and psychological reasons. It is clearly unacceptable to the developing countries to be told that after what our science has done for us in many different dimensions, it will be withheld from the developing countries so they can undertake an undisturbed search for the pie in the sky. Whether true or not, such a policy would appear indistinguishable from scientific imperialism and condescending paternalism, and would arouse feelings of a type that in today's world need little stimulus to be aroused anyway. For an illustration of such feelings in the context of nuclear science and technology, see Moravcsik (1972).

In view of these considerations, I cannot help feeling that claims of a "new science" for developing countries, and the policy directives arising from such claims, are unsubstantiated by past and present, extremely unlikely in terms of any extrapolation into the future, and on top of all this, practically unrealistic and unfeasible. As a result, they must be strongly rejected and neutralized so they do not interfere with the heavy task of worldwide science building.

Acknowledgements

This article grew out of many discussions I had over the years with numerous colleagues interested in science development. Specifically, I want to thank Drs. Roger A. Blais and Leopold A. Heindl, both of AGID, for their respective recent letters which gave the immediate impetus toward the articulation of these ideas. I am also grateful to an unknown referee of the first version of this paper who suggested some useful additions.

REFERENCES

Ben-David J. (1971): *The Scientist's Role in Society; A Comparative Study,* Prentice Hall, Englewood Cliffs, New Jersey, particularly chapter 9.

Eisemon, T. (1976) and Rabkin, Y: Academic Engineers in Quebec: The Myth of Linguistic Demarcation, Proceedings of the First Annual Meeting of the Society for the Social Studies of Science, Ithaca, N. Y. (to be published).

Elkana, Y. (1971): The Problem of Knowledge, *Studium Generale* 24: 1426-1439.

Forman, P. (1971): Weimar Culture, Causality, and Quantum Theory, 1918-1927; Adaptation by German Physicists and Mathematicians to a Hostile Intellectual Environment, *Historical Studies in the Physical Sciences III,* University of Pennsylvania Press, Philadelphia, pp. 1-115.

Moravcsik, M. (1972): Science Development in the Framework of International Relations, Southern California Arms Control and Foreign Policy Seminar, Santa Monica.

Moravcsik, M. (1975a): Motivations of Physicists, *Physics Today* 28: 10, 9.

Moravcsik, M. (1975b): Scientists and Development, *Journal of the Science Society of Thailand* 1: 89-95.

Moravcsik, M. (1976a): *Science Development,* PASITAM, Bloomington, Indiana (Second Printing).

Moravcsik, M. (1976b): The Context of creative Science, *Interciencia* 1: 50.

Moravcsik, M. (1977): Applied Scientific Research and Developing Countries, (to be published).

Price, D. de S. (1965): Is Technology Historically Independent of Science? A Study in Statistical Historiography, *Technology and Culture* 6: 553- 568.

Price, D. de S. (1969): Who's Who in the History of Science: A Survey of a Profession, *Technology and Society* 5: 2, 52-55 (1969).

Rahman, A. (1975): Goals for Basic and Applied Research in Different National and Cultural Contexts *Journal of Scientific and Industrial Research (of India),* 34: 1-7

Raman, V. V. (1976): Universalism in Science and Technology, *Philosophy and Social Action* 2: 1-2, 79-81.

Weinberg, A. (1963): Criteria for Scientific Choice, *Minerva* 1: 159-171.

Ziman, J. (1968): *Public Knowledge,* Cambridge University Press, Cambridge.

C. The Bridging of the Gap

Preprint, published in *Science and Culture*
39, No. 5 (1973) 205.

A CHANCE TO CLOSE THE GAP?

Michael J. Moravcsik

I.

The existence of a scientific and technological gap is one of the dominant factors of our world in the second half of the twentieth century. Scientific research, both pure and applied is overwhelmingly concentrated in a few countries, and the resulting technological innovations and know-how are also primarily confined to those countries. At the same time, these are also the countries with the highest per capita gross national product exceeding up to twenty-fold those of the countries where science and technology are non-existent or in their infancy.

Furthermore, the gap between the "advanced" and the "developing' countries has continued to increase in absolute terms. Clearly, even if both types of countries develop their scientific and technological resources at the same percentage rate, the gap will nevertheless increase. To make matters worse, the developing countries have been investing, even relatively speaking, less effort in research and development than the advanced countries, thus increasing the gap even more.

This state of affairs, if extrapolated into the future, leads to a rather pessimistic prediction for the state of the world some decades hence. It is not only the purely material aspects of life that are involved here. Undoubtedly, a growing disparity in science and technology between a few frontrunners and the rest of the world will accentuate the difference in the material standard of living between the haves and the have-nots. But the effect goes far beyond that. Being steeped in the scientific way of doing things also has a profound influence on the general outlook in life. It is not our purpose here to argue whether this is for the better or worse. It is enough to state that the influence is drastic, and hence the difference between countries where science and technology are known only to a handful and countries where science and technology pervades every walk of life is dramatic. This influence makes itself felt in social and political matters as well as economic ones, and hence equalization in this respect could establish a common basis from which the dialogue between countries can proceed.

But the effect goes even beyond that. The incorporation of technology into the lives of people results in the creation of leisure time, and a sizeable fraction of the life of every individual needs no longer be spent on work for the absolute necessities for survival. Thus, cultural, spiritual, and recreational opportunities, heretofore absent, gradually emerge, and this again has a profound effect on the view of life.

These are all generally recognized facts, which are emphasized here only to underline the crucial importance the closing the gap has on our future, and to demonstrate again that by the traditional analysis the gap appears to be bound to get wider instead of narrower.

The aim of this essay is to argue that, due to a relatively recent and perhaps unexpected development, there might be a distinct possibility to close the gap after all. The argument is admittedly quite speculative, and in many other

contexts it would be idle pastime to consider it. Because of the enormous
portent of the present context, however, it might in fact be worth while to
discuss it somewhat more seriously, and derive from it guidelines for present
action if it is to be exploited.

II.

Risking some oversimplification, we can list four essential ingredients
for the scientific and technological development of a country, society,
community, or group: Trained manpower, favorable circumstances for the
deployment of this manpower, capital, and high morale. A brief exposition of these
might be in order.

One of the most important elements is trained manpower. In science and much
of technology, productivity per capita is the dominant factor in success, and to
increase that, one needs highly trained, competent researchers, teachers, and
technicians. Furthermore, since science and technology often progress through
"breakthroughs", and these breakthroughs are due to the efforts of a very few
people at the apex of scientific or technological creativity, there must be
great emphasis on the quality of manpower as well as on the quantity. Second
best in science and technology is often only tenth best when it is converted into
its effect on the economic development of a country.

In addition, science and technology are prime examples of collaborative
undertakings, in which the weakest link might determine how effectively science
and technology can be embedded into the economic and cultural life of the country.
Thus, it is essential not only to develop some absolutely first class scientists
or technologists, but also to set a rigid minimum standard for all those participating in scientific or technological activities.

The second requirement is the presence of favorable circumstances to deploy
such manpower. This involves a variety of factors, such as political non-inter-
ference in matters of science and technology, a competent administrative and
managerial framework to provide support for science and technology, and an
adequate social climate in which research and development activities are at least
condoned. This last element is particularly important. There must be a general
level of "scientific point of view" on the part of the whole society, a general
readiness to encourage science and technology for what they are, to understand and
value their benefits and so give support and status to them.

The third ingredient is capital for investment in science and technology.
This is perhaps the most obvious and more often discussed of the factors
affecting development, and as such hardly needs more exposition here.

The last of the four requirements, on the other hand, is very seldom
discussed, and much less obvious. Nonetheless, it is perhaps the most crucial
of all. It is what I call high morale, and because of the vagueness in the
term, this does need elaboration.

Let us for example consider the physicist from one of the developing
countries who was trained in an advanced country, and after obtaining his Ph.D.
spent a few years in another advanced country. There he was connected with a
smaller research institution of second rank, where he had access to a few

equally competent colleagues, and where the facilities were adequate for productive research. Indeed, he remained quite active, and generated a number of interesting research results.

He then returned to his own country, where he was placed in one of the best research institutions that the country had to offer. There he was also surrounded by a few equally competent colleagues, and had equally adequate facilities for research. In addition, his salary, compared to his surroundings, was much more generous than it had been at his previous place of employment, his social standing was higher, and last but not least, he lived in his home environment, one that he has always treasured even during his years away from it.

Nevertheless, his activity in research came to an almost complete stop. Since he was a personal friend, and since I was familiar with and interested in science development in that country, I inquired about the reasons for this change. His reply was that, as everybody knew, one just could not do physics in that country. Not satisfied with this, I tried to pinpoint the tangible reasons for this, but as we surveyed the various factors, it became evident that in terms of a rational analysis, his present place was in no way inferior to his former one. And yet, when I then repeated the original question, the answer remained the same. It was a shining example of what I mean by low morale.

In more general terms, I mean by morale a motive force, a sense of purpose and faith in what one is doing, a philosophical and psychological conviction about certain values and goals. It includes what in modern terms is called "being turned on". Naturally, it is often verbalized and dressed up in rational justifications, but these rationalizations are convincing only if they are rooted in a much less rational soil, and this is what I wish to call high morale.

To return now to the comparison of advanced and developing countries, in terms of the traditional analysis the latter lag behind the former on all four counts. Scientific manpower in most developing countries is minuscule, fragmented, not always well trained, and lack the critical mass and coherence to be effective. Furthermore, there are great difficulties in training future manpower, because indigenous education is often not strong enough, and training abroad brings with it a myriad of new, and, in the final analysis extraneous, problems.

As to the circumstances for deployment, the picture is again less than favorable. Political interference is often present, and political instability prevents long range planning. Scientific and technological managerial skills in developing countries are frequently even less developed than the direct scientific skills. Furthermore, the society is often not attuned to the appreciation of such skills, and might consider scientists or technologists irrelevant parasites on the "useful" activities of society. There is usually also a shortage of capital, particularly in foreign exchange.

Finally, and most importantly, morale is often very low. It is partly a matter of osmosis from the static quality of the society in which these scientists or engineers are embedded, and partly a feeling of hopelessness when it comes to an extrapolated comparison with the more advanced countries. But it goes deeper: as nothing succeeds as well as success, also nothing prevents success so much as lack of success.

III.

We have now come to the main point of this discussion, which is the following assertion.

There appears to have developed recently a serious loss of morale among scientists and technologists in some of the most advanced countries. If this indeed turns into a trend, it will cripple the further development of the advanced countries, and thus possibly turn the tide away from the widening of the gap. This in turn should be utilized by the developing countries to redouble their effort, since now the closing of the gap becomes a distinct possibility.

The discussion must start with some support for the contention that a loss of morale by scientists and technologists is in fact occurring in some advanced countries.

The most superficial investigation would at first give the contrary impression. It would appear that in fact there is a lessening of enthusiasm for science and technology on the part of the public and some governments, which is manifested in terms of decreased financial support, but that scientists and technologists are desperately fighting to turn back this trend. If one penetrates somewhat deeper, however, the picture changes drastically.

Because it is exactly among scientists and technologists that one finds a feeling of confusion in goals, a sense of uncertainty in values, and a deep emotional involvement with doubts and fears. To be sure, most (though not even all) of them insist on continued support for their own particular scientific activities, but at the same time, they often insist that the technology resulting from their work be suppressed unless absolute guarantees are given that no effects, presently known or eventually discovered, will be harmful. What would appear to many as a basic incompatibility between these two stands does not seem to strike them as such. Others would be all in favor of government expenditure for pure scientific research, but then decry any spending on technological uses until all social problems are solved. It seems rather evident that these convictions must stem from much deeper reasons that these rationalizations would suggest, since a purely rational analysis would instantly reveal that absolute iron-clad guarantees for only beneficial effects of a technology are impossible, and that waiting until all social problems are resolved means postponement forever.

There are many specific examples of issues where these attitudes surface. Stands in connection with nuclear power, supersonic aircraft, peaceful uses of nuclear explosives and space exploration, are some of the most recent ones. It is important to emphasize that whether these are legitimate issues or not is not in dispute. They certainly are, and there is much that scientists can contribute to their discussion. What is more pertinent here is the method of discussion. A balanced and quantitative investigation of issues in the spirit of tolerance toward those with a different point of view is one thing. Quite another thing, however, is an all-or-nothing attitude, a fear-ridden craving for the absolutely secure, coupled with a view of the world as conspiracy filled against the noble and enlightened. Such a stance, a sharp contrast to the way scientists would treat a purely scientific matter, gives strong hints of much deeper psychological underpinnings.

Furthermore, even the question of whether all this alarm and emotionalism is "justified" or not is irrelevant for the present discussion. For let us assume for a moment, that indeed ecological, psychological, social, and political considerations of the future would force on us an absolute ban on further technological development. Even in this admittedly completely extreme case, one should argue that the advanced countries must stop first, and give an opportunity to the developing ones to catch up before a worldwide ban goes into effect. It may be hoped that in the process of this equalization the latecomers might "profit" from what some consider "mistakes" of the now advanced countries and hence reroute their own development so as to avoid these "mistakes". This, however, is a decision to be made by the developing countries themselves and not by those in the advanced countries who claim to "know better".

All in all, therefore, what is relevant here is the fact that, whether "justified" or "unjustified", there seems to be a loss of morale among some scientists in some advanced countries. This loss of morale is manifested in terms of a fear of the future and of the unknown, a sense of dizziness in the increasing pace of development, a loss of faith in the philosophical and moral bases for scientific research, a gripping sense of uncertainty about social and political questions, a drop in self-confidence and an introverted guilt feeling about past accomplishments. As is often the case, such feelings are then coupled with a desire to withdraw from the world as it is, and try to move in model worlds with simplified problems and a clear cut line-up of "good" and "bad". The result is a lessening of creative power, a cut in the fraction of mental energy devoted to science and development, and, quite importantly, a huge decrease in credibility when the case for science and technology is presented to the public for financial support.

The main point of our discussion is that this halt in the progress of science and technology in some advanced countries gives an unparalleled opportunity to the developing countries to catch up. We should therefore turn to the question of how this opportunity can be exploited.

IV.

The first order of business is for scientists, technologists, and scientific managers in the developing countries to take note of this possible opportunity, and publicize it as much as possible. High morale is a collective phenomenon, and it is greatly enhanced by a coherent effect among individuals. Thus an explicit community awareness must be created, an awakening to the idea that because of these new developments, many old preconceived notions might become obsolete and that in effect the game goes on with new rules, much more favorable to the developing countries. To be sure, this awareness might not be enough in itself to create high morale, but it should be an enormous contribution toward it.

Secondly, developing countries should find now much more inducement to make a supreme effort for development of science and technology. Even in terms of percentage of gross national product, the developing countries have been quite remiss in devoting sufficient resources to these problems. The knowledge that investment in these areas might now pay off more promptly than in the past should stimulate interest.

It would be beyond the framework of this article to discuss in detail the specific steps this effort to accelerate the development of science and technology in the emerging countries should take. In any case, there is a large literature on the subject, and many meetings and conferences address themselves to this question. One particular example, however, might be useful since it specifically pertains to the situation hypothesized in our discussion. The decrease in morale in some advanced countries might very well herald the beginning of a reversed brain drain, in which scientists from the advanced countries might be induced to move to the less developed countries. Particular effort should be made to lure back those who were originally nationals of the latter but emigrated because of increased scientific opportunities.

In addition to such special opportunities, increased effort must be directed toward the development of indigenous advanced technical and scientific education, since such education might be increasingly more difficult to acquire in the advanced countries. A long range policy committement toward a fixed fraction of the GNP toward science and technology would also help in assuring a smooth and well integrated development.

Thirdly, the scientific and technological personnel must maintain this newly found high morale even in the face of erosive effects due to the lowered morale on the part of their colleagues in the advanced countries. There are a number of such colleagues already, who oppose scientific and technological development of the less advanced countries on the grounds that they should not go astray on the same "disastrous" route the advanced countries have. Although this attitude, on the surface, is almost indistinguishable in its practical implications from economic colonialism, it must be recognized that it is meant in all sincerity and good will. Nevertheless, it must be strenuously resisted, and it must not be allowed to interfere with efforts for scientific development in the emerging countries.

It might be well to emphasize again that I do not advocate the disregard of all ecological, philosophical, moral, or social implications of science and technology in the developing countries. On the contrary, since these are long range considerations which take a long time to mature properly, they should be initiated at the very outset. What must, however, be avoided at all cost is the creation of a low morale situation where abstract and speculative negativism can cripple the faith and purposefulness that is so much needed for making headway.

An intensified program of science and technology in the developing countries is also dictated by two other considerations, both potential outgrowths of the new situation outlined above.

The first is the conjecture that the loss of morale and the concomitant introversion in some advanced countries undoubtedly will result in decreased technical and scientific assistance by the advanced countries to the developing ones. Thus, the latter countries will have to "go it alone" to an even larger extent than in the past. The extent of such assistance was never large enough for the needs, but in the future it is expected to decrease even further. This means that a strong self-sufficient program of science and technology development becomes even more of a must for each country. It will require greatly intensified effort and increased financial investment by the developing countries.

Second, indications are that the loss of morale in advanced countries is not a universal phenomenon in all such countries or that at least the extent varies greatly from country to country. Thus it would not be unreasonable to expect that if the demoralization suggested in this article is indeed a long range trend and the countries with a loss of morale decline in comparison to others more fortunate, the balance of power in world politics will shift substantially. In particular, it might become impossible in the future for smaller developing countries to navigate securely in the shadow of a balance of large powers, since such balance might no longer exist. In such a new political environment self-sufficiency and strength for developing countries will become even more of a sine qua non condition for their continued survival and prosperity. Thus the development of indigenous science and technology might take on a crucial importance.

History plays strange tricks with nations, and leading countries turn into insignificant entities while small, unnoticed countries burst into trailblazers. At any given time, it is difficult to believe that the giants of today might in fact be overtaken by the poor and backward. I have tried to suggest that in our age, it might perhaps be disparity in scientific morale that will trigger such a reversal of roles. If there is any chance of this prediction being correct, it must be incorporated into the planning of the future of the now underdeveloped countries.

The Transmission
of A Scientific Civilization

One observer perceives waning public confidence in science and low morale among scientists as possible indications that the decline of scientific civilization in the West might have begun. He contends that because science is closely related to objective reality, its "civilization" might be transmitted, relatively undistorted, to receptive societies in developing countries. Dr. Moravcsik is a theoretical physicist at the Institute of Theoretical Science and the department of physics, University of Oregon.

MICHAEL J. MORAVCSIK

That ours is the scientific age is by now a truism. The most common, and perhaps most superficial, interpretation of this truism is in terms of the products of science which surround us and have a profound influence on our lives. The statement, however, is true also in a somewhat deeper sense. We live in the scientific age because science also has represented over the past few decades a focus of our aspirations and a strong force on our philosophical and ideological thinking. In this sense, we can call our milieu the scientific civilization. Indeed, one can argue convincingly that a good way to characterize a civilization is by its abstract concerns, those of its motivations which in the most recent terminology would be called "irrelevant."

From this point of view, it is not difficult to demonstrate that we do live in a scientific civilization. A relatively large number of our contemporaries spend their lives doing science. To be sure, they are not supported by society because the majority of citizens are sufficiently devoted to the ideals of science to be willing to provide for the maintenance of so many trained practitioners of these ideals. Instead, motivation for the public support of science comes more from utilitarian roots. People believe, rightly or wrongly, that their own lives will be made more comfortable and more enjoyable if science is practiced extensively. People might also believe, rightly or wrongly, that eminence in science is a matter of prestige, personal or national, and they may attach great importance to this. But these more utilitarian motivations for the support of science do not differ greatly from the motivations of the masses for their support of the ideals of previous civilizations. Large-scale support of religiously inspired activities, for example, stemmed in part from a utilitarian concern about afterlife.

Perhaps more importantly, the actual practitioners of science are in fact motivated by the ideals of sciences. Very few scientists would list as their main incentive for doing scientific work the desire to contribute to the utilitarian needs of society. Instead, their main motivation is the excitement of scientific discovery and the aesthetic satisfaction of understanding the laws of nature. These pleasures fuel the scientific creativity which they convert into research work.

It may be objected that while science is indeed a strong element in our activities, it alone does not make a whole civilization. Although this is undoubtedly true to some extent, science in fact has a strong influence also on the other ingredients which make up our civilization. It is not at all farfetched to argue that one of the major forces in shaping the development of twentieth century music has been science. The trend toward abstraction, and in fact toward abstraction of the mathematical type, is a case in point. Twelve-tone music and computer generated "random" music are examples. To this one can add the science-aided exploration of completely new musical (?) sounds (the so-called "electronic" music), as well as the very profound effect on music of the phonograph record. Analogous trends can also be discerned in fine arts, where abstractions are again the rule, and the geometrical shapes and combinations play a very important role. The creation of mobiles and moving displays also owe their origin in part to science. In literature, the concern with the more analytical is undoubtedly also connected with the scientific method.

Transcendental Brainwash

It would be very simpleminded to claim that everything in our present civilization is a product of the upsurge of science. Human affairs are much too complex to be explained, at any time, in terms of a single influence. For the present discussion, however, it is sufficient to argue that our contemporary civilization is very strongly influenced by science, and that science is in fact one of the dominant factors in it. In this sense, therefore, we can refer to our environment as a scientific civilization.

Throughout the history of mankind civilizations have risen, reigned and fallen. This interweaving pattern of waves of cultures has been so invariant and so steady that one is tempted to postulate it to be a necessary law of human behavior which therefore can also be extended into the future. In order to do this, however, it might be advisable to describe in somewhat greater detail how these waves of civilization interact.

In the birth of new waves of civilizations, perhaps the most important factor is morale. A group of people, however small, "backward," and "uncultured," can originate a new wave if they are imbued with confidence, with high morale, with a deep-seated belief. The new civilization then can spread not because it offers anything "superior" to what had transpired before, but because of the ability of its proponents to transmit their enthusiasm for and their belief in these new ideals. Naturally, confidence alone does not make a trend; and many movements, fueled by enthusiasm or even fanaticism, have vanished without making a civilization because the worth of their wares, measured in terms of somewhat more objective criteria than pure enthusiasm, has proven to be slight or even negative. Nevertheless, the importance of high morale, of the ability to brainwash in this transcendental sense, and of the deep-seated personal identification with some ideals cannot be overestimated.

Ideals and Civilizations

A new wave of civilization, as it begins to roll, naturally encounters other, already existing civilizations. The encounter often turns out to be a conflict, since the two sets of ideals frequently appear to be mutually exclusive. What determines the outcome of such a collision of civilizations? Would the long-standing, well-rooted culture not have an edge over the new wave, which is still in the process of trying to establish itself? The answer is in the negative: often the established civilization succumbs because of old age. It is likely that the main aging factor in existing civilizations can again be traced back to morale. As a trend gets entrenched in a certain segment of humanity it becomes somewhat stale after a while, not because its potentialities have actually been exhausted, nor because another trend is necessarily "proven" (by objective standards) to be superior to it, but rather because its practitioners lose interest in it, and hence lose their morale and incentive for devoting themselves to it with true enthusiasm. The visible shells of practice might linger on for some time thereafter, but in fact the civilization is dead, and ready to be swept away by the next wave.

We might wonder why the ideals of a civilization could not be taken over by another group, previously uncontaminated and hence not demoralized in the above sense. If this could happen, one would see an empire with a certain civilization being swept away by a dynamic wave of "barbarians" who, in turn, would take over that civilization and carry it on essentially unchanged. This in fact does not happen. To be sure, civilizations do not vanish without exerting some influence on the conquerors, and the new trends often exhibit discernible elements of the old and dead which they have replaced. But whatever is carried on undergoes a strong transmutation before being incorporated into the new civilization.

Such a transmutation occurs because the laws regulating civilizations are arbitrary. Standards of beauty in the Greco-Roman civilization cannot be compared in an absolute way with standards of beauty in, for example, Chinese civilization. Each had sufficient force to fascinate vast numbers of people for long periods of time, but it would be impossible to establish a nonsubjective set of criteria by which one could be ranked superior to the other. Thus each group of people, as it gathers conviction and belief in a different set of ideals, develops its own set of criteria. When the remnants of an old civilization are incorporated into a new wave, they are selected, interpreted and transmuted according to the new set of criteria. The result is a basically new civilization and, therefore, it cannot be claimed that the old one survived at all in a direct sense.

Beginning of the End?

Can we now apply these concepts about changing trends to our own, scientific civilization? One might ask first why the eagerness to talk about the subject? Is our scientific civilization not strong and vigorous, and does it not continue to reign unchallenged by new movements? I would like to argue that the answer begins to be in the negative, and therefore the topic is of considerable timeliness.

What are the signs that the scientific civilization in its present locale is beginning to decline? First, it might be pointed out that this locale is mainly in the so-called Western countries, that is, Europe and North America, with the addition of Japan. To be sure, some rudimentary scientific activity is also beginning in other parts of the world, but the level of the activity is not very significant and it does not influence, at the present time, more than a tiny fraction of the population. Thus, for present purposes, we shall discuss science in that limited area of the world.

The most visible sign of the beginning of the decline of scientific civilization in these countries is the slackening of public support for science. The fact that this decreased support is rationalized in terms of various mellifluous phrases, such as "reassessment of priorities," "redirected relevance," and the like, should not obscure the fact that such an attitude by governments and individuals is an expression of a decreasing identification with the ideals of the scientific civilization. It is, naturally, much too early to tell whether the present signs constitute a trend or only a passing whim, but at their face values they could be construed as a possible indication for a decline of the scientific civilization.

A much more profound sign can be found in the intellectual attitude toward science by many people. It is increasingly said that science is irrelevant to the important problems of our age, or that science, in its applications, is predominantly harmful and hence should be banned somehow. The arguments, both pro and con, eventually boil down to value judgments; hence a change in the balance of the arguments represents a declining morale and waning confidence in the scientific civilization. At the same time, activities with markedly anti-scientific or at least non-scientific attitudes seem to gain new adherents, as the recent revival of astrology demonstrates.

The most important sign of the possible decline of the scientific civilization is, however, the attitude of scientists themselves. When the high priests of a culture themselves begin to loose their motivation, the cause is in serious trouble. I want to suggest that there are signs that this is actually beginning to happen now. A small but increasing number of scientists themselves begin to question the relevance of science to our age, and begin to be riddled with fears of all sorts. This manifests itself not only in some negativistic contributions to the philosophy of science on an academic level, but also in attitudes toward the more applied scientific-technological developments. Some examples are evident in discussions of nuclear power, genetic engineering, pesticides, supersonic transportation, space exploration and many others.

Fear-ridden Spirit

Often these attitudes surface in the context of international affairs or ecology, but these subjects themselves are only incidental, since it is not the mere investigation of the pros and cons of these problems that is relevant here. Instead, the heart of the matter is the fear-ridden, super-conservative, security-craving spirit in which these matters are sometimes looked at, with the implication that unless absolute guarantees can be given for the absence of detrimental effects, presently known or eventually to be discovered, a new undertaking should not be allowed, or, that such new developments should be postponed until all "relevant" problems of the world have been solved. In other words, the analysis is carried out as if scientific and technological developments had no inherent value (or even a negative inherent value) beyond the narrowly utilitarian consequences. It is not difficult to see what a devastating effect such a mentality would have had on the history of mankind had it always persisted.

Some will claim that the above examples pertain only to the uses of science, and that attitudes of scientists toward science itself has not changed. Such a claim has very little credibility. It is not possible to separate science and its consequences in such a schizophrenic fashion. Even though a scientist might not be motivated primarily by the effects of science on society, if he is convinced that science is basically "bad" (even if delightful), he is not likely to pursue it with the high degree of dedication needed to maintain the scientific civilization, especially when it is under attack from the outside.

To one not affected by this malaise, the trend appears to be dominated by the crippling shackles of doubt and a lack of purposefulness which make small minds out of potentially great ones and which constrain individuals to fall below their own capability rather than transcend it. Undoubtedly, these attitudes appear in a quite different light from the inside, but that is by and large irrelevant in the present context. Similarly, the question of whether these doubts and fears will turn out to be well founded is also void of an objective meaning in our present discussion. For our conjecture of the decline of the scientific civilization, the important point is that such a loss of morale appears to be gaining ground within the scientific community itself.

Negative Trends

It is, therefore, not completely farfetched to argue that perhaps what we are witnessing today is the beginning of the decline of the scientific civilization in the advanced countries. To be sure, no viable alternative civilization appears to take its place, since all other trends on the horizon are purely negativistic in their conceptions, and such trends very seldom become waves of the future. The possibility, however, that the present scientific civilization might soon become ripe for a possible overthrow is in itself an incentive to discuss its future now.

A crucial question is whether the scientific civilization could perhaps be transmitted to other parts of humanity, not at the present steeped in it and therefore not plagued by declining morale. On the basis of our discussion of the general problems of transmissions of cultures, the answer would be in the negative. I will now argue, however, that the scientific civilization has a certain feature which makes it unique among all previous civilizations, and this feature might hold out hope that such a transmission in this case might be successful. I contend that because science is closely related to "objective" reality, its civilization might be transmitted from group to group with very little transmutation.

We have seen earlier that one of the main obstacles to the transmission of a civilization is the subjective nature of its criteria. If, therefore, we can demonstrate that in science these criteria are more universal than they were ever before, the main contention is certainly made more plausible.

Indeed, science appears to have such an added objectivity. It pertains to the outside world, and it is generally possible to bring about a considerable consensus concerning what the facts are about that outside world, at least if all participants subscribe to the scientific method as a means of investigation.

One might counter this by saying that the scientific method is by no means the only way to look at the outside world, and that by making people subscribe to that method we have artificially brought about an apparent consensus, which will be shattered if others refuse to subscribe to this particular method of inquiry. This is an undoubtedly valid objection, and indicates that science has no absolute universality either, and hence would not be able to bring about a completely unambiguous and immutable transmission of its culture. But compared to other civilizations, science is still in a favorable position. In many other cultures, even when the methods of inquiry are agreed on, there can be very fundamental differences in the resulting criteria. The main reason for this is that other cultures do not have the objective experiment as an arbitrator of disputes. In science not only are the criteria universally agreed upon, but these criteria are also unambiguous.

This universal quality of science is very evident to its practitioners who often have contact with scientists from all over the world, many of whom come from backgrounds and traditions very different from our own Western civilization. These differences, however, vanish instantly when the communication turns to matters of science, and general agreement can usually be achieved even if deep political and cultural distinctions or even antagonisms otherwise exist between the communicants.

Objective Criteria

To be completely honest, one should add, of course, that even in the sciences the objective criteria do not always work instantly. For one thing, experiments on occasion can be wrong or indecisive, and it sometimes takes a little time to apply the criteria of the scientific method to a particular situation so that a possible dispute is resolved to everybody's satisfaction. In the long run, however, all such disputes are in fact settled. Similarly there can be animosity among scientists on account of quarrels over priority, though this is not strictly a scientific dispute. But on the whole, in the sciences, much more than in other areas, an objective standard exists.

One can then suggest that with these objective criteria, the transmission of a scientific civilization might be done successfully and without major distortions. Granting this, and assuming that the analysis about the beginning decline of the Western scientific civilization is correct, one should then explore ways in which this civilization can in fact be transmitted to new, dynamic, "uncontaminated" societies.

There are, of course, many candidates, since the present scientific civilization is in the bloodstream of only a very small fraction of the world. In particular, all of the developing countries are such candidates. Most likely, only a very few of them will have the other necessary characteristics to become the carrier of a new wave of scientific civilization, and it would be very difficult indeed to predict which ones are in this category. For those of us who believe that our scientific civilization is worth saving, and who have not been demoralized, bringing science to the developing countries— a necessary and worthwhile task in itself—takes on an added dimension: that of an attempt to transmit this civilization before its complete decline in the West.

DOCUMENT

Intercultural Transmission
of Science

Michael J. Moravcsik

Modern science (i.e. the natural sciences since the 16th century) first developed almost exclusively in the "West", that is, in Europe and North America (1). Starting perhaps a half a century ago, however, and since then at an increasing pace, the practice of modern science has been spreading throughout the rest of the world. With Japan perhaps leading the way (2), contributions to science began to appear in various Asian countries, notably India and China and in some Latin American countries. After the second World War, this universalization of science accelerated to include other Asian and American countries, and later also African nations. To be sure, even today the scale of this "extra-Western" contribution to science is relatively modest, amounting (if we exclude Japanese work) to barely 10 % of the "Western" effort. Yet it is clear that we are, happily, on our way to make the practice of science a common heritage of all humanity.

It is also clear that there are a number of obstacles to and difficulties in this process of intercultural transmission of science. One need not have to visit the developing countries for a long time before becoming aware of the substantial problems in the path of this spreading of science. Many

(1) For a concise historical account of this, see George Basalla, "The Spread of Western Science", Science *156*, 611 (1967).

(2) A graphic illustration of the origins of modern science in Japan is given in Derick de Solla Price, *Little Science, Big Science*, Columbia University Press, New York (1963), p. 99.

MICHAEL J. MORAVCSIK

of these problems are material or economic in nature and therefore can be pinpointed relatively easily. They can also be solved relatively easily and speedily, provided that the one necessary ingredient, money, comes forth from some beneficial source.

A much larger set of problems, however, has nothing or little to do with money, and in fact cannot be remedied quickly even if huge sums of monies are available. These problems are perhaps less evident, and more complex in nature. They are, consequently, also more interesting intellectually, and more challenging to a pragmatist who wants to get things done.

In this discussion I want to concentrate on these more complex obstacles in the path of the evolution of science in developing countries. My approach will *not* be that of a theoretical analyst presenting a logical and causal framework of philosophical, sociological, and anthropological factors (3). Instead, I want to rely on my longstanding personal contact with the *de facto* problems scientists encounter in developing countries and on my past opportunities to learn from a large number of scientific friends and colleagues in many developing countries who have poured out their hearts in conversations, discussion, or articles.

In fact, I want to structure the discussion in terms of a chronological survey of the life of a scientist in a developing country, pointing out, step by step, the factors which impede him in converting his talent in science into actual scientific productivity.

We might begin with a possibly sensitive question, that of innate ability. Hard information on possible differences in innate ability between various races for one or another human activity is completely missing, and work on this subject is practically impossible in the present overcharged atmosphere. We will, therefore, assume henceforth in this discussion that there are no such innate differences. Just on general statistical grounds, as well as on the basis of obvious physiological differences between races (in terms of size, weight, body build, etc.), it is highly unlikely that future research will not discover differences also in innate ability with respect to one or another human endeavor. At the present, however, the magnitude of such differences is unknown, and it is also unknown which races will show greater ability in which undertakings.

Starting, therefore, from the assumption that a child in any part of the world is born, on the average, with equal innate ability toward scientific activities, let us first concentrate on early childhood experiences. There are many aspects in which the environment of a child in a develop-

---

(3) For a broader theoretical essay on this subject, see for example Yehuda Elkana, "The Theory and Practice of Cross-Cultural Contacts in Science: Queries and Presuppositions" (to be published, preprint available from the author) (1978).

INTERCULTURAL TRANSMISSION OF SCIENCE

ing country is different from that of a child in an industrialized country. I want to mention two (4) in particular which are likely to affect the child's future scientific activities.

The first of these pertains to the playing environment of the child. The toys children play with in the industrialized countries increasingly incorporate sophisticated technology, which accustoms the child to the operation of machines and to the idea that man's inventiveness can bring about a growing assortment of "wonders", some going far beyond what nature itself can provide. More importantly, however, these toys arouse, quite early, the child's inquisitiveness as to how things work. "Western" parents hear an unending stream of "why?"'s and "how?"'s from the lips of their children, questions which are basic to scientific inquiry of any kind. The parent may not be able to answer these questions in all instances, but in any case it is recognized by the child and parent alike that the answer is known, and that it is a desirable trait for a person to be able to answer these questions. Such an environment is generally absent for a child brought up in a developing country.

The second important environmental effect is the early exposure to abstract thinking. A ready, though by no means unique example of this is the acquaintance with the idea of maps. As research has shown, "Western" children are, at an early age, familiar wtih the abstractions that go into the drawing of a map to scale, while children from developing countries of the same age visualize a map as a conjunction of realistic images arranged without regard to spacial or temporal relationships. Since abstract thinking is a cornerstone of at least the more sophisticated of the modern sciences, we encounter here another factor that might be construed as an obstacle in the path of the transmission of science to the developing countries.

It might be interjected at this point that our investigation may be *a priori* slanted in a direction that predetermines its outcome. It seems that we are asking about the establishment of "Western" science in non-Western societies. Is it not possible that non-Western societies will develop their own science, quite different from "Western" science, and this development will not be hampered by the factors that could obstruct the spread of "Western" science.

This question, namely whether the developing countries can have their "own" science or not, is often discussed in various contexts . My own stand on this, at least until I am persuaded otherwise, was recently stated in an article (5) dealing entirely with this problem. It is my con-

(4) For these effects at the grassroots, see the writings of Francis Dart, in particular Francis Dart and Panna Lal Pradhan, "Cross-Cultural Teaching in Science", Science 155, 649 (1967), and Francis Dart, "Science and the Worldview", Physics Today 25, 6:48 (1972).

(5) Michael J. Moravcsik, "Do Less Developed Countries Have a Special Science of their Own?", Interciencia 3, § 1, 8 (1978).

MICHAEL J. MORAVCSIK

clusion that there is no "Western" science and "Third World" science, but that basically we deal with only one kind of science, which happens to have started in the "Western" countries but is much more universal in its character. As a result, I will assume throughout this discussion that we are in fact investigating the obstacles to the spread of this one and unique science.

So let us return to the child's environment, and take a look at it beyond the specific toys or concepts the child is exposed to. Some very substantial influencing factors exist in the prevailing epistemology (4) in which the child of a developing country is likely to be submerged. Two aspects are of particular importance. One is the simultaneous, peaceful coexistence of several explanations for the same natural phenomena, some resting on traditional mythology or religious concepts, some on popular versions of scientific theories. Thus earthquakes are interpreted, simultaneously and by the same person, as due to forces of volcanism in the inside of the earth as well as to the shifting of the turtle on the back of which the earth rests. Though it would be an exaggeration to claim that in "Western" culture all phenomena have only one, universally accepted cause, the all-prevailing pluralism of some non-Western civilizations is different from the general "Weltanschauung" that a child is brought up in, in most industrialized countries.

The second epistemological element is of even greater importance, since it deals with the origin of our knowledge. It has been shown that in various non-Western cultures the origin of knowledge may be thought to reside in the heads of old men, in books, or in the powers of a sorcerer or guru, but not in the process of inquiry itself. Consequently, in the former views knowledge is often thought of as a closed domain: Everything that is to be known has been discovered already and is to be memorized through contact with books, old men, or magicians, rather than by asking new questions. It is not necessary to point out what a fundamental effect such a world view will have on the fostering of scientific inquiry in a country.

Our child from the developing country is, therefore, already handicapped when he enters school at the age of 5 or 6. Here he is confronted with new blows. Partly due to the factors discussed earlier, he will be taught by teachers to whom science is not a living method of inquiry but a dead set of facts. School will demand rote learning, and success at examinations will depend on the quick and verbatim recall of memorized material. Schools will also be short of demonstration equipment, let alone laboratory opportunities for the student himself to experiment. This in turn will lend an even more bookish demeanor to the whole business of learning science (6).

---

(6) For a discussion and references, see Michael J. Moravcsik, *Science Development*, PASITAM, Indiana University, Bloomington, Second Edition (1976), Chapter 2.

INTERCULTURAL TRANSMISSION OF SCIENCE

Schools will also be very "democratic", which in practice translates into mindless egalitarianism. All students learn exactly the same material, and little attention will be paid to individual interests and capabilities. The prevailing external (and hence often centralized) examination system originally conceived to pull up the schools of lowest quality, will also strengthen this trend by pulling down the gifted and the motivated.

The next problem arises in connection with career choices. Here three elements often found in non-Western societies play a dominant role: gerontocracy, a closely knit family structure, and an absence of individualism. Thus career choices for the young will be strongly influenced by the older members of the family and will not be selected by taking into account internal, individual motivations, but rather by external criteria, societal prestige, and the adherence to the prevailing norms.

Since the pursuit of science is still minuscule in developing countries, the societal image of a scientist is either non-existent, or vague and hence not yet ranked. As a result, the external pressures on the young student in choosing a career will very seldom steer him toward becoming a scientist. Among the intellectual professions medicine and engineering are far preferred, and often only those who have failed to become a doctor or an engineer end up choosing science.

There are, of course, always exceptions. But the exceptional student, whose internal drive and curiosity won out against all the opposing forces, and who chose science, will find himself a freak. He will encounter few other students to share his excitement and wonder about science, and often few professors to be recognized by.

There are few professors, because college level science education suffers from the same defects of rote learning and formalistic approach. Added to these is the tendency toward premature specialization, that is, the lack of a broad education in a large area of science. Instead, a student, a mere four years after completing secondary school, is allowed to narrow down his studies into an exceedingly specialized small domain of one of the sciences (e.g. spectroscopy of a certain class of molecules). Sometimes this specialty is an obsolescent one to begin with, but even if it is not, it is likely to become one in a decade or so, at which time the student will lack the broad foundations to switch fields.

In most developing countries educational opportunities in the sciences exist locally only up to a college (Bachelor's) degree. If more advanced education is desired (and in most sciences the opportunities for original work are exceedingly limited without such advanced degrees), it must be acquired abroad.

At this stage another set of obstacles looms ahead (7). For one

---

(7) Michael J. Moravcsik, "Foreign Students in the Natural Science: A Growing Challenge", International Educational and Cultural Exchange 9, § 1, 45 (1973).

MICHAEL J. MORAVCSIK

thing, obtaining such advanced education is not at all easy. Admission and financial support to graduate schools in "Western" institutions is not given out very easily. The evaluation procedures used by such institutions to select students are rather haphazard, favoring students whose transcripts are interpretable, and students coming from institutions previously yielding other successful students. As a result, many countries are *a priori* handicapped, and mostly those where science is in a particularly rudimentary state. Examinations like GRE and TOEFL, required by many of the "Western" institutions as proof of competence, are costly to take, are given only in certain places and at certain times which might be inaccessible to the student, and are subject to corruption. These tests are also strongly biased in favor of students who were previously educated in a "Western" environment. Applications fees, negligible by "Western" standards, are huge by local standards and are often unattainable because of foreign currency regulations. Travel funds to the locale of advanced education are scarce, even if the student has an assistantship to maintain him once he begins his advanced studies. The list is almost endless. This is an aspect of science development, however, which has been discussed amply in the past, even though very little has been done to improve the situation.

But the problem is not only in the logistics of obtaining such an advanced education abroad. There are also the more substantial problems of acclimatization to a new educational environment, and the question of whether good advanced science education as designed for "Western" students constitutes a complete and good science education also for students who will return to a developing country to do science.

In the area of acclimatization, the faculty adviser plays a crucial role. Yet this aspect of "Western" education is often very much neglected. The advising of students in general is not a strong point of most "Western" educational institutions, and the advising of foreign students is even worse, particularly because many foreign students are used to not talking to the professor unless being talked to. In addition to having general interest, however, faculty advisors would also need to know something about the educational environment of developing countries in order to give wise advice to the entering foreign student. There are by now a sufficient number of "Western" scientists with personal experience in developing countries so that such expertise could be made available in almost any science department of almost any "Western" university. The problem, however, has not been generally understood and the corresponding minimal organizational steps have not been taken.

With respect to the appropriateness of "Western" education for the student from a developing country, one can say that on the whole, the education is good but not complete. It is incomplete since it does not make allowance for the fact that, unlike his fellow student from the host country, the student from a developing country, when he returns home,

INTERCULTURAL TRANSMISSION OF SCIENCE

will be faced simultaneously with the task of doing science *and* the task of helping to create the conditions under which science can be done. No education is provided for the student for this latter task. Though there are relatively easy ways of providing some supplementary preparation for the students in this direction, nothing has been done in practice.

Let us now continue to follow our potential scientist from the developing country. He is in his late 20's, and, having received an advanced degree in one of the sciences, he now returns home.

The greatest obstacle confronting him now is isolation (8). Science is a very collective undertaking in which interaction with other scientists in various parts of the world is essential. Of particular importance is such an interaction right after a Ph.D. when the young scientist is generally not ready to act as a research leader, though he might be quite capable of performing research if he has the advice and collaboration of other, more experienced scientists.

But in his country he is unlikely to have such an interaction, and he will also be short of communication through journals, reports, conferences, proceedings of meetings, short term visitors, books, review, and all other vehicles of scientific interaction. This is also an area that has been discussed amply in the literature (9) of science development, even though very little has been done to relieve this isolation.

The isolation affects not only the scientist's direct ability to pursue his work, but also has psychological implications, effects on his morale. Receiving external recognition is an important element in a scientist's knowledge of being on the right tract, in his knowledge of his contributing to the universal and timeless edifice of scientific knowledge. Thus isolation can result in discouragement, or in flight into esoteric problems the solution of which is of no interest to other scientists or to technologists.

Apart from being isolated from the worldwide scientific community, our scientist who just returned home may find himself also lacking comprehension or recognition by the society that surrounds him. For the reason discussed earlier, the image of science in the society of a developing country is often not clear and hence the prestige and interest surrounding the profession is undeveloped.

More importantly, however, this lack of comprehension of what science does, needs, and provides carries over also into the governmental bureaucratic structure. Thus our scientist encounters indifference, ignorance, or even hostility in the governmental offices whose function should

---

(8) For a very eloquent picture of this isolation, see Abdus Salam, "The Isolation of the Scientist in Developing Countries", Minerva *4*, 461 (1966), reprinted in Edward Shils (Ed.), *Criteria for Scientific Development: Public Policy and National Goals*, MIT Press, Cambridge, Mass. (1968).

(9) See Reference 6, Chapter 4.

MICHAEL J. MORAVCSIK

be to foster scientific activity in the country. In addition to these explicitly negative factors, we have the general and all-prevailing deadly influence of bureaucracy *per se*. The burgeoning of bureaucracy is perhaps the one single aspect of contemporary life in which the developing countries have successfully competed with the industrialized countries — and often even won the competition. Thus our scientist has to fight endless and debilitating battles with interminable red tape, robotminded regulations, administrative inability, and a disregard of the importance of the time element in the making of decisions.

Superimposed on this, we find, in many countries, a blatant interference of politics into scientific matters. In the most extreme cases, ideological criteria dictate the selection of personnel, the awarding of research money, or even the choice of scientific problems or methods. In such an environment, in which the government lives by the motto of "If you are not with me, you are against me", scientists are constantly forced to spend their time and integrity reaffirming their loyalty to the prevailing ideology. Furthermore, political weapons are then also used for personal squabbles among scientists, thus making those squabbles potentially fatal, so that it becomes a matter of personal survival for the scientist to waste his time and energy in fighting those squabbles.

The situation is somewhat better in the case of regimes where the motto is "If you are not against me, you are with me", that is, where neutralism is not construed as a threat to the government. In such a situation it becomes possible to establish a position for science which will be supported by any regime, somewhat akin to the support by any regime of the supply of clean water to a large city. But even in such a situation, the temptation of political corruption is great, resulting in the appointment of science policy makers on the basis of party affiliation rather than technical competence.

It might appear that much of what I mentioned in connection with the social isolation and political status of the scientist pertains to his relationship to people outside the scientific community. But this is not necessarily so. If a young scientist returns home, full of capability and knowledge to participate in modern scientific activities, he might very well find that his older colleagues at home, who (often through no fault of their own) have not had the opportunity to acquire this potential toward a productive scientific career, will greet him with envy, bitterness, and hostility. Thus even in his own university, research institution, or laboratory, the scientist might feel unwanted and isolated.

There are also purely material factors adding to his difficulties. His salary may be insufficient (as compared to his peers in other professions) so that he has to spend time to supplement the income through secondary jobs. Experimental equipment may be lacking, or the repair of such equipment may prove to be impossible or lengthy. Technicians and

INTERCULTURAL TRANSMISSION OF SCIENCE

assistants may be non-existent, libraries rudimentary, stockrooms primitive.

From another direction, a different danger threatens him. Because of the rampant bureaucracy in most developing countries, bureaucratic positions are many and quite often attractive. It is, therefore, extremely tempting for the young scientist to throw in the towel as far as science is concerned and switch to a bureaucratic position in what is euphemistically called "science policy", or in educational administration. It is most regrettable that even some international organizations, such as UNESCO, encourage this process by sponsoring the proliferation of a multitude of national, "subregional", regional, and worldwide councils, committees, offices, and divisions.

There is, of course, great need for some scientists in the developing countries to make an effective case to the government for the needs of science. But the divorce of scientific activity from this advocacy, and the siphoning of many of the potentially most creative scientists into formalistic and unnecessary bureaucratic manipulations is something else again.

Thus we come to the end of a very sketchy cataloging of some of the influences and factors that make an intercultural transmission of science difficult. Some of these factors are purely cultural and conceptual, others are connected with societal characteristics. Again others are a consequence of science being new in some countries. Added up, however, the collection of all these factors is so stupendous as to suggest that the transmission of science into the cultures of developing countries is in fact an impossibility.

I do not believe that this is so, and I think I can both argue theoretically and demonstrate empirically that this is not so.

On the theoretical level, one can devise remedies for every single one of these adverse factors. In fact, such an attempt at therapy is not even a purely theoretical undertaking: Many of the suggestions have been tried, at least on a small scale, and were found to work (10). A tremendous amount remains to be done, not so much because remedies have not been suggested, but because action has not followed these suggestions. As a result, one can be sure that the transmission of science into non-Western cultures will be a prolonged process, the impatience notwithstanding that is exhibited by many orators from the developing countries who, understandably, would like to see instant but painless transformations.

---

(10) For an example, see Francis Dart, Michael J. Moravcsik, Andrew de Rocco, and Michael D. Scadron, "Observations on an Obstacle Course", International Education and Cultural Exchange *11*, § 2, 29 (1975) and the references therein.

MICHAEL J. MORAVCSIK

Perhaps more convincingly, however, one can point at developing countries where science was practically non-existent 2-3 decades ago but which are now well on their way toward the development of a creative and self-propelled indigenous scientific infrastructure. The examples are yet few, and sometimes temporary reverses occur due to political interference or other factors. Yet, if one's perspective used a characteristic time of several decades, one cannot be but optimistic that eventually the pursuit of science will become a truly worldwide activity. How long it will take to reach this state depends very much on the understanding, interest, and willingness for active collaboration that the "Western" countries, and their scientific communities in particular, exhibit. The general outlines of the path to be taken have been demarcated. What we need now is less rhetoric and more action.

MICHAEL J. MORAVCSIK

Preparing the new pioneers

It is very likely that within the next century a significant fraction of the world's population will have left Earth as a domicile and will have moved into space colonies. The technology to do so already exists, and soonor or later will undoubtedly be used for that purpose.

Which country, or which group of people will take the initiative on a sufficiently large scale to give the decisive impetus to this movement is not yet known. In the past, the bold, the motivated, the imaginative, the innovative, and the energetic—sometimes arising from quite unexpected quarters—changed the course of the human race.

The last few times there was such a drastic expansion in the horizon of a segment of humanity, it brought great benefits to some but had at best

Michael J. Moravcsik, physicist, is in the Institute of Theoretical Science at the University of Oregon in Eugene. He is the author of *Science Development* (1975).

mixed consequences for many others. When Europe expanded into the New World at the beginning of the sixteenth century, the event had a stupendous impact on the life of most Europeans, generating a development that took them from the marginal Middle Ages toward a greatly enriched and broadened life style. The expansion meant slavery, however, for countless Africans and hard work in an inferior status for many Asians, all of whom were transported to the newly discovered continent. The move also meant a virtual end of the less dynamic civilizations that already existed in the New World.

At the time of the discovery, there were large differences within the human population in terms of attitudes, skills, motivations, energy, and determination. It is not surprising that the strong, the ones with broad vision, the ones skilled in the crafts and techniques of the day, came to dictate the direction and the structure of these new developments, and that the rest were simply dragged into it, serving in inferior positions and functions.

The explorations and discoveries centuries ago generally came as surprises that fell upon humanity unexpectedly, and there was no way to make preparations to meet the challenge and to avoid the pitfalls of disparity and inequity.

Today the situation is different. We have advance warning that we are about to embark on an expansion of our horizon, on an enlargement of our living space—for a generation of pioneers—and we know that we still have a number of decades to prepare for it. We must therefore take advantage of this foresight and see to it that the circumstances that brought about the disparities during the last pioneering period are not reproduced.

Again, we will need new skills, crafts, technologies, and knowledge to be successful pioneers, but this time on an even more sophisticated level, demanding even more from those who participate. The new age will demand an equally significant broadening of our attitudes, a major psychological widening of our view of the world and our environment. The pioneers experienced strange plants, strange animals and strange natives, but the same sky, the same day and night, the same air, and the same soil. This time, it will be different.

We must, therefore, see to it that as many of us as possible are prepared for this challenge, and have the opportunity to undergo the growth and development needed to be at least potential pioneers.

In this process of broadening science will undoubtedly play a crucial part. Many of the specific skills, crafts, knowledge, and understanding will be science-based. That does not mean, of course, that only PhD scientists would survive in a space colony. On the contrary, life there will be specifically beneficial to the layman who has a variety of interests and capabilities. Nevertheless, even such laymen would have to count on possessing an enhanced amount of scientific knowledge and understanding in order to utilize the new environment optimally.

Science will also play a significant role in forming the new spirit needed to thrive in the Newest World. The belief that man, through knowledge, can control his environment and can affect his fate, is the foundation of modern science and was created by the development of modern science. The problem-solving attitude of science will be pivotal in the development of a new environment.

Life in space colonies will not be coldly scientific. On the contrary, people will gain added opportunities and color in terms of an almost unlimited spectrum of channels through which our non-material aspirations and interests can be satisfied. But in addition to such rich outlets for non-scientific activities, the scientific side of our being will have plenty of incentive to come to the fore to enhance the quality of life.

If science is likely to play such a pivotal role in the preparation of the new pioneers, and if we want to avoid creating potential disparities when the new migration begins on a large scale, we must assure that science as a craft and source of knowledge and understanding, as well as science as a cultural force, is fairly evenly shared among the various groups of humans and the different countries of the world. Unfortunately, this is not at all so today.

Some 95 percent of the new scientific knowledge is created by scientists who come from countries representing a mere one-quarter of humanity. The other three-quarters are passive spectators in this development, catching the material morsels of science only occasionally and partaking in the cultural impact of science only in most superficial ways, and then at second– or third–hand.

Some of the causes of this disparity are not difficult to understand. The scientific revolution started in Europe, and people descending from Western civilization therefore have been living with modern science for at least 200 years. In the rest of the world, modern science is at best decades old, and the scientific infrastructure is still weak and small.

Thus, the worldwide development of science will take some time, since it is not easy to force the tempo of a broad cultural evolution. Nevertheless, those who are active in science and have been for some time must make sure that the rest of humanity shares in the treasure of science.

Aspects of science like science education, communication among scientists, as well as the transmission of science as a powerful force in the life of a common citizen require international cooperation. Instead, the fledgling scientific communities in the developing countries are very much isolated from their more fortunate colleagues, and international and national organizations continue to provide fish to the hungry instead of teaching them how to fish.

Thus, we seem to be preparing to repeat the past, either to leave the non-scientific cultures behind when we head for happier pastures, or to drag them along as menial laborers, bringing them into a world which they cannot comprehend and in which they have little chance to succeed.

It is still not too late. We still have half a century at the least, in which the present situation can be corrected. This appears to be the time span required for a country to proceed from rudimentary use of science to a self-propelled and productive scientific infrastructure, coupled with a general population that has managed to integrate science into its own culture. The year of mass migration could see pioneers from all around the world and not only from a few privileged countries.□

Preprint, published in the *Proceedings of the Symposium on the Future of the Arabian Gulf Universities* Bahrain, 1983.

THE WORLD IN THE 21st CENTURY --- THE 2020's IN THE CONTEXT OF EDUCATION

Michael J. Moravcsik
Institute of Theoretical Science
University of Oregon
Eugene, Oregon 97402
USA

Preface

Talking about the future is both risky and tempting. It is risky because Man's ability to foretell even the next few decades is very weak indeed. The weave of events is a very complex and very non-linear system, and unexpected discoveries propagate throughout this system rapidly and make a mockery out of feeble extrapolations of the present.

Yet talking about the future is also a temptation, in spite of the unencouraging precedents. It tickles one's fantasy, and it also appeals to a feeling that by forecasting certain trends for the future one might bring about self-fulfilling prophecies. Also, forecasting far ahead prevents one from being proven wrong in his lifetime.

Talking about the future at this meeting is, nevertheless, not a mere idle play but is done with a definite and serious purpose in mind, namely to take some concrete steps toward making education more effective on a broadly international level. For this reason, I will delimit my task in two respects.

First, I will aim at the future only some 40 years ahead. I strongly feel that forecasting farther than that is not functional because of the large uncertainties involved in the future more than a half a century ahead. For example, in the field of science and technology, even the scientific base of the technology that will govern the world 50 years from now is unknown. The second constraint I will impose on myself is to cover only predictions which have a direct relationship to education. One might claim, of course, that everything is related to education, but I will construe such relationships in a strict way.

My discussion is divided into two parts. First, I will list nine elements of the future that I feel are both important and closely related to education. The second part of the discussion will then draw seven conclusions from these elements which, I hope, will have a direct impact on our formulation of a specific set of activities to promote educational objectives on a worldwide basis, and particularly in the developing countries.

Nine Elements of the Future

As mentioned, this section deals with nine aspects of the future which have direct bearing on educational policy.

1. The Systems Approach

The first element is an epistemological one. Our efforts so far to learn about the world around us have been structured predominantly in a "fundamentalist" way, namely complex phenomena have been disected into fundamental and elementary components and we have strived to understand these components one by one. It was then assumed that once the components are understood, one can easily synthesize the complex phenomenon out of these components.

This epistemological method has been very successful in many areas of human knowledge. Almost all areas of mathematics, physics, chemistry, geology, etc. have so far been conquered by the use of this fundamentalist approach. Some progress was also made in this way in some areas of biology, and slight traces of progress might even be perceived in the so-called social sciences.

Nevertheless, in recent years, there have been increasing signs that in many areas in which we want to acquire new knowledge this fundamentalist approach needs to be supplemented by a different method which we may call the "systems approach". It is possible that the behavior of many systems is not just a simple sum of the laws governing the components, but the system itself has certain laws of its own which need to be combined with the component behavior in order to understand the system.

This general situation manifests itself in many ways in different fields and is described by various labels. In physics, we deal with strongly non-linear systems, or talk about strong-coupling interactions. In biology, we develop phenomenological ways of describing large systems. In the social sciences, the isolation of components continues to be a utopistic goal. There is also an increased awareness that even in the "hardest" of sciences, the interaction of the observer with the phenomenon being observed is an inseparable and essential part of the picture, and hence, in that sense, it is impossible to talk about an elementary and fundamental component. This is a complex and still nascent subject which could not be discussed here in detail, but for our purposes it is enough to say that the "systems" approach to knowledge shows a strong ascendancy. I expect this trend to continue at a rapid pace, and thus, by the 2020's, our scientific and scholarly pursuits should be very strongly influenced by this development.

2. Rapid change

As the science-technology chain becomes shorter and a scientific discovery becomes increasingly faster utilized in technology, as our means of communication and transportation become increasingly more rapid, significant changes in our lives will occur faster and faster. Thus, in a human lifetime (which itself is increasing in length) the world is transformed very much more drastically than it was the case for our grandparents. This is something that has been the subject of a large number of recent writings, ranging from psychological essays to popular books, so this phenomenon probably does not require elaboration.

3. Increased sophistication of employment

Perhaps the greatest benefit of technology for human beings is our liberation from routine labor and the manifold multiplication of our capabilities through machines and automation. At the beginning this applied mainly to physical work, such as pushing a cart, digging a ditch, drilling a hole, moving the needle when sewing. With the advent of more and more "intelligent" machines, a similar trend has also begun with respect to mental labor. Record keeping by computers, manufacturing by robots, keeping track of air planes by automated radar and piloting systems are just a few examples. Contrary to the expectations of the ludites, human employment has not been curtailed as a result of these developments. The nature of the employment that is available has, however, changed. The positions available and needed have increased greatly in sophistication, in the necessary education prerequisites, in the amount of knowledge needed to discharge work satisfactorily. By no means is this change complete yet, and many jobs remain which, although perhaps more comfortable than their predecessors, are still routine supervision of machines for which people can be trained relatively easily. In a way this is fortunate, since there will always be people who are incapable to hold sophisticated responsibilities. Yet, an increasing fraction of the labor force is placed in relatively sophisticated positions. I expect this phenomenon to accelerate significantly in the next 40 years, so that by the 2020's well above half the positions will require a degree of sophistication and knowledge which was needed, a century prior, only by a tiny fraction of the positions then available.

4. Several careers in a lifetime

Partly because of the more rapid changes, I expect that by the 2020's a significant fraction of the labor force will have more than one careers in the lifetime of the individual. There will be several factors bringing this about. Some vocations will simply become obsolete when the individual is 45, even though they were quite viable when the person entered the labor force at the age of 20. The proverbial fireman on a Diesel locomotive is a relatively trivial example, which also illustrates how labor unions cling to old-fashioned ways by striving for life-long job security for a person in an obsolete vocation. In some other cases the switch in careers can be precipitated by a personal desire for change, by a decreased enjoyment of a line of work maintained for 2-3 decades. In again other cases the prospects of rapid advancement and increased material rewards in new fields will constitute the stimulus. Such a switching of careers occurs already today, but it is relatively rare. It will, I expect, become more the rule than the exception 40 years from now.

5. The role of extraordinary individuals

I expect that as time goes on, the leverage, impact, and influence of extraordinary and creative individuals, in whatever field, will grow. As discoveries and inventions have an increasingly radical impact on our lives, the rare people who are capable of being the creators and implementers of such breakthroughs will also gain in importance. To say this in a different and perhaps less extreme way, I expect the quality of manpower to gain in importance as time goes on, and, in particular, the upper "tail" of the quality distribution will play a particularly crucial role.

6. Decentralization

I expect the world to be considerably more decentralized in the 2020's than it is today. I call a system decentralized when it consists of many units which, although in contact and cooperation with each other, can make independent and substantial decisions, can experiment and progress without being too much tied to others, and which thereby can contribute to an overall pluralism of ideas, methods, procedures, and achievements. I strongly feel that the bankrupcy of the hierarchial and centralized structures in administration, government, and production has been sufficiently demonstrated to precipitate a new approach, much more decentralized than the present. As mentioned above, decentralization does not imply anarchy or even a lack of cooperation, but a cooperation of a varied assortment of parts on an equal footing is quite different from one huge bureaucratic and operational hierarchy.

7. Raw materials less important

The world is increasingly influenced and dominated by products and services in which the raw material component is subordinate to the input in the form of brainpower. Although high-technology products often depend on "special materials", these are special because of the processing they require and not because the raw material out of which they are made is scarce or unevenly distributed. This trend is clearly evident, for example, in energy production, where the old-fashioned sources like wood, oil, coal, and hydroelectric are being increasingly replaced by fission-nuclear, solar, fusion-nuclear and other, yet to be invented sources, all of which are very much more independent of scarce raw materials than were the old fashioned methods.

This increasing freedom from scarce or unevenly distributed raw materials is a most powerful force in providing "equal opportunity" to all countries around the world when it comes to competing in economic development. I expect this factor to increase considerably by the 2020's. As we all know from numerous examples, equal opportunity is by no means tantamount to equal results, and there are and will be many other factors beside raw materials that will determine whether a country is successful in economic competition. Yet, the decrease of the importance of raw materials will shift the attention to the competition in developed brain power, in inventiveness, and in motivation.

8. Expansion into space

I expect that a hundred years from now a signficant fraction of humanity will live not on the surface of the earth but in space colonies. This evolution will not yet have reached a very signficant stage 40 years from now, but I expect to have some permanent large space colonies (with, say, 10,000 inhabitants each) in place by then, and the outlook into the future at that time will be strongly oriented toward planning for life in space colonies. Space colonies will provide an opportunity for a much greater variety of life styles, since the different colonies will be independent from each other and selfsufficient, but at the same time they will also raise the requirements in education and skills for the population.

9. A changing of the guard in world leadership

It is common to view the world of today as a division into two separate and unchangable groups of countries, the "developed" and "less developed" countries, with a huge gap between them which is thought to be unbridgable. I believe that the projection of this model into the future is completely erroneous, and that staggering changes in this respect will occur in the next four decades.

There have been definite signs in the last decade or two that the now leading countries have reached the zenith of their cycle of development and have entered the declining phase of their civilization. It is very difficult to diagnose such an event, since we mistake fluctuations for trends and vice versa, and because historians, well versed in the factual details of past human events, are quite incapable of giving us answers to the broad questions of history, such as why certain civilizations rise and fall. There are some insightful exceptions, such as the beautiful book by Kenneth Clark, "Civilizations". According to Clark, the decline is marked by feelings of fear, uncertainty, and purposelessness, and these result in a loss of morale, a loss of motivation to engage in novel undertakings, to take risks, to explore the unknown.

These symptoms have certainly been in evidence lately in the now leading countries of the world. Fear of diseases, of the environment, of the dangers of acquiring new knowledge, of risk taking, of novel explorations have been mounting, and these insecurities show up in a broad spectrum of phenomena starting from economic slumps to the crime rate, or from endless and pointless legal battles to a deterioration of interpersonal relations. How far this deterioration will progress by the 2020's is hard to gauge, but in a pessimistic scenario, these countries will, by then, be tied up in knots, their energy resources stifled by the halt in the development of hydro-electric resources so that the three-toed two-beaked yellow warbler could have its entire territory and by the halt in fission- and fusion energy development because of fears of unknown or unevaluable side effects, their raw materials locked up so that the affluent could have a playground in wilderness areas, the educational system ineffective on account of unenforced standards of quality in a quest of mindless egalitarianism, their labor force crippled because of an adherence to job security as the supreme criterion and hence because of a lack of opportunity to modernize, their industry and productive sector paralyzed by bureaucratization both by the government and by industry itself, their transportation system in shambles because of an obsession with "absolute" safety, and their governmental coffers overstrained in the support of an increasingly helpless population which demands more and more but gives less and less,

These predictions may appear too severe today, but since the effect is a psychological one which is highly non-linear and can spread like an epidemic, the above picture is certainly a possible one for 40 years from now.

I therefore forsee a definite shift in the leadership in the world, in as much as a different set of countries may emerge as leaders 40 years hence, many of which may very well be among the "less developed" countries today. There is a crucial "if" in this expectation, however. Some of the now trailing countries can rise to world leadership in 40 years <u>if</u> they manage to transfer, adapt, absorb, and utilize the educational, scientific, technological, and cultural legacy of the now leading countries <u>without also accepting the decadent value system in which this legacy is enveloped</u>. This is a most difficult assignment. At the present, the developing countries, in their interaction with national, regional, and international assistance and cooperation agencies, receive both valuable ingredients for their own development and the distorted value systems, the subjective points of view which constitute the context in the developed countries for those ingredients. To separate the two, to take the first and reject the second is supremely difficult, though it is not without precedent in history. Whenever "barbarians" invaded a highly developed but decadent civilization, they generally managed, without being conscious of this, to learn from that civilization and at the same time not catch the malaise of decadence. Perhaps this can occur again, even if the improved tools of communication work against such a separation.

Seven Consequences for Education

The nine elements of the future described in the previous section carry some consequences for education in the coming decades worldwide, and particularly in the developing countries. Manpower is perhaps the foremost ingredient in the strength of countries, and education is the creation of an appropriate manpower. Thus an educational system must be tailored to respond to the needs some decades hence--a difficult and challenging task.

A. Education as a potential-builder

Since the epistemological basis of knowledge will signficantly change, since changes in general will accelerate, since jobs will become increasingly more sophisticated and will use new knowledge, and since a given individual can be expected to have more than one careers during his lifetime, education must be increasingly broad, aiming not at "vocational" training but at developing a potential for further and continuous learning. In particular, two presently very common defects of educational systems, namely rote learning and premature specialization, must be deemphasized in the various educational systems. The attitudes and expectations of students should be shaped accordingly, and the curricula be arranged to serve this goal.

B. Change of careers

Because the science-technology chain is shortening and hence rapid changes become the rule, since a person can expect to pursue several careers in his life time, and since drastic changes in our lifestlye (like the moving into space colonies) can be expected, â university must accommodate the lifelong education of people and must serve individuals who want to change careers. Special curricula and opportunities will be needed to accomodate such people, whose educational base is different from that of the regular students, and whose educational needs are also different. Special workshops, seminars, short courses, and conferences will be needed for those who want to catch up on developments in their fields, supplemented by introductory courses for those switching fields. By 2020 I would expect about an equal number of "new" students and "old" students at a university.

C. The role of inner motivation

Since jobs will become more sophisticated and hence more strenuous to prepare for, since extraordinary individuals will be even more valuable in the future, and since the full utilization of brain power rather than the possession of raw materials will be the deciding factor in success, the educational system must place a very great emphasis on the identification, encouragment, and education of people with inner motivation to get involved in a field. Among the various forces influencing the career choice of young people, perhaps the strongest and the one most difficult to destroy is an inner motivation, that is, an innate interest, curiosity, and esthetic affinity. It is not easy to spot people with such motivations, but thorough personal contact with young students usually reveals them. Today's internally motivated student is quite likely to turn into the exceptional individual later on who makes a gigantic imprint on his domain of interest.

D. Self-reliance and self-education

Virtually all the elements mentioned in the previous section play a role here. In the future we will increasingly need professionals who are self-reliant, who can chart their own course of work, and who can self-educate themselves both during their university career and, even more importantly, thereafter. There are many measures a university can take in its educational methods and structure to encourage the development of these traits. Above all, the university must demonstrate to the student that knowledge is open, that we know yet very little of all there is to be known, that even the professors know only a very limited amount, but that there is a method of asking new questions and finding new answers which can constantly increase our understanding. An important element in giving this impression at a university is the constant involvement of the faculty in scientific and scholarly research, so that when they talk about the perpetual quest for new knowledge, they do not only preach, but also constitute an example. This is one of the many reasons why higher education and research are inseparable.

E. "Interdisciplinary" education

The label "interdisciplinary" is, like most of the words we use, ambiguous. It is sometimes used to describe specific and often narrow problems of inquiry which happen to lie in between two traditional disciplines. The other and more significant meaning of the word describes a study which involves elements from a large number of the traditional disciplines, these elements being joined by the particular subject that is being investigated. Perhaps "transdisciplinary" would be a better description of this situation. It is my strong recommendation that such transdisciplinary fields be given an important role at any new university from the very start.

Specifically, in the area of science and technology, I would like to urge that educational systems include a separately demarcated Department for the Context of Science, performing research and offering education in the very new and fast growing transdisciplinary field sometimes called "the science of science", or at other times "science policy", which includes elements and representatives of many of the traditional disciplines, such as the history of science, the philosophy of science, the sociology of science, psychology of scientists, economics of science, policy making and management of science, etc. This new field which is both intellectually very exciting and practically most essential in a world of growing scientific activities, has so far been recognized only in a few universities around the world (usually the leading ones), so any other university now becoming involved in this field would have the advantage of being a pioneer in this investigation right at the start. Forty years from now, I anticipate, hardly any university will be able to exist without activity in this field.

F. Education for the gifted

Since the extraordinary people can be expected to play an increasingly important role in development, a university must cater specifically to the very gifted students. Special programs, "honor" programs, individual interaction with faculty through reading and research courses, extracurricular groupings of gifted students with common interest are among the ways to accomplish this. To spot such gifted students, assessment of entering students in depth will be needed. Programs for the gifted should receive formal recognition in the university's adminstrative structure, perhaps in the form of a dean for programs for gifted students.

G. An indigenous educational system

It is of course true that forming an indigenous educational system requires significant financial resources in order to establish an infrastructure that would make it indigenous and independent. Money, however, is by no means the most important aspect of independence. The most important goal for the independence of the new university is along intellectual, scientific, and scholarly lines. As I mentioned in the previous section, knowledge transferred from other parts of the world usually comes with a coating of values, preferences, and psychological elements. Only a faculty of outstanding, penetrating, critical, and independent minds can assure that the new university does not simply parrot what other universities teach, but that the new institution develops its own intellectual tradition and integrity.

It will take a considerable amount of time, perhaps all there is between now and 2020 to achieve such a status. Great universities never sprung up overnight. In the meantime, very careful selection of faculty is needed to assure the eventual attainment of this goal. A fatal error is to compromise quality at the outset, in the hope that this can be corrected later. It cannot be. It is much better for the new university to evolve slowly, perhaps even slower than anticipated, but to maintain high quality at all times. Assessing quality is not easy, and will involve peer judgement by the worldwide scientific and scholarly community.

In the formative years of a new educational system it is likely that the new universities will have to rely, in part, on engaging faculty from outside the country. Indeed, such influx of expertise is desirable for the initial momentum of the university and for setting high standards for it. Five-year contracts for such faculty are generally sufficient for effectiveness and yet are not too long. As time goes on, and nationals from the country return from abroad with an advanced degree and with professional experience, the indigenization of the educational system can gradually proceed. Indeed, most great universities throughout history had international faculties thus injecting variety as well as excellence. The criteria for selecting faculty must rely more on quality and on independence of mind than on national origin.

Epilogue

It is probably evident from the above discussion that, from the point of view of the presently scientifically less developed countries, my forecast for the future is a very optimistic one. I claim that the competition in the future among countries will be decided not by who has how much raw material, but by who can develop its brain power best. I also predict that the presently less developed countries will not have to contend with the large gap that now exists between the leading countries and many countries in the Third World, because by the 2020's many of the leading countries will have declined substantially. Thus the opportunity exists for new countries to assume a leading position in the world.

D. The Personal Angle

Communication & Cognition
Vol. 13, No 2/3 (1980), pp. 237–248

WHAT MOTIVATES THE DEVELOPING COUNTRIES TO DO SCIENCE AND TECHNOLOGY ?

Michael J. Moravcsik
Institute of Theoretical Science
University of Oregon

Motivational research is notoriously tricky. No matter what methodology is followed, it is subject to serious and valid criticism. Perhaps the simplest method is to ask people about their motivations. That this might not be too reliable is due to the fact that people's knowledge of themselves is imperfect, and what people say and what they actually do in concrete situations might therefore be quite different. Another method is to confront people with hypothetical situations and ask them how they would behave. This is similarly unrealiable since people's professed action in hypothetical situations might be quite different from the way they would act when placed in a real-life quandry. A third way of proceeding would be to record which motivation is most expressed in controversies and debates arising in actual practice. This method suffers from the shortcoming that certain very deeply held motivations may never become challenged from the outside and hence never appear in public controversies. For example, scientists experience an esthetic thrill when making scientific discoveries, but since thrill cannot be taken away from the outside, it has no reason to appear in conflicts between scientists and the rest of the world, or between one scientist and another.

In view of this lack of a reliable methodology for motivation research, it is not unreasonable to fall back on the traditional process of gathering "wisdom" in a problem, and simply explore the personal, intuitive, accumulated experience of people who have been, for some time, in contact with the situation to be investigated. Indeed, in a wide variety of human areas of interest, such informal sources provide most of the knowledge and information that exists.

It is in this vein that I am offering my view on one of the most fascinating and crucial motivational questions of our age, namely why there is such a worldwide movement on the part of the developing countries (constituting three-quarters of the world's population) to participate in scientific and technological activities. My qualifications to do so are an active and close involvement in this problem for almost 20 years, and, as part of it, contact and friendship with hundreds of scientists and technologists from the developing countries located over all continents. I have had the good fortune of being able to travel in many of these countries and experience things first hand. In addition, I have also had the privilege of knowing many other scientists and technologists from the developed and the developing countries who have also been involved in this problem are for a long time, and who have also traveled widely. Their experience and opinion was folded into mine during many interesting discussions and debates, and what I am suggesting in this talk is therefore a synthesis of all these influences.

The first pivotal point one has to make is that motivations are extremely rarely one-dimensional. The pluralism of causes of most human activities, actions, motivations, and aspirations is a fundamental law of the world. This is not surprising since the same also holds for the non-human domains of nature. In fact, one of the most difficult tasks of the "scientific method" is to create conditions for scientific experiments in which the usual pluralism of causes is reduced to the one single cause one wishes to study in that experiment.

In view of the fact that pluralism, multidimensionality is such a pervasive feature of our world, it is astounding that when people attempt to analyze the causes of some event, or the motivations for some occurrence, they almost invariably lapse into "one-dimensional" thinking and search for *the* cause, *the* motivation. This way of thinking frequently gives rise to various kinds of "conspiracy theories" of the world. "If only the oil companies could be brought under control, the energy crisis would vanish", goes the saying, or "If only the labor unions could be eliminated, inefficiency and featherbedding in our technological age would be a thing of the past".

In this talk, therefore, I will consistently maintain the attitude that motivations for doing science and technology are a complex, multidimensional set of aspirations, and correspondingly science policies aiming at the evolution of science in the developing countries must be equally pluralistic. This point is very much worth emphasizing since until recently the prevailing view in the developed countries and also in international agencies has been that the only and sole valid justification of science and technology in the developing world was the short term improvement of the most basic material conditions of life : Food, health, shelter, etc. The picture has now begun to change, and a more realistic and sophisticated conceptual framework is beginning to emerge which comes closer to coincide with the viewpoint I will take in this talk.

Briefly, my train of thought will be as follows : First I will summarize the motivations that *generally* exist in the world to undertake research and development in science and technology. Having such a summary at hand, I will explore the extent to which the developing countries feel in tone with this list of motivations, and will try to establish an order of importance among these motivations as seen from the vantage point of developing countries. Finally, I will try to draw some policy conclusions from the previous results.

So let us begin with a summary of the various motivations one encounters for pursuing science and technology. There are a multitude of these, reflecting the great heterogeneity of viewpoints among various professions and various personal convictions.

Not only are these motivations multitudinous, but they are sometimes conflicting with each other also. The detective and the criminal might both rejoice in the new possibilities scientific and technological developments offer, even though they have rather opposite points of view. The doctor may like science and technology because they make the art of healing more effective and more systematic, while the scientist himself may enjoy the esthetic pleasure of learning about new ideas and is not particularly impressed by the effect of all this on medical activities. The general, on the other hand, values science and technology for producing better weapons. Patriots may look upon science and technology as excellent areas to exhibit national superiority, while others may stress the international aspects of science and technology and their influence in helping to erase national differences. One could go on for some time with such an itemized listing of the attractions of science and technology for various types of people.

All this, however, can be classified and synthesized into three broad areas in which science and technology affect us and hence in which they generate motivations for us to pursue them.

First, science is the basis of modern technology and modern technology has made our lives longer, easier, richer, and broader in a large variety of ways. This point is generally fairly well recognized, and in fact this kind of impact of science and technology is the one that is first perceived by a casual or superficial observer. It is generally acknowledged by now that while technology in the "old days" was mainly based on empirical trial-and-error and hence was to a large extent divorced from science, in the last 150 years or so both science and technology have begun to explore and utilize phenomena that are not directly observable through our human senses of hearing, seeing, touching, tasting, etc. and which pertain to ranges of the physical variables (length, time, temperature, mass, etc.) which are very different from those of human beings. This exploration of the new domain cannot be done any longer by "trial and error", but instead technology must use knowledge developed by science as a basis for development, inventions, and processes. Furthermore, more and more of the important items of production are characterized by a large dose of incorporated know-how compared to the small amount of raw material needed for its production. It is in this way that countries poor in raw materials, like Japan, Singapore, or Switzerland, can forge ahead into the ranks of the most effluent and, for their size, most influential countries in the world.

Since this first motivation for doing science and technology is so well accepted, we can quickly turn to the second one, less obtrusively evident but no less important for that : Science and technology are human aspirations in the 20th century.

Aspirations have been the hallmark of all ages, all civilizations, and all individuals, and are the driving froce for all of us. People or peoples who have lost their aspirations, who cannot see a purpose in their existence, who are void of aims and directions to work toward soon die, either in fact or for all practical purposes. The type of human aspirations vary from person to person, from group to group, fro age to age. Many of these aspirations are material, involving bodily gratification in food, shelter, luxuries, sex, clothes, and many other items. An even larger number of aspirations, however, are non-material, having to do with religion, esthetics, interpersonal relations, political power, conversion of talent into achievements, ideology, and a myriad of other facets of our mind and soul. throughout history such non-material aspirations loomed high and served as one of the primary or dominant motivations for people to act. As an example, a look at wars or near wars presently existing between countries reveal in each case that *a*, or perhaps *the* dominant cause lies in a conflict between the non-material aspirations of the feuding parties.

The inner urge to explore nature and to use this new knowledge for the broadening of our horizon and spectrum of actions is definitely a strong aspiration in the 20th century. It was not always such a predominant aspiration. For example, in the Middle Ages in Europe this kind of activity was not at all foremost in people's minds, who instead found pleasure, delight, fulfillment and selfrealization in other kinds of aspirations. But in the present century, science and technology do play such a prominent role. That this is so is well indicated by a multitude of elements in our lives : Popular interest in the latest discoveries of science, fascination with technological feats like traveling to the moon, the great popularity of science fiction, or the fierce competition among nations in areas of scientific research and technological development even apart from practical utilization. The race to accumulate Nobel Prizes, the national pride in technological excellence all suggest that being a "great" country is seen to include being "great" in science and technology.

To the narrow minded economist such a preoccupation with "useless" feelings and stimulants often appears incomprehensible or at least despicable, even though a quick look into the psychological mirror would convince our economist friend that he himself, as a human being, is also predominantly driven by such "useless" forces. Indeed, from a certain point of view, one could deplore the existence of non-material aspirations, since a world entirely catalyzed by material considerations would be a much easier arena to adjudicate and negotiate in. It is relatively easy to arrange "compromises" if only material stakes are at issue, but even the idea of a "compromise" may be totally repugnant to somebody with strongly held religious, ideological, national, and other non-material aspirations.

In any case, the aspirational weight of science and technology constitute motivation for undertaking science and technology. Having established this, we can now turn to the third and last major area, namely the motivation for doing science and technology because these activities have a profound influence on Man's view of the world and of himself.

If you live in what in a cultural sense is called the Western civilization, it is by now hard to be fully aware of the impact the scientific revolution had even on the "common man" in this civilization. One must read a vivid and "true-to-life" story of life in the Middle Ages, of people's conceptions, mentality, attitudes, and beliefs, in order to grasp the enormous change that has taken place in the Western civilization as a result of the scientific revolution. Today even the person with little if any formal scientific educational would show signs of having been influenced by this revolution. His attitude toward change, toward the way we acquire new information about phenomena, toward the extent to which Man can influence his fate, and toward the possibility of turning aspirations into reality would all bear the hallmark of tenets of the "scientific thinking", even if he is unable to grasp the law of gravitation and does not know what a logarithm is. I believe I do not have to dwell on this point before an audience sensitized to the theory of knowledge.

At the same time, it is important to emphasize, even to such a sophisticated audience, that these attitudes are *not* universal around the world, and that much of the world's population, for better or worse, has not been significantly touched yet by the scientific revolution as a cultural force, even though transistor radios or motorcycles may be ubiquitous among them. It is also important to realize that some 400 years ago, when nobody in the world was yet affected by the scientific revolution, the differences in material standard of living as well as in world outlook and attitudes between, say, Western Europe and the rest of the world were very much smaller than they were 300 years later and than they are even today. Indeed, the huge gap between the "advanced" countries and the developing countries which exists today is a relatively recent phenomenon and can be traced directly to the scientific revolution. When confronted with such a situation, one becomes acutely aware of this third justification for engaging in science and technology, namely their impact on Man's view of the world and of himself.

In the advanced countries these three major areas of motivation for doing science and technology carry roughly equal weights. Whether this is the case also in the developing countries and from their own point of view is our next topic of discussion.

The impression one acquires after extensive contact with this problem from the vantage point of the developing countries themselves is that the third type of motivation does not play a very large role. This is not altogether surprising. A very large fraction of the people in the developing countries, even including many decision makers, having themselves not been basically influenced by the scientific revolution as a cultural force, cannot appreciate its power in molding individuals. To be sure, one hears complaints from time to time about how the population in the country is "backward in its thinking", "not ready to participate in development activities", etc., and such opinions are frequently offered by people from the developing countries themselves. Very seldom have I heard, however, somebody connecting such complaints with specific causes and, even less, to specific remedies.

In particular, educational systems in the developing countries seldom if ever emphasize science and technology as a cultural force, and practically never would a student in a developing country be exposed to a discussion of the methodology, the conceptual foundations, the attitudinal innovations of science and technology. On the contrary, science and technology education in developing countries tends to be extremely narrow, focused only on those pieces of factual knowledge which are, rightly or wrongly, thought to be directly connnected with the designated activity and job description of the person to be educated. Unfortunately, such a narrow focus in education does not even result in functionality within a limited scope, since while the syllabus is thus focused, the method of teaching is far too much oriented toward rote learning to be useful for problem solving.

That science and technology as cultural forces are not well recognized from within the developing countries need not mean, however, that this aspect of science and technology should not play a major role in motivating international assistance by the advanced countries. Indeed, there appears to be a steadily growing realization in the international development community that if the countries receiving such assistance are to be helped toward a self-sufficient state in which they have their own internal capacity to acquire, generate, and utilize the science and technology they need, then a true scientific revolution must take place in those countries to provide the roots and foundations for such self-sufficiency. The meetings auxiliary to the recent United Nations Conference on Science and Technology for Development, which took palce recently in Vienna, showed clear evidence that after many years of unsuccessful groping in the dark, this pivotal point in development philosophy begins to emerge.

This leaves us now with the other two main kinds of motivations for science and technology, namely their material effects and their serving as human aspirations. Are these two important in the eyes of the developing countries ?

Conventional wisdom has it that the first of these motivations, namely the material and economic considerations, are predominant in the developing countries, and that the second type plays a negligible role. It is often added that this is indeed rightly so, because the developing countries must first close the material gap between themselves and the advanced countries before they have the "right" to be concerned with "luxuries" such as non-material aspirations.

What evidence is there for this conventional wisdom ? Some of it simply comes from the condescending attitude toward the developing countries that some people nurture (more confidentially than openly). But if that were the only evidence, one could dismiss it immediately. In fact, there is extensive additional evidence that appears to support the contention that in developing countries the main motivation for doing science and technology is the desire for material, economic improvement.

For one thing, national development plans handle science and technology exclusively as tools toward material, economic progress. This appears to be a weighty piece of evidence until we remember that such development plans are invariably written by economists and hence are naturally documents for only economic development. I have yet to see a national development plan which seriously addresses itself to development in the full sense of the word, including cultural, conceptual, spiritual, and general human development. Thus the fact that science and technology are justified in development plans in terms of economics is merely a piece of tautology and not evidence.

One might be inclined to take, as another piece of evidence, the fact that the overwhelming fraction of research and development expenditure in developing countries is devoted to institutions carrying the proud names of Natural Agricultural Institute, Federal Research Centre for Forestry, National Applied Scientific and Technological Laboratory, and alike. It would appear indeed as if the material applications were the sole justification for spending on science and technology.

On second thought, this evidence also fades considerably in the degree to which it can convince us. for one thing, one remembers that in *all* countries around the world, whether "advanced" or developing, the lion share of funds set aside for science and technology is spent on technological development, and the second largest share on applied scientific research. These two consume, in virtually all countries, at least 80–85 % of the total research and development funds. Then, it may also occur to us that the shiny names may not appear only, or not even primarily, in order to identify the institution with economic, material objectives, but to symbolize the involvement in science and technology by the country so as to satisfy her aspiration to join this central human activity of the 20th century.

One could cite many other examples, but by now it should be evident that my own personal view tends toward doubting that the economic, material motivation is the predominant element in the raison d'etre of science and technology from the vantage point of the developing countries themselves. Indeed, I now want to offer evidence that much stronger than such material motivations is, in the developing countries, the desire to do science and technology in order to become, thereby, an equal to the "advanced" countries in the practice of an important preoccupation of the 20th century.

To support this view, let me first describe what to me appears to be the central element in the state of being "underdeveloped", as viewed by the developing countries. It is not primarily the matter of having a per capita gross national product of $ 200 instead of $ 7,000. Much more pervasive and burdensome is the fact that the developing country is, in virtually all respects, completely dependent on outside influences and forces. Whether we consider cultural, economic, political, scientific, or technological matters, a developing country has little if any choice : The innovations, new trends, influential decisions and activities are all generated abroad, in the advanced countries, and the developing country, to an overwhelming extent, is simply a puppet that must respond to what occurs in other parts of the world.

Total dependence, a frustrating existence without self-expression is thus the most debilitating and humiliating aspect of being "underdeveloped". In previous times this was formalized, for some countries, in an actual colonial status, but even for countries that were never colonies, the situation was hardly different. Today the age of colonies has almost completely passed, but erasing the name did little to alter the deeper causes of the situation that had brought about colonial status. These deeper causes have to do with the huge gap that developed between countries since the dawn of the scientific revolution, a gap we discussed earlier. This gap cannot be eliminated by purely political action, and there is no United Nations, no amount of resource transfer, no New Economic Order that can alter it in the short run. It is the basic aim of the slow and long term process called development to bring about the fundamental changes in the developing countries themselves which will then "liberate" them from this subservience.

It is therefore hardly surprizing that as the Third World walks along the slow road of development, it constantly looks for signs of progress, for indications of having narrowed the gap, for evidence that at least in some respect the country is now on an equal footing with the more advanced countries. Indeed, such signs offer encouragement, a proof that progress is possible, and a pride in what has already been accomplished. Signs might be found in all walks of life, whether in a hockey match or a Nobel Prize in literature, in a medal at the olympics or perhaps in a mountain climbing feat. In some cases it might also be in an industrial achievement, perhaps a successful instance of competition with one of the leading countries in some special branch of production.

Science and technology, in many ways, offer a promising area for such signs of progress. For one thing, modern science and contemporary technology by their very nature symbolize being in the forefront, being among the leading countries of the world. Also, while it is unlikely that a developing country would, in the near future, be in the position of competing with the leading countries in, say, steel production or in computer design, it is quite possible for such countries to make contributions to science which are noted and recognized worldwide.

But in fact, the making of actual contributions to science and technology is not even necessarily needed for bringing about encouragement or high morals. For that even the externalities of science and technology might be sufficient : A National Science Council, a "science policy", shiny new buildings with large letters announcing its being a "research center", etc. Thus we find that the motivation for doing science and technology, or at least for performing the externalities of doing science and technology, could very well come from this second general justification of science and technology, namely that they constitute a human aspiration in the 20th century countries on an equal footing.

So far this is a mere speculation. What are the actual pieces of evidence that would support such a contention ?

There are quite a few. Perhaps the most striking one is the fact that in developing countries extremely rarely does one find an attempt to evaluate or assess the scientific and technological work already performed, and to measure the extent to which they have contributed to the economic state of the country. Development plans are silent about it, funding agencies are not interested in doing so, and even within the scientific and technological community itself there is little activity of evaluation or assessment. It is indeed difficult to avoid the strong impression that the main objective is *to be involved* in science and technology and not to produce results in these areas.

There are many other signs that science and technology are valued in the developing contries for their symbolic significance more than for their functional contributions. Formal organizational manipulations (forming "National Research Council"'s, planning rather than making decisions or implementing, etc.) take precedent over functional arrangements. Scientists and technologists often scurry into the flashy and high-profiled administrative positions instead of remaining in the laboratory or at the bench to do research or development. New and shiny equipment (often far too complex for the task from a purely functional point of view) is purchased with gusto and then is left unused, on exhibit. When searching for scientific visitors, elderly Nobel laureates are heavily favored over those who are presently the front-line contributors to the scientific areas of interest. In selecting personnel, formal qualifications like degrees are given weight over the *de facto* functional potential. Governmental technological institutions often select research projects for showiness and not for easily transmittability to local users. The list is indeed endless.

This state of affairs is sometimes well recognized by observers in the advanced countries and is considered by some of them an annoying situation which proves the incompetence and lack of development in those countries. This view might, however, be on a completely wrong track, since it assumes a motivation for doing science and technology (namely to produce knowledge useful in economic activities) which may not be the dominant one in the developing country itself.

Naturally, I do not mean to say that developing countries have the *sole* motivation of wanting to appear on an equal footing with the scientifically and technologically leading countries. Here as in all other situations, pluralism reigns, and the overall motivations are composite and complex. All I wish to stress is that in the developing countries one of the dominant motivations for engaging in science and technology is one that is often completely overlooked or disparaged in developmental circles.

Assuming this to be the case, what policy consequences can we derive from it ?

First, the "advanced" countries should openly recognize the "legitimacy" of this type of motivation, and cease claiming that only the rich has the right to the "luxury" of activities undertaken for other than short-term material reasons.

Second, advanced countries should try to influence the developing countries in the direction of *combining* the externalities of science and technology with productive accomplishments in these areas. After all, there is no basic conflict between being proud of the externalities of indigenous science and technology and being also productive and creative in the essence of science, technology, and their applications. Thus international assistance should offer cooperation for the full spectrum of scientific and technological activities the developing countries wish to undertake, and let the motivations for these activities be chosen and judged by those countries themselves.

The "advanced" countries nowadays are under heavy political attack by the developing countries who claim that they are not understood and listened to. if what I tried to suggest in this talk is true, it shows that, indeed, there are some basic misunderstandings between people in the "advanced" countries and in those of the Third World. The matter therefore deserves further investigation and exploration by scholars and policy makers alike.

Institute of Theoretical Science
University of Oregon
Eugene, Oregon 97403 (USA)

THE SCIENTIST IN A DEVELOPING COUNTRY AND SOCIAL RESPONSIBILITY *

MICHAEL J. MORAVCSIK

SUMMARY   Instead of the customary discussion of this topic in the form of a model sermon, an "operational" analysis is offered. I ask the questions "How does an observer determine whether the behavior of a scientist is in accordance with social responsibility or not?" and then divide this into two steps: 1. The observer determines the objectives of society. 2. The observer compares the scientist's behavior to see whether it is in accordance with the results of step 1 or not.

The first step must be looked at along five dimensions. First, the scientist belongs to many societies simultaneously, and so we have to decide the objectives of which society we are concerned with. The problem is, of course, that the objectives of the various societies the scientist belongs to are likely to be in conflict with each other. Second, which observer should we have in mind? The scientist is judged by a great variety of people, all with different values and different ways of relating to science and the scientist. Thirdly, responsibilities might vary drastically depending on whether we focus on short term or long term considerations. In practice, short term perspectives often win out, but not necessarily because they are more important. Fourthly, there is the choice of broad problems for which the scientist is not fully trained versus the specific problem areas in which the scientist can fully use his competence and background. This aspect touches on the question of the involvement of scientists in politics, which is sketched in the Latin American context. Finally and fifthly, even in a given category along the above dimensions, there are many different challenges and the scientist cannot face all of them equally successfully.

The second step is then analyzed along four dimensions. First, should we judge a scientist on what he says or on what he does? This is the old problem of ideals versus practice. Second, how can we construct a reliable measure for the output of a scientist? This is a general conceptual and theoretical problem in the science of science which is yet to be solved satisfactorily. Third, a certain action can be prompt-

ed by a variety of motivations, some conflicting with others, and so the question arises whether the scientist is "socially responsible" or not if he does the "right" things but due to the "wrong" motivations. Finally and fourthly, social goals and individual aspirations may not be mutually exclusive and incompatible, and hence we might wonder whether a scientist should be called "socially responsible" when he simply does things to satisfy his personal, individual aspirations which happen to coincide with those of society.

This apparently theoretical analysis has practical utility since it turns a fuzzy question into a well-defined one, it reminds us that the simple form of the question is meaningless unless narrowed down through specifiers and qualifiers, and warns us not to make sweeping, collective judgments but evaluate each case on its own merit. These guide lines can be extremely clucial in public discussions of this subject, even though this article so far provides only a framework for the analysis of the social responsibility of scientists. To contribute to the discussion of the substance of this problem also, four elements are suggested which are likely to be valid no matter which of the above discussed dimensions is chosen as a basis. These four elements are as follows.

First, the responsible scientist strives for the highest possible degree of scientific expertise and understanding. Second, in his public pronouncements, he tries to separate his views given as a scientist on aspects of the problem which are scientific, from statements he makes in non-scientific areas as an amateur. Third, a responsible scientist realizes that working on interdisciplinary problems involving also non-scientific components requires a thorough retraining on his part, which consumes years of time and much effort, and hence he assumes a modest and humble attitude toward such problems. Finally, he is willing to admit publicly if he was wrong in his previous stands on some public issues, just as he would have to retract his views on scientific issues in the face of contrary experimental evidence.

Customarily articles on this subject are written in the format of a moral sermon, containing an exposition of what ought to be done by scientists and how scientists should respond to the responsibilities society imposes on them.

Since that approach is by now well represented in the literature, I will offer a completely different type of an analysis of this problem area, one that might be called phenomenological. Assume that we have a scientist and an "observer". How does the observer de-

termine whether the behavior of the scientist is in accordance with social responsibility or not? This is the "oper-

* Written for the NGO (Non-Governmental Organization) Forum: Science and Technology for Development, Vienna, August 19-29, 1979.

Michael J. Moravcsik is Professor in the Department of Physics at the Institute of Theoretical Science of the University of Oregon. His research in theoretical physics has been concerned with elementary particles and nuclear problems. For the past 15 years he has also been involved in problems of science building in the developing countries, and is the author of the book "Science Development", as well as over 160 articles in theoretical physics, science policy, and the science of science. Address: University of Oregon, Eugene, Oregon 79403, U.S.A.

STEP 1: DETERMINATION OF THE SOCIETAL OBJECTIVES

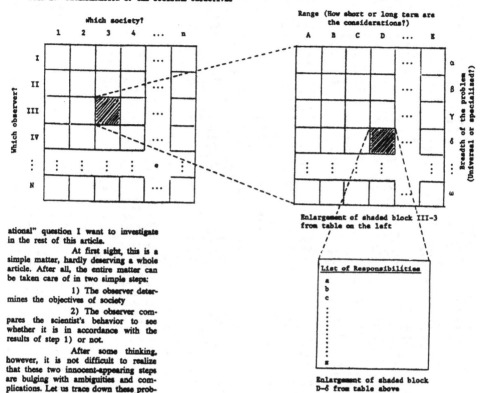

Enlargement of shaded block III-3
from table on the left

List of Responsibilities

a
b
c
:
:
z

Enlargement of shaded block
D-δ from table above

ational" question I want to investigate in the rest of this article.

At first sight, this is a simple matter, hardly deserving a whole article. After all, the entire matter can be taken care of in two simple steps:

1) The observer determines the objectives of society

2) The observer compares the scientist's behavior to see whether it is in accordance with the results of step 1) or not.

After some thinking, however, it is not difficult to realize that these two innocent-appearing steps are bulging with ambiguities and complications. Let us trace down these problems one by one.

Which society, which observer?

For a start, let me analyze the first step, that of the observer determining the objectives of society. There are five dimensions here that deserve to be pointed out.

First, which society? A scientist in a developing country belongs to many societies at once, the interests, objectives, values, and aspirations of which may be quite different from each other. To list these in an order of increasing size, the scientist has a family which has its own interests and aims. For example, the welfare of his children is a societal matter, and whether these children will be able to live a productive and contented life is of societal concern. In many developing countries the concept of a family goes much beyond spouse and children, and includes dozens

or scores of other people, to whom traditional obligations exist through a network of family ties which constitute an important element in societal fabric and values.

The scientist, at the same time, is part of the society of the town he lives in, the region he belongs to, the ethnic group from which he comes, the religious environment he was raised in, the country he is a citizen of, the subcontinent or continent he is located at, and finally of humanity as a whole. All of these groupings are societal and contribute in important ways to the workings of society, and yet there may be an infinite number of conflicting aspirations, values, responsibilities, and activities within the overall conglomeration or these groupings.

Then the scientist is part of the scientific community, and

this membership also exists on various levels: His place of work, his local professional society, the worldwide community of specialists, etc. All these entities will impose some responsibilities on the scientist, and it is almost certain that some of these will be in conflict with each other.

Second, which observer? In a heterogeneous society (and after all, most societies are heterogeneous to some extent), the definition of what constitutes a responsibility will depend considerably on whose judgment we take. In many developing countries, the judgment which is often most prominent is that of the "government", but since abstract bodies like a "government" have a definite opinion only in principle and not in practice, we must further inquire as to who in the government is making the judgment: The President in a speech on Independence Day, the Minister of Science on the occasion of opening a new institute, the economist who writes the national development plan, the superior of the scientist who, for extraneous reasons, is mad at him, or perhaps the scientist himself? Furthermore, are we to use these judgments as made in 1955, or in 1975? They may very well differ drastically, especially in countries where ideology likes to make 180° turns. Or should we instead rely on observers abroad, perhaps in "neutral" universities of great excellence, presumably containing wisdom not available to the uninitiated? Or perhaps the United Nations documents are the proper source of comparison? The list is clearly endless and confusing.

Third, responsibilities might differ very much according to whether they focus on the short term situation or on long term considerations. Perhaps Newton should have been directed to worry not about the laws of motion but about food production in England in the 17th century which, undoubtedly, had left much to be desired. This dichotomy of short vs. long term perspective arises especially often in discussions of aims, objectives, and hence responsibilities. It is human nature to focus much more acutely on immediate, short term problems and visions, and as a result be confronted with recurring crises and challenges which appear unexpectedly because of the neglect of adequate long term preparation for it. It is most unfortunate that in the development strategy of the developing countries, as far as science and technology is concerned, short term technological manipulations take undue precedent over the long term preparation of the country's scientific and technological infrastructure, and hence the latter is perpetually found to be inadequate as new challenges appear. As a result of this unbalanced view, scientists with strong feelings of responsibility toward long term goals are often chided for not having social responsibility at all, simply because their main efforts are directed toward deeper and more long range developmental activities.

Fourth, there is the dichotomy of big responsibilities, the handling of which the scientist is not trained for and not particularly competent in, versus the preoccupation with more specialized problems for the solving of which the scientist has a unique or rare talent and capability. This is also a common point of conflict in various developing countries. An example of it is, for example, the often debated

question of the extent to which a scientist, say, in Latin America, should be involved in politics. One school of thought (especially in Latin America) would hold that the general political issues are so paramount as to eclipse any specific responsibilities which may conflict with them, and hence all scientists must be deeply involved in politics even though they may be poorly equipped to be effective in it. The opposing view, on the other hand, claims that in order to be able to build the country's scientific and technological infrastructure (a long and difficult process needing long term stability and support), scientists should forego political involvement in order to preserve their neutrality and freedom of action that will permit them to devote themselves to infrastructure building for which they have unique or rare qualifications. There is no "logical" way to resolve this conflict, though one can point out, in a pragmatic spirit, that one of the main reasons why Latin American science has been underproducing compared to its potential is the frequent and debilitating blows science and the scientific community there suffers through the instability of scientific manpower which, due to its political involvement, has to lose half of its membership every time the government changes.

Fifthly and finally, even in a given category of short or long range objectives, or general or special responsibilities, there are many different tasks, and it is impossible for a given scientist to cater to all of them. The weighting of these tasks, in order of priority, however, is a matter very much dependent on personal value judgments, and so ambiguities, accusations, and confusion are again possible.

How to compare behavior

Although the above five dimensions by no means exhaust the possible ambiguities inherent in determining the aspirations and objectives of society and hence the responsibilities of scientists, the list should suffice for demonstration purposes and so we can now turn to a similar analysis of the second step, namely the comparison between the behavior of the scientist and the responsibilities specified in the first step. Again, we find that there are a number of ambiguities and dichotomies to confuse us, of which I was to discuss four.

First, in judging the scientist's behavior, do we consider what he says or what he does? After all, most of us human beings are far from

being self-consistent, and hence what we say often does not harmonize with what we do. In some cases this may be for sinister and intentional reasons, but in the overwhelming fraction of the cases it is much more likely to be a matter of a gap between intentions and capabilities, between a future vision and de facto practice, or between an ideal and an imperfect realization of it. Should we then give credit to the scientist for what he says his intentions are, or should we simply judge him by his actual accomplishments?

Second, even if we decide to judge by accomplishments, it is far from trivial to assess the achievements of a scientist. The measuring of scientific output is a thorny theoretical question in the science of science and a vexing problem in practical science policy, so that even the best available knowledge used by the most knowledgeable and expert people provides only fragmentary and questionable information, at least in a contemporaneous sense (and after all, that is what we are usually interested in for evaluation procedures). Even less reliable is then the judgment of a ministry bureaucrat with little scientific background when, under political pressure, he is asked to "Evaluate" the activities of some scientists in the country. From the type of evaluations I have often seen in various countries (judging the success of the research project on the sole basis of whether the "progress report" written by the researcher himself contains the right key words or not), I feel strongly that accusations made on the strength of such evaluations are meaningless, and in fact harmful, and hence they should not be made at all if they cannot be done much better.

Thirdly, I want to point out that the motivation of a scientist to do something is one thing, and whether his activity is in accordance with some objective or not is a quite different thing. The point here is that a person's motivations are virtually always pluralistic, that is, a person usually has many reasons for wanting to do something. A scientist pursues science because he is curious, because he wants to convert his talent into accomplishments, because he likes to compete, because he is esthetically thrilled with the beauty of the subject, because he wants to help humanity, because he likes to be well paid, because he likes to get external peer recognition, etc., and all these motivations may and generally will coexist in the same scientist.

STEP 2: COMPARISON OF THE BEHAVIOR OF THE SCIENTIST WITH SOCIETAL OBJECTIVES

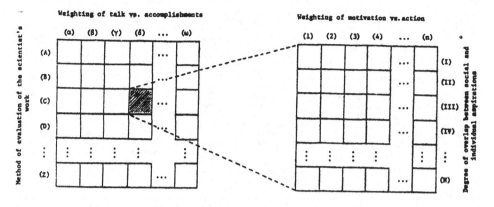

Enlargement of shaded block (C)–(δ) from table on the left

Conversely, a given activity will have many different motivations and justifications in a society. In fact, the most successful societal undertakings are likely to have many and often conflicting justifications, thus soliciting support from a wide variety of the members of the society.

Thus it might very well happen that from the observer's point of view, the scientist's activity is the "right" one as far as societal responsibility is concerned, but for the "wrong" reasons. For example, let us assume that a country is plagued with a river constantly changing its bed and hence making agriculture in the area unpredictable and inefficient. The observer, being a practical soul, places a high premium on any activity that works in the direction of remedying this situation, and hence a scientist working on this problem will be judged by him to be socially responsible. In fact, however, it might happen that the scientist could not care less about the practical problem of agriculture, but instead he is fascinated by hydrological aspects of geology because his interest is in explaining the strange patterns on the moon and on Mars which some people have claimed to be results of once existing waterflows. Our scientist therefore makes great progress in understanding the cause of the meandering of the river, and thus is proclaimed socially highly responsible by the observer, even though "in reality" (e.g. judging by motivations), the scientist may be the most "selfish" and "so-cially irresponsible" creature on earth.

Fourthly and finally, social goals and individual aspirations may not at all be mutually exclusive and incompatible. It is deplorable one-dimensional thinking to assume that a given goal has only one path leading to it, and that therefore social goals can be reached only by the discharge of social responsibilities which are opposite to individual strivings. Much of society's aspirations and aims are simply coherent superpositions of individual aspirations and aims, all pointing more or less in the same direction. For example, the conversion of talent into achievement is something one finds commonly among the motive forces of individuals, and at the same time it serves as a source of societal (e.g. national) pride also. Should we then call a scientist socially responsible merely because his personal aims and aspirations happen to coincide with some of the societal ones we happen to focus on?

What is the use?

So far I seem to have made progress only in the direction of increased obscurity and an enhanced obfuscation of the question we are discussing. What was, at the beginning, an apparently simple question with a simple answer has turned into a maze of additional questions, each of which is very difficult to answer simply and unambiguously. Am I claiming then that the situation is hopeless, the original question meaningless, and the determination of whether the scientist is socially responsible or not impossible?

Not so. What I tried to show only is that before one can answer the original question, namely whether an observer considers a scientist socially responsible or not, we have to specify the question more thoroughly, attaching to it some specifiers and qualifiers, which in turn depend, in part, on some personal value judgments. Once those particulars are specified, it may very well be possible to answer the question, even in practice.

This analytical exercise, therefore, is useful for several reasons.

First, it allows us to turn the original fuzzy question into a much better defined one that may be answerable in practice.

Second, it helps to remind us that any answer given to the original question is meaningless unless the specifiers and qualifiers used to narrow down the question are stated together with the answer.

Thirdly, it reminds us that it is impossible to make sweeping, collective judgments about social responsibilities of scientists, and rash pronouncements on this subject are also unwarranted. Each case must be carefully evaluated on its own merit, and then appraised in terms of the set of choices made earlier in terms of the criteria for social responsibility.

These three elements can be extraordinarily helpful in real

situations. Public discussions on most issues, including that of the social responsibility of the scientist, suffer greatly because of the ambiguity in the statement of the problem under discussion, and acrimonious debates with a great deal of hostility are generated by two people actually discussing two different problems while under the impression that they are debating the same subject. Similarly, debates can be much more bitter when each party believes that his answer is an absolute one, value-free, and valid under all circumstances, rather than one that is contingent on the specifiers and qualifiers attached to the question and hence steeped in value judgments.

At this point it might be objected that in spite of all these glittering semantic manipulations, the discussion has failed to attack the main problem and to answer the question: Is a given scientist under given circumstances socially responsible or not? The criticism is well taken, and in fact the present discussion is not intended to go all the way in reaching an answer. The aim was to establish a *framework* in which the question can be treated much more easily, much more sharply, and much more operationally. Thus everything said in this article is simply a prelude to the "real" discussion. But then, the years of exercises, scales, proper finger positions, appropriate bow technique, and the like that, say, Yehudi Menuhin had to go through was "only" a prelude to his interpretation of the Brahms Violin Concerto, and yet they made the difference between *his* rendering of this composition and that of a person who picks up the violin for the first time.

Rules of conduct

It would, however, be somewhat anticlimactic to end a discourse on the social responsibility of scientists on such a rarified and technical level. So, in conclusion, let me offer four rules of conduct for scientists which, in my opinion, will be beneficial no matter which of the innumerable alternatives we choose for judging the social responsibility of scientists.

1. It is the social responsibility of a scientist to strive for the highest possible degree of scientific expertise and understanding. This may take years of work, and there is no shortcut to this process, no matter how tempting certain apparent alternatives are. A person who is, say, a nuclear physicist, and then wants to become a truly productive and effective contributor to the public debate on the pros and cons of nuclear power should expect to have to devote a very large amount of time, effort, and humility in learning the areas of biology, chemistry, economics, engineering, and other fields that are significant components in the problem. This rule is very frequently violated by scientists who have a penchant for talking off the top of their heads on public issues of all varieties, and doing this *as if* they were speaking in the role of scientists.

2. Related to the previous point, scientists, when working on interdisciplinary problems of social importance, should be extremely careful to separate the scientific components of the problems from the other ones, and emphasize when they talk on the basis of significant professional expertise in areas of science, and when they say something as amateurs in fields in which they have no training or experience.

3. Further related to the previous two, when scientists get involved in socially significant issues, most of which are, by nature, "interdisciplinary", they should be particularly prepared to devote a long time and much effort to getting acquainted with the non-scientific issues in the problem area. An outstanding biologist, when faced with issues of, say, political science, economics, or philosophy, is certainly not better off at the start (and perhaps worse off), than a starting graduate student in that field, and hence even five years of full time study will take him only up to the threshold of the ability to contribute. For this reason it is often advisable for a scientist to choose less broad, more modest and limited social problems to contribute to, rather than the cosmic, universal, and difficult ones which are highly interdisciplinary and where the chances of making a significant contribution are slim and the opportunities of doing damage by rash judgments and by the advocation of poorly thought-out action are stupendous.

Fourthly and finally, if the scientist recognizes that he was wrong in his stand on a public issue, he should back down and change his position in the same way as he is used to doing in a purely scientific debate when the power of experimental facts demonstrates that his position was in error. Scientists find such a retreat in public issues very difficult, partly since, in his innocent arrogance, the scientist frequently tends to "go out on the limb" and take an unduly definitive stand, unwarranted either by the facts available or by the depth of his understanding of the problem. As a corollary of this, beware of scientists (or others) who claim that their view of an issue represents a "moral" argument. The implications of such a stand are vicious: It implies that the person disagreeing with the scientist is immoral and hence does not deserve consideration or a decent treatment, and it also implies that the stand is not open to change or compromise since, after all, nobody can compromise his morals. Thus, I conclude, it is best to keep "morals" out of the explicit discussions on the social responsibilities of scientists.

E. Some Benefits

SCIENCE AND A CRUCIAL STAGE OF ECONOMIC DEVELOPMENT

There is a great variety of scenaries among the economic development of countries around the world. Yet, with only a minimum of oversimplification, three different phases can be discerned in such a development.

MICHAEL J. MORAVCSIK

Institute of Theoretical Science,
University of Oregon,
Eugene, Oregon 97403, USA.

In the first and most rudimentary stage, the country's exports consist mainly of raw materials. In some cases this might be food stuff (e.g. rice in Thailand), in other cases mineral resources (oil in Nigeria), in again other instances agricultural products other than food (jute in Bangladesh), etc.

With a few exceptions of countries with small populations having large quantities of very valuable raw material (oil in Kuwait), this stage of economic development suffices only to produce a GNP per capita in the lowest range, say below $300 per annum. The point is simply that the specific value (i.e. value per unit weight) of such raw materials is usually fairly low (compared with the kind of products we will discuss later), and so even fairly large bulks of export will produce only modest amounts of financial returns.

The second stage of economic development is based on the country being able to export technological products of varying degrees of sophistication which were researched and developed abroad but which the country can now imitatively produce. At such a stage the country needs skilled and yet fairly lowly paid workers who can equal the quality of products made abroad and yet undersell them.

This second stage of development can carry the country far beyond the $300 per annum for the GNP per capita. There are examples of countries thriving on this mode with a figure approaching $2,000, though the more common amount is half that much.

Finally, in the third stage of economic development, the country has the indigenous capability to create new technological products and manufacture them. If such products incorporate modern, sophisticated technologies, their export will catapult the country to the highest existing values of GNP per capita, in the $6,000 range.

In countries well along in the second stage of development, there is considerable pressure to make the transition to the third. For one thing, the mechanism of the second stage can raise the standard of living only up to a certain point, and there will be public demand to continue the growth. More crucially, however, there will be outside pressure to stimulate the transition. The rising wages in the country will make it increasingly more difficult to undersell the competitor who is already in the third stage, and, from the other side, new competitors will arise from countries just entering the second stage in which the wages are still low enough to undersell the country we are considering.

What is the role of science in each of these stages? In the first stage, science has only an indirect, a covert part in the economic development process. At this stage the country uses, if at all, mainly relatively simple and traditional technology which may need to be adapted for local use. The technological activities surrounding such adaption are also relatively simple, but still they need to be based on some scientific activity, both in terms of the education of the personnel and in the research and development field itself. This indirect role, however, is often not well recognized by the decision makers of economic policy (consisting mainly of politicians and economists). Hence science frequently is deemphasized, and what there is of it is permitted only in areas which to the uninitated appear to be "relevant" to the economic task at hand. As a result, the country's development is retarded. Nevertheless, the production of the raw materials keeps limping on somehow.

In the second stage, science still plays an indirect role in the economy, although its importance begins to be more evident. In the education of manpower with more evident. In the education of manpower with more sophisticated skills, as well as in the creation of complex production processes with quality control, the scientific background is essential. Still, science is only used one step removed from the actual production process, and hence it still tends to be deemphasized and neglected. Some of it undoubtedly evolves somehow, but the balance between science and technology is less than perfect even from the point of view of the present stage of development.

The crisis really strikes, however, when the country attempts to make the transition from the second to the third stage. Nowadays almost all innovative technology is strongly science-based, and the competition is so great that doing innovative technology on borrowed science represents, even at best, a debilitating time-delay. At this point, the country should have a broadly based, productive, and self-generating scientific infrastructure of active researchers with a research tradition in a wide variety of scientific disciplines. Modern technological products are strongly inter-disciplinary, requiring, simultaneously, expertise in physics, chemistry, computer science, material science, and others.

But why is this a cause for a crisis? The reason is simple: It takes several decades, at the least, to evolve such a productive and stable scientific infrastructure, and hence the process should have begun 40 years earlier, when the country was still in the first stage of economic development. In the overwhelming majority of the cases, this was, however, not done, or not done on a sufficiently energetic scale so as to provide the country with the scientific base when it is explicitly needed, when the impact of science on the country's economic development surfaces from an indirect, covert link to a primary, explicit factor.

These observations can be verified now on a number of emerging examples. The positive example is Japan, which started its scientific development over a hundred years ago and which, when 80 years later it reached the third stage, had the necessary scientific base to go ahead full steam.

On the other hand, negative examples appear to be provided by a number of countries at the present (South Korea and Singapore being two) which are about to enter the third stage of economic development—or, more precisely, they would enter if they had a sufficiently strong scientific base to do so. At the moment they do not have it, and the reason can be traced directly to the neglect, during the past two decades, of science in favor of imitative technology. In South Korea this is now recognized, and very recently governmental policy has switched to a much more vigorous support of indigenous science. it is, however, too

late by now—not forever, but too late not to produce a considerable delay in the country's entrance into the third stage.

It is unfortunate but true that the blame for this oversight in most of the developing countries cannot be ascribed solely to the local decision makers of "science policy". There is a long string of international accomplices to this shortsightedness, most of whom are merrily continuing with their misguided policy. Many United Nations agencies, most of the large national agencies for international assistance, as well as other organizations have subscribed to the view according to which the scientific and technological structure of a country must be shaped to the present economic needs, and hence, during the first and second stages, science should be considered "relevant" only in the context of the narrow criteria of contemporary economics. In other words, it is claimed that first the country must have a set of economic objectives and only then should it be concerned with the development of its scientific infrastructure which then must be closely dovetailed into this economic plan.

Quite apart from the fact that such a view completely ignores the non-economic impact of science on a country, it is nonfunctional even from a purely economic point of view. In fact, in practice this view even failed to promote sufficient development in the first two stages, and it certainly fails altogether when we come to entrance into the third stage. The whole situation reminds one of the father who, at the time of the birth of his son, decides to orient the son's upbringing strictly toward the task of becoming a locomotive fireman—only to find 20 years later that not only has the son become a non-achiever and a delinquent on account of the confining childhood, but in the meantime the job of a locomotive fireman has been annihilated by changing technology.

What is then needed to avoid such problems in the future? First, objective, independent, and thorough research, preferably concentrating on case studies in depth, should be carried out on the situation of the countries which are now trying to enter the third stage of development and are in trouble. The analyses of these studies should then be widely publicized in order to influence the still upcoming other developing countries and the international assistance agencies. There are still very many developing countries around the world which are in a sufficiently early phase of economic development (that is, in the first stage or at the outset of the second stage) that an appropriate emphasis by them on science now might be able to avoid the crisis later.

The change that is needed is not very large in terms of the rearrangement of financial and manpower resources. The development of science in a country is a relatively inexpensive undertaking. If a total of 0.2% of the GNP is directed toward scientific research (whether "basic" or "applied" is not so important from this point of view, and in fact is somewhat of a semantic question is any case), then the desired scientific infrastructure has a good prospect to mate-

rialize. Furthermore, the needed change is much more a matter of understanding and attitude than money. The "new look" needs to start with science education at the very beginning of schooling, where science must be understood as a method of asking and answering questions about the world around us, questions to which the answers have not been known previously, rather than a set of rules to be memorized. The new attitude must pervade the exposure of science and technology in the mass communication media to show that the basic methods and beliefs of science can also be used by nonscientists in their everyday lives. It must carry over to colleges and universities to demonstrate that science is an exciting, open-ended field and not just a last resort for those who failed to get admission in medical schools, law schools, or engineering schools. It should envelop also the practising scientists in the country, particularly in the huge governmental research establishments which consume the lion's share of the country's money earmarked for research, so that the work there turns into creative scientific research ("basic" or "applied") and not low level imitative technical manipulations, as is often the case.

All this must take place at the very beginning of the country's economic development, even at a time when the effect of science on the country's economy is mainly the indirect kind discussed earlier. Questions of "relevance" must be judged with a much broader outlook in mind, taking into account present and future, economics and the other aspirations of the country, already evident as well as only vaguely anticipated trends in the needs of the country.

The so-called "developed" countries in most cases had several centuries at their disposal to absorb the impact of the scientific revolution as a cultural force, and to integrate science and technology into their economic development. The developing countries want to move faster, perhaps in only a few decades, and hence they cannot take the leisurely road of making mistakes and hence losing time. Thus conceptual foresight is essential. This is the true "planning" process, not the drawing of organization charts and the preparation of unrealistically detailed blueprints of exactly what kind of activity a given institution should be engaged in ten years hence. In such a conceptual "planning", the recognition of the direct role of science at the third stage of economic development, and of the necessity of preparing for that role decades ahead is crucial.

The role of science in technology transfer

Michael J. MORAVCSIK

Institute of Theoretical Science, University of Oregon, Eugene, OR 97403, U.S.A.

Final version received April 1983

The discussion begins by listing three meanings of technology transfer (TT), involving importation of technological products (TTA), the ability to imitatively produce technological products (TTB), and the ability to create new technology (TTC). This is followed by an investigation of the relationship between science and technology, using a multidimensional conceptual framework in which one cannot ask "What is *the* cause of something else?", but in which one can trace the contributory and the necessary conditions for something to take place. In this light it is shown that in modern times science is a necessary condition for technology and *vice versa*, the links being of a variety of sorts, partly epistemological, partly attitudinal, partly infrastructural. The discussion then turns to technological development, and four stages of that are distinguished: stage 1 rests on traditional trial-and-error-based technology, stage 2 on imitative low-level technology locally performed, stage 3 on imitative high-level technology locally produced, and stage 4 on the creation of novel technology. An anomalous stage involving small oil-rich countries is also mentioned. The role of science in each of these four stages is complex, involving both production and importation of technology, forming both direct and indirect links, having both simultaneous and delayed impacts, and acting also though attitudinal influences. Combining the previous conclusions it is then found that science has an essential role in all three meanings of technology transfer. It is, consequently, urged that organizations for international scientific and technological assistance and cooperation take a broad view and orient their activities to deal with the whole interlocking and interwoven complex of science and technology in the Third World.

1. Introduction

Ours is an age of technology, but it is also an age of egalitarianism. With regard to the second, although large differences around the world continue to exist in many respects, our *Zeitgeist* appears to contain an urge to level those differences and bring about a uniform distribution of whatever resources and benefits we have.

It is nor surprising, therefore, that we also strive toward a more uniform distribution of technology, one of the most conspicuous resources of our century. As in most other walks of life, in technology also, uniform creation, generation, and production is at present not possible, and so the goal is approached by a redistribution of technology, transferring it from one part of the world to another, from the creators to those unable to create it. Hence technology transfer is a very prominent subject of discussion nowadays.

There are many facets of technology transfer (henceforth denoted as TT). A huge literature has been generated on how TT can be done, how much of it is feasible, what kind of technology can be transferred, what financial arrangements need to be made, how one should prepare the receptive end for such a transfer, etc.

In this article I want to focus on only one particular element in this complex picture, one that has perhaps not been very widely and thoroughly discussed. It is the role science plays in TT. I will attempt to make the discussion concise, precise, and, at least in its outlines, reasonable comprehensive. To achieve this objective, some preparatory analysis will be necessary. In section 2 the various meanings of the phrase "TT" will be distinguished and defined. This will be followed, in section 3, by a summary of the quite subtle relationship between science and technology. I will then turn, in section 4, to a listing of the various stages of technological development, and of the role of science in these various stages. After these preparations, the stage becomes ready, in section 5, to outline the various roles science has in the various meanings of TT. Section 6 will try to summarize the situation and draw some policy conclusions from it.

** The writing of this article was stimulated by a discussion I was asked to lead at the Office of Technology Assessment in Washington. I am indebted to Martha Harris for the invitation to that discussion.

Research Policy 12 (1983) 287–296
North-Holland

2. Meanings of "technology transfer" (TT)

Like most terms we use in everyday conversation, TT also has several meanings which are seldom carefully differentiated. Some of them appear to stretch somewhat the "natural" meaning of the phrase, others may be more directly related to the most obvious connotations. In our discussion it will be essential to make a distinction among these various meanings, since the relationship of science to these different kinds of TT will turn out to be quite different. In particular, I want to distinguish three different types of TT, all referred to in everyday parlance only as TT.

(1) The first type of TT I will denote as TTA. It denotes a transfer of finished technological products from one country to another. On the simplest level, selling completely manufactured and all-assembled transistor radios to Chad is a TTA. On a somewhat more complicated level, selling F5 fighter planes to Bahrain, or selling isotope manufacturing facilities to Thailand are also instances of TTA. In such cases we deal with simple trade of agricultural or industrial products which happen to have involved technology in their manufacturing process which took place in the seller (or donor) country.

(2) The second type, denoted as TTB, involves the transfer of the know-how to make use of technology already invented and established in other countries. The establishment of a car factory in India to make cars designed abroad may be an example. In fact, if this car is designed in India following well established principles and practices of car-making and containing no significant innovations in such technology, the process still falls under TTB.

(3) The third and most sophisticated type of TT, designated as TTC, involves the transfer of the ability to create new technology. For example, Japan, in its contact with Europe and, to a lesser extent, with the United States, transferred to itself, during the first half of the 20th century, the ability to create innovations in technology which in the last two decades resulted in Japan taking a lead in a number of high-technology areas, such as some fields of electronics, robotry, automobile manufacture, etc.

The eventual goal of every country is to acquire TTC, at least in some areas of technology. This is,

however, a long range goal, difficult and laborious to achieve. On the way to it, countries seem to pass through TTA and TTB, as will be discussed presently. In fact, TTA and TTB remain important elements of the economic life of a country even after TTC has been attained in some areas.

It is perhaps unnecessary to add to the above classification that discrete categorization always represents a certain degree of simplification, since the real situation is more likely to be a continuum with peaks and valleys. Indeed, there are various borderline situations between the above three TTs. Consider the sale of a car to a country with a climate of high humidity which therefore has to modify slightly the carburator system of the car to account for the difference in environment. Is this TTA or TTB? When Bahrain buys fighter planes and then trains its own technicians to service, maintain, and repair the aircraft, are we not passing from TTA to TTB? These are valid, but on the whole, inessential questions. If we treat definitions not with the mind of a pedestrian bureaucrat but in the spirit of trying to gain understanding, the applicability of such definitions in more than three-quarters of the interesting cases will suffice to prove their utility and value.

People with different interests may focus on different types of TT. For example, to a scientist or technologist, TTA hardly seems to earn the right to be called TT. In contrast, to a sociologist or anthropologist, TTA is of great importance because it can cause large scale and profound effects of the types sociologists or anthropologists like to study. The same holds for economists, though they are also equally interested in TTB and TTC.

Sometimes it is uncertain whether a given act of TT belongs to one or the other category. Some years ago there was a discussion among experts in the United States about the advisability of selling some advanced computers to the USSR. Some argued that this would be a TTB and perhaps even TTC, since once the USSR had in its possession a sample of such a computer, it could manufacture others by copying it (TTB), and perhaps even gain the ability to make further innovations on them (TTC). Others argued that the state of auxiliary technologies (e.g. materials science) in the USSR was so low that even in possession of a prototype, the USSR would be unable to copy it (TTB), let alone improve on it (TTC). The example shows that the impact of an instance of TT can depend on factors and circumstances going far beyond the particular technical subfield which that specific TT refers to.

3. The relationship between science and technology

An immense amount has been written on this topic, and hence this brief account should not be taken to represent an encyclopedic review of the subject, or even an impersonal, "objective" resumé. I want to present instead a personal view in a form which will be suitable for the subsequent discussion.

First, some definitions [1]. Science is a human activity, using a particular methodology whose objective is to explore and understand the world around us. The product of science is knowledge. Science can be classified in several ways. One way is by the motivation of the researcher or of the sponsor of the research. If the primary emphasis is on acquiring knowledge for general epistemological, intellectual, or esthetic reasons, the research is tagged as being "basic", whereas if the primary motivation is in the possibility of applying the knowledge to other human undertakings, the research is tagged "applied". This dichotomy is very ambiguous and vague, because the classification depends on whether we do it according to the researcher's motivation or the sponsor's motivation, because the classification depends on whether we consider short or long time perspectives, because both motivations can coexist in the mind of the same assessor, and for many other reasons. For science policy purposes, however, it is possible to make somewhat arbitrary but fairly well defined distinctions between "basic" and "applied" research.

A second way of classifying scientific research is by the methods used by the researcher. Research in which already acquired observations are organized and explained through new concepts and theories is called "theoretical" research, while "experimental" research is the process in which new observations are collected through scientific experiments.

The two dichotomies of "basic" versus "applied", and "theoretical" versus "experimental" have nothing to do with each other. Accordingly, there can be four different situations if we con-

sider both classifications: basic theoretical, basic experimental, applied theoretical, and applied experimental research.

Technology is a different kind of human activity in which prototypes of gadgets, or patents, or procedures for making things are created, using methods of trial and error and/or results of science. We see that the way to distinguish between science and technology is by the product: science produces knowledge (an intangible), while technology produces something more tangible in the form of a gadget or a process to make gadgets. While these definitions are somewhat rough, they suffice for our purposes.

Before we turn to investigating the relationship between science and technology, a word about general methodology. Virtually all real life situations, including that of science and technology, are multidimensional [2], that is, their description must be made in a space of many dimensions, each representing one aspect of the problem. This is in contrast to one-dimensional models, in which cause and effect are well defined and are unique, in which events and entities can be lined up in a chainlike fashion, in which superlatives ("largest", "greatest", "best", etc.) make sense, and in which one side's benefit is the other side's detriment.

In a multidimensional situation it makes no sense to ask what is *the* cause of something else. Since the points denoting entities and events are dispersed in the multidimensional space and connecting lines criss-cross in large numbers among these points, everything has many causes an many effects. Since the assessment of an event point is a composite of assessments along many different axes, an overall rank will depend on the relative weights attributed to the various axes. The relationships in such a multidimensional space are therefore very much more complex than in one-dimensional models.

There are, however, more modest questions that we are allowed to ask even in a multidimensional model. For example, one can ask whether A is a *necessary* condition for B or not, that is, whether B can occur without A previously occurring. That A is not a *sufficient* condition of B is virtually certain because of the multi-dimensional character of the space.

All this may sound obscure and esoteric, and quite unnecessary for the consideration of "practical" questions like the relationship between science and technology, but in fact the situation is quite the opposite. I recently encountered [3]. for example, a learned essay on whether the transistor was a consequence of modern quantum mechanics or of prior rectifier technology. Regardless of what answer this essay happened to arrive at, its very way of posing the question is meaningless in the framework of a multidimensional view of the situation.

In view of all this, I want to discuss four questions which can arise in thinking about the relationship between science and technology.

(1) Is science a (contributory) condition for technology?
(2) Is science a necessary condition for technology?
(3) Is technology a (contributory) condition for science?
(4) Is technology a necessary condition for science?

The first question is relatively easy to answer. All we have to do is to name one example where science contributed to the development of a technology. The nuclear power reactor is such an example. The fission of uranium was discovered by scientists, and that laid the basis for the construction of a nuclear power reactor.

The second question is very much more involved. It asks whether a given technological product could be developed without a scientific input.

There is little doubt that 400 years ago, in most instances, technological gadgets did develop without recourse to scientific input, partly because such scientific input was not available, and partly because those gadgets could be easier developed by trial and error, by tinkering of the every day sort. Thus at that time, in almost all cases, science was *not* a necessary condition for technology.

Nowadays, the situation is quite different, for reasons which need not be discussed here. Virtually all technological innovations of any importance contain necessary elements supplied by science. To be sure, they also contain elements obtained by trial and error, but nowadays even trial-and-error is guided by some knowledge of science, which narrows down the number of possibilities that need to be tried. In some cases the "application" of newly found scientific results is direct and immediate, such as in the above example of nuclear reactors. In other cases, the necessary ingredient is

"old" science, scientific knowledge established decades ago. In some cases science manifests itself by supplying materials to be used in the innovation, materials which had been created with the help of scientific knowledge. In one way or another, however, the hallmark of science is clearly present in one or many of the necessary conditions for the birth of the new technology.

Let me stress again that science being a necessary condition of a new technology does not preclude the possibility (in fact high probability) that some previous technology is also a contributory or even necessary condition for that technology. In the example mentioned above, rectifier technology undoubtedly had an important and perhaps even necessary part in the development of transistors. Form the point of view of the science–technology relationship, however, all we have to convince ourselves of is that modern quantum mechanics was also a necessary condition for the transistor, that is, an event without which the transistor would not have come about, no matter how stimulating the influence of rectifier technology may have been.

So far I have considered only the role of science in the form of providing knowledge on the basis of which new technology can be evolved. But science has other effects also besides supplying knowledge on which technology is based. In general, the impact of science can be classified in three main effects [4]: science as a basis of technology, science as a human aspiration, and science as an influence on the view of man of the world and of his role in it. I will not discuss here the second of these, but the third one is very pertinent indeed to the science–technology link. The scientific revolution as a cultural force has injected into Man's thinking the notion that by following a particular methodology (called sometimes "the scientific method") man can learn increasingly more about the world around him and then can use this understanding to influence his fate in the world. This is an attitudinal revolution, compared to the much more passive. lethargic, and static view that often prevailed prior to the scientific revolution [5].

This new attitude has a· profound effect on technology. It is not enough that a given technological innovation is, in principle, made possible by new scientific knowledge and/or by the direct opportunity for tinkering or working in a trial-and-error fashion. What is also needed to turn this opportunity into accomplishments is the interest, the will, the motivation to use the knowledge, to tinker, to engage in trial-and-error. The man stimulated by the broad horizon, the promising future, the opening up of possibilities that the scientific revolution as an attitudinal force projects, will have such a motivation and will-power to pursue the opportunities.

Let me now turn to the third question: Is technology a (contributory) condition for science [6]? The answers to this question (as well as to the last one) parallel closely those given to the first and second questions, thus exhibiting the high degree of symmetry and mutuality that has evolved between science and technology. In particular, it is clear that technology contributes to scientific research: Most modern scientific research laboratories are filled with technological equipment used to create the conditions to be studied, to record the signals obtained from these experiments, and convert them to the type of signals our human sense can directly perceive.

Similarly, when we ask whether technology is a necessary condition for science, the answer is again the same as the one we had for the second question. Some 400 years ago, technology was quite separate from science, and when science first concerned itself with phenomena which were directly perceptible to the human senses, technology played a relatively small role in fostering science. But as science increasingly expanded into phenomena more remote from everyday experience and reach, starting with telescopic observation of stars and quickly progressing to electric phenomena, atomic physics, microbiology, molecular biology, nuclear physics, elementary particles, radioastronomy, and many other fields, technology played a more and more indispensible role. Again, as in the case of science, it was sometimes "old" and sometimes "new" technology that was needed, sometimes it was in the form of new materials produced by technology that proved essential (e.g. in the case of particle detectors). In every case technology was not the *only* ingredient in the development of science, but it increasingly became a necessary one.

The "spirit" of technology also contributes to science just as the reverse is true. Those excited by the achievements of gigantic and complex technological systems (for example, the Saturn-V rockets) will also have a drive, will-power, patience, imag-

ination, and boldness to design and build circular particle accelerators several miles in diameter. This is also an attitudinal effect, less tangible and measurable than the direct material link between technology and science, and yet hardly less important.

Finally, there is also the even more indirect but equally crucial link between science and technology, both ways, via the overall rapid increase in the economic resources of humanity. In order to spend $300 million on a particle accelerator, or on a space probe, or in order to afford flying scores of Boeing 747's costing $50 million a piece, society as a whole must be rich, and indeed it is in comparison with 400 years ago. This affluence also has many causes, among them the rapid development of science and technology in the past 400 years, and the evidence is very persuasive indeed that science and technology have in fact been necessary conditions for the development of such an affluence.

4. Stages of technological development

Our next step in preparing for the discussion of the link between science and TT is a description of the consecutive stages of technological development in a country. As before, it is convenient, though somewhat oversimplifying, to describe such a development in terms of four distinct stages [8].

In stage 1 the country has some rudimentary agriculture, sufficient for domestic consumption or sometimes even sufficient for some export, and there is also cottage industry, the products of which also satisfy local needs and can be used for international trade. Both agriculture and cottage industry rest on traditional trial-and-error-based technology. Countries at this stage of development usually subsist on a gross national product per capita (GNP/capita) of the order of $300–500 or less. Since at this stage the local technology is mostly indigenous traditional craft, TT at this stage has little if any role for the technology *produced* in the country.

Stage 2 represents a somewhat enhanced level of technological activity. At this stage the country is able to produce locally some items of imitative and low level technology, using them for import substitution and perhaps also for low-priced export. In the latter context the success is due to the very low domestic wages which enable the country to sell these goods for a very low price, low enough that they are bought even if the quality of the goods is not very high.

Stage 3 is faintly similar to the previous stage except that by now the country can produce high technology products also, on an imitative basis. This is often done through subsidiaries of multinational firms, but it can be also by genuinely local companies. At this stage the manpower must be more sophisticated and more highly skilled, and the wages can be considerably higher without precluding competition with the advanced countries themselves. Countries like Singapore, with a GNP/capita of over $2000, thrive to a large extent on such stage 3 activities.

Finally we have stage 4, at which the country is able to create new frontline technology, and hence becomes a leading country, an "advanced" country, a "developed" country. At this stage the wages are usually sufficiently high for the country to be virtually forced to engage mainly in novel, frontline technological activities, because in imitative activities at the lower level it has strong competition from countries at the lower stages of technological development, and in such a competition the stage 4 country tends to lose unless it resorts to governmental subsidies, artificial trade barriers, and other tricks which in the long run prove to be untenable.

To what extent and in what way is science needed in these stages of development?

It would appear from the above description of stage 1 that in it there is no need for science. This, however, is not so. For one thing, even the traditional methods of agriculture can be made more productive by the infusion of a bit of help from modern science and technology, and hence agricultural research is usually needed even at stage 1. The same can be said in connection with matters of health. Furthermore, even though the country's *production* does not directly involve science since it consists of the traditional trial-and-error-based technology, the country's *consumption*, through its imports, will undoubtedly include articles produced by science-based technology and used in science-related situations. The pertinent science may be elementary and "old" science, but it will still be present.

To illustrate this point, here are three examples, taken verbatim from a previous study [9] devoted specifically to the status of "everyday" science in developing countries.

Example 1

In urban buildings in tropical areas, built using modern techniques, the top floor is often unbearably hot. The reason is simple: the usually flat roof absorbs solar radiation, and no provisions are made either to insulate the roof from the ceiling of the top floor, or to design the building with a simple double roof with air space between the two for ventilation. Ironically, such features are often included in the traditional, older, and rural structures, but since the scientific reason for such features is not comprehended, builders cannot distinguish between those features of traditional architecture which are functional and those which are ceremonial or due to limitations of construction technology.

Example 2

In numerous developing countries tin-box ovens used on charcoal stoves are common. They simply consist of a usual thin-walled surplus tin-box, placed on top of a regular iron charcoal stove (similar to the hibachis popular in Western countries for use on outings and backyard barbecues).

It does not take very much science to realize that the efficiency of such an oven, and also the temperature attainable inside, could be increased if all sides of the tin box except the one in contact with the stove were insulated. A little more science tells us that such a saving in fuel would be quite substantial, very rapidly returning the small initial investment in one of the commonly available insulating materials.

Example 3

Recently, in one of the developing countries, the supply of tap water became quite intermittent for several months. It was then found that during those hours of the day when the water supply did not operate, much subsequent loss of water was initiated by taps being left in an "on" position during the time when the supply was off. People would try the tap, see that no water was coming, and then leave the tap without turning it to the "off" position. When the water supply returned, the water kept flowing sometimes during the whole day until somebody happened to return to it and turn it off.

Water being expensive, one of the important contributing factors in this apparent neglect is likely to be the complete absence of an image of a water flow with a tap being a gadget to transmit or obstruct this flow. When this simple point is not understood, there is no more "reason" for turning the tap to the "off" position when the water fails to emerge. This is science on such an elementary level that most of us might find it difficult even to realize that it is science.

These considerations also apply to stage 2, in fact even more so. Even if an old and simple technology is copied, its essence must be understood by the copier in order to be successful, and this essence almost always involves science of some sort. Perhaps this science is only on a level that a good elementary or secondary school should be able to get across, but such lower level schools will not exist in a country which is void of scientific activities of any kind. We see, indeed, that the connections between science and the technological state of the country are complex but strong.

The scientific requirements become increasingly more stringent as we proceed to the more advanced stages of technological development. When in stage 3, a country is involved in high technology production, even if only on an imitative basis, the production of special materials through sensitive processes, the importance of quality control, the complexity of the articles produced will increasingly demand at least some knowledge and understanding of the scientific basis of these products. In high technology items the particular way of proceeding during manufacture is increasingly more crucial, and even one apparently "small" mistake can lead to failure. In following the proper path, a scientific understanding of at least the basic principles pertaining to the product is extremely helpful. Without such an understanding, following the manufacturing instructions for a product is akin to copying a page of a book written in a script that we cannot read or write.

It is needless to say after the preceding discussion in previous sections of this analysis that in stage 4 frontline scientific activity within the country is absolutely crucial. Japan's recent lead in electronics, robotry, and other frontline technologies would not have been possible without a highly developed scientific infrastructure in that country. Significantly, this does not mean that *all* pioneering

research and development has to be performed in a given country before that country can be a leader in the technological exploitation of a new branch of science. On the contrary, in robotry, for example [10], most of the pioneering research was originally done in the United States. Yet, there are 20 times as many robots in operation in Japan than in the United States, or 40 times more per capita, and Japanese manufacture of robots is at least as far ahead of the United States. This is not surprizing, in view of our multidimensional model, since the flourishing of a given technology in a given country has many contributory causes (and in fact more than one necessary condition also), and it is the combined workings of all those conditions that regulate success or failure. For our discussion, however, we only need to stress that a well developed scientific infrastructure is *one* of the necessary conditions for such technological leadership.

In addition to the above four stages, I have to mention yet another and somewhat anomalous stage of technological development, which applies to relatively few countries which, however, are very conspicuous. In most previous stages discussed, the modest state of the *productive* technology in the country places an appropriately modest limit on the *imported* technology also, due to the rough balance that has to exist between the economic size of imports and exports. This balance is, however, radically different in those countries which happen to have a very valuable raw material to export—such countries which are small *and* oil-rich. Here the productive technology may be rudimentary and yet huge resources exist for importing large quantities of very sophisticated. frontline technology. Many of the problems discussed above, pertaining to the scientific level needed to utilize sophisticated technology even if it is merely imported are greatly accentuated in such an environment. Another special feature in such countries is the fact that the rate of indigenous technological development is in no way limited by the availability of funds, and hence the other limiting factors (which play a crucial and in fact decisive part in other countries also but are sometimes camouflaged by the lack of funds) now become truly central. Among these limiting factors the most important is the lack of appropriately educated manpower, and, specifically, scientific manpower.

To conclude this section, I must mention two more and somewhat related aspects of the connection between science and the stages of technological development. Even when the country is at a beginning stage of such a development, it generally undertakes planning and decision making with respect to the future development of the country. National development plans, decisions on university development, the establishment of governmental research and development institutions are common in all countries. In these activities background in the sciences is essential. To be sure, the number of people required for such policy making functions in a given country is not large, but those needed must be of high quality. Experience has shown that it is rare to encounter successful science policy makers who have not had themselves a substantial personal involvement with scientific research, preferably in the not too distant past. Thus we see here yet another link between science and technological development, indirect and yet very important indeed.

When making projections for the scientific and technological development of a country, one must realize that to attain a self-propelled, creative, and functional scientific infrastructure in a country takes several decades, starting from the time of inception. If, therefore, it is projected that the country will need such a strong infrastructure, say, 50 years from now, the beginnings of such a infrastructure must be laid today, in terms of the fostering of quality institutions of higher learning, in terms of sending manpower abroad for advanced education, in terms of laying the foundations of local research activities, in terms of establishing a good communication system for the scientists and technologists in the country, in terms of ensuring the flow of talented and motivated young people into scientific and technological professions, etc. The more sophisticated and extensive an infrastructure is needed, the farther ahead work on it must be started. From this point of view, therefore, science in the country must be related not only to the *present* technological stage of the country but also to the projected *future* stages. In practice this means, in most cases, work on the building of the scientific and technological infrastructure that, judged merely by the requirements of the *present* stage of technological development of the country, appears far fetched, "irrelevant" and wasteful. The failure to understand this point has been one of the most glaring defects in the work of most national and international scientific and technological assistance and cooperative agencies.

Let me summarize the results of this section. We found that there were *four different stages* of the technological development, *plus an anomalous stage* represented by the small oil-rich countries. The *requirements of technology* in each stage pertain to two different aspects, namely the *production* of technology and technologically based products and the *consumption* of these, including the consumption and utilization of imports. In tracing the *link* between science and these stages of technological development, we have to distinguish among various types of effects of science on technological development. Along one dimension, there are *direct* lines in which scientific knowledge is needed for the generation, or adaptation, or utilization of technology, and *indirect* links, such as the effect of science on the education process or on the planning and management of technology, both of which then have an effect on the technology development as a whole. Along another dimension, we can distinguish between *simultaneous* impacts and *delayed* impacts. In the first, the presence of science at a given point in time impacts on technology at that same point in time. The delayed effects consists of science initiated and fostered over a number of decades resulting in the presence of a scientific infrastructure at the end of that period which then has an influence on technological activities. Finally, we must also take into account the even more indirect, *attitudinal* effect of science as a cultural force, in that the outlook science presents tends to set up an environment which is motivationally favorable for technological development.

5. Linking science with technology transfer

We now have all the background we need to deal with the link between science and the three kinds of TT we discussed in section 2.

We see that even in the case of the most rudimentary of the three meanings of TT, namely for TTA, science has a number of roles. In TTA we are talking only about the importation of products from abroad which were manufactured abroad with the help of some scientific input. We saw, however, that the utilization of such technological products itself involves some scientific knowledge and background. Such knowledge is also needed to make traditional agriculture more effective. We also saw that the management and planning of local technological processes, including decisions on what to buy from abroad, also involve scientific background. All this presupposes some functional infrastructure in the country in science education, which in turn requires some indigenous scientific activity. The motivation and will to use technological products in lieu of traditional ways of doing things require a type of world view which is fostered by science as a cultural force, and this in turn is likely to act only if there are some local scientists as carriers of this point of view. Finally, there are the delayed effects of science, effects that will be needed in later stages of development but which must be initiated at an early time, when TTA is dominant, in order to have them ready decades later when they are needed.

In TTB, the previous factors are all in effect, but some additional ones are in operation. In TTB the country needs the ability to imitate technology created abroad, and to imitate successfully a goodly amount of scientific knowledge is needed to grasp the principles on which the technology to be imitated operates. Much of the science needed for this purpose may be "old" science, fairly elementary science, "simple" science, but to acquire even such science in a functional way the country must have a perceptible scientific infrastructure.

Finally, In TTC the direct role of science becomes even more central. As we have seen, almost all present day technology is science-based in the sense that science is a necessary condition for the creation of such technology, and hence TTC, which absorbs the capability to create new technology is unthinkable in the absence of a strong scientific base. As I mentioned, it is not essential that all pertinent novel scientific research pertaining to a new branch of technology be also carried out in the same country. In fact, such a requirement would be impossible. Even the presently leading country in the science, namely the United States, produces only about 30–40% percent of the new science created around the world. What is needed, however, is that the country undertaking TTC have the scientific expertise to participate in research work in the general areas pertinent to the new technology, so that scientific and technological developments originating elsewhere can be un-

derstood, assessed, and utilized promptly by the country in question.

6. Conclusions

Several kinds of conclusions can be drawn from the foregoing discussions.

We have seen that in the framework of the multidimensional model for the relationships among science, technology, technological development, and technology transfer, certain questions are interesting, pragmatic, and answerable, while others are void of any meaning. In particular, we must not ask what is *the* cause of something else, but we can usefully inquire whether something is a contributory or necessary condition of something else, always keeping in mind that even if the answer is in the affirmative, there might be many other contributory or necessary causes of the same event, all of which together must act to bring about a certain happening.

Specifically with respect to TT, the conclusions we can draw point at the necessity of initiating, fostering, stimulating, and supporting indigenous science activities in a country at the earliest possible time. We have seen that even when the country is involved exclusively in TTA, science has many necessary roles both for the present and toward development in the future. The impact of science on TT is complex even at the TTA stage, and we must avoid taking a simplistic view lest we mutilate the complex environment needed to make TT successful. Such an overly simplistic, narrowly utilitarian, and excessively short-ranged attitude has been common in the policy decisions of most national and international scientific and technological assistance agencies, resulting often in failures and low efficiencies in TT processes they became involved with. If the developed countries really mean to promote effective TT to the Third World, they cannot avoid dealing with the whole interlocking and interwoven complex of science and technology.

Notes and references

[1] For some particularly articulate discussion of the definitions of science and of technology, see the writings of Derek de Solla Price, e.g. *The Difference Between Science and Technology*, address at the International Edison Birth-

158

day Celebration, Thomas Alva Edison Foundation, February , 1968.

[2] I recently analyzed the methodology of the multidimensional model of events in M.J. Moravcsik, *Life in a Multidimensional World (Scientometrics*, to be published).

[3] M. Gibbons and C. Johnson, Science, technology, and the development of the transistor, In: B. Barnes and D. Edge (eds.), *Science in Context* (MIT Press, Cambridge, MA, 1982) pp. 177–185.

[4] See, for example, M.J. Moravcsik, *How to Grow Science* (Universe Books, New York, 1980) Ch. 2.

[5] A pioneering, penetrating, and yet not sufficiently known researcher of the differences in world view concerning science in the various cultures was the late Francis Dart. See, for example, F. Dart, The Rub of Cultures, *Foreign Affairs* (1963) 360; F. Dart and P.L. Pradhan, Cross-Cultural Teaching of Science, *Science* 155, (1967) 649; F. Dart, Science and the Worldview, *Physics Today* 25, (1972) 48.

[6] A most eloquent though perhaps overly zealous advocate of the effect of technology on science has been Derek de Solla Price, see for example, *The Science/Technology Relationship, the Craft of Experimental Science, and Policy for the Improvement of High Technology Innovation.* Final report for the Division for Policy Research Analysis (National Science Foundation, Washington, DC 1982).

[7] For a recent discussion of a number of subtleties in the relationship between science and technology, see M.J. Moravcsik, Reflections on the Science–Technology Link, *Journal of the Scientific Society of Thailand* 8, (1982) 77.

[8] For a discussion of these stages in the context of two specific countries, see M.J. Moravcsik, Science and a Crucial Stage of Economic Development, *Bulletin of the Institute of Physics, Singapore* 7 (1) (1979) 1; and M.J. Moravcsik, Science and Technology, 'Basic' or 'Applied' – How to Balance Economic Development, *Science and Technology (S. Korea)* 14 (6) (1981) 26. (in Korean).

[9] See R.H.B. Exell and M.J. Moravcsik, Everyday Science in Developing Countries, *Nature* 276 (1978) 315, which was published, without the authors' consent to the change in title, under the racy, journalistic, and somewhat misleading title of "Third World Needs 'Barefoot' Science". The article also contains additional examples.

[10] Interesting data on robotry is contained in *Science Indicators 1982* (National Science Foundation, Washington DC), in the chapter on international science (to be published).

Reprinted by permission of the *Bulletin of Research and Information* 1 (1981) 16.

Global scientific and technological co-operation

Michael J. Moravcsik*

TODAY many national, regional, and international organizations and agencies are involved in programmes labelled "international scientific and technological co-operation", and a substantial number of people as well as hundreds of millions of dollars are devoted to such tasks. At the same time, such activities are not infrequently viewed with a certain amount of suspicion, questioning, uneasiness, and criticism, and in fact from various and often diametrically opposed points of view.

On the part of some of the primary recipients of such international co-operation, charges of "neo-colonialism" are heard, while some of the primarily donor countries and organizations express reservations on the grounds that such co-operation amounts to a hand-out to freeloaders, or that it creates future competitors. Such an atmosphere of suspicion, especially if unwarranted or grossly exaggerated, can be very damaging to the cause of such co-operation, and hence it might be useful to analyse the balance of such a co-operation. Who benefits from it and for whom is it detrimental?

SOME PREREQUSITES

In setting up the framework of the discussion, I would like to discuss briefly two observations which will be important in the analysis to follow.

The first of these can be summarised by saying that all real-life problems and situations are multi-dimensional and must be discussed as such. I elaborated on this point in greater detail in a recent book of mine. Very briefly, the point is that when one comes to analysing the causes of a situation, the motivations of people in it, the alternatives available for the future, or even the present status of a situation, one must always work in terms of a multitude of causes, a variety of motivations, a whole range of future possibilities, and even a number of different aspects and facets of the present situation.

For example, the answer to the question that forms the title of this article cannot possibly be given by naming one person, or

*Institute of Theoretical Science
University of Oregon
Eugene, Oregon 97403, USA*

organization or country, or faction. Similarly, when we ask: Why does country so-and-so participate in international scientific cooperation, there is no single answer to such a question.

The answer is inherently multi-dimensional, and hence all such problems must always be considered in such a multi-dimensional framework. Only in the natural sciences do we deal with problems and questions which are one-dimensional and the answer to which can be given in 2—3 words, and even in those cases the scientist has to work very hard to isolate the problem, to abstract and idealise it, so that it becomes, at least approximately, one-dimensional.

There is a specific consequence of this multi-dimensionality that we will use in our analysis. It pertains to the fact that while in one-dimensional thinking everything is visualised to occur along a line, and hence, in a dispute, if one party gains the other must necessarily lose, in multi-dimensional thinking this is not at all necessarily so: It is possible to have·courses of action which benefit both protagonists simultaenously.

The second observation we will agree on in our analysis is that since we talk about a real-life situation with real people as characters in it, we must describe people as they really are: Neither as pure "angels", nor as pure "devils". Fortunately our multi-dimensional thinking allows this, though it does not follow from it, and hence this second requirement must be stressed separately.

In discussing motivations of people we will find among them some which are altruistic, some which point toward one course of action and others that point toward an opposite one. All of these together form the picture, and any of them when isolated and unrelated to the others is likely to be extremely misleading and hence useless.

SOME MOTIVATIONS

An occurrence which promotes the realisation of some aspiration is considered a benefit for the aspirant. Thus, in deciding who benefits from international scientific and technological co-operation, we must first discuss the aspirations and the motivations of the participants in this co-operation.

At the risk of some over-simplification, we can say that such co-operation has two types of participants: The developing countries, where the scientific and technological infrastructure has not evolved far enough to allow the country to function on a more-or-less equal footing with other countries, and advanced countries, with a substantial scientific and technical tradition, and a well developed scientific and technological infra-structure which is well connected both to the cultural life and to the productive sector of the country. Such sharp dichotomy, of course, does not really exist in the world, and the spectrum is in reality a continuous one, but this particular simplification will not affect our discussion substantially.

Let us first discuss some of the motivations of the developing countries in participating in scientific and technological co-operation. There are a number of these.

First, since science, and to a somewhat lesser degree, technology, are collective and universal pursuits, if a country wishes to work in the sciences and in technology, it must be "plugged into" the worldwide scientific and technological community. These disciplines depend on a researcher being aware of the work of others, being subjected to the critique of others in the field, having a good communication system with the other practioners of the discipline.

In fact, isolation is probably the most severe single handicap in the pursuit of science or technology in a developing country, ranking even higher than the absence of financial resources. Hence one of the important motivations for a developing country in participating in international co-operation is in assuring that its own science and technology remain healthy. I am placing this motivation first on the list because it pertains equally to developing or "developed" countries.

The second dimension for the motivation is the fact that when a country wants to evolve its scientific and technological capability, it must, at least in the initial stages, rely on assistance from other cuntries. The education of scientists and engineers, the formation of "schools" of research and development, equipment and its maintenance for research and development all demand such a co-operation.

The third motivation for such co-operation is the fact that successful participation in such co-operation alleviates the feeling of dependence which is perhaps the most burdensome element of being a developing country. As I tried to analyse elsewhere recently[2], such dependence is partly "real" and partly a psychological but in either case an undesirable situation.

International co-operation in the sciences and in technology, especially if it is a functional one with actual give-and-take and interaction with other countries, but to some extent even if it is a nominal one through a paper membership in some international organization, can contribute to a change in such a dependent status.

One might even go one step further and ask why a developing country should want to have its own science and technology? The many answers to this question can be gathered under three broad headings: One, an important influence on Man's view of the universe and his place in it. I have discussed this question in greater detail elsewhere[3]. In our present discussion it suffices to list them as forming the basis for a desire to participate in international co-operation in science and technology.

Since science and technology is part of the aspiration of the 20th century Man, they have become objects of "national prestige". There are numerous instances[4] of Third World leaders pronouncing science and technology being a crucial component in the prestige of the country, and thus a motivation of a developing country for participating in science and technology may be also "to be great". It may very well be that such a motivation, in fact, originates mainly with one particular leading person, a politician or a scientist, whose energy, influence, and hard work manifests itself in the activities of the whole country.

Thus personal aspirations also become mixed with collective and societal feelings. In such instances co-operation on an international level provides a good arena in which to exhibit the country's progress in in science and technology.

The picture would not be complete without mentioning some motivations and feelings in developing countries which disfavour participation in international co-operation science and technology. One of these is national pride which is sometimes interpreted as insisting on going it alone, on doing things without external help, and hence it looks on co-operation as an admission of weakness or failure.

A second motivation stems from a fear of being "corrupted", culturally, or religiously, or socially, by contact with external trends and currents of thought and action. The third involves a suspicion that in any international contact the country will be taken advantage of, will be exploited, will be manipulated to the advantage of the other countries. All such feelings argue against international co-operation, against internationalism, for isolation and for a regressive world view which abhors change.

The above discussion of the motivations of developing countries to participate in scientific and technological co-operation is far from complete, but even so, it exhibits a broad variety of various arguments and trends. The list is by no means less varied and heterogeneous in the case of the analogous motivations in the case of the "advanced" countries.

In discussing that, it is helpful to consider separately the two main types of organizations in the "advanced" countries which are involved in such co-operation. namely the government and the private sector. To be sure, in some relatively "advanced" countries the private sector is not a significant force compared to the governmental sector, but in most of the countries which have a leading position in international scientific and technological interaction, the private sector plays an important role.

Turning first to the governmental sector, one motivation to participate in scientific and technological co-operation is that it forms a good field in which active relations among countries can be established. "Good relations" with other countries is an objective for most countries, for political and other reasons. In science and technology the manpower to be used is there, and sometimes even plentiful, as during a period when the creation of scientific and technological manpower exceeds the momentary domestic demand for such people.

It has happened more than once that when two heads of State met and could not get very far on resolving problems and conflicts of a political and social nature, they fell back on establishing some scientific and technological agreement for co-operation between the two countries.

It is believed on the part of the "advanced" countries that scientific and technological co-operation allows a general political and cultural interaction with the recipient country and hence is advantageous from the point of view of political and cultural influence. For this reason much of such co-operation is often under the jurisdiction of the "advanced" country's foreign affairs ministry rather than under its scientific institutions.

A different motivation for scientific and technological cooperation with developing countries is given in terms of trade and international markets. Most "advanced" countries depend, for their economic well being, on as large an international trade as possible, and on as extensive international markets as possible. Countries which have a low gross national product per capita constitute poor trade partners because they cannot afford to buy, if they buy they cannot afford to concentrate on the advanced type of products that many of the "advanced" countries are good in producing, and because they do not have much to sell either. It is, therefore, of great importance to an "advanced" country to build up the developing countries economically so that productive trade with them becomes possible.

A yet another motivation is basically a moral one. Whenever cultures or civilisation have risen to considerable heights and have accomplished much in some area, they tended to assume a messianic, missionary demeanor and to consider it their moral duty to convert the outsiders to the practices and beliefs of their culture. This phenomenon is a quite controversial one, since what may appear to be a moral urge from one side can appear as an intolerably arrogant and materialistic force from another side. The question of what this "really" is has no meaning. Instead, we simply have to list the phenomenon itself in our multi-dimensional syllabus of factors.

There is also a more general motivation for the "advanced" countries to evolve science and technology in the developing countries. At the present, virtually all of the new science and technology in the world is created by countries with an aggregate population of about one quarter for the world's population. The remaining three-quarters are, therefore, only a latent potential.

If this latent potential for scientific and technological development could also be utilised, it is virtually certain that much more rapid progress could be attained in science and technology, something that is likely to benefit everybody, including the "advanced" countries.

There are also feelings on the part of the governments of the "advanced" countries which run against international cooperation. Isolationism exists in this context just as it exists in the developing countries, though perhaps for somewhat different reasons: There are those in the "advanced" countries who believe that their countries would do better by isolating themselves from the problems of the world and paying attention only to the domestic concerns.

It is unfortunate that in some countries the labour unions have been especially vocal in voicing such sentiments. A somewhat related feeling is that by engaging in cooperation with the developing countries, future competitors are being groomed which will eventually decrease the economic and political power of the "advanced" countries.

Such arguments are based on the "fixed-pie" model of the world, in which the assets and resources are finite and the question, therefore, is only on how to distribute the total of fixed size. Such a view of the world is quite at odds with our historical experience. The total is far from fixed, and the aggregate wealth of the world can grow by leaps and bounds. The most important resource is the brainpower and inventiveness of humans, and that appears to be limitless.

Such creativity can circumvent shortages of minerals and of raw materials by substitution, and make products of fantastic value out of materials of great abundance, using ever more powerful energy sources harnessed by the same creativity and inventiveness. Even our living space expands constantly, and we are on the threshold of space colonisation thus entering into a new era of human habits.

There is one fear of the isolationists which is justified but largely irrelevant. It is certainly true that throughout history countries, civilisations, and world leaders have risen and fallen, and this will undoubtedly continue. Thus the presently leading countries of the world will also decline, and their places will be taken by others. This consideration is irrelevant, however, since such a decline has primarily internal origins and isolationism cannot postpone the decline substantially.

Let us now turn to the private sector of the "advanced" countries. Their aim is to succeed in a business sense. One, but by no means the only, indicator of such a success is immediate profit. Others are stability, public image, adequate share of markets, harmonious relationship between the workers, the management, and the stock holders, a diversified base of operations, etc. In all these respects the private sector has incentives to be active in scientific and technological cooperation with the developing countries, where there are new markets, important raw materials, excellent potential manpower for workers and management, important potential sources of new inventions and patents, a potentially more comfortable business climate, opportunities for new inventions and products according to the geographical locations of the developing countries, etc.

Especially companies with a home base in a declining "advanced" country find it increasingly more advantageous to operate in a "young" country with more energy, drive, and imagination, where the fear of the unknown and the fear of risks have not ruined the climate for creative production. Many of the developing countries represents pioneer land with unlimited opportunities.

The above discussion pertained primarily to co-operation between an "advanced" and a developing country. Similar arguments can be listed for the scientific and technological co-operation between two "advanced", or two developing countries, using the motivational elements outlined above.

CAN THE MOTIVATIONS BE RECONCILED?

In deciding about the benefits of scientific and technological cooperation, we have to decide now whether the multitude of motivations we discussed which fuel the activities of the developing and the "advanced" countries can be reconciled with each other. whether they can coexist to their mutual benefits, whether they are compatible with each other or not.

It is my inescapable conclusion from looking at that list of motivations, that, on the whole, they can be very well reconciled with each other, and that it is very possible to have a cooperation in which most of the above motivations are simultaneously satisfied and hence all parties can benefit. Looking at the problem with a multi-dimensional eye, this is hardly surprising. To be sure, there might be and will be instances where conflicts will arise, and where negotiations must, therefore, take place, resulting in some kind of a compromise solution in which each party both takes and gives. This conviction is certainly strengthened by our experience in seeing a number of developing countries evolve fast through such an international cooperation.

From the point of view of the developing country, success in cooperation in science and technology rests on two very important foundations.

First, the country must at all stages strive to have as strong, broad, and active an indigenous scientific and technological infrastructure as possible. Cooperation, negotiations, collaborative activities, joint planning and implementation of projects all require expertise on the part of both the participating developing and the participating "advanced" country. In the absence of such expertise in the developing country, that country will not be able to utilize cooperative possibilities, even if the other, "advanced" country has the best of intentions and the most altruistic stance. It is, therefore, very much more important for a developing country to evolve its own infrastructure than to be the recipient of assistance projects in which the results of science and technology are simply imported into the country.

Second, the developing country must have a maximally pragmatic attitude toward such international cooperation. Instead of worrying about whether the "advanced" country participating in the cooperation will benefit from it or not, the emphasis should be on whether the developing country will benefit or not, and if it will, the fact that the other country also benefits should make the cooperation only more attractive.

Cooperative projects should be assessed on the basis of real scientific and technological benefit to the developing country, and not in terms of appearances, semblances and fashions. Such an assessment is not easy and requires the broad and active scientific and technological infrastructure mentioned earlier.

In a sense the question in the title of this article is an academic one. No country has ever been able to evolve itself scientifically and technologically without international cooperation, and hence the question concerning international collaboration is not so much "whether" but "how". The aim of this article was to show that even the question of "how" is not as controversial as some may think, since many of the motivations for such cooperation held by various parties around the world point in the same direction. The problem, therefore, is not an ideological one, but only one of skill, understanding, masterly of negotiations and management and patience. As such, the solution of the problem is definitely possible and can be arranged to the mutual satisfaction of almost all.

REFERENCES:
1. M.J. Moravcsik, *How to Grow Science*, Universe Books, New York (1980), Appendix A.
2. M.J. Moravcsik, *Dependence* (submited for publication) (1981)
3. M.J. Moravcsik, *The Context of Creative Science*, Intersciencie 1;2, 72 (1976).
4. For example: "It is an inherent obligation of a great country like India, with its tradition of scholarship and original thinking and its great cultural heritage, to participate fully in the march of science, which is probably mankind's greatest enterprise today." (Jawharwahl Nehru)

F. Research on Science

Preprint, published in *4S Review* 3:3 (1985) 2.

Science in the Developing Countries: An Unexplored and Fruitful Area for Research in Science Studies

Michael J. Moravcsik
Institute of Theoretical Science
University of Oregon, Eugene, Oregon

I. INTRODUCTION

What we know about the context of science and about the scientific activity as a human endeavor comes almost exclusively from two sources: The investigation of the contemporary status of science in the scientifically advanced countries, and the study of the history of science in the Western civilization. Yet even today, the scientifically advanced countries comprise only about one-quarter of the world's population. What about contemporary science among the remaining three quarters of humanity? Information on this score is extremely fragmentary. There are, of course, exceptions but so few in number that they just tend to strengthen the rule. Research by the late Francis Dart[1] into the concept of nature in non-Western societies stands out as pioneer classic. Some of the work of Derek de Solla Price[2] includes quantitative indicators for developing countries also. More recent work by Davidson Frame and others[3] provides insight into distribution and patterns of scientific disciplines in those countries. Research by the late Olga Gasparini[4] on the Venezuelan scientific community offers a most valuable source not only for "facts" but also for the self-image of scientists in a developing country. Then we have the study of the Useems.[5] Though these efforts do not exhaust the list,[6] the total body of such research is very tiny compared to the fast growing literature dealing with science in the scientifically active countries.

I want, therefore, to present a case for undertaking much more research in science studies which pertain to developing countries. I will first give some reasons why such research would be attractive and valuable, and how one can arrive at a list of research topics that would be fruitful and interesting. I will then proceed to offer a partial list of such topics. Finally, I will discuss some of the methodological and logistic problems connected with such research. This is frankly a note about the future, about things to be done, rather than about results already obtained and conclusions drawn. I am quite certain, however, that at least some of you will be impressed by the vast potential and great promise of the work that awaits us in this domain.

II. WHY DO SCIENCE STUDIES ON THE DEVELOPING COUNTRIES?

There are basically two kinds of motivation for learning about science in the developing countries, just as there are similarly two aspects of science studies in general. First, the developing countries offer a domain for research in the science of science which is basically different from what is available to us otherwise. We can say in effect that the developing countries offer a contemporary opportunity to study the history of science. I am fully aware that by putting the matter in those terms I imply a basic assumption, namely that there is only one kind of science[7] in which some countries have progressed to a larger extent, and others to a lesser extent. This assumption itself is testable through research on developing countries. If true, these countries give us an opportunity to view the development of science at stages which in the Western civilization lie in the past and hence are accessible only by historical methods. Furthermore, since many developing countries building their science are fairly isolated from each other, one can even explore several variants of the same development, something that is impossible when we deal with one fixed and concentrated history. In other words, the study of science in developing countries promises an independent method to test our ideas about the science of science and to deepen our understanding of the processes that take place when science is practiced.

The second motivation is a more pragmatic or functional one. Most of the developing countries around the world are in the process of building their scientific infrastructure, with skill and intensity that may vary greatly from country to country. At the same time, there are many international and national organizations and agencies engaged in international assistance in the areas of science or related fields. In order to build science successfully or to assist effectively in this process, we need factual information on the problems and circumstances that exist in this regard in the developing countries. Considering

the magnitude of the effort around the world, and the great stakes involved, even a minor amount of improvement in science development would have a large impact.

In this respect it is particularly regrettable that expertise, interest, and activity in science studies in the developing countries themselves are even more sporadic than expertise, interest, and activity in the sciences. It has been stated[8] that while about 92% of the world's scientists are in the scientifically advanced countries, 96% of the historians of science are there, and the percentage is probably even higher if we consider everybody working in science studies. Thus it is likely that, for a long time to come, the developing countries will not be in the position to do the necessary research in science studies to supply the facts and understanding needed to analyze their own problems in science building. It is, therefore, (or should be) part of international scientific assistance to carry out research work in the scientifically advanced countries on topics that pertain specifically to the developing countries.

II. HOW TO MAKE UP AN AGENDA FOR RESEARCH?
Apart from the very sporadic instances of solid research mentioned earlier, information on science in the developing countries comes from two sources.

The first of these consists of governmental or international statistics and reports. On the whole, these suffer from several deficiencies. First, the statistical information is often unreliable because of ambiguities in definition, because of faults in the methodology of collecting it, and because of the constraints imposed by national pride, political compromises, and other irrelevant factors. Second, the content of the statistics and the reports often tends to be formalistic, concentrating on logistic and bureaucratic questions which are of no great interest either to the scholar in science studies or to the practical science policy maker. This is not to say that this source is entirely without value, but it certainly covers only a small corner of the whole domain we wish to explore.

The second source of information on science in the developing countries contains empirical, anecdotal, personal accounts by various indigenous and visiting observers, and some generalizations drawn on the basis of these observations. This source is much more colorful, interesting, and functional than the first, but even so, opinions and individual views are not the same as systematic research of well delineated aspects of the overall problem. Thus much work remains to be done.

Nevertheless, these personal accounts can very well serve, at the initial stages, to compose a list of topics that appear to be particularly worthwhile for research in the context of the developing countries. The credibility of such a list is strengthened by the fact that there appears to be a great amount of unanimity among observers who have had a substantial amount of personal experience with science in the developing countries concerning what the characteristics are and what the problems tend to be.

I would like, therefore, to set myself up as a representative of this observer group, and offer you a shopping list for research topics. I have have been involved in science in the developing countries on a part time basis but to a significant

extent for about 25 years, and have had the good fortune to be able to observe, first hand, science in a large number of developing countries spread over all continents. To be sure, perception and insight are not acquired like seniority in the civil service system, and hence the mere fact of a prolonged and extensive exposure does not guarantee the wisdom of what I have to say. Since, however, very few of the above mentioned observer groups are simultaneously also active in our Society, my note can be construed as one of the first steps in establishing a link between the concerns of our society and the arena of developing countries.

IV. WHAT TO RESEARCH IN DEVELOPING COUNTRIES

As an introductory note, let me stress again that it appears to many observers that the science development problems in developing countries are, to a remarkable extent, the same the world over, and independent of the historical, cultural, racial, religious, economic, ideological, and social characteristics of the individual countries. What differences exist between countries can be interpreted more as different stages of the same development than altogether different directions of development.

This statement itself needs further research and verification. For the moment, however, I will assume that it holds, and hence from now on I will not mention the possibility that all the topics to be listed may need to be researched in various environments separately, thus expanding the list by a large factor.

In offering the areas of research, I will use a set of subheadings that I have found useful before in discussing science in developing countries.[9]

1. Motivation and justification of science

There are three groups whose attitudes toward science play an important role in science policy: The scientists themselves, the people who make decisions about science policy (that is, about providing for science), and the population as a whole. The questions below therefore can be asked about each of these three groups.

What is the prevailing concept of nature, laws of nature, and the way events in the world occur?

What is the prevailing concept of what scientific activity is, that is, of what the objectives, methods, and needs of this activity are?

What are the individual motivations for scientists to engage in scientific work?

What are the societal justifications for the support and development of scientific activity?

What are the expectations with respect to the outcome of such scientific activities?

In researching these questions, care must be used to distinguish between conspicuous rhetoric on the one hand, and actual attitudes on the other.

These questions might appear academic and abstract, but in practice they are of crucial importance in science development. Ignorance about the nature of science, confusion about motivations and justification, misconceptions about the requirements for productive scientific work are among the primary retardants of the establishment of indigenous science in many countries. Measures to correct these deficiencies have not

been effective and numerous partly because there has been no systematic diagnosis of the nature and extent of the malady.

2. Science education

How do prevailing concepts of nature and of laws of nature reflect in the type of science education offered?

To what extent are scientists in developing countries effective in influencing the population as a whole in the direction of "scientific thinking"? In other words, is indigenous scientific activity helpful in bringing the conceptual scientific revolution to the masses of the country?

What brings students in developing countries into science? What part does the existing scientific community play in stimulating the next potential generation of scientists into choosing science as a profession?

It is said that in many developing countries abstract and theoretical considerations of science are preferred by scientists to empirical and experimental work. Is this claim true, and if so, what are its causes?

It is said that scientists of breadth and versatility are, even relatively speaking, in shorter supply in the developing countries than in the scientifically advanced countries. Is this claim true, and if so, what are its causes?

What are the origins of the two major defects of science education in many developing countries, namely rote learning and premature specialization?

It is said that the concept of lifelong, continuous education for a scientist is disregarded in developing countries, as compared to scientifically advanced countries. Is this claim correct, and if so, what are its causes?

What are the advantages and handicaps of having been educated abroad as compared to indigenous education, from the point of view of a working scientist, his productivity, his role and influence in the indigenous scientific community, his choice of research problems, and his ability to participate in organizational activities?

To what extent does foreign education produce schizophrenia in cultural affiliation, in aspirations and motivations, and in scientific affiliations?

In retrospect (that is, after having returned home from abroad with an advanced degree and having worked at home as a scientist for 2-3 years), what are in the opinion of foreign-educated scientists, the most beneficial and the most detrimental elements of advanced scientific education at American graduate schools from the point of view of somebody whose scientific career will take place in a developing country?

It would appear that some of these questions could be relegated to specialists in education and would not have to be researched by people well versed in broad areas of science studies. This is, however, not so. In order to have the power criteria for gauging scientific performance and attitude, experience in general science studies is most desirable.

3. Manpower

The development and maintenance of scientific manpower is the pivotal element in science development which determines the rate at which the country can build its indigenous scientific infrastructure. Research pertaining to it, therefore, is of special importance.

Is the productivity distribution of scientists in developing countries the same as in the scientifically advanced countries? Is Lotka's law valid in any country?

Is there a correlation between the productivity of scientific communities and the prevalent language of the country?

Is the productive lifetime of scientists in developing countries different from those in the scientifically advanced countries?

Is the attrition of scientific manpower with time due to switching into nonscientific professions different in developing countries, and if so, what are the patterns?

Is the extent to which scientists fall victims of administrative and bureaucratic duties and hence cease to be scientifically productive greater in developing countries than in the scientifically advanced ones?

What to do with "aging" scientists is a worldwide problem. With scientific productivity often peaking relatively early in life, finding adequate outlets and gratification for older scientists is a challenge. Is this more or less severe in developing countries and what are the options used there to cope with the problem?

Some countries intentionally over produce scientific manpower with the aim of letting the surplus be "stored" abroad for a time, to be "recalled" later when the country's opportunities and requirements demand it. Does such a procedure work in practice, and how do such scientists who resided abroad for a long time integrate into the indigenous scientific community?

How large is a "critical mass" of scientists that can be productive in a developing country? Is the size different from the one in scientifically advanced countries? What parameters influence the size?

Are scientists in developing countries more or less "competitive" than scientists in the scientifically advanced countries? Is the scientific community there more divisive, more strife ridden?

What is the status of social studies of science among the scientific communities of developing countries?

4. Communication among scientists

As we know, the communication patterns within the worldwide scientific community are determined primarily by the scientists themselves. The system that thus evolved is based on the desire of maximizing scientific research output in the very near future. A consequence of this principle is that in scientific communications, the rich is getting constantly richer, and the poor constantly poorer. The documentation of the inequitability of scientific communication patterns handicapping the developing countries, and the search for ways to remedy the situation are therefore very important objectives in which research in social studies of science can play a very substantial role. There is, consequently, a long list of interesting research topics available in this area, both with respect to the internal scientific communication within a given country, and the international communication that links the scientists of that country with the rest of the world.

How do the internal scientific communication patterns within a given developing country differ from those in a scientifically advanced country? What function do the communication media (domestic journals, scientific meetings, seminars, exchange visits, letters, telephone calls, etc.) play?

We know something about the relative importance of the various scientific communication modes (journals, preprints, letters, meetings, seminars, informal travel, telephone, etc.) for scientists in the scientifically advanced countries. Is this hierarchy different for scientists in developing countries?

In what way are domestic journals in developing countries different from international journals, in terms of standards, subject matter, readership, prestige, worldwide citability, and social functions?

In what way are professional societies different in developing countries from their counterparts in the scientifically advanced countries?

What are the patterns of communication between universities and governmental research laboratories in developing countries, and how do these differ from similar patterns in scientifically advanced countries?

What is the communicative structure in developing countries between scientists and technologists? Between science based industry and universities?

What is the intensity and pattern of communication of scientists in developing countries with those who make national decisions for science policy, that is, with politicians, bureaucrats, civil service personnel, etc.?

Given a fixed amount of money to be spent on improving domestic communication patterns within the scientific community of a developing country, which mode or modes of communication provide optimal investment of these funds?

In what way and to what extent does isolation (i.e. lack of adequate communication with the worldwide scientific community) influence the choice of research topics, the standards of research, the pace of research work, and the utilization of the research results in developing countries?

Is there a correlation between the scientific productivity of scientists in developing countries and the strength of their communication ties with the domestic and/or international scientific communities?

In what way and to what extent are scientists in developing countries handicapped in publishing their papers in international journals, due to the page charge requirements, refereeing systems, and editorial policies of these journals?

What are the citation patterns of and by scientists in developing countries, and can possible differences between these patterns and those of scientists in the scientifically advanced countries be attributed to particular causes?

Is science in the developing countries underrepresented at international scientific conferences?

Are scientists in the developing countries underrepresented among the speakers at international scientific conferences?

Are scientists in the developing countries underrepresented in decision making bodies of the international scientific community, such as editorial boards of international journals, executive councils of international scientific organizations, organizing committees of scientific conferences, etc.?

To what extent have scientists in the developing countries communicated their special problems to the members of the worldwide scientific community?

5. Scientific research
Is there, in any sense, a special type of science for the developing countries, different from that of "Western" science, and if so, what are its characteristics?

In what way is the choice of research problems and the balance among various scientific fields different in developing countries from the problems researched in the scientifically advanced countries?

There is frequent discussion about, and statistical information on, "basic" vs. "applied" scientific research in developing countries. There are also some international guidelines for composing such statistics. What is the de facto status of the definitions and distinctions made between "basic" and "applied" scientific research in developing countries?

Is there a difference between the research patterns and research productivity of scientists from a given developing country who work in that country, as compared to nationals of the same country working in a scientifically advanced country?

Many official policy documents in developing countries call for scientific research "relevant to the needs of the country." What is meant by this in principle, and in practice, what criteria are used to judge the relevance of research, and how successful is this objective in terms of the country's development?

While the predominant fraction of resources devoted to scientific research in developing countries go to governmental research institutions and only a small fraction to universities, scientific productivity in the universities appears to be much higher than in the research institutions. Can this impression be documented, and if so, what are the causes of this discrepancy?

What are the patterns of research cooperation between university and governmental research in the developing countries, and how does this compare with such cooperation in the scientifically advanced countries?

What is the relationship, if any, of technological research and development and scientific research within a given developing country?

What is the success rate of "applied" scientific research in developing countries, measured in terms of successful transfer to and use in technological development?

How successful is the scientist in a developing country in serving as a conveyer of scientific information generated abroad to the technological developmental activities in the country?

Is there a correlation in the developing countries between the involvement of a scientist in research and his effectiveness and impact in indigenous science education?

What is the impact of research performed by international research institutions located in certain developing countries (e.g. IITA in Iabadan, IRRI in Los Banos, ICIPE in Nairobi, etc.) on the research activities of indigenous institutions in the same country?

6. Scientific organization

In most developing countries the participation of scientists in decisions pertaining to providing for science is minimal. Thus there are two distinct groups, the scientists and the administrators, who often have only tenuous communications with each other, and whose background and framework of thinking are far apart.[10] The result tends to be a psychological environment for science which is discouraging from the very start, and which continually demands a sizable fraction of the scientists' physical and mental energies for the purpose of assuring the protection of even the minimal elements necessary for the productive pursuance of scientific activities. It is against this backdrop that the questions listed below emerge as relevant ones.

What is the educational and professional background of science policy decision makers in developing countries?

What is the attitude toward science of high level governmental officials in developing countries? What are their expectations with respect to scientific activities?

In many countries' development plans there are sections pertaining to science development. What is the relationship between these plans and the actual scientific activities that take place in the country during the time period planned for?

What is the extent of the influence of indigenous scientific communities in developing countries on science policy in those countries, and in what respects is it most effective?

To what extent are scientific communities in developing countries knowledgeable about matters of social studies of science and of science policy, and to what extent are problems of national science policy discussed within these scientific communities?

What are the scientific productivity patterns of governmental research institutions in developing countries as a function of size, time since establishment, area of research, and location?

What are the characteristics of academic institutions which are most productive in scientific research?

What is the specific cost of research in developing countries, by area of science, institution of the researcher, and other parameters, and how do these figures compare with analogous figures in the scientifically developed countries?

What is the de facto method used in developing countries to decide the allocation of resources for various scientific projects? De facto here refers to what actually takes place in decision making bodies, in contrast to what the formal rules and procedures may be.

What evaluation procedures if any, are used to assess the success or failure of completed scientific research projects?

7. International aspects

Some of the questions of international scope were listed in earlier sections. There are, however, some additional aspects also, mainly connected with international assistance and cooperation.

What has been the impact, if any, of the UN agencies (UNESCO, IAEA, FAO, WHO, UNDP, etc.) on the building of scientific infrastructures in developing countries? To what extent do such UN programs affect the work of a research scientist in those countries?

What has been the impact, if any, of specific national foreign aid agencies (AID in the US, CIDA in Canada, SIDA in Sweden, etc.) on the building of scientific infrastructures in developing countries? To what extent do such programs affect the work of a research scientist in those countries?

To what extent have regional scientific cooperative schemes been successful in stimulating the indigenous scientific communities?

To what extent are bilateral links between a group of scientists in a developing country and a counterpart group in a scientifically advanced country successful? What are the characteristics of the successful links as compared to the less successful ones?

What is the attitude of professional scientific societies in the United States toward cooperation in science with developing countries? How is this evident in the membership, in the leadership, and in the administrative staff of these societies?

What are the areas in which the scientific communities of the developing countries would be most interested in seeing action by international agencies in science and technology?

8. General remarks

The above questions, intended only as samples of the type of problems which appear to be of interest in science development, could all be answered in terms of opinions offered by people with extensive personal experience in this field. In fact, it is likely that asking any of these questions will prompt such an "expert" to launch into an extensive discussion of his theory and view on science in the developing countries. Such opinions, however, have neither the scholarly standing nor the practical persuasive power that systematic and extensive studies would possess. Governmental officials, international servants, science administrators can be impressed by statistics, quantitative assessments, and analytical studies, especially if those studies are sufficiently well circulated so as to have an effect on the external image of the country or community they describe.

It should also be mentioned that in addition to studies pertaining to the questions listed above, there is also a great need for an integrated study of science in a given country, since this would also reveal something about the interrelationship of the many factors appearing in the previous list. To the best of my knowledge, there has been no extensive study of a whole country's scientific life which goes beyond giving statistics and listing formal organizations. Such a case study would be most revealing in that it would clearly exhibit the nature of the science gap that exists today in the world and would definitely point at particular measures to help closing this gap.

V. HOW TO DO SUCH RESEARCH?

There are several special difficulties in pursuing research in the social studies of science in developing countries.

First, almost all such research would involve travel to and fairly extensive stay in one or several developing countries. To be sure, some research could be done through questionnaires, or through the use of the Citation Index or similar statistical data bases. Most research, however, would require some on-the-spot investigation. This raises the problems of acquiring travel funds (for foreign travel, which is difficult), and adjusting to a different environment over a short enough time so that effective research can be done.

Second, and related to the first, there are relatively few organizations ready to support such research. There are obvious candidates for the source of such support, such as AID, UNESCO, etc., but, for a variety of reasons, in practice one cannot count on these organizations to any substantial extent. They are either interested only in technology and not in science, or, for political reasons, shy away from "controversial" subjects. NSF appears to be an ideal source for research, and indeed, NSF does sponsor some research in science policy and in the science of science, but the focus there is overwhelmingly on domestic science, or on science in the other scientifically well developed countries.

Another source is private foundations, though they have not been overly interested in supporting projects connected with science, and are often more action than research oriented.

All in all, it is clear, however, that new opportunities for research support are needed. In this respect, I think it would be most appropriate and perhaps even effective if our Society created a well-argued and well-documented policy statement urging more research opportunities in this field, and then transmitted this statement, coupled with person-to-person discussions, to the appropriate key people in national, regional, and international organizations active in world-wide scientific cooperation, and to private foundations with an interest in international science.

Finally, such research should also make some contribution toward building the local infrastructure of social studies of science, and hence should, if at all possible, be carried out in cooperation with some local people who are active in such social studies, or at least have an incipient interest in such studies. In the long run, special problems of developing countries should be researched, analyzed, and remedied by these countries themselves, even though international cooperation in such studies will remain a desirable objective at all times.

Appendix about the bibliography of this field

The literature on science in the developing countries is very diffuse, appearing in a multitude of journals, reports, and conference proceedings. An attempt was made to offer an extensive though not necessarily comprehensive bibliography in reference 8. That book covers the literature only up to about 1972. A supplement of that bibliography, with a closing date of Summer 1977, is given in M.J. Moravcsik, Indigenous Science-A Kingpin in Selfpropelled Development, a report prepared for the Policy Research Analysis Division of the National Science Foundation in September, 1977. It is high time for a more up-to-date bibliography of the field to appear.

References

1. Francis E. Dart, "The Rub of Cultures," Foreign Affairs 41, 360 (1963); Francis E. Dart and Panna Lal Pradhan, "Cross-Cultural Teaching of Science," Science 155, 649 (1967).

2. Derek J. de Solla Price and Suha Gursey, "Some Statistical Results for the Numbers of Authors in the States of the United States and the Nations of the World," Preface to Who Is Publishing in Science 1975 Annual, Institute of Scientific Information, Philadelphia, Pa. (1975).

3. J. Davidson Frame, "Mainstream Research in Latin America and the Caribbean," Interciencia 2, 143 (1977); J. Davidson Frame and Francis Narin, "The International Distribution of Biomedical Publications," Federation Proceedings 36, 1790 (1977); J. Davidson Frame, Francis Narin, and Mark P. Carpenter, "The Distribution of World Science," Social Studies of Science, 7, 501(1977); J. Davidson Frame, "National Economic Resources and the Production of Research in Lesser Developed Countries," Social Studies of Science 9, 233(1979); Herbert J. Inhaber, "Distribution of World Science," Geoforum 6, 231(1975).

4. Olga Gasparini, La investigacion en Venezuela: Condiciones de su desarrollo, IVIC, Caracas (1969).

5. John Useem, Ruth Hill Useem, Abu Hassan Othman, and Florence E. McCarthy, "Transnational Networks and Related Third Cultures: A Comparison of Two Southeast Asian Scientific Communities." Pp. 283-316 in K. Kriesberg (Ed.), Research in Social Movements, Conflict, and Change. Vol. 4. JAI Press (1981).

6. For a few more recent

references, many of them by authors from developing countries, see S. Arunachalam and K.C. Garg, "A Small Country in a World of Big Science: A Preliminary Bibliometric Study of Science in Singapore," Scientometrics 8, 301 (1985); Thomas O. Eisemon, The Science Profession in the Third World, Praeger Special Studies, New York, (1982); Larissa Lomnitz, "Hierarchy and Peripherality: The Organization of a Mexican Research Institute," Minerva 17, 627 (1979); Lewis Pyenson, "The Incomplete Transmission of a European Image: Physics at Greater Buenos Aires and Montreal, 1890-1920," Proc. Amer. Philos. Soc. 122, 92 (1978); A.B. Zahlan, Science and Science Policy in the Arab World. St. Martin's Press, New York (1980).

7. Michael J. Moravcsik, "Do Less Developed Countries Have a Special Science of Their Own? Interciencia 3:1, 8 (1978).

8. Derek J. de Solla Price, "Who's Who in the History of Science: A Survey of a Profession," Technology and Society 5:2, 52 (1969).

9. Michael J. Moravcsik, Science Development--The Building of Science in Less Developed Countries, PASITAM, Bloomington, Ind., Second Edition 1976.

10. Michael J. Moravcsik, "The Missing Dialogue--An Obstacle in Science Development," International Development Review: Focus 18:3, 20 (1976).

II. THE PROBLEMS

A. The Nature of the Problem

Dependence

Michael J. MORAVCSIK

Institute of Theoretical Science, University of Oregon, Eugene, OR 97403, U.S.A.

The aim of this paper is to analyze the concept of dependence, particularly in the context of international scientific and technological interactions. It is argued that dependence is more than mere interaction or transaction, and it consists of a crippling loss of freedom of action on the part of the dependent because of his lack of skill and capacity, because of his feeling of indebtedness, and because of the limited opportunities for independent decision making. It is then pointed out that the actual onset of the feeling of dependence is due primarily to psychological factors. Such feeling of dependence creates frustration and hatred toward the country on which the dependent country depends. Dependence also creates a feeling of lethargy and pessimism. It is finally suggested that efforts to end dependence by legislation are futile, and that the only effective road toward lessening dependence is the evolution of an indigenous infrastructure in the dependent countries.

Keywords: Dependence, international relations, scientific assistance

Michael Julius Moravcsik was born in Hungary and attended the University of Budapest, Harvard University, and Cornell University, receiving the Ph.D. degree in theoretical physics from Cornell. He has worked at Brookhaven National Laboratory, the Lawrence Radiation Laboratory of the University of California, Livermore, and since 1967, with the Department of Physics and Institute of Theoretical Science at the University of Oregon. He has held a number of Visiting Professorships in the U.S., Pakistan, Japan, and England. He has lectured in many other countries and participated in numerous symposia on science, technology, and development planning. He is the author of three books (*The Two-Nucleon Interaction*, 1963; *Science Development*, 1975; *How to Grow Science*, 1980) and some 220 articles dealing with physics and science organization & policy, particularly in developing nations.

I am indebted to Professor Edith Moravcsik for a critique of the first draft of this manuscript.

North-Holland Publishing Company
Human Systems Management 2 (1981) 268–274

1. Introduction

Dependence is one of the pivotal words of our age. Colonialism, imperialism, and domination all imply dependence. The often heard 'liberation' suggests a riddance from dependence. The policy called energy independence similarly means a lack of dependence on foreign oil. In the area of science and technology, the dependence of the developing countries on the advanced countries in these areas motivates the worldwide move toward the building of indigenous science and technology in those countries so as to terminate the present dependence.

Being so closely tied to phrases and issues which are so emotion-laden and politically so explosive, the word 'dependence' is heavy with ambiguities, tendencious uses, serious implications, important effects, and significant consequences. It therefore needs clarification, discussion, and resolution with respect to its meaning and import and with respect to how one can lessen its detrimental impact. This will be the aim of this article, especially with regard to the areas of science and technology. In that context one of the recurrent symptoms of dependence is the periodic emergence of the 'two-science' theory, that is, of claims that the developing countries have, or should have, a science of their own which is, or should be, different from the 'Western' science. It is my view that to a large extent this phoney discussion of several sciences is a consequence of the feeling of and resentment against dependence, and that once this dependence is lessened or eliminated, such discussions of several sciences will fade out. To hasten this is one of the motivations for this article.

Our discussion will begin with a listing of what dependence is *not*. This will be followed by a survey of the 'objective' elements of dependence, after which some psychological elements will be analyzed. Having by then clarified the nature of dependence, we can turn to its effects, followed by the investigation of some imagined and real remedies.

2. What dependence is *not*

Because dependence is so closely connected with emotions, and since we want to arrive at a reasonably functional and practically meaningful definition of it, it might be helpful first to point at some possible uses of the word which should be excluded.

'To depend on' is not a synonym of 'to interact with'. In practical and political terms, this means that isolationism is not a proper reaction to dependence. In the areas of science and technology, all countries of the world are interacting with each other, and yet this interaction, *per se*, does not engender a feeling of dependence in all or even most cases.

Neither is 'to depend on' a synonym of 'to get from'. In the areas of science and technology this is particularly evident. For example, the scientifically leading country of the world, the United States, produces only about 25% of the new science and technology in the world. Thus the United States not only gets science and technology from abroad, but in fact gets three times as much from abroad than it produces itself. For other developed countries this ratio is even larger. Yet, this situation is rarely described in terms of the word 'dependence'. Among the scientifically advanced countries this dispersion of the source of science and technology is construed more as an international team work, an association of countries in which each of the individual members both contributes and benefits.

Dependence therefore must mean more than just communicating and trading. To be sure, in any situation in which we talk about dependence there is always some interaction and transaction. But just because interaction and transaction occur, we cannot conclude that dependence exists. There are some additional elements in dependence which we will analyze in the next two sections.

3. 'Objective' elements of dependence

First let us address ourselves to the more tangible, the more objective and universally perceived elements in dependence. 'Material' would not be an appropriate term to use in this context since, as we shall see, some of these elements are not in the realm of food and shelter, or even money and luxury articles. And yet, these elements are material in the sense that they are not exclusively a matter of perception but manifest themselves in the material aspects of life.

Dependence means that the dependent cannot function, cannot develop unless he receives cooperation and, in some cases, permission by the source he depends on. For example, if country X in Africa wishes to construct a new transmission line from town A to B, it cannot do so unless it receives the cooperation of a technologically more developed country which supplies the equipment, the knowhow, the maintenance, etc. If the African country is rich, financial obstacles may not be substantial, and yet, the dependence still remains since the country is incapable of achieving the goal in question, should individuals, organizations, or governments in other countries refuse to cooperate.

To be sure, all countries in the world depend on each other one way or another. To take a ludicrous example, the United States depends on other countries for pandas in its zoos, since pandas do not live in the United States and presumably cannot be raised there outside zoos. One might call such a situation natural dependence, since no country can be entirely self-sufficient in everything. To exclude such inessential instances of sporadic 'local' dependence, we have come to use 'dependence' to denote the situation when a country depends on others for much or most of its needs, economic or otherwise. Indeed, the needs may not be consumption goods. It may be instead scientific knowledge, or novels to read, or popular tunes, or movies, or philosophical frameworks, or political systems.

A second and related aspect of dependence is the state of indebtedness of the dependent to outside entities. This indebtedness may be financial, but in many cases it is of a more intangible but not less noticeable kind. For example, a 'borrowed' political system, an adoption of popular tunes, local currency or local stamps manufactured in a foreign country may not result in a large amount of financial debt and yet imposes on the country a feeling of owing to another country. Such indebtedness can automatically signal a feeling of dependence.

A third aspect of dependence is a greatly reduced degree of indigenous decision making power. It is true that in todays world which, as the cliche would have it, is highly 'interdependent', nobody can make decisions all the time and in all respects without interaction with others. But here again, it is the overwhelming degree of reduction in decision making opportunities that characterizes dependence. When a country finds that in economics, in politics, in cultural matters, in science, in technology, and in other fields its future course is laid out by events, people, developments, and forces that are located outside the country, it will conclude that it is depressingly dependent on others, and discouragingly unable to make its own decisions and shape its own fate. It is this realization that the phrase 'neocolonialism' arises from, since it is easy to jump from this state of affairs to the conclusion that sinister conspiratorial forces are at work.

A yet another way to phrase this example is to point out that a dependent person has a very limited opportunity to realize his own aspirations. One way of characterizing the lives of people or of peoples is to say that we strive toward the realization of our aspirations, whatever they may be. Indeed, throughout history, and even today, as we view the different countries and the different individuals within countries, we encounter a myriad of different aspirations, material, cultural, social, spiritual. Perhaps one way of describing the elusive concept of 'happiness' is to equate it with the realization of aspirations.

It is, therefore, a sad, frustrating, and potentially dangerous state for an individual or a country when its aspirations are blocked or when, without such intentional blocking, events simply run in a direction in variance with the realization of such aspirations. Yet that is likely to be the fate of a dependent whose decision making power is limited, whose actions are determined by outside forces, who owes its resources and tools to external sources. Afterall, the outside entities will generally pursue their own aspirations, which will generally be in directions other than those of the dependent, and the latter can expect to make progress toward its own aspirations only when either they happen to coincide with those of others, or when the forces he depends on condescend to the charitable act of helping the dependent on his way. That such a situation involves an acute sense of dependence is hardly surprising.

All this may be easier to bear if the situation were significantly reciprocal, that is, if the dependent, in turn, could also act, at least in some important respects, as a determining outside force in the context of some other countries or groups. But this is seldom the case in the situation when such dependence exists and is perceived. Thus yet another way of depicting dependence is to say that it is a state in which there is a gross mismatch, an overwhelming imbalance between the influence the dependent can exert on others and the influence others exert on the dependent.

In summary, we might say that all three aspects of the 'objective' hallmarks of dependence have their roots in a common situation: a crippling loss of freedom of action on the part of the dependent. He is unable to act freely because of his lack of skill and capability, because he is inhibited by a humiliating feeling of obligation, and because of his very limited opportunity for decision making.

4. Psychological elements of dependence

From the previous section it may appear that the symptoms of dependence are so numerous and so clear that one can measure, using concrete and universal indicators, when, where, and how strongly dependence exists. In this section I want to argue that this is by no means so. On the contrary, while the qualitative existence of certain ingredients of dependence can be objectively ascertained, the determination of when these ingredients are strong and overwhelming enough to constitute a perception of dependence is to an overwhelming extent a matter of the state of mind, a psychological factor. In other words, using popular language, dependence is almost entirely 'in your head'.

That psychological factors can have such a great influence in human affairs hardly needs documentation. In the classic and ubiquitous example in which ingredients of dependence are certainly present, namely the relationship between a child or juvenile and his parents, the perception of dependence and the ensuing rebellion against the parents' values, lifestyle, advice, and aspirations is not a result of a factual evaluation of the degree, nature, and desireability of the dependence but of a spontaneous strive toward breaking off, toward becoming independent, toward grabbing self-determination. On the political scene, to take a particularly conspicous example, the creating of Bangladesh by its breaking away from Pakistan was clearly not brought about as a result of an economic study of self-interest. In fact, by breaking away Bangladesh lowered its GNP/capita by at least a factor of 3 and brought about social and political difficulties which far exceed those which existed before. Yet, such a breaking away was undoubtedly inevitable and unavoidable since the psychological forces at play were enormously stronger than 'rational' economic considerations.

Similarly, Iran's take over by the Khomeini group and the subsequent hostage episode, on a rational, material, and factual level, brought about near-collapse economically, an invasion from outside forces, and social and political chaos. How could this happen? It is not difficult to understand: The new masters of the country are possessed by psychological forces of such magnitude that material considerations appear to them to be insignificant in comparison.

A final example is that of the colonial history of most developing countries. Many of these countries were colonies for decades or centuries, and were brought by their colonial masters into the age of science, technology, and other aspects of what is generally regarded as the character of the late twentieth century. It is difficult to imagine that had these countries been left isolated and left to their own resources to this day, they would have developed faster, or developed even substantially in the sense of becoming equal partners to the 'Western' countries in matters of economics, standard of living, technological capacity, scientific knowledge, social systems, and other aspects. This does not mean that the colonists helped these countries to the best and most possible extent. On the contrary, especially viewed with hindsight, we can confidently say that they did not. Yet, the alternative of not having been colonies most likely would have lead to an even more dependent state today. The few countries which were never colonies or were such only for a very short time seem to confirm this feeling.

Yet, and entirely understandably, there is worldwide resentment against colonialism not only in the present, but also as a historical phenomenon. This resentment is not justified by objective studies showing the inferiority of this path compared to others. Instead, the source of the resentment is in the psychological feeling that people and peoples must learn on their own and must be allowed to make their own mistakes, whether this is 'good for them' or not.

In view of such examples, some of which explicitly involved dependence, it should not be difficult to accept the contention that dependence is primarily a psychological matter. We might say that while the factors described in the previous section characterize qualitatively the elements of dependence, the qualitative boundary between dependence and being a partner is determined by psychological elements. In all factors discussed in the previous section, it was the overwhelming nature of the loss in decision making, of the indebtedness, of the constraint in achieving aspirations that precipitated the feeling of dependence. Exactly when such elements become over-

whelming is, however, a matter of opinion, and in fact a matter of the psychological state of the individual or country.

Perhaps an analogy might be helpful here, one that is much in the focus of attention these days. The concept of a handicapped person has some objective hallmarks, in terms of a physical disability of the person. Yet what counts primarily in terms of the person's life and of his standing in society is his psychological attitude toward such a handicap, his attitude toward his place in the world. The current movement toward self-sufficiency, independence, and an active and energetic life for the handicapped realizes this crucial point and thereby transforms the crushing, essential handicap into a relatively minor, purely 'external' impairment. This situation has a very close analogy in the realm of dependence in international relations. Regardless of the presence or absence of certain 'objective' symptoms of a status of dependence, what matters most is whether the country *acts* as a dependent or not. This is predominantly a psychological question.

For example, we mentioned that no country can make decisions by itself in all matters and to an unlimited extent. But even though all countries are thus affected by a decrease in their decision making power, in many cases this does not lead to a prevalent feeling of dependence. Whether a certain factual situation is construed as dependence will depend on such psychological factors as self-confidence, serenity, optimism, fatalism, envy, impatience, aggressiveness, lethargy, and many others.

Similarly, we could turn to the other two characteristics of dependence that were discussed in the previous section, and demonstrate how psychological elements draw the line between what is natural and tolerable, and what is outrageous and burdensome.

It is crucial to realize the pivotal importance of these psychological factors when making recommendations in developmental matters. As an example, it has often been suggested that developing countries should import their science and technological know-how rather than undertake the much more difficult task of creating their own. Let us for the moment ignore the fact that such importation is not possible in the absence of some indigenous science and technology. Even if we do so, the recommendation becomes completely absurd and unrealistic because it is oblivious to the psychological factors that fuel countries to have their own science and technology, even if at a given stage such an endeavor is not 'cost-effective'.

So far in this discussion we attempted to define what dependence is, and survey its material and psychological bases. In the next section we will take the dependence of many countries as a fact and will describe the effects that such feeling of dependence has.

5. The effects of dependence in international relations

Whether the feeling of dependence is due to 'real' factors or whether it is a mental condition, it has bad effects in either case. I want to emphasize here, in particular, two main ways in which such a feeling of dependence affects the dependent in an adverse way. They are natural consequences of the criteria and symptoms of dependence discussed in the last two sections. The first such effect is the creation of frustration and hatred in the dependent. The perception is that of a hopeless situation which the dependent is powerless to change (or at least change quickly), and hence this frustration is vented in a hatred of the outside forces which are felt to cause the dependence. Such a state of mind is often a vicious circle in that frustration creates more dependence and hence more hatred. It is for this reason that I always found it utterly naive when American policy listed, as one of the justification of the international assistance, the hope that thereby the United States will earn the gratitude of the countries helped. To be sure, there are, from time to time, some superficial expressions of gratitude from the recipient countries. But underneath one cannot mistake a much deeper feeling of frustration, humiliation, and hatred which is directed toward the only obvious target, namely the donor country.

The same sentiment is behind the various theories of conspiracy, in which people in the assisted countries speculate about the 'real' motives of the donor country, and go far in 'tracing down' the secret ways in which the assistance in actuality damaged the recipient country rather than helped it. In all such theorizing, it is always assumed that in such transactions only one party can benefit, and hence if the donor country in any way benefited from the act of donation, then, *ipso facto,* the recipient must have suffered.

An example of such resentment is the feeling in many developing countries toward multinational companies present in those countries. These companies are obvious symbols of the existence of countries on which the developing country is heavily dependent, as exemplified by the presence of the multinational. The multinationals obviously find it to their benefit to do business in the developing countries—otherwise they would not be there. At the same time, they contribute significantly to the educational structure of the country, to the training of manpower, to labor welfare standards, and to communal and social responsibility. This aspect of their presence is, however, often ignored, and their entire performance is assessed only in the sole dimension of their being symbols of 'imperialism' (i.e. dependence).

Much of the instability, conflict, and hostility in the world today can be, at least in part, attributed to such feelings of frustration and hatred engendered by feelings of dependence, and hence the lessening of such real or imagined dependent relationships is a task of high priority.

The second effect of dependence is the creation of impotence. Dependent countries, even if they vent their frustration and hatred in various symbolic, sterile, and sometimes violent ways, often are deeply pessimistic and impotent with respect to action than can truly begin to reverse the status of dependence. Such actions are, almost without exception, of long range nature, slow, difficult, and uncertain, requiring perseverence, patience, quiet and hard work instead of conspicuous grand-standing, and such requirements cannot be easily reconciled with the burning desire to end the dependent status. Thus often these truly remedial alternatives are spurned in favor of make-believe remedies. Local industries are prematurely nationalized and then ruined, foreign embassies are attacked, hostages taken, airliners hijacked, 'educated' manpower is 'created' by decree or by paper targets of five-year plans, 'instant' research institutions are formed where a shiny building shelters unprepared, inactive, and misemployed people, backyard steel mills are ordered, cadres of 'red and ready' 'scientists' and 'technologists' are unleashed on the country, creating havoc and setting back the real scientific developments of the country by several decades.

At the same time, the feeling of hopeless dependence creates lethargy toward constructive action. Not only does it distort judgment and suggests nonsolutions, but it also dulls the sense of experimentation, lessens the willingness to take risks in novel undertakings, and tends to shift the blame away from the dependent and onto some outside force. Such a mentality is exemplified by a bureaucrat working in a huge, catatonic organization in which the individual is utterly dependent and helpless, and where the aim has ceased to be to get the system to move, and has become the mere mechanical survival in a formal sense.

An illustration of this kind of lethargy and pessimism can be found in the way the results of local, indigenous scientific research and technological development are often handled in developing countries. The utilization of such results is often meager or non-existent and instead foreign technology is imported to the country at great cost, even if such an imported technology is unadapted to local needs and conditions. The underlying feeling is that *since* the local technology was created in the country itself, it cannot possibly be good, and can therefore not compete with the similar technology created abroad. Thus the hated dependence is just accerbated and prolonged by the feeling of helplessness that is a consequence of it.

6. What to do about dependence

Some modes of action follow from the previous discussion. One mode of action is to vent the frustration through basically destructive and pointless actions, examples of which were given above. It is not infrequently that such a mode is chosen. This is purely a relief measure of the dependent itself and hence needs no further descussion.

The second mode, also much in evidence, apparently more constructive, but in fact equally futile, is to legislate away dependence. The United Nations and some of its agencies have been particularly fond of such methods. Declarations are made about "New International Economic Order", and endless conferences are held to generate further pronouncements of legislation about how the world should be. 'Codes of conduct', international treaties allegedly regulating the relationships between the dependents and the outside forces, resounding votes in international assemblies and agencies are some of the other tools of such legislation. None of this affects any of the real causes of dependence, and neither does it seem to ameliorate the state of mind of the dependents or to reduce the feeling of dependence or the imagined instances of dependence.

The third mode is the only one that is realistic and which therefore must be exercised: We must redress the worldwide imbalance in the know-how and concurrently, by encouraging some real instances of success and true signs of independence in the Third World, we must foster the self-confidence, the optimism, the self-respect that will eventually erase the imagined dependencies and the mental state of feeling dependent.

This is, of course, easier said than done, and the implementation of such an exhortation involves in practice innumerable small steps in many different areas of activity. Let me nevertheless suggest two more general 'recipes' which are likely to further the eventual achievement of the above outlined objectives.

First, it is very much more important to assist a country in achieving capability in a certain field than to import into the country ready made products pertaining to that field. Indigenous infrastructure building is of foremost importance, even if it appears to be very long range or even anachronistically misplaced at a time when the country lacks many essential commodities. It is foremost because it is a real contribution to the lessening of dependence, both *de facto* and psychologically.

Second, from the point of view of shedding the psychological burden of dependence, specific examples of success are of pivotal importance, even if they are achieved in areas which do not bring instant material improvement to the country. In science and technology, the launching of a satellite, a Nobel prize, or even a home-made cyclotron or atomic bomb can have and has had such an electrifying influence on morale. Without pride in oneself, preferably related to an actual achievement, it is hard to shake off a feeling of inferiority or dependence.

The world used to be more equitable in the distribution of know-how, and the parts of the world which were behind were at the same time sufficiently isolated from the rest of the world that 'they did not matter'. It was about 400 years ago when the participants of the 'Western civilization' suddenly embarked on a rapid pace of development spurred by the evolution of modern science and technology. By 1900 the differences thus created in know-how were enormous. As the century progressed, these differences remained the same and at the same time became increasingly more conspicuous as communication and transportation improved and the intercomparison of all parts of the world became feasible. Thus we suddenly found that the world consisted of two groups: Three-quarters of the world became heavily dependent on one quarter.

In recent years there are beginning to be signs that the disparity might lessen, or at least that we move toward only two-thirds depending on one third. Some countries (like Brazil and South Korea), 40 years ago heavily dependent, are finding their self-confidence and are making considerable strides toward becoming a partner rather than a dependent. At the same time, some of the countries only recently in the lead appear to falter, appear to lose their self-confidence, their optimism, their élan, and become more dependent not out of necessity but due to their own decline.

A large part of the world, nevertheless, remains in a deeply dependent status, and hence their transformation into partners must continue to have high priority. The main aim of this discussion was to show that such a creation of an indigenous capability, know-how, skill, vitality, and energy must be the primary objective of international development assistence, playing at least as large a role as the ubiquitous short term aid delivering ready-made relief supplies. A short discussion like this is not the place to go into details of how a development assistance can complement and enhance local efforts to create such an infrastructure. In the specific area of science and technology, I have tried in the past to direct attention to ways of accomplishing it, and so have a few others. The present discussion thus can serve as an additional foundation for such efforts, justifying them in terms of the desireability of lessening the present heavy dependence of three-quarters of the world on one-quarter.

RELEVANCE:
An Analysis Illustrated on Science Education in the Third World
MICHAEL J. MORAVCSIK

ords are often used as weapons. In a debate, words can be thrown around in a tendentious, value-laden way so as to prejudice the discussion from the very start. This is facilitated by words having many meanings, and by words acquiring certain implied meanings in certain contexts which can then be used for innuendos in other contexts. It is thought to be especially effective to use words the opposite of which is clearly damning. "Relevant" is such a word, since who would want to be tagged "irrelevant"? "Sane" is another, with its opposite being "insane."

Rigidly ideological organizations particularly delight in titling themselves with such value-laden words. Paradoxically, the actual character of such organizations can very often be described by taking the opposite of such headings. The Daughters of the American Revolution consists of very conservative people opposed to any revolution. The Americans for Democratic Action consists of an intellectual elite engaged in talking but no implementation. People's democracies the world over are dictatorships in which a tiny, self-selected group rules the people. Various organizations claiming to be for a "sane world" advocate harebrained schemes which would plunge the world into chaos and conflict. Organizations

heralding themselves as being for "policy alternatives" try to resuscitate old policies already tried and discarded as unworkable. Organizations self-tagged "public interest groups" advocate policies favored by some small minority and ignored or even opposed by the majority of the public. The list is long and illuminating.

The purpose of this essay is to take one such overused, distorted, and violated word, namely "relevance," and try to reinstate its meaning and its proper use. In doing so, I do not mind taking the risk of being labeled "irrelevant." As we will see, much of the proper use and meaning of "relevance" involves being "irrelevant" as judged by the present, crippled use of "relevance".

The aim of this analysis is, however, not purely abstract. "Relevance" is used profusely in the context of development problems and, in particular, when discussing the evolution of science and technology in the Third World. I want, therefore, to apply the general considerations to this particular example to generate some criteria for determining whether some proposed action is or is not relevant to the evolution of science and technology in the developing countries. The general analysis will be contained in Section II, and the application in Section III.

General Analysis
(Section II).

There are five important factors in the analysis of a word like "relevance," without which its use is meaningless, can only vent emotions, and cannot communicate rationally.

1. Context

The meaning of a word must be judged in a context, especially a word like "relevant" which goes with the prepositional phrase "relevant to X." In non-relational adjectives, the context is contained in the word itself. "His head was spherical" is an example. But "relevant" is very much "relational" since to make sense it must be followed by the thing to be relevant to. We cannot say that a hammer is relevant, only that it is relevant to driving in a nail. Another example: "Flashcards are relevant to education." (Flashcards serve to memorize arithmetics or foreign vocabulary.)

2. Purpose

To say whether A is relevant to B, we must know the purpose, the aim of B. This is a special part of the context deserving special emphasis. For example, in general, "A working engine is relevant to an airplane" is questionable, since an engine is relevant if the plane is to fly, but not if it is

Michael J. Moravcsik is Professor in the Department of Physics and the Institute of Theoretical Science of the University of Oregon. His research in theoretical physics has been concerned with elementary particles and nuclear problems. For the past 15 years he has also been involved in problems of science building in the developing countries, and is the author of the book "Science Development", as well as 160 articles in theoretical physics, science policy, and the science of science. Address: University of Oregon, Eugene, Oregon 97403, USA.

exhibited in a museum. Another example: a functional evaluation and assessment system may or may not be relevant to a research institute in a developing country. If the purpose is to produce good science, it is relevant. If the purpose is to proclaim to the world that the country is scientifically on an equal footing with the "advanced" countries, it may not be relevant, since the judgement may be done on the basis of the input and the material structures alone. Similarly, flashcards are or are not relevant depending on whether memorization is a goal of education or not.

3. A word is operational

To judge relevance we also must have in mind a way to attain the stated purpose. Example: Are X-rays relevant to medical diagnostics if, for example, the purpose is to ascertain whether a bone is cracked or not? Prior to the discovery of the X-rays, the method of diagnosis was to "feel" by manual touch or to open up the patient. The discovery of X-rays introduced a completely novel procedure (namely, taking a photograph by transmitted X-rays), and thus the previously apparently irrelevant research on X-rays suddenly became most relevant.

Or consider pest control. When this was done by killing individual members of the species by chemical poison, research into the reproductive processes of the pest were deemed irrelevant. Such research, however, produced a completely new *procedure* for pest control, namely interference with reproduction, and thus suddenly became most relevant.

4. The word is embedded in a multidimensional world

In making the determinations outlined above, we operate in a multidimensional world (Moravcsik 1984). There are, simultaneously, several contexts, several objectives, several ways of achieving objectives. When building a new highway, the motivation may be a need for local transportation, a need for long-range transportation during a military emergency, the rewarding of a congressman for support on a different issue, the creation of jobs, etc. Something may be relevant to some of these motivations but not to others. For example, in attempting to improve the education and job-preparedness of minority students in the United States, the procedure may include the education of the parents of these students, remedial

instruction for students below the average, the exposure of students to model individuals from the same minority, the distribution of books containing biographies of successful individuals, etc. No single program would be relevant to all of these procedures.

Shorter vs. longer term objectives form a particularly important dichotomy. Human nature tends to focus on the former, and crises are dealt with in a stop-gap and inefficient manner. For example, molecular biology in the 1950s was judged irrelevant (Moravcsik 1983) for the developing countries and hence was not supported. Now when there are multibillion dollar industries based on that field, some support begins to exist, but building capacity in such a field takes 30 to 50 years, so the efforts are hopelessly too late.

Contexts, objectives, and methods also depend on the individual or organization involved in the process. In judging relevance for science development, a UNESCO official in contact only with ministerial officials in the Third World may come to one conclusion, while a currently active scientist, in contact with his colleagues in the developing countries, may reach the opposite conclusion. Indeed, it might be thought that forming a consensus on relevance is an impossible task.

In reality this is not so, due to a fortunate circumstance. A specific project can often enlist the support of a large number of people, each cooperating for a different reason, stemming from *his* particular view of relevance. Theoretical debates can continue while practical action can be instituted. This is not Machiavellian but simply pragmatic, and its validity originates in the multidimensional nature of relevance.

5. A priori or a posteriori judgement?

Most judgements of relevance must be made a priori, with respect to something that lies in the future. Some other adjectives express the same situation. We talk about a "wise investment," a "fortunate decision," a "happy choice" for a spouse. In each

case the decision was made relevant to something in the future. But can we, in fact, judge relevance a priori?

Yes and no. Some predictions can be made a priori, by extrapolation. Other aspects are impossible to assess ahead of time without knowing the results of the activity to be undertaken. Without knowing about X-rays and their properties, it would have been impossible to predict their relevance to medical diagnostics. On the other hand, in the development of thermonuclear fusion energy sources, which has been proceeding for 25 years, it is relatively easy to judge what is relevant, and hence progress has been steady and predictable.

In general, predictability in technological activities is much higher

---

Indeed, it might be thought that forming a consensus
on relevance is an impossible task.

---

than in scientific activities. Within science, the evolution of "normal science" (Kuhn 1970) is more predictable than the occurrence of "breakthroughs." But since science is a constant mixture of "normal science" and "paradigm changes" (on smaller or larger scales), and the breakthroughs often are the most important elements, science on the whole is quite unpredictable.

Actually, what is unpredictable in science is where, when, by whom, and through what method a breakthrough will be achieved, and what its content will be. If we ask *whether* a breakthrough will be achieved if we provide the high quality input into scientific activities, we can, statistically, reply in the affirmative with considerable confidence. We can also point at scientific disciplines in which such breakthroughs in the not-too-distant future are very likely. As the literature on "criteria for scientific choices" (Weinberg 1963, Moravcsik 1974, etc.) indicates, the broader scientific research area we consider, the more certain the breakthrough is and the greater impact it will have in applications.

In striking a balance between such "normal science" activities and work on broader scientific problems, developing countries are frequently very deficient. There a pedestrian and distorted sense of "relevance" dictates only "normal science" activities, and in fact often not even that but instead imitative technological manipulations. There is thus need for clarifying what "relevance" means.

In summary, we found that while the outcome of science cannot be predicted, the relevance of certain policies and measures pertinent to scientific activity and organization can often be judged with some confidence.

Application: Science Development in the Third World (Section III).

The general considerations of Section II will now be applied to the area of science development in the Third World. In particular, we will explore the aspects of science education that are "relevant" to the task of building science in the developing countries.

Aims

The above paragraph defines the context in which we want to determine the relevance of education. In doing so, following the "recipe" given in Section II, our first task is to establish the aims of science in a developing country, keeping in mind the multidimensional nature of such teleological questions.

The aims of science and the related question of the motivations for doing science have been discussed before (Moravcsik 1980), and so I will just summarize the conclusions briefly. There are a great many justifications for engaging in scientific activities, both on an individual level and on the part of society as a whole. Depending on the occupational, philosophical, and social background of the respondent, one can obtain scores of different answers to the question: "Why should we pursue science?" A doctor will refer to the scientific basis of medicine, the general will stress the scientific aspects of defense, the economist will explain the extent to which science contributes to the economy, the scientist may tell about the excitement of scientific discoveries, the criminologist will praise science for making the investigation of crime easier, the transportation expert will emphasize the revolution in travel due to scientific discoveries, the poet may cite science's role in providing him with more spare time to devote to poetry, etc., etc. Yet these many answers can be concisely collected under three principal headings. These are: a) "Science is a basis of technology," b) "Science is a human aspiration in the 20th century," and c) "Science is a powerful influence on Man's view of

the world and of his role in it." In the context of the developing countries, the relative weights of these three motivations are roughly equal, though the empirical determination of these weights is not easy, since explicit statements, words left unspoken, and deeds must all be combined to deduce such weights, and since the answer depends on whom one asks, when, and under what circumstances.

Following our recipe, as our next step, after having established the aims of the thing to which relevance is sought, we now have to specify the procedures, the methods, the ways in which these goals of science will be attained, and hence, in which science education can serve the process. Since the set of goals is multidimensional, our

analysis of the procedures must also follow that format.

Science as a basis of technology

Let us start then with science being the basis of technology. This statement means that scientific knowledge and understanding is a necessary condition to technological progress in our times, though, of course, not a sufficient condition also. There is a multitude of other factors which are also necessary conditions. Yet science, sometimes "old", sometimes "new," sometimes in a conceptual way, sometimes only as a measuring tool, is now indispensable to technological activities even on as rudimentary a level as the usage of imported products of technology.

So the Third World country we are considering will need a base of scientific knowledge and understanding in order to have a technological base. The science that is needed can be classed into three categories, according to the procedure by which we want science to serve technology.

First, there is "old" science, that is, results of past scientific research and the body of established scientific knowledge. For example, the scientific knowledge involving most of the common electric phenomena which are so completely ubiquitous in almost any technological device nowadays is certainly "old science," having been discover-

ed and researched mainly in the 19th century. This is the scientific base of the training of engineers, and hence is mainly fed into the S&T system of the developing country through its engineering education. It has to be taught, in that context, in a problem-solving way, that is, as a tool toward the solving of new problems, and not as a collection of facts to be memorized. In order to present such "old science" in a live and problem-solving spirit, the transmitters of such science need to remain in constant contact with science as a problem-solving activity, that is, they must continue to be involved in scientific research. The topic of that research is, from this point of view, not very important, as long as the research is of high quality as judged by the usual internal and ex-

---

In general, predictability in technological activities is much higher than in scientific activities.

---

ternal criteria used in the assessment of scientific research.

Second, there is "new" science, that is, the results of current scientific research. For example, science utilized in biotechnology (a very rapidly expanding novel technological area) is primarily of very recent vintage, having been discovered and researched in the last decade or so. In the case of any Third World country, the overwhelming fraction on such current scientific research will be performed by scientists in other countries, simply because a given Third World country is scientifically tiny and contributes only a fraction of a percent to the overall worldwide growth of science (Price 1969, Frame 1979, Vlachy 1979). From this point of view, the role of a scientist in the Third World country is to serve as a funnel of such scientific information into the country. He must be aware of important developments around the world, must have the experience of assessing the importance of developments, must have the expertise to make selections among the huge body of new results, must be able to understand and assimilate such new knowledge and must be able to transmit it to those in the country who need it, including technologists, other scientists, administrators, decision makers, policy makers, etc.

Finally, there is science that has not been generated previously, and which is not being generated abroad, partly because the number of scientific problems awaiting research is always much larger than what existing scientists can tackle at a given time, and partly because the new science required might be connected with some technological situations which are very specific to the particular Third World country and hence is of no great interest elsewhere. Such scientific problems arise preferentially in the life sciences, in medical research, and in some areas of geology and astronomy, which are all more geography-specific than, say, physics or chemistry are. Research in tropical diseases or in tropical agriculture serve as specific instances. This "local new science" therefore has to be researched by the scientists in the country in question. Although important in its own right, in the overall picture of "old science," "international new science," and "local new science," the last one occupies only a small fraction in terms of the total needs of the country, with the first two categories being very much larger.

In order for the scientist in the Third World country to be successful in all three of these aspects of science, he must continue to carry on his personal research. I already mentioned this in connection with "old science." The emphasis is even more acute in connection with "new international science," since the only known way to remain a member of the international scientific community, and the only way to maintain capability in serving as a funnel of scientific knowledge and understanding into the country, is to maintain personal research. Finally, personal research capability is clearly needed also for creating "local new science" in the country.

For the first two categories of science, the particular topic of the personal research the scientist conducts is not very important. Working on some topic enables the scientist to keep in contact with a much larger area in science. For the third category, namely research on the special scientific problems of the country, special expertise in certain areas is advisable, though practice has shown that a good scientist is very versatile in his problem solving ability. In the United States, for example, the record shows that scientists (and particularly physicists) make crucial contributions in a very broad assortment of tasks and activities, going far beyond physics or even science. Future employers, and particularly those in centers of applied scientific research, pay much more attention to the overall quality of candidates than to the past disciplinary experience of such candidates, something that is confirmed through conversations with many science managers in industrial research.

So far, therefore, viewing relevance from the standpoint of science as a basis of technology, we can conclude that the educational requirements should stress a very broad education, with emphasis not on memorization but on problem solving, and on the development of an ability for further self-education so as to keep up with the rapid evolution of science throughout the person's professional lifetime.

Science as human aspiration

Let us now turn to the second heading for the aims of science, creation of new scientific research results ("basic" or "applied") which are recognized, respected, and utilized by the worldwide scientific community. To be able to do this, the country needs a scientific infrastructure of very high *quality*, no matter how small it may be. The educational requirements for this are many. A good selection system for students of high promise, encouragement of such students in their schooling and in their career choices, an educational system of high standards and a stimulating and active faculty, opportunities for contact with outstanding scientists, exposure to scientific meetings and lectures are some of these requirements. After obtaining the degree, the young scientist needs further encouragement, collaborative opportunities, channels for con-

But since science is a constant mixture of "normal science" and "paradigm changes" (on smaller of larger scales), and the breakthroughs often are the most important elements, science on the whole is quite unpredictable.

namely the aspirational one. Science being one of the conspicuous aspirations of our times, the knowledge and the manifestation that a country participates in scientific discoveries on an equal footing with other countries is a strong motivation for pursuing scientific activities. This is particularly important for the developing countries (Moravcsik 1981) which suffer anyway from the feeling of dependence, the feeling that the country is not its own master and is not in the position of determining its own fate. A particularly eloquent expression of this motivation comes from the late Prime Minister Nehru of India, who said: "It is an inherent obligation of a great country like India, with its traditions of scholarship and original thinking and its great cultural heritage, to participate fully in the march of science, which is probably mankind's greatest enterprise today."

Some of this motivation manifests itself in mere externalities: The erection of showy buildings, the formation of national research councils and similar bodies of high visibility, etc. For this there are no educational requirements, but then these are the unimportant and superficial manifestations of this motivation. The more substantive realization of such a motivation is in the tinuous self-education, etc. It is known empirically that highly productive scientists are very rare: The number of scientists producing n papers each during their lifetime is proportional to $1/n^2$ (for a bibliography of this "Lotka's Law," see Vlachy 1978). If, therefore, the aspirational motivation is to be satisfied, and if the country wishes to produce really first class science ("basic" or "applied," it does not matter), the entire educational and research infrastructure must be geared toward the encouragement of exceptionally talented and productive scientists.

A particularly essential element in attaining excellence is a functional assessment system for evaluating people, research proposals, and research already completed. More about this later.

Science as a world view

Finally, the third rationale for science is its effect on Man's view of the world and of his place in it. Here we are considering the scientific revolution as a cultural force, and hence we deal with the impact of science on the population as a whole. In a country where science is a new activity, the people who can authentically transmit the impact of science best are the scientists themselves. For a scientist to be effective in this, the educational system in which he was brought up must be broad, dealing not only with the technical details of science but also with the context of science. In addition, the scientist needs to develop personality traits which facilitate interaction with non-scientists, with public communication media personnel, and with people at large.

In addition to requirements pertaining to the three aspects of science, there is an additional, very general one also when we consider science in the environment of the developing countries. In such countries a scientist is always faced with the *double task* of doing science *and* creating the circumstances under which science can be done. Thus the scientist also needs to be knowledgeable about the infrastructural aspects of science and about ways science is built, supported, and evolved. This is a very crucial educational requirement, without which the scientist in the Third World will be discouraged, hampered, demoralized, diverted, driven to emigration, and scientifically destroyed.

Let us summarize, then, the educational requirements which we found to be relevant to being a scientist in the Third World. A breadth in interest and coverage, high quality scientific training, an inquisitive and problem solving attitude which considers science an open field rather than a set bundle of facts, a concern with the contextual and infrastructural problems of science, a personality which successfully interacts with technologists, with policy makers, with public communication media personnel, and with people in general are high on the list. A concern with specialization is to be downgraded, and instead versatility and an ability for self-education are to be emphasized.

Comparison with the statu-quo

How do these requirements compare with the reality of the education of scientists in the Third World? It is, of course, risky to generalize about as diverse a subject as science education in the Third World. Practices differ from country to country, from institution to institution, and, most importantly, from individual to individual. Indeed, I have, during my two decades of acquaintance with science in the developing countries, seen numerous exceptions to the general image I will

now describe. These exceptions notwithstanding, I believe it is possible to discern certain general and prevalent patterns in science education in the Third World which govern the large majority of educational efforts there. It is in that sense that the following statements are to be interpreted.

It would be highly welcome if quantitative data could be exhibited to prove the statement on science education in the developing countries which follow. Unfortunately, I know of no such comparative studies. The conclusions are arrived at in one international conference after the other, and the consensus of knowledgeable observers is considerable, yet the conclusions remain based mainly on peer review rather than data-based studies. To

be sure, there is some information, for example, on the scientific productivity *per scientist* in the developing countries, but, due to the multidimensional nature of the scientific infrastructure, it is difficult to establish a causal relationship between such productivities and science education.

It is indeed ironic that, in the name of "relevance," the actual educational system in the Third World tends to be shaped to be almost totally irrelevant to the requirements of science in the Third World, given its goals. In terms of our five factors, the system is constructed partly without a context, partly in a one-dimensional context with an incomplete and simplistic set of purposes, using much of the time the wrong procedures to attain them, even though a very much more effective system can be instituted even *a priori*. In particular, education tends to be narrow, memory-ridden, and with premature specialization. The subject matters are often constrained with "relevance" in mind, thus eliminating most of the scientific areas truly relevant to the country's future development. Furthermore, relevance is judged almost exclusively in terms of subject matter and not in terms of the other parameters which we found to be much more important in our dis-

cussion. Contextual education is altogether neglected, in line with the narrowness of the education offered. It is even more unfortunate that many of these elements also appear in the education of Third World students at universities in the scientifically advanced countries. An additional touch of irony is in the fact that such distorted education is even more detrimental for the Third World's capacity to perform "applied" scientific research than it is for their "basic" scientific research efforts. It is worth emphasizing this point. The more advanced science becomes, and the more "high" the accompanying technology becomes, the more crucial it becomes that relevance be judged appropriately, and the more sophisticated and difficult this judgement becomes. In

> . . .the only known way to remain a member of the international
> scientific community, and the only way to maintain
> capability in serving as a funnel of scientific knowledge and
> understanding into the country, is to maintain personal research.

this sense the remarks in this analysis are equally applicable to the pursuit of high technology, whether by developing or by "advanced" countries. Since high technology represents a complex and strongly interwoven system of activities, relevance must be judged successfully for all components of the system in order to be successful.

Indeed, in developing countries applied science and technological development work suffer more due to the misjudged relevance of science education than the basic scientific activities do. Since, however, functional assessment systems to gauge the effectiveness of science are practically nonexistent in the Third World, the detrimental nature of the existing educational system cannot be easily demonstrated.

We have, thus, an excellent illustration of the importance of using "relevance" thoughtfully. Indeed, the example is startling: The pedestrian slogan of "relevance" can produce completely "irrelevant" results and thus bring about waste and delay in an area of activities which is crucial to a large number of people. The aim of the methodology offered in this essay is therefore to correct such situations and to stimulate "relevant" analyses of relevance.

REFERENCES

Frame, J. D. (1979): National Economic Resources and Production of Research in Lesser Developing Countries, Soc. Stud. of Sci. *9*, 244-246.

Kuhn, T. (1970): *The Structure of Scientific Revolutions*, 2nd ed., Univ. of Chicago Press, Chicago.

Moravcsik, M. J. (1974): A Refinement of Intrinsic Criteria for Scientific Choice, Res. Policy. *3:1*, 88-97.

Moravcsik, M. J. (1980): *How to Grow Science*, Universe Books, New York, Chapter 2.

Moravcsik, M. J. (1981): What is a Developing Country?, casmn Journ. *2:1* 16-27.

Moravcsik, M. J. (1983): Generating Innovative Capabilities in Science and Technology in Developing Nations, Lecture at the Wilson Center, Washington, D. C. (to be published).

Moravcsik, M. J. (1984): Life in a Multidimensional World, Scientometrics *6:*, 75-86.

Price, D. de S. (1969): Measuring the Size of Science, Proc. of the Israeli Acad. of Sci. and Humanit. *6*, 98-111.

Vlachy, J. (1978): Frequency Distribution of Scientific Performance — A Bibliography of Lotka's Law and Related Phenomena, Scientometrics *1*, 107-130.

Vlachy, J. (1979): Publication Output of World Physics, Czech. Journ. of Phys. *B29*, 475-480.

Weinberg, A. (1963): Criteria for Scientific Choice, Minerva *1*, 159-171.

GENERATING INNOVATIVE CAPABILITIES IN SCIENCE AND TECHNOLOGY IN DEVELOPING NATIONS

Michael J. Moravcsik
Institute of Theoretical Science
University of Oregon
Eugene, Oregon 97403

Discussion presented at the Woodrow Wilson International Center for Scholars, Washington,D.C.
December 9,1982

Genetic engineering and, more generally, biotechnology have, in the last five years, exploded from nothing to an industry representing billions of dollars. Indeed, it represents a qualitatively new step in our ability to alter the world according to our will, desire, and presumed advantage. Previously, throughout the last two centuries, our science-based technology has brought about increasingly complex and novel _inanimate_ tools and gadgets with which to extend our capabilities and to change the "natural" course of events in the world. With the onset of biotechnology, our ability to create tools for this purpose has been extended to _animate_ objects also — a qualitatively new step indeed.

This new technological revolution has been taking place in the United States, in Europe, in Japan — in other words, in the so-called developed world, containing about one-quarter of the world's population. The other three-quarters, labeled as the Third World or as the community of developing countries, once again failed to participate in this novel event at its outset, just as they have missed the previous technological revolutions during the last 300 years. To be sure, there will now undoubtedly be some attempts in some of the most advanced of the developing countries to get on the bandwagon, but such an attempt is likely to bring fruition only very slowly and only in an imitative sense.

It is most interesting and instructive to ask why the Third World missed the boat this time. For a start, one might resort to the usual set of reasons and excuses recited to explain the backwardness of the developing countries. Let us take these one by one.
1) "The developing countries are poor in raw materials and hence cannot compete with the advanced countries in technological ventures." Since biotechnology requires only materials that are ubiquitously and abundantly available, this argument has no bearing on this case.
2) "The developing countries are poor in capital and hence cannot compete with the advanced countries in technological ventures." The biotechnology industry required very little capital indeed at the outset. Some of the early founders who started with $500 from their own pockets, are now worth millions. The capital requirements of this industry have so far been well within what is available in virtually any country.
3) "The developing countries have been prevented from competing in technological ventures by the multinational corporations." In this industry, at the outset, multinational corporations played very little role. Instead, specialized small companies, formed recently, have been carrying the torch.

These and other similar arguments for the backwardness of the developing countries all fail in the face of the characteristics of this new technological revolution. What are, then, the real reasons for the developing countries failing to participate in these novel enterprizes?

As usual, there are undoubtedly in this case also a number of reasons contributing. Two, however, stand out as major, crucial, and in fact probably predominant. They are the lack of manpower appropriate for the task, and the lack of entrepreneurship to connect the scientific and technological advances to the productive sector of the economy.

Let me, in particular, dwell on the first one of these in some detail. In order for a developing country to have now a significant group of scientists at the forefronts of this new area of biology, the seeds should have been planted some 30-40 years ago. With a very minor allocation of resources, even measured in terms of the means available in a developing country, a small group of biologists would have had to be established in the country, and supported sufficiently to allow productive research activities. The span of 30-40 years is a reasonable period over which some tradition, excellence, and strength in productive research could have been achieved. Alternatively, perhaps a regional center for such research could have been set up, encompassing several countries with an aggregate population of, say, 100 million people, where the small national resources could have been pooled to form a critical mass of researchers

In reality, however, nowhere in the developing world was such a research group created, or such a laboratory established. Indeed, had somebody, 40 years ago, walked into the office of any science manager in any developing country with a request for funds for a small group in frontline modern biology, he would have been laughed out of the office. This may possibly be excused by remembering that, especially 40 years ago, science managers in developing countries were singularly ill equipped in terms of education and background to make well informed policy decisions in science and technology, and that this deficiency was a natural outgrowth of the rudimentary state of development of the country.

It is, however, much less excusable, and in fact bordering on the scandalous, that had some-body 40 years ago walked into the office of an official of our Agency of International Develop-ment (or of its predecessor) with the same request, he would have been equally promptly laughed out of that office. The lack of an opportunity to acquire the proper education and background cannot be used in this case as an excuse. Our officials had the opportunity of growing up in a society in which science as a cultural force has been in operation for centuries, and where innumerable case studies are readily available to base science policy decisions on.

Indeed, it is the "thesis" of my short presentation today that both the science policy of the developing countries and the international assistance policy of the United States have been extraordinarily myopic, pedestrian, and hence ineffective in creating indigenous capabili-ties in science and technology in the developing countries, and as a result, these countries, inspite of their occasional successes in some imitative technological activities, have remained in subservient position to the advanced countries in their innovative capacity and in their ability to participate in the frontline events of novel technology. It is furthermore my claim that much of this deficiency is due to the simpleminded and pedestrian interpretation of "re-levance" and to the neglect in the development of a strong science base in these countries.

If we lived 400 years ago, this would not matter much. Then much of technology was still trial-and-error-based, since science was in its infancy, and because technology was then utilizing laws of nature that are directly observable through our natural senses of seeing, hearing, touching, etc. In other words, the objects then observed and dealt with matched in the various dimensions (length, weight, time duration, etc.) the rough scale of our own body, and hence could be directly perceived by our body.

Since about 150 years ago, however, we have been increasingly moving toward the perception and exploration of laws of nature that pertain to realms of the universe that are more remote from the dimensions of our own bodies. We deal with very high or very low temperatures, with astronomical distances as well as with atomic dimensions, with time spans of millions or billions of years or of a billionth of a second or less, with very high speeds and with very high pressures, etc. Coincidentally, we have more or less exhausted the new technology that can be fashioned out of everyday objects and are now involving the same remote objects into our technological innovations that are the subjects of our scientific investigations. In doing so, our direct senses do not suffice, our everyday intuition (acquired since birth through dealing with everyday objects) does not help, and we are therefore involved increasingly more deeply into scientific research and technological development work which is both conceptually and operation-ally very complex and indirect. Indeed, science and technology have fused to an overwhelming extent, and practically all of novel technology has become science-based. This does not mean, of course, that trial-and-error has no place at all any more anywhere in the lengthy process of technology and production, but it does mean that science has become an absolutely necessary ingredient in technology. Sometimes the science needed is not frontline science but that of 30 or even 50 years ago, but the scientific ingredient is there.

I would like to offer two concrete examples of developing countries which have recently become aware of this close linkage between science and technology and are in the process of making belated efforts to strengthen their science base. They are South Korea and Singapore.

In South Korea technology has been in the focus of governmental plans for several decades. The Korean Institute of Science and Technology (KIST) (which, despite its name, has been a purely technological institution) was founded some 20 years ago, and subsequently the Korean Advanced Institute of Science (KAIS) was added (now merged with KIST), which performed research and advanced education in some areas of science closely related to current technology. During this period, however, the Korean universities were grossly neglected, and hence the broad scientific base of the country failed to evolve significantly, inspite of the few islands of activity due to some exceptional and devoted individuals.

It was only a very few years ago when the Korean government finally realized that severe deficiencies existed in the country's science base. By this time Korea was doing quite well in imitative routine technology, in which it could compete because of the well disciplined working force and the wages which were lower than those of some competitors. But life on imitative technology is one of short duration. As the local wages rise, as other countries evolve into competitors in imitative technology, as technology itself extends into new areas, and as the country's own aspirations and ambitions grow, the urge to have frontline innovative capacity increases. So South Korea suddenly found itself in need of such capacity, but at the same time had a lack of the scientific foundation that is a prerequisite for such a capacity.

Suddenly, therefore, an extensive program was initiated to upgrade, expand, and deepen scienc education and scientific research. As an example, Seoul National University, on its new campus outside Seoul, undertook a multimillion dollar project to strengthen its scientific base, including in areas that would be deemed useless by the usual bureaucratic determinations. While there is some progress already, it will take at least a decade or two for this process to mature sufficiently to provide South Korea with the scientific base needed for innovative capacity in frontline technologies.

The case of Singapore is somewhat similar, though less favorable. Singapore's development patterns have been characterized by what is called pragmatism by some, but which, from another point of view, appears to be nearsighted utilitarianism. Singapore has also been thriving on imitative technology, and on the even more rudimentary activity of high technology manufacturing through foreign subsidiaries. With one of the highest gross national products per capita in Asia, and with new competitors springing up (such as Malaysia), Singapore would also be in need of an upgraded innovative capacity. There are some signs that this need is being tended to in some respects. The new campus of the National University of Singapore, which is a fusion of the previously separate University of Singapore and Nanyang University, and which represents a huge investment for the small country, is one such sign. The establishment, a few years ago, of the Singapore Science Center, imposing in size as well as laudable in its operation, is also such a piece of evidence. Yet in Singapore one has much less of an impression than in South Korea that the policy makers have understood the importance of long range thinking in the evolution of local technology. It is my guess that Singapore's economic development will be substantially hampered in the next two decades by its failure to deal with these deeper, more subtle, and longer range problems of technological development.

What is then the bottom line, the policy recommendation that emerges from this analysis? To avoid any misunderstanding, let me state that my recommendation is not that we should turn the international assistance program of the United States upside down, and provide massive billions for the support of the study of general relativity theory in Chad. In the overall spectrum of our scientific assistance and collaboration program with the developing countries, much of the effort and funds could continue to go to more immediate, short term, and more rudimentary technological activities and to that part of the scientific activities which are directly and closely related to them. To be sure, I would make considerable changes in the way these funds are administered, but that is a topic for another discussion. For the present one it suffices to say that I would not re-earmark the majority of the efforts and funds now devoted to such international assistance and cooperation in the sciences and in technology.

I would, however, separate about 10-15% of these funds, and use them for the support of the building up of the scientific infrastructure in the developing countries across the board, not using the customary definitions of what is "useful" and what is "useless". In fact, I would pay special attention to the cutting edges of the various scientific forefronts. Why 10-15%? Because that is roughly the fraction of research and development funds most countries devote to the nebulously defined category of "basic scientific research". Although there is no theoretical determination that this is in fact the optimal fraction to spend for this purpose, the success of the scientifically leading countries suggests that it is at least a workable fraction, and most developing countries (urged by international bodies) have followed this division of resources.

Administratively, such a reorientation of a small part of the funds could best be done through a small and specially designed organization which is independent of the rest of our foreign aid machinery. An attempt in this direction was made a few years ago, with a proposal for an Institute for Scientific and Technological Cooperation. This is not the place to describe the sad tale of this proposal, which finally died in Congress, having been ill formulated from the beginning, mishandled politically, and finally compromised almost beyond recognition.

It is worth noting that many other developed countries, much smaller but more active and imaginative than the United States in this respect, have such an organization. Canada and Sweden are examples. It can be argued successfully that even the organizations in those two countries are not farsighted and imaginative enough, but at least they provide a vehicle for the kind of activity I am suggesting.

A second and perhaps more promising way to work in the same direction is through the education of the hundreds of thousands of students from the Third World who are getting their advanced degrees in the sciences at American universities. Since eventually these students will constitute the forces for progress in their countries, given appropriate exposure and education in the United States, they may be able to bring about a more farsighted and effective science policy in their own countries.

Let me end this discussion, appropriately enough, on a long range note. As cultures and civilizations rise and fall, and as the leadership in the world shifts from one group of countries to another, much of the world's progress at any time depends on the activities and capabilities of the then most advanced and leading countries. It is most likely that over the next 50 years, major shifts will again occur, and that some of the countries now designated as "developing" will be among the leaders. It is therefore in our own self interest to prepare these countries well for this future leadership role.

Some of the elements of this discussion are discussed in greater detail in the following writings of mine:

Science Development: The Building of Science in Less Developed Countries, PASITAM, Indiana University, Bloomington, Ind., 1976 (Second printing)

How to Grow Science, Universe Books, New York, 1980

"Paradesia and Dominatia: Science and the Developing World" (with John Ziman) Foreign Affairs 53, 699 (1975)

"Science and a Crucial Stage of Economic Development" Bull. of the Inst. of Physics, Singapore 7:1, 1 (1979)

"Science and Technology, 'Basic" or 'Applied' —How to Balance Economic Development" Science and Technology (South Korea) 14:6, 26 (1981) (In Korean)

"Intercultural Transmission of Science" Civilisations 29:1/2, 127 (1979)

Preprint, published with permission from *Approtech*
5 (1982) 70.

SCIENCE DEVELOPMENT - A NETWORK OF VICIOUS CIRCLES

BY MICHAEL J. MORAVCSIK

INTRODUCTION

The events of the last half a century have made it quite evident
that the building of science in the Third World is a difficult, laborious
and slow process. There are many ways of conceptualizing this state of
affairs. In this discussion I want to visualize science development as
a complex and multidimensional network of vicious circles, in each of
which to attain A one needs B, but to attain B one needs A. In the
absence of interaction with the outside, the creation of A or B, therefore, is very difficult, and depends on the chance occurrence of a situation which can circumvent this vicious circle type of a relationship.
If, in addition, we also consider that the process by which A creates B
is slow and has a small leverage ratio, we can appreciate the reasons
why science development takes so long and is so strenuous, with periods
of stagnancy and even relapses.

The model, like any model, should not be taken completely literally,
lest it deviate from reality. There are also crucial aspects of science
development which are not necessarily describable in terms of vicious
circles. Furthermore, the image can also be embellished by transforming
it from a network of vicious circles to a set of nested vicious circles,
that is, to point out that the relationship of vicious circle "a" (containing A and B) and vicious circle "b" (containing A' and B') is itself
in the form of a vicious circle. These additions, however, are not
necessary for the purposes of this discussion, since the aim here will
be to point at the remedy to being mired in vicious circles, and that
remedy is equally valid whether we consider the simpler model or whether
we add the supplements which bring it even closer to reality.

THE CIRCLES

I will now offer a dozen illustrations of these vicious circles in
science development. It would not be difficult to find three times as
many. In every instance when one studies some aspect of the evolution
of science in the Third World, one encounters examples of such vicious
circles. Since the ones I have chosen come from a variety of aspects
of science development, however, they should suffice to make the point.

Deciding to Have Science

At the very beginning of science development, the initial decision
has to be made to build science in the country. To make this decision

---

*Michael J. Moravcsik is with the Institute of Theoretical Science,
University of Oregon, Eugene, Oregon 97403 USA.*

there must be <u>some</u> acquaintence in the country with science, however
superficial and tenuous it may be. Somebody must realize that in order
to have material development science needs to be included in the country's
infrastructure, or that science is a preoccupation of the 20th century
that contributes to the self image and national pride of a country, or
that having science in the country has a positive effect on the world view
of the people, or some other justification for science. If there is no
science education in the country at all, and nobody is sent abroad for
education either, it is difficult to find people in the country with
sufficient vision, determination and know-how to establish science locally.
But such people are not created unless science is already established.

Local Educational Institutions

Manpower is pivotal to science so to produce scientists and science
managers the country must have a good educational system. This means
having good science teachers. Good science teachers need to be educated
in good educational institutions, thus closing the vicious circle. Again
without at one point resorting to either education abroad or teachers
imported from abroad, it is difficult to see how such a self interacting
system of teachers and students can be started.

Education Abroad

Even if education abroad is resorted to, vicious circles continue
to plague us. In order to select institutions abroad to apply to, in
order to decide the areas of education sought, in order to have credibility in the eyes of universities abroad so that students from the
country are offered admission and financial aid for study abroad, the
country must have some people who have been educated in the sciences
and in fact, educated abroad. Such people versed, at least to some
extent, in the sciences, have the vantage point to select institutions
for the subsequent generation of students to apply to, they can decide
on the areas of education and they constitute a precedent in the experience of universities abroad on the basis of which subsequent generations
of students receive favorable consideration. But how would the <u>first</u>
such students handle these problems, in the absence of help from
previous generations? Herein lies the obstacle of a vicious circle again.

Attracting Good People to Science

In a country where being a scientist is not a known profession, it
is difficult to pinpoint, attract and persuade young people to become
scientists, especially in countries where social pressures, parental
will, peer judgement, gerontocracy and the force of traditions far out-
weigh personal aspirations in the choice of profession. In fact, just
about the only group that can tilt the balance in favor of science is
the already existing scientists and science teachers. But how would
they have been attracted into science at a time when scientists and
science teachers did not yet exist in the country?

The Training of Manpower

The most crucial part of a scientist's education consists of his
early association with other, older and more experienced scientists.
Most autobiographies and similar recollections of famous scientists
allocate predominant space to this type of apprenticeship, and, in
comparison, hardly mention at all the more formal institutional education. But how would young scientists in developing countries benefit
from such an opportunity to apprentice if there is no previous genera-

tion of scientists in the country?

Creating a Scientific Tradition

To create a scientific tradition in a research group, research laboratory or institution, or in a country as a whole is a lengthy process. Once such tradition exists, it can be maintained and strengthened by the new generation of scientists. But how does such a tradition begin? Not infrequently it is through the personality, talent, determination and influence of a truly exceptional scientist, the type that is in very short supply in any scientific community and especially in the tiny ones in the Third World. To evolve the talent and capability of such exceptional leaders, early immersion in a group or institution of some other tradition is a very important element, thus closing the vicious circle between creating scientific tradition and educating people who can create such a tradition.

The Management of Science

Especially in a context in which most of the material support for scientific work comes from governmental sources, the management of science calls for the existence of science managers with personal experience in scientific activities as well as in science policy and management. Such personnel can hardly be created in the absence of indigenous science, but on the other hand, indigenous science can hardly be evolved in the absence of appropriate science managers.

Becoming Part of the International Scientific Communication System

The communication system within the international community of scientists is geared preferentially toward those who have been prominent and successful in the recent past in producing scientific research results. This holds for all modes of scientific communication: access to preprints, reprints, and reports; access to publication in prominent journals; access to the back volumes and current issues of scientific journals; access to scientific conferences; access to international scientific visitors; access to influence in the large international scientific organizations which largely determine the allocation of communication resources; etc. But to become prominent and successful in the recent past in the production of scientific research results, being part of this international scientific communication network is a crucial element, without which becoming prominent and successful is very much more difficult. Since in most developing countries the scientific productive system is just being evolved and hence has not had the opportunity yet to become prominent and successful, such groups are disadvantaged or even practically excluded from participation in the communication system. Another vicious circle.

Competing for Research Grants

Research grants are usually awarded, in part, on the basis of demonstrated past ability to deliver research results. Furthermore, the writing of research grant applications is a technique that is acquired by experience in writing research grant proposals. In a situation, therefore, in which scientific groups from the Third World have to compete with groups from the scientifically advanced countries for research grants, the former are at a considerable disadvantage because of the lack of past accomplishments in research and the lack of past experience with writing research grant proposals.

Contact Between Scientific Research Results and Their Users

In the scientifically well developed countries, the link between the results of scientific research and the users of these results in the productive sectors of the country is secured by both a "push" and a "pull" mechanism. There are established ways of calling such results to the attention of potential users, and, in reverse, potential users are looking for new scientific advances that can be utilized, and in fact often themselves sponsor the generation of such new scientific results. In contrast, in most countries in the Third World the productive sector is largely unfamiliar with and unaccustomed to using results of scientific research and instead, imitates already existing technologies developed abroad. As a result, the performance of what is intended to be applied scientific research is fueled only by a "push" mechanism by the government, but the direction of this "push" and the method of performing applied scientific research fail to benefit from a feedback from the users and hence often wander off into completely unproductive and useless directions. As a result, local potential users become even more convinced that local science has no utility for them. The vicious circle thus lacks both the development of good applied research and the evolution of productive local industry and agriculture.

Inputting Science into the Local World View

One of the three and perhaps, in the long run, also the most important influence of science is on the world view of the population of a country. Science as a cultural force has been affecting Western civilization for such a long time that those living in it tend to ignore or at least underestimate the impact science, as an element in people's view of the world and of themselves in it, has made. In many Third World societies where such an impact has not yet occurred the lack of it is very much in evidence.

Here again, we encounter a vicious circle. To make such an impact on the population at large, the country must have some scientists who are trained not only in a technical sense but who have also assimilated and internalized the cultural, epistemological and ontological elements of science. But it is very difficult to have such scientists in a country in which the scientific component is to a large extent missing from the social, intellectual and cultural environment.

The Maintenance of Local Scientific Journals

As a final example, here is a specific case from the area of science management. The internal communication within a local scientific community is considerably helped by the existence of locally published scientific journals, which at the same time also encourage research on locally important research topics, offer inexpensive channels for publication, promote the development of a scientific vocabulary in the local language, etc. Such journals, nevertheless, must maintain high scientific standards if they want to be accepted by the worldwide scientific community as respectable places of publication. Thus these journals must have a good refereeing system. Since the size of the local scientific community is so small, a good refereeing system must involve referees from around the world. Here is where the vicious circle appears. In order to know how to select good referees for an article in a given field, the editor or his advisors must be acquainted with the people working in that field, which in turn can be acquired mainly by having done research in that field for some time. But scientists in countries where the local journals are just beginning tend not to have had such experience. Thus again, the two elements of research experience and international connections, on the one hand, and having good local journals, on the other, represent two ingredients in the vicious circle which depend on each other.

THE SOLUTION

As has been illustrated by the above examples, it is very difficult or even impossible to break these vicious circles that appear in science development without going outside the scientific infrastructure of the particular country in question. Hence the solution of the dilemma is to go outside, that is, to resort to international assistance and cooperation. In fact, I want to claim the breaking of these vicious circles and the transformation of them into growing spirals is the most important and perhaps even the only beneficial aim of international scientific assistance to the Third World.

Let me make a distinction in terminology between assistance and cooperation. The latter is a natural part of doing science which routinely and regularly occurs even among countries which are all scientifically well developed. Science is a collective and universal activity and such collaboration among countries is simply part of this activity.

In contrast, assistance denotes a situation in which the "donor" offers more than the "recipient" and the latter benefits to a greater extent. Naturally, good assistance programs bring benefit to donor and recipient alike, but the primary objective of an assistance type of a program is to help the recipient with its scientific development. The above claim refers to assistance.

The claim in question can be supported in many ways. Let us compare it, for example, to the presently much more prevalent forms of assistance. One of these is the solving of a scientific problem "for" the developing country through research done in the donor country by scientists from the donor country. The other prevalent way is to support an isolated and specific research activity with a definite short term (and generally technology-oriented) goal. In both cases, isolated short term successes may be scored, but little if anything is being contributed to the breaking of any of the vicious circles inherent in the country's scientific infrastructure, and hence little if anything is contributed to the scientific development of the country, to the building of a permanent scientific infrastructure of the country, to the ability to avoid similar crises in the future.

Such short term assistance programs tend to be very expensive, particularly in the long run, since they do not make use of the large leverage factors latent in the local scientific community which can be catalyzed resulting in the transformation of vicious circles into growing spirals. Let me offer an appropriate example. One of the vicious circles I discussed was that of obtaining good and appropriate scientific education abroad, and the problems of selection and "authentification of credentials" that are connected. A short term, "let's do it for them," an expensive "solution" to this problem is one often pursued, namely to offer the countries free training in areas decided by the donor country to students selected by the recipient country by its own methods of selection. The result is a group of students, some of whom are ill prepared for the particular educational experience, being brought to institutions in the donor country, and hence being placed in

a situation in which the students are not only ill prepared but possibly
also disinclined in terms of personal interest. Furthermore, the price
tag of this program is huge and is borne by the assistance agency in the
donor country.

In contrast, the appropriate program to relieve the difficulties of
placing students into foreign institutions of learning is exemplified
by the Physics Interviewing Project (now also adopted by chemists), in
which a small team of scientists from the donor country visits the
potential recipient countries and interviews students who are considering
getting advanced scientific education in the donor country. All the
interview does is to generate standardized evaluations about each student
interviewed concerning his background, potential and ability to receive
advanced education in that particular scientific discipline. The interviewers also distribute factual information on how to make a choice of
schools to apply to. The rest is left to the student and to the universities in the donor countries which then, on the basis of the information generated by the interviews, can make decisions about the student
applications and can use their existing teaching and research assistantships to support the students they offer admission to. The process is
very inexpensive and leaves almost all processes in the local infrastructure to operate and flourish. The various latent resources and
interest in the infrastructure of both the developing and the donor
countries are utilized and only the activity needed to link up these
latent resources is actually performed.

Similar examples could also be given in connection with the other
eleven examples given above.

In other words, I claim that the purpose of international scientific
assistance to the Third World is mainly to catalyze the transformation
of the network of vicious circles into an interwoven pattern of growing
spirals, fueled by feedback and "bootstrap" from the infrastructure of
the recipient country itself. To achieve this would unfortunately mean
a rather drastic reorientation of most present scientific assistance
programs of most of the scientifically well developed countries. I
firmly believe that eventually this change will have to come about,
because it will become increasingly evident that the present system of
short-term assistance programs leaves the recipient countries in a continued state of dependence and scientific backwardness. Perhaps the
kind of conceptualization I tried to offer in this discussion will
advance the time when such a change will occur.

B. Research and Its Applications

Applied scientific research and the developing countries

Michael J Moravcsik
Institute of Theoretical Science, University
of Oregon, Eugene, Oregon, USA

The author discusses concepts of basic
science, applied science, and technology, and
of their relationships, and reviews the
status of applied scientific research in
developing countries. He finds that there is
very little applied scientific research in these
countries; this is *not* caused by 'too much'
basic scientific research; the predominant
mode of activity under the heading of
'science and technology' is imitative technological manipulations based on trial and
error, which is mainly ineffective. The
causes of this state of affairs can be traced
to a combination of ten factors. Some
remedial action is suggested to improve the
situation.

The status and role of applied research in less
developed countries is a frequent subject of
discussions, pronouncements, reports, and
policy documents, and in a variety of contexts.[1] It is sometimes said to be the primary
aim of research in those countries.[2] On
other occasions, it is charged that applied
research is non-existent in the developing
world.[3] Then again, in national science
plans one finds the overwhelming fraction
of the research budget earmarked for it.[4]

In viewing this situation the observer
quickly comes to two conclusions: there
appears to be a fair amount of confusion
about terminology and facts concerning
applied research, and, consequently, also
about any measures that are to be taken to
remedy the situation; a clarification of this

confusion deserves high priority because of
the amount and intensity of attention
focused on this issue.

Science, technology and production

In discussing applied research, one has to
be clear about the much broader areas of
science, technology, production, and their
relationship to each other. It is therefore
advisable to summarize briefly these concepts and their connections.

In making such classifications, one always
faces charges that a continuous spectrum of
possibilities cannot and should not be simplified into discrete classes. In situations,
however, like the present one, when the
continuous spectrum conspicuously bunches
into discrete clumps, with only tenuous
wisps between them, the discrete conceptualization is an excellent approximation
and a great practical help.

In this spirit, we then can say that the
distinction between science, technology, and
production can best be made[5] by the *product*
of each activity. The purpose of science is
to provide knowledge about natural phenomena. (In this discussion I will talk only
about the natural sciences; the social sciences
in their present state would need a different
type of classification and analysis).

Technology, on the other hand, produces
processes, procedures, prototype gadgets,
methods for producing something material.

Finally, production aims at actually creating material objects or services, usually on a large scale, in industry, agriculture, or other sectors.

The people involved in these three types of activities are scientists, technologists, and entrepreneurs, respectively. On the whole (with remarkably few exceptions in any part or the world), these three types of people are quite different from each other: They have different motivations, different talents, different methods, different skills, and, last but not least, very different educational and professional backgrounds.

Though it is sometimes discussed, the question of whether any of these activities are superior to any of the others is a rather meaningless one. They are interrelated in a way so that all three are crucial for the survival of the other two, and the rest is up to personal taste and interest.

Technology, by its very nature, is always applied in that it is always oriented toward being used in production. As far as science is concerned, the situation is different. One often hears the distinction made between fundamental, pure, or basic science on the one hand, and applied on the other.[6]

The distinction between the two can be made not, as in the case of the distinction between science and technology, on the basis of the product, but instead according to the *intention* of the scientist or of the sponsor of science.

Accordingly, basic science is performed with the intention of acquiring knowledge about nature simply because we find such knowledge interesting, beautiful, and thus an objective in itself. On the other hand, applied science is practised with the intention of providing knowledge that might be useful for serving as a basis of the work of the technologist.

This distinction cannot be made according to the product of the activity because historical evidence indicates that virtually every good piece of scientific research eventually finds application somewhere. Furthermore, the time and place of such a future application is often unknown at the time when the research is performed.

Classification criteria

If, therefore, one would classify science into basic and applied by its product, virtually all old science would be called applied, and much of contemporary science basic, resulting in an absurd and useless dichotomy.

The classification according to intention, however, is also far from being free of ambiguities. For one thing, intentions and motivations are seldom single-channeled, and hence 'basic' and 'applied' are not mutually exclusive, but in fact may often coexist side by side. Second, a given scientific project may appear primarily basic to the scientist who carries it out, while at the same time it will seem applied to the agency or organization supporting it financially.

Whether it appears mainly basic or primarily applied will also depend on the time span contemplated. A project may be justified in the short run mainly in terms of its basic merit, but will appear worthwhile by applied criteria in the perspective of the next 30 years.

The ambiguity of the dichotomy between basic and applied science does not mean, however, that in practical science policy decisions one should not use these terms at all. On the contrary, they are useful for attaining, or at least aiming for, some kind of a balance among various scientific activities in a country, an organization, a a laboratory, a group.

For example, a government might decide that that it will call basic research a scientific activity which will probably produce knowledge of the type which is unlikely to be directly usable by technologists in a certain specified area during the next 5 years. This

applied scientific research

sounds like an excessively circumscribed definition, but is is probably the best one can offer without plunging into serious ambiguities.

I want to emphasize that the definition of basic research is *not* scientific activity which is useless. This definition, which is often used explicitly or implicitly, would include applied research of such a low quality that it does not produce knowledge or technique at all - and hence it cannot be used.

It is this factor which contributes, above all, to the impression that a large fraction of research in the less developed countries is basic. People who claim this usually lump genuine basic research together with unsuccessful applied research - and in this merger the latter not infrequently swamps the former.

In fact, the element of quality enters into the picture in other ways also. The word 'knowledge', without further qualifications can be misused in an overly broad manner.

For example, the measurement of the distance flown by one particular hawk in Northern Iran on October 15, 1968 is a piece of knowledge in a way, but its *value* from the scientific (basic *or* applied) point of view is either infinitesimally small or in fact zero. What this value is for any particular piece of knowledge can be decided only by the scientific community itself through its consensus.[7]

While this criterion raises further conceptual and methodological questions far too complex to be treated here, on the whole, this consensual assessment of the

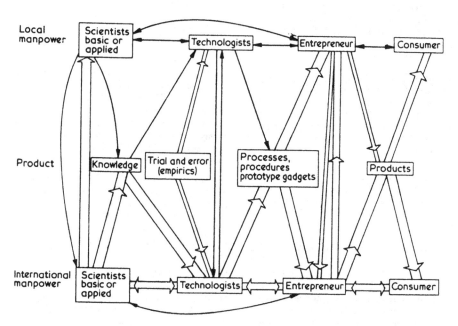

Figure 1. Schematic diagram of basic and applied science, technology, production, and consumers, and their relationships. The bottom line represents the worldwide communities, the top line the local communities in a developing country. The widths of the connections indicate the strength of the coupling

ranking of scientific knowledge works extremely well in practice. Thus, when I talk about producing knowledge, I implicitly mean the type of knowledge that is considered valuable by these criteria.

From the point of view of the discussion in this article, one of the most pivotal points to emphasize is that *applied scientific research is not the same as technological research*. They have different intentions, different products, different methods, and are carried out by different types of people.

Naturally, there is a connection between the two, as there is a connection among all the elements of science, technology, and production, but that does not make them identical or interchangeable.

In fact, these relationships are shown in figure 1 in a schematic way. With our later discussion in mind, the figure separates the man-power into those within the country of interest and those outside it.

In all cases (including that of the United States), the former is a smaller group than the the latter. In the case of most of the developing countries, the former is very much smaller than the latter. The arrows, sometimes monodirectional and sometimes ambidirectional, indicate interactions, and their widths intend to show relative importance. These relationships are shown in a somewhat idealized case. As we will see in a later section, in real life some of the arrows are even weaker than shown in the figure.

For a *given* piece of scientific knowledge, its subsequent technological utilization, and the ensuing production, the chart can be taken as a chronological description of happenings, though the time scale is left undecided by this chart. In the *overall* activities of a country, however, *all* parts of this chart occur simultaneously. Correspondingly, in any plans for the future development of a country, *all* parts of this chart must be included and made provisions for.

Status of applied research

In providing information about applied research research one has to accept the definitions of applied and basic as used in such statistics. There is some ambiguity about this, and it might even vary from country to country.

On the whole, however, it agrees with the one given in the previous section in that the distinction is made on the basis of intention rather than product, no so much because of the conceptual superiority of the former way, but rather because, a) in development plans and research reports (where most of the statistics come from) the product, even in principle, is not known yet, and hence only intention or expectation can be stated, and b) because in developing countries countries completed research projects very seldom are evaluated in any case with respect to their products.[8]

Basic expenditure

Taking such definitions, therefore, the first point to note is that in virtually all countries around the world, the expenditure for basic scientific research is between 10 per cent and 20 per cent of the total budget for research and development.[9] There is perhaps a very slight tendency for this figure to be *lower* in less developed countries than in the scientifically more advanced ones, but the effect is small.

Thus, the remaining 80–90 per cent is earmarked for applied scientific and applied technological research and for development. For reasons evident below, no further subdivisions are usually made to indicate the amount for each of these three categories separately.

Our first conclusion then is that if we assume that the fraction of financial resources to be devoted to basic scientific research out of the total research and development budget should be more or less the same for any country, regardless of its stage of scientific development, then we see that *if there is a shortcoming in the area of applied scientific research in less developed*

applied scientific research

countries, this is not a result of too much
being spent on basic scientific research.

But is there such a shortcoming? In order
to answer this question, one must assess
applied research (scientific and technological) in less developed countries. Unfortunately, this cannot be done very well from
documentary evidence, because, as mentioned before, it is not the custom in less
developed countries to *evaluate* ongoing
or completed scientific or technological
projects in other than a perfunctory way.

For example, applied scientific projects
are practically never subjected to the peer
judgement of the international applied
scientific community, and their assessment
by the visible results of stimulated technology is greatly beclouded by the weakness
of the interaction between scientists and
technologists, or even between technologists
and the production sector.

Data limitations

Therefore, one must rely primarily on the
much more unsatisfactory method of
personal, anecdotal impressions gained
during actual visits to applied research
centers in such countries and through conversations with scientists in those countries.
Accordingly, the evaluation I give here is
admittedly a personal one, but it seems to
be corroborated by a sufficient number of
other personal experiences that I think the
burden of proof is on those who wish to
prove this view incorrect.

In any case, the conclusion is two-fold:
In most of the developing countries, there is
practically no applied *scientific* research at
all; and there is a huge amount of *technological* research, but most of it based on trial-
and-error rather than scientific knowledge
(see Figure 1), and much of it is of low
quality or at least low efficiency.

One must add immediately that the absence
of *scientific* applied research is not intended.
On the contrary, huge institutions,

laboratories, and centres are proclaimed to
have the mission of applied scientific
research,[10] and large sums are appropriated
for this purpose. The actual activity,
however, consists, in a large majority of
the cases, of technological manipulations
aiming at producing a gadget or process.

Trial and error

In viewing this, the realization comes easily
that at least one element that is missing is
the understanding of what applied scientific
research is in the first place, in other words,
that there is much *scientific knowledge* that
needs to be developed and/or transmitted
and transplanted in connection with technological problems of the developing
countries.

Since this realization is missing, one cannot
be too surprised that in the existing applied
technological research also, the use of
scientific knowledge (developed elsewhere)
is greatly suppressed compared to the trial-
and-error method. The main trouble with
this attitude is that the trial-and-error method
in most cases is very much less efficient,
if applied to contemporary technological
problems, than the 'scientific' route.

Is is undoubtedly true that in each technological problem there always remains an
element of trial and error, to help in
situations when scientific knowledge is not
available, or the system is too complicated
or too specialized for scientific knowledge
to be applied efficiently. Yet, such an
attitude should be reserved for the time
when all known scientific knowledge has
been inspected and either applied or found
inapplicable in this case.

Ten causes

Assuming that the above assessment is
correct, let us now inquire into some of the
causes of this state of affairs.

1. The first cause was already mentioned:
There is often a lack of understanding
on the part of research managers and

governmental officials about what applied scientific research consists of. This results in misplaced allocation, administration, and expectation.

2. The low efficiency of any applied research is also due to the two primary shortcomings of scientific educational systems in many developing countries: rote learning and premature specialization. Ironically, the latter is to a great extent the result of a desire to aim at applied research.

The indigenous 'conventional wisdom' in science policy often claims that 'we cannot afford to give a broad education to our scientists since we want them to work mainly in applied research', thus forgetting that *applied scientific research requires a considerably broader educational background than basic research,*[11] because in the former the problems are formulated externally and hence are almost always 'interdisciplinary'.

3. Another factor is the lack of special attention to students from developing countries when they are receiving advanced scientific education in scientifically advanced countries. If such education were *supplemented* by elements not immediately needed by the counterpart local science students in the advanced country (and which therefore are not taught at those educational institutions), the scientists returning to their home countries would be better prepared for doing science (basic or applied) in the context of their own scientific environment.[12]

4. There is an overall mismatch and misuse of scientific and technological personnel in many developing countries, which goes as follows: There is usually a shortage of competent technicians in those countries. As a result, people with full technological training (eg engineers) are used for work that should be done by technicians.

On the other hand, applied technological research then remains unattended, partly because engineers are doing the technician's work, and partly because the engineers, due to the narrowness of their education, are often incapable of performing useful technological *research* (though they might be able to perform routine technological tasks).

Thus scientists are called on to do technological research. Partly as a result, scientific research (basic or applied) remains undone or is underemphasized.

This system ensures that a) everybody is used for something other than what he was educated for, what he has experience in, and what he has talent for; b) everybody is unmotivated since he is assigned to something other than what he is interested in.

5. In many developing countries there is a sharp division, or even emnity, between the academic and the non-academic segments of the scientific and technological communities.[13] Since basic research, if done at all, is likely to be in academic institutions, while the governmental research has a primarily applied mission, the schism between academic and non-academic is also a gulf between basic and applied research.

As a result, it is easier for the applied research community to drift altogether away from science and lapse into pure technological manipulations.

6. There is a lack of appropriate facilities for applied research. By appropriate I mean facilities in which applied scientific research could be in close contact with technology and then with production.

The governmental research places are often isolated from industrial and agricultural organizations, and indigenous private industry itself is uninterested even in technological research, let alone applied scientific research. Finally, the subsidiaries of multinational corporations very seldom do their research locally.

applied scientific research

7. There is a lack of external stimulus through international interaction in areas of applied scientific research.

Scientific visitors to developing countries, if they come at all, are likely to be mainly academic people with a stress on basic research, and at the other end of the scale technological visitors tend to be either industrial representatives with a specific business mission, or AID-type narrow specialists interested in getting a specific project done and not in transmitting or transplanting technological skills.

Some programmes, like the AID-NSF SEED project, begin to make the right kind of a contribution to remedy this, but their extent is still minuscule, and the selection criteria often still too narrow.

8. Social traditions often run counter to a functional applied research programme, in that doing something for use tends to be considered less noble than doing it for spiritual enrichment, and also 'theoretical' efforts are often preferred to 'manual' operations.

9. Applied research, because of its technologically oriented nature, and also because of its interdisciplinary character, needs more careful administrative and organizational support than basic research. On the other hand, organizational capability and experience is, in many developing countries, in even shorter supply than is scientific expertise.

10. Applied scientific research in developing countries also suffers from the lack of a sufficiently broad market that could utilize its products.

If, as a consequence, applied research is organized by selecting a general area of presumed importance, the results (which, even in applied research, do not arise in a very predictable way, at a predictable time, and in a predictable place) might not find an immediate user among the few industrial or agricultural organizations in the country,

even if the results themselves, by international criteria, are meritorious.

On the other hand, if first a specific narrow problem is selected for a 'technological fix', doing research 'for it' in this reverse way is often not nearly as efficient and successful as it would be the other way around (ie by doing the research first and then finding a specific application for it).

The above list by no means exhausts the diagnostics of applied research in developing countries, but it will suffice for some constructive suggestions in the following section.

Pointers for action

Suggestions for action to try to correct some of th deficiencies listed in the previous section follow naturally from that listing. They can be grouped into three main areas: Conceptual clarification, education and communication, and indigenous research facilities and structure.

It might, at first sight, appear unnecessary to call for conceptual clarification with respect to the problems discussed in this article. Have we not had scores of impressive conferences, seminary, reports, and declarations during the past 30 years,[14] all dealing with science policy and the developing world? Is it not true that we have huge divisions in international organizations dealing with such problems?

Do we not see in almost every country piles of science policy documents, and do we not hear science policy terms used generously in speeches and statements on the role of science and technology in development? What more is needed after all this?

In trying to assess the practical effects of all this effort on the actual provision for science in less developed countries, I am inclined to believe that what is really needed is a brief, vividly written, practically oriented, simply worded brochure, very broadly distributed and hence very cheaply produced, containing many illustrative examples, and starting from the most basic concepts and terminology, which could serve as a common denominator and solid basis for the everyday decision making of science managers and scientific leaders all around the world.

In it, for example, there would be a discussion of the difference between technology and applied science, one of the elements in the present article. It would also deal with education, funding, organizational structures, communication, manpower factors, and a number of other ingredients of the environment of science-making.

The brochure would be void of the rhetoric so abundant in proceedings of international conferences on science policy, and it would also avoid a braggard-like exhibition of sophisticated-sounding terminology. The aim would be to transmit the elementary ideas and tools needed for a functional creation of a context for science.

The whole brochure need not be more than 10 000 words, and it should be sufficiently intriguing, educational, and entertaining so that every fairly educated person, whether he has a formal scientific background or not, and whether he works for a scientific institution, industry, government, or is still a student, would find it engaging. In this way, it would not end up on bookshelves or in filing cabinets, gathering dust unread as many other science policy documents.

Education and communication

Turning now to the field of education and communication, a much larger assortment of promising approaches open up.

For example, the education of science students from developing countries at the universities of the more advanced countries leaves much to be desired.[15] Of the many aspects of this problem that have been discussed elsewhere, I want to underline one in particular: the need for a programme to provide such students with summer jobs in

applied scientific research in the outstanding institutions of the country where they are being educated.

To be sure, the type of applied research they are likely to encounter in their own country upon return will most likely be quite different from what they have contact with during these summer jobs. Nevertheless, they will at least have some personal experience with what applied research itself is, something they are not likely to acquire at the universities where they are being educated.

The investment needed for such a programme is indeed rather modest, especially since it would not represent new funds, as most of these students (at least on the graduate level) are employed already during the summers at their universities. An experimental pilot project, handling 100 such students a year, could be financed on less than $ 200 000 a year.

At a more advanced level, there is a great need to involve the applied scientists of both the advanced and developing world into the already existing exchange programme of scientific personnel. Those utilizing sabbatical leaves, Fulbright fellowships, and similar opportunities are predominantly academic scientists. I do not mean to imply that there are too many of these participating at the present. On the contrary, channels for these types of scientists must also be expanded. But the channels for applied scientists are at the present even more meager.

At the other end, there are ample though perhaps inefficiently structured channels for technologists through AID projects, international organizations, and other vehicles. It is the representative of work which is, in some sense, in between basic science and technology, who suffers the worst.

In the reverse direction, channels must be opened up for productive, promising, and able scientists from the developing countries to

applied scientific research

spend periodic extended visits in centres of research in the advanced countries. In this respect also, applied areas suffer badly.

The opportunities to visit academic institutions are very scarce but not altogether absent, and, at the other end, AID and similar organizations have large programme to retrain, update, refresh, and circulate people involved in direct technological operations. But opportunities to visit applied *scientific* research centres in advanced countries are practically non-existent.

New role for institutions

In this respect there is a powerful role to be had, and at the present almost completely neglected, by the large, usually governmental, research institutions and laboratories which exist in most advanced countries. For example, the national laboratories in the United States[16] at the present are virtually completely void of such an involvement, apart from sporadic small programmes, initiated by interested individuals who have to fight the prevailing trends to get anywhere.

One of the difficulties here is the funding structure of these national laboratories which, at least until very recently, has been mainly from a single governmental agency, generally with narrowly circumscribed interests. This has made diversification in general difficult for these national laboratories, and excursions into what many would consider to be under the jurisdicition of the State Department (namely international science) have been particularly difficult to make palatable.

In the last two to three years these laboratories have begun to get used to being funded on a multiagency basis, simply because not enough money was available from a single-purpose sponsor, especially if this purpose lost some of its former popularity. This would be the time,

therefore, to bring international science explicitly within the 'mission' of these laboratories.

Such a broadened scope of activity for these laboratories might also help to cure a very serious 'disease' of these institutions, which is getting graver every year: what to do with aging scientists?

While at universities there are graceful and useful ways of making use of the talents of scientists who are no longer as much involved in research itself as they used to be, in national laboratories most of these channels are missing. Collaboration in research, and particularly applied research with developing countries, would suddenly open up a new horizon for some (though not all) of this to scientists, and would give new meaning to their activities.

To conclude the aspect of education and communication, I would like to mention the need for scientific personalities who have been been eminently successful in applied scientific research (whether in advanced or developing countries) to be sent on a lecture tour to the developing world so they can publicize their experience.

Again, the investment needed would be modest. A one-month lecture tour, including salary, transportation, and accommodations, for 20 top level scientists of this type could be financed on less than $ 200 000.

Not only would such an appearance make an impact on the existing science managerial infrastructure, but it is also likely to influence and reorient future generations by being accessible to student audiences also. People selected for such a tour must have the ability to explain simply, clearly, and effectively what applied scientific research really is, preferably using local opportunities as examples.

Turning finally to the question of the indigenous research facilities and structure, we first encounter the necessity for local research by multinational companies.[17]

While the standard arguments of efficiency might still dictate a central research laboratory in the home country, factors of public relations, future stability, and even specialized markets and products tend to argue the other way. In any case, the setting up of a small research group locally represents such a negligible expenditure compared with the overall size of the operations, that experimenting with it might be worth the risk.

But the onus is not entirely on the multinationals themselves. Very few countries have regulations requiring such companies to do some indigenous research, and furthermore, local, government-owned industrial operations mostly also lack such research facilities.

The closing of the gap between the applied and the academic sector of the scientific communities in developing countries encounters a number of difficulties. Beside strong traditional values and sentiments, there is also the danger that an effective consulting type arrangement between university faculty and applied research institutions will result in the faculty member 'moonlighting' too much and neglecting his university duties.

While I heard this complaint voiced in more than one country, at the present, I think, the danger is still mainly in the opposite direction, namely in the sharp chasm between universities and applied research institutions, and hence I would strongly support further development of such consulting arrangements.

The interaction however, should also work in reverse. Significant figures from applied research facilities in the developing countries should be invited to lecture part time at universities, and take on research students. This probably runs into even more opposition, since universities often have formalistic and unrealistic visions of grandeur about themselves, which precludes collaborating or sharing teaching with 'external' elements. Yet, examples exist for such university instruction by applied scientists from the outside, though the experience has not been

long enough to form a conclusion about its effects.

Finally, I want to add that I have said nothing in this article about the deficiencies of applied research performed by advanced countries which directly benefits the less developed countries. Included in this topic would be the supernational organizations like the International Institute for Tropical Agriculture in Ibadan, which, though located in a developing country, are most of the time quite isolated from their immediate surroundings.

They do not employ significant numbers of indigenous staff, their interaction with neighboring local institutions is weak, and even the transmission of their research results to the surrounding technological personnel is tenuous.

The usual justification for this mode of operation is in terms of the efficiency and standards of the research itself, and this might might very well be valid as far as it goes. Yet, one cannot but regret that some of these admittedly high quality institutions, extensive in size and spectrum of concerns, have so little impact on the stimulation of the local scientific infrastructure, and particularly on the development of indigenous applied scientific research.

References and notes

1. For references, see the bibliographies in J Spaey *et al, Science for Development,* UNESCO, Paris, 1971 (particularly pp 37-44); G Jones, *The Role of Science and Technology in Developin Countries,* Oxford University Press, London, 1971; M J Moravcsik, *Science Development,* PASITAM, Bloomington, Indiana, Second Printing, 1976 (particularly pp 32-34, 96, 105-107 and 137).

2. Some examples are quoted in M J Moravcsik, *Science Development* (see reference 1), p 106.

3. This point of view is implicit in some of the references listed in the bibliographies referred to in references 1 and 2, and is also often voiced as an anecdotal impression by scientific visitors to less developed countries.

applied scientific research

4. Even a brief glance at almost any national development plan which contains a 'research and development' section will confirm such a claim. See eg M J Moravcsik, *Science and Technology in National Development Plans: Some Case Studies,* Agency for International Development, April 1973 (unpublished but available in report form).

5. For an eloquent exposition of this subject, see the writings of Derek de Solla Price, eg *The Relations Between Science and Technology and Their Implications for Policy Formation,* FOA (Research Institute of National Defense), P Rapport B 8018-M5, Sweden, 1972.

6. For a discussion and references, see M J Moravcsik, *Science Development,* see reference 1, pp 95-97.

7. The role of consensus in scientific methodology is the central theme in J M Ziman, *Public Knowledge,* Cambridge University Press, London London (1968).

8. This is discussed in M J Moravcsik, 'The context of creative science', *Interciencia* Vol 1, No 71, 1976.

9. For statistics see the discussion in M J Moravcsik, *Science Development,* (see reference 1), pp 107-109.

10. As an example, in a number of Asian countries there is a CSIR (Council of Scientific and Industrial Research) with one or even a whole network of laboratories. Another example is presented by the Federal Institutes in Nigeria, for Industrial Research (Lagos), and for Forestry Research and Agricultural Research (Ibadan).

11. For some compelling testimonial on this subject, see National Academy of Sciences, 'Applied science and technological progress', *Report to the Committee on Science and Astronautics,* Washington, 1967.

12. M J Moravcsik, 'Some modest proposals', *Minerva,* Vol 9, No 55, 1971.

13. As an example, see University Grants Commission Commission and National Council for Science Education, *Physics in India: Challenges and Opportunities,* New Delhi, 1970, which is a record of a conference on this subject held in Srinagar, Kashmire, at which, however, virtually only academis scientists appeared, in spite of the all-inclusive title.

14. UNESCO alone must have had at least a dozen large conferences on this subject during the past 20 years. For references see the entry of UNESCO in the bibliography of M J Moravcsik, *Science Development* (see reference 1).

15. See eg M J Moravcsik, 'Foreign students in the natural sciences: a growing challenge'. *International Educational and Cultural Exchange* Vol 9, No 1, p 45, 1973.

16. Perhaps the most active in this respect has been the National Bureau of Standards. For a review, see M J Moravcsik, *Science Development* (see reference 1), pp 168, 186, and for a discussion of some of the problems in these national laboratories, M J Moravcsik, 'Reflections on national laboratories', *Bulletin of the Atomic Scientists,* Vol 26, No 2, p 11, 1970.

17. For a recent study of multinational companies in the context of science and technology in the developing countries, see National Academy of Sciences, *US International Firms and R, D, and E in Developing Countries,* Washington, 1973. In general, the literature concerning the role of multinational companies in the *technology* of less developed countries is huge; a corresponding treatment of the role of multinationals in the *science* of less developed countries is minuscule.

Reprinted by permission from *Nature* **276** (1978)
315 © 1978 Macmillan Journals Limited

Third World needs 'barefoot' science

How does a water tap work?
Michael Moravcsik and
R. H. B. Exell explain why knowing the answer can be a breakthrough for an African village.

The practical need for science in developing countries is rarely discussed. Debate usually rises to a general, analytical, or theoretical level. There is no question that such discussions help in understanding the role of science in those countries, and assist the proponents in convincing incredulous officials in national, international, and other decision making organisations, that science is necessary. And they also boost the morale of the scientists in the developing countries—who otherwise have many reasons for being discouraged and frustrated. But one should also be aware that the need for science can be much more down-to-earth.

In this brief note, therefore, we want to present a few examples where exposure to science as a way of thinking and as a method of inquiry and problem solving could play an important role in very simple everyday situations.

The examples are taken from "real life" situations that we have encountered during our stays in developing countries. Colleagues of ours could doubtless add dozens of such examples to the list.

Michael Moravcsik is from the Institute of Theoretical Science, University of Oregon, USA; R. H. B. Exell is from the Asian Institute of Technology, Bangkok, Thailand.

Keeping cool: In urban buildings in tropical areas, built using modern techniques, the top floor is often unbearably hot. The reason is simple: the usually flat roof absorbs solar radiation, and no provisions are made to either insulate the roof from the ceiling of the top floor, or to design the building with a simple double roof with air space between the two for ventilation. Ironically, such features are often included in the traditional, older, and rural structures, but since the scientific reason for such features is not comprehended, builders cannot distinguish between those features of traditional architecture which are functional and those which are ceremonial or due to limitations of construction technology.

Keeping hot: In numerous developing countries tin-box ovens used on charcoal stoves are common. They simply consist of a usual thin-walled surplus tin-box, placed on top of a regular iron charcoal stove (similar to the hibachis popular in Western countries for use in backyard barbecues).

It does not take very much science to realize that the efficiency of such an oven, and also the attainable temperature inside, could be increased if all sides of the tin box except the one in contact with the stove were insulated. A little more science tells us that such a saving in fuel would be

quite substantial, very rapidly returning the small initial investment in one of the available insulating materials.

Current problems: Electrical wiring in developing countries is often installed very flimsily and haphazardly, so that breaks in the wiring and other types of damage occur frequently. In such instances there are many accidents as well as further damage due to incorrect rewiring. For example, we have seen instances when the break in a two-strand wire was "repaired" by joining all four dangling wires in one knot.

Even the most qualitative and simplest *conceptual* understanding of current as a flow (and hence a schematic knowledge of what bears the unnecessarily forbidding title of Kirchhoff's laws) would do much toward preventing such occurrences.

Turning on: Recently, in one of the developing countries, the supply of tap water became quite intermittent for several months. It was then found that during those hours of the day when the water supply did not operate, much subsequent loss of water was initiated by taps being left in an "on" position during the time when the supply was off. People would try the tap, see that no water was coming, and then leave the tap without turning it to the "off" position. When the water supply returned, the water kept flowing sometimes during the whole day until somebody happened to return and turn it off.

Water being expensive, one of the important contributing factors in this apparent neglect is likely to be the complete absence of an image of a water flow with a tap being a gadget to transmit or obstruct this flow. When this simple point is not understood, there is no more "reason" for turning the tap to the "off" position when the water fails to emerge. This is science on such an elementary level that most of us might find it difficult even to realise that it is science.

Counting squares : The next two examples are perhaps more mathematical than physical, but they again stress the importance of understanding concepts rather than following recipes.

A junior lecturer at a university, engaged in an "applied" problem in botany, was confronted with the task of measuring the surface area of leaves. He kept postponing the measurements, however, because an expensive planimeter ordered from abroad had not arrived and, according to some textbook, was *the* instrument to use for such measurements. He was amazed that tracing the leaf on to readily available graph paper with a square millimetre network and then simply counting the squares within the traced figure yields the desired answer with as great (or greater) precision than would have been obtained with the planimeter.

Saving time : In mathematical calculations related to applied scientific or engineering problems, the lack of a conceptual understanding of mathematics often prevents the practitioners from using approximation methods (particularly numerical ones) which yield the answer to the required degree of precision much easier than fancy analytical methods or computer evaluations of the "exact" problem. When in a calculus of variations problem, the shape of a curve has to be optimised, the approximation of the smooth curve by a chain of straight-line segments can reduce the problem to the solution of a simple set of algebraic equations.

The same is true in complicated boundary value problems in differential equations of heat transfer and other problems. Countless hours of precious computer time could be saved on expensive and scarce electronic computers in developing countries if their users were better trained in the fundamental concepts of science and mathematics and relied less on recipies from enginering handbooks or on imitating calculations described in esoteric books

Roofing in the tropics: applying simple scientific principles can make a hot house cool.

dealing with different problems and different tolerance requirements.

Hot water : A solar water heater is to be designed. Even if the basic qualitative principles of such a design are understood, much will depend on the ability to carry out a fairly reliable estimate of heat losses, conversion efficiencies, and the like. For this the scientific principles underlying these phenomena must be clearly understood, because engineering handbooks or the lecture notes saved from university training will hardly give automatic recipes for such "non-standard" situations.

If an engineer claims that he could make steam with such a solar heater, he must also be able to estimate with sufficient accuracy the temperature and the amount of such steam. In the absence of such ability, his project remains an academic exercise and cannot be introduced to "real life" situations where the economic considerations are important and where a quantitative assessment is essential.

A similar situation exists when a claim is made for a design of a practical solar-heat dryer of agricultural products. In this case the useability of the resulting products as well as their quantity will depend crucially on the quantitative factors in the design which, as before, must be estimated on the basis of the scientific principles underlying the engineering applications.

So much for the examples. In seeing them, the reader may advance various objections, at least two of which should be briefly discussed.

The first objection may be that our examples depend not so much on the lack of science but on the lack of "common sense". It is relatively easy to dispose of this objection. Indeed, to a person imbued with some conceptual background in science and technology, the examples may appear a matter of common sense. This is so, however, because he has internalised simple concepts of science and mathe-

matics to such an extent that viewing the world in that light now appears to him a matter of common sense and no longer a matter of science and mathematics.

This is the main point we wish to make: in developing countries such internalisation and fusion of scientific and mathematical conceptual elements with the customary empirical view of the world is very rare, whereas it is much more frequent (though far from universal) in the more advanced countries. This shaping of man's world view is, in fact, one of the primary goals of science in developing countries, and yet it is one that is much too infrequently included in the discussions that prevail throughout "development" literature.

The second objection may be that the missing elements in each of our examples could be provided by a well-trained technologist and engineer and hence science is not really directly involved. In some idealised world this statement may carry some validity. In practice, however, the technological and engineering education, particularly in the developing countries, is such that a functional understanding of the conceptual scientific basis of technology is extremely rare among the graduates of technological and engineering institutions. This is partly so due to directives based on a total misunderstanding of the nature of applied scientific research and of technological development.

According to such directives, engineers and technologists should be given a narrow education in which "unnecessary" abstractions are weeded out and training with respect to a very narrow set of specific projects is the goal. In part, however, the lack of conceptual understanding is due to the lack of contact of these students with teachers who are, themselves, involved in some scientific work and who therefore have had a personal experience with how science is actually done., as distinct from how it appears from the textbooks. Thus we conclude that we need a much greater attention to education.

Scientometrics, Vol. 4. No. 2 (1982) 135–169

SCIENTIFIC OUTPUT IN THE THIRD WORLD*

J. BLICKENSTAFF, M. J. MORAVCSIK

*Institute of Theoretical Science, University of Oregon Eugene,
Oregon 97403 (USA)*

(Received June 10, 1981 in revised form August 25, 1981)

Although such indicators exhibit only certain aspects of the contribution of science to a country, the number of scientific authors in a given year is plotted for every year between 1971 and 1976, inclusive, and the number of scientific authors divided by the population of the country is also given for those years. The number of scientific authors is the number of scientists who published at least one article in a journal in that given year. The data were taken from a survey which, although it covers only about 4000 scientific journals, includes a large fraction of all articles published.

The results are given in 43 graphs, the first 17 of which show the number of authors and the second 16 the authors per capita. The graphs are divided according to geographical areas: Latin America, Africa, the Middle East, and Asia, and within each region countries with roughly comparable output or output per capita are grouped together.

The last ten graphs show the growth rates of authors and of authors per capita, compared to the 1971 values, for groups of countries aggregated according to various parameters with which correlation is being investigated. Continent, size of population literacy rate about 25 years before, the percentage of gross national product spent on military expenditures, and colonial past.

*This article was first suggested by and submitted to *Interciencia*, but upon its completion Interciencia was hesitant to publish it because of the coverage by the ISI compilations of Latin American scientific journals. It was felt by Interciencia that the coverage was so skimpy as to fail to do adequate justice to the scientific output of Latin American countries. By implication, the same may be said about the coverage of other parts of the "Third World".

On the one hand, as explained in the article, this criticism does not invalidate the findings of this article if they are formulated in a sufficiently careful language. On the other hand, the criticism has merit since there is a great need for a much more complete coverage of the scientific publications, authors, publications, and citations in the Third World. For reasons which are quite legitimate in their own right, it is unlikely that an organization like ISI can undertake a coverage of the thousands of journals published in the Third World. I have therefore been suggesting for some time to various appropriate people in the developing countries that these countries, singly or in groups, initiate their own computerized compilation of journals, authors, publications and perhaps even citations. Such a program would be of great value not only in an international context, but also in terms of national and regional science policies, in terms of studies in the science of science of efforts in the developing countries, in terms of evaluative and assessing efforts in those countries, etc. I hope that this article and the reaction it may produce will accelerate the beginning of such an effort.

As far as the numbers of authors are concerned, there are many countries with outputs so small (less than 20 authors) that the statistical fluctuations drown out any possible trends. The remaining countries are grouped into moderate producers (below 80 authors) and large producers (about 80 authors). The developing countries, with the exception of Israel (which is hardly a developing country anyway) and India, all have less than 1200 scientific authors, while, in comparison, the United States has 160 000.

As to the rate of growth, countries can again be divided into three groups: Stagnant or even declining countries, moderately growing countries (up to 60% growth in the period investigated), and fast growing countries. In comparison, the average growth for the scientifically advanced countries during that period was 60%. Thus the percentage gap is narrowing for the fastest growing developing countries (which, however, represent the minority of the developing countries), and is widening for the rest. It also appears as if there were relatively fewer very fast growing countries in Latin America than in the other areas. This might be connected with the interference of politics into science and of scientists into politics in many Latin American countries. In fact, in general there appears to exists a positive correlation between political stability and the growth rate of science, but it does not appear to make any difference what the exact nature of the stable regime is. It is hypothesized that in countries where politics is not mixed with science, the latter can prosper.

The authors per capita figures show a division of countries also into three groups: The lowest group has authors per capita less than 2 per million of population, the middle group from 2 to 10, and the highest group between 10 and 60. The exception is Israel where this number is 1000, exceeding the 750 for the United States. There appears to be little correlation between the number of authors and the authors per capita numbers, so that there is no indication that small countries could not make up for their small size by a larger concentration of scientists.

The results of the aggregated graphs show that, (a) the Americas and Africa grew faster than the Middle East or Asia, (b) small and medium size countries grew slower than most large ones, (c) the highest literacy rate correlated with the slowest growth of science, (d) the less the country spent on the military the faster its science grew, (e) former French colonies grew somewhat slower than former colonies of other countries. Many of these conclusions, however, may have been caused not by the parameters indicated above but by parameters which happened to be correlated with the above ones.

There are many features in these graphs, however, which do not lend themselves to an immediately obvious explanation: Sudden one-year peaks, steady declines, fluctuations, etc. Thus the data presented in this paper clearly constitute a basis for much further research.

Introduction

In looking for indicators of scientific output, one has to be modest and emphasize the limitations of any single indicator, especially if it is a strictly quantifiable one which is therefore very convenient and accessible. There are many ways science contributes to the development of a country. For example, science has an effect on the general world-view of the population which in turn can influence attitudes and activities in areas completely remote from science. It can generate a feeling of national pride and accomplishment which is the crux of the state of mind of nations the world over, but which cannot be strictly quantified. Yet, the primary objective of science remains the generation of new knowledge and understanding about the world, whether for the purpose of intellectual or spiritual growth ("basic research") or for the purpose of application in technology ("applied research"). Thus, in measuring the scientific output of a country, it is reasonable to concentrate on measuring the extent to which scientists in that country contribute to the growth of scientific knowledge.

Whether or not this growth can be measured by articles in scientific journals is the next question we need to answer. There are, after all, other ways the results of scientific research are propagated: informal reports, lectures, informal discussions, books, conferences, etc. These are all modes of communications, and one assumes that all scientific research does get cummunicated to other scientists. This last assumption is correct almost by definition. Research, the results of which remain with the researcher who produced them fails to become part of the body of scientific knowledge or of the basis on which technology is built, and hence is useless. As to the other modes mentioned above, it is perhaps not too incorrect to assume that all scientific research *eventually* makes a trace in a journal publication, even if it first appeared in lectures, reports, or oral discussions. Contrary to some impressions, both "basic" and "applied" research appears in the form of journal publications. Technological development work does not, but that is not the subject of this study.

The next problem that arises is the coverage of scientific journals. Altogether there are some 70 000 such journals around the world. In contrast, all of the available surveys of the scientific literature (including the one we will use) cover only a small fraction of these journals. The omitted journals, however, are so small and infrequently published that scanning only that small fraction of the number of journals will catch an overwhelming fraction of the total number of scientific articles produced around the world. Furthermore, since we are mainly interested in a *comparative* study of the developing countries, it is reasonable to assume that the incomplete coverage affects all developing countries the same way and hence cancels out.

Nevertheless scientometric research based on such surveys has been criticized, sometimes quite vehemently, by scientists in the developing countries, and particularly in Latin America, and it has been claimed that their results are useless. We of course do not believe this to be the case, and claim that information based on the best available sources is always valuable, even if these sources are in some respect deficient. At the same time the complaint of the developing countries about the incomplete coverage has some merit and could be remedied in one of two ways. One would be for the organizations preparing the surveys and compilations to extend their coverage drastically to cover a very much larger number of journals. This is most likely to be financially unfeasible unless heavily subsidized from the outside. The other possibility would be for the developing countries themselves, individually or, preferably, regionally, to organize surveys and compilations of their scientific literature in a computerized form so that scientometric research on the material is easily possible. Both possibilities need to face some conceptual problems, such as the definition of a respectable scientific journal, language, double publication, etc. We nevertheless urge those who complain about the present coverage of the scientific journals of developing countries to turn their energies toward the exploration of such constructive solutions of the problem.

Finally, one might wonder whether a mere counting of scientific articles is reasonable, since such counts do not take into account the size or quality of the articles. This is a longstanding and quite valid critique of article counts which is generally acknowledged by scientometricians. The objection can be partially eliminated by using citation counts to weigh publication counts, but that is a more cumbersome procedure and at the same time still not entirely objection-free.

All this suggests that the results of this article should be interpreted with caution and with the full realization of the methodological limitations. At the same time, quantitative indicators have proven very useful and productive in various realms of science policy, and their utilization should be encouraged.

The specific author data for this article were taken from the volumes of *Who Is Publishing in Science*[1] for 1972–77, inclusive. This series of surveys is produced by the Institute for Scientific Information in Philadelphia, the same organization that publishes the *Science Citation Index.* In 1979, they scanned about 3850 scientific journals the world over, of which about 85 were published in developing countries. The list of the latter is given in Appendix. The ratio of journals from the developing countries to the total number of journals is about 2.2%. This is not very different from the ratio of developing country authors to the total number of authors (5.5%), especially if we consider that authors from developing countries often publish in journals published in the scientifically advanced countries, but the converse seldom happens. It is possible, however, that both percentages are biased in the same way. In any case, however, as said earlier, such biases are likely to be greatly reduced when we view the *relative* numbers of publications in developing countries from country to country.

Articles are assigned to countries according to the byline of the author which indicates his institutional affiliation. Joint papers are split. All authors of a multiple-authored article are recorded.

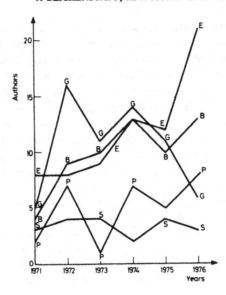

Fig. 1. The number of scientific authors as a function of time for South American countries. Dashed lines represent actual figures divided by 10. Dotted lines represent actual figures divided by 100. NA = figures not available. B = Bolivia, E = Ecuador, G = Guyana, P = Paraguay, S = Surinam

Results

The results are shown in Figures 1–43. They are divided into two groups. The first group, Figures 1–17, give the number of scientific authors in a particular year (i.e. the number of people who published a scientific paper at all in that year), *versus* the year, for each country. The countries are grouped geographically, but within each broad geographical area (i.e. Latin America, Africa, the Middle East, and Asia) the countries are arranged so that the ones with approximately similar outputs appear on the same figure. For easier intercomparison, the scales on the ordinates of the figures are coordinated so that the total number of different scales be reduced to a minimum. This objective resulted in some countries, with relatively large outputs, being plotted as multiplied by a reduction factor of 10 or 100, and they are indicated by broken or dotted lines. This is explained in more detail in the captions.

The second group of graphs, namely Figures 18–33, show the number of scientific authors per capita *versus* the year, for the different countries. The popu-

J. BLICKENSTAFF, M. J. MORAVCSIK: SCIENTIFIC OUTPUT

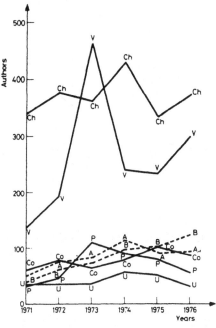

Fig. 2. The number of scientific authors as a function of time for South American countries. For further explanation see Fig. 1. A = Argentina, B = Brazil, Ch = Chile, Co = Colombia, P = Peru, U = Uruguay, V = Venezuela

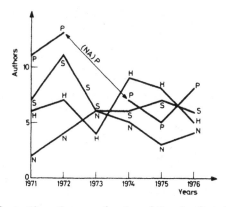

Fig. 3. The number of scientific authors as a function of time for Central American countries. For further explanation see Fig. 1. H = Honduras, N = Nicaragua, P = Panama, S = El Salvador

226

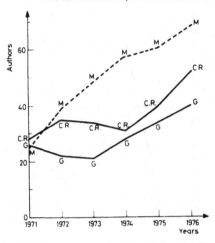

Fig. 4. The number of scientific authors as a function of time for Central American countries. For further explanation see Fig. 1. C.R. = Costa Rica, G = Guatemala, M = Mexico

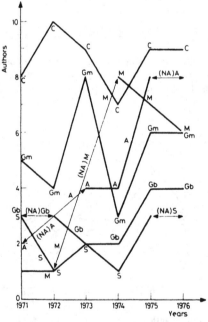

Fig. 5. The number of scientific authors as a function of time for African countries. For further explanation see Fig. 1. A = Angola, C = Chad, Gb = Gabon, M = Mali, S = Swaziland

J. BLICKENSTAFF, M. J. MORAVCSIK: SCIENTIFIC OUTPUT

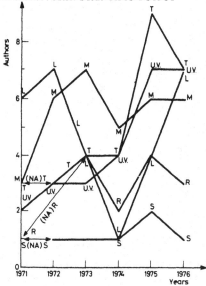

Fig. 6. The number of scientific authors as a function of time for African countries. For further explanation see Fig. 1. L = Liberia, M = Mauritius, R = Rwanda, S = Somalia, T = Togo, U. V. = Upper Volta

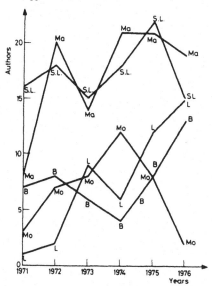

Fig. 7. The number of scientific authors as a function of time for African countries. For further explanation see Fig. 1. B = Botswana, L = Lesotho, Ma = Malawi, Mo = Mozambique, S. L. = Sierra Leone

J. BLICKENSTAFF, M. J. MORAVCSIK: SCIENTIFIC OUTPUT

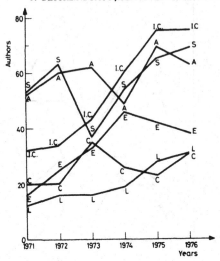

Fig. 8. The number of scientific authors as a function of time for African countries. For further explanation see Fig. 1. A = Algeria, C = Cameroon, E = Ethiopia, I. C. = Ivory Coast, L = Libya, S = Senegal

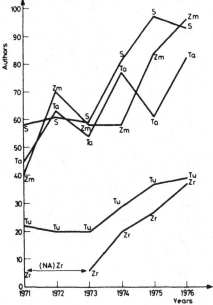

Fig. 9. The number of scientific authors as a function of time for African countries. For further explanation see Fig. 1. S = Sudan, Ta = Tanzania, Tu = Tunesia, Zm = Zambia, Zr = Zaire

J. BLICKENSTAFF, M. J. MORAVCSIK: SCIENTIFIC OUTPUT

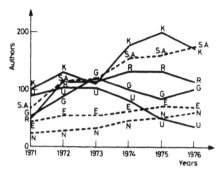

Fig. 10. The number of scientific authors as a function of time for African countries. For
further explanation see Fig. 1. E = Egypt, G = Ghana, K = Kenya, N = Nigeria,
R = Rhodesia, S. A. = South Africa, U = Uganda

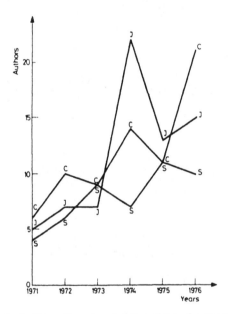

Fig. 11. The number of scientific authors as a function of time for Middle Eastern countries.
For further explanation see Fig. 1. C = Cyprus, J = Jordan, S = Syria

144

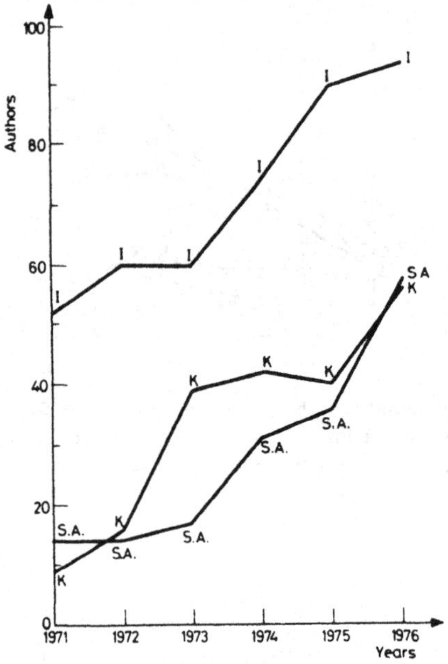

J. BLICKENSTAFF, M. J. MORAVCSIK: SCIENTIFIC OUTPUT

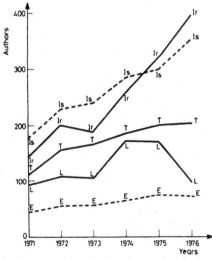

Fig. 12. The number of scientific authors as a function of time for Middle Eastern countries. For further explanation see Fig. 1. I = Iraq, K = Kuwait, S. A. = Saudi Arabia

Fig. 13. The number of scientific authors as a function of time for Middle Eastern countries. For further explanation see Fig. 1. E = Egypt, Ir = Iran, Is = Israel, L = Lebanon, T = Turkey

J. BLICKENSTAFF, M. J. MORAVCSIK: SCIENTIFIC OUTPUT

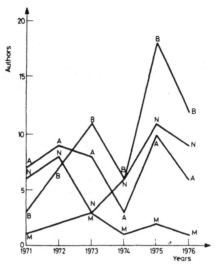

Fig. 14. The number of scientific authors as a function of time for Asian countries. For further explanation see Fig. 1. A = Afghanistan, B = Burma, M = Mongolia, N = Nepal

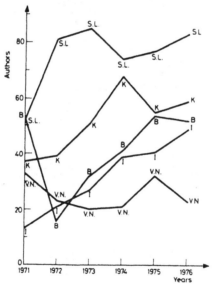

Fig. 15. The number of scientific authors as a function of time for Asian countries. For further explanation see Fig. 1. B = Bangladesh, I = Indonesia, K = Korea, S. L. = Sri Lanka, V. N. = Viet Nam

J. BLICKENSTAFF, M. J. MORAVCSIK: SCIENTIFIC OUTPUT

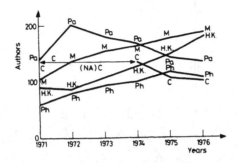

Fig. 16. The number of scientific authors as a function of time for Asian countries. For further explanation see Fig. 1. H. K. = Hong Kong, M = Malaysia, Pa = Pakistan, Ph = Philippines

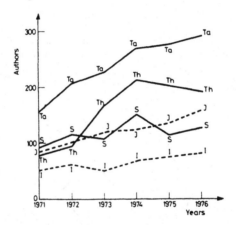

Fig. 17. The number of scientific authors as a function of time for Asian countries. For further explanation see Fig. 1. I = India, J = Japan, S = Singapore, Ta = Taiwan, Th = Thailand

J. BLICKENSTAFF, M. J. MORAVCSIK: SCIENTIFIC OUTPUT

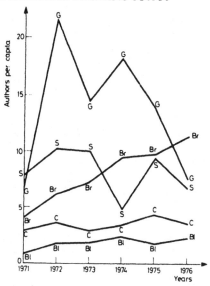

Fig. 18. The number of scientific authors per capita as a function of time for South American
countries. Dashed lines represent actual figures divided by 10. NA = figures not
available. Actual figures are given in number of authors per million of population.
Bl = Bolivia, Br = Brazil, C = Colombia, G = Guyana, S = Surinam

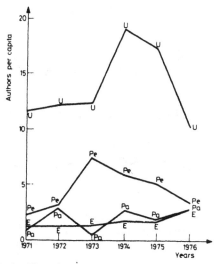

Fig. 19. The number of scientific authors per capita as a function of time for South American countries. For further explanation see Fig. 18. E = Ecuador, Pa = Paraguay,
Pe = Peru, U = Uruguay

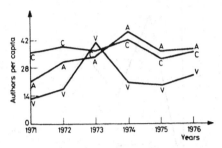

Fig. 20. The number of scientific authors per capita as a function of time for South American countries. For further explanation see Fig. 18. A = Argentina, C = Chile, V = Venezuela

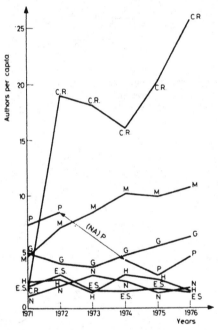

Fig. 21. The number of scientific authors per capita as a function of time for Central American countries. For further explanation see Fig. 18. C. R. = Costa Rica, E. S. = El Salvador, G = Guatemala, H = Honduras, M = Mexico, N = Nicaragua, P = Panama

In Figure 21, the 1971 value
for Costa Rica should be 15.5
rather than the 1.55 shown.

J. BLICKENSTAFF, M. J. MORAVCSIK: SCIENTIFIC OUTPUT

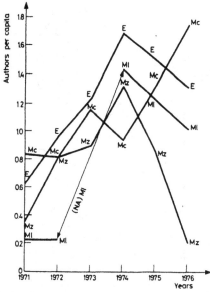

Fig. 22. The number of scientific authors per capita as a function of time for African countries. For further explanation see Fig. 18. E = Ethiopia, Mc = Marocco, Ml = Mali, Mz = Mozambique

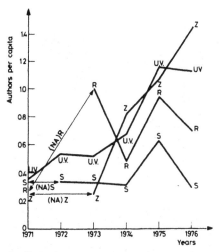

Fig. 23. The number of scientific authors per capita as a function of time for African countries. For further explanation see Fig. 18. R = Rwanda, S = Somalia, U. V. = Upper Volta, Z = Zaire

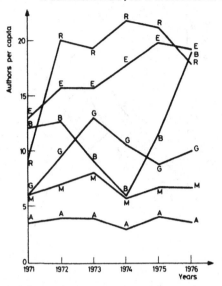

Fig. 24. The number of scientific authors per capita as a function of time for African countries. For further explanation see Fig. 18. A = Algeria, B = Botswana, E = Egypt,
G = Ghana, M = Mauritius, R = Rhodesia

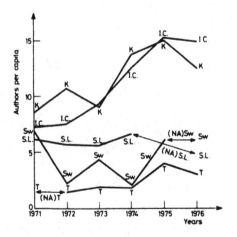

Fig. 25. The number of scientific authors per capita as a function of time for African countries. For further explanation see Fig. 18. I. C. = Ivory Coast, K = Kenya, S. L. =
Sierra Leone, Sw = Swaziland, T = Togo

5

J. BLICKENSTAFF, M. J. MORAVCSIK: SCIENTIFIC OUTPUT

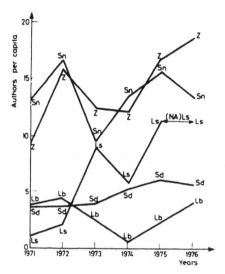

Fig. 26. The number of scientific authors per capita as a function of time for African countries. For further explanation see Fig. 18. Lb = Liberia, Ls = Lesotho, Sd = Sudan,
Sn = Senegal, Z = Zambia

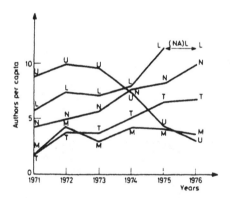

Fig. 27. The number of scientific authors per capita as a function of time for African countries. For further explanation see Fig. 18. L = Libya, M = Malawi, N = Nigeria,
T = Tunesia, U = Uganda

238

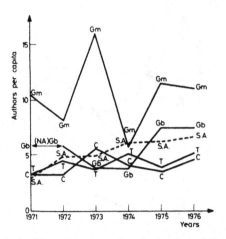

Fig. 28. The number of scientific authors per capita as a function of time for African countries. For further explanation see Fig. 18. C = Cameroon, Gb = Gabon, Gm = Gambia, S. A. = South Africa, T = Tanzania

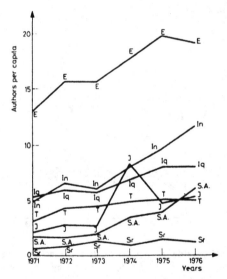

Fig. 29. The number of scientific authors per capita as a function of time for Middle Eastern countries. For further explanation see Fig. 18. E = Egypt, In = Iran, Iq = Iraq, J = Jordan, S. A. = Saudi Arabia, Sr = Syria, T = Turkey

J. BLICKENSTAFF, M. J. MORAVCSIK: SCIENTIFIC OUTPUT

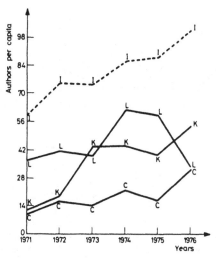

Fig. 30. The number of scientific authors per capita as a function of time for Middle Eastern countries. For further explanation see Fig. 18. C = Cyprus, I = Israel, K = Kuwait, L = Lebanon

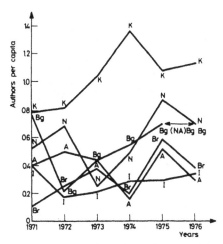

Fig. 31. The number of scientific authors per capita as a function of time for Asian countries. For further explanation see Fig. 18. A = Afghanistan, Bg = Bangladesh, Br = Burma, I = Indonesia, K = Korea, N = Nepal

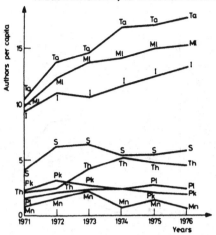

Fig. 32. The number of scientific authors per capita as a function of time for Asian countries. For further explanation see Fig. 18. I = India, Ml = Malaysia, Mn = Mongolia, Pk = Pakistan, Pl = Philippines, S = Sri Lanka, Ta = Taiwan, Th = Thailand

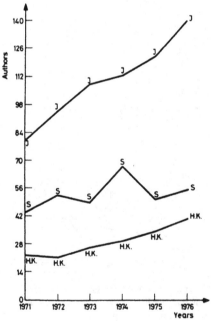

Fig. 33. The number of scientific authors per capita as a function of time for Asian countries. For further explanation see Fig. 18. H. K. = Hong Kong, J = Japan, S = Singapore

J. BLICKENSTAFF, M. J. MORAVCSIK: SCIENTIFIC OUTPUT

lation figures were taken from the U.N. Statistical Yearbook[2] of 1977. Here also, the same guidelines were used for geographical groupings and for the choice of scales.

The third group of graphs, namely Figures 34–43, show the number of authors and the number of authors per capita, aggregated according to some parameter the correlation with which we want to explore. In particular, Figures 34 and 35 show these quantities aggregated according to continent. Figures 36 and 37 show aggregation by the size of the country's population, with S = small denoting countries up to five million, M = medium designating countries with a population between five and twenty million, and L = large including countries with populations over 20 million.

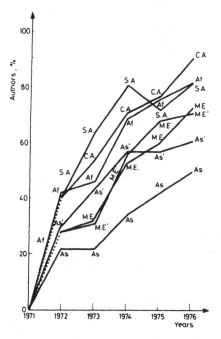

Fig. 34. The compound percentage change of scientific authors as a function of time for various continents. Figures for Africa and the Middle East both contain data for Egypt. Figures for Asia do not include Japan. Countries without complete data were omitted. Dashed lines represent shared values. Af = Africa, As = Asia, C. A. = Central America, M. E. = Middle East, S. A. = South America. As' = Asia without India, M. E.' = Middle East without Israel

242

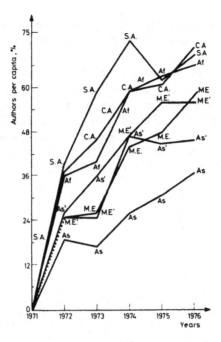

Fig. 35. The compound percentage change of scintific authors per capita as a function of time for various continents. For explanation see Fig. 34.

Figures 38 and 39 show the number of authors and the number of authors per capita aggregated according to the literacy rate of the country[4] in the year closest to 1950 for which statistics is available. The three categories of S.M. and L include such rates from 0 to 33%, from 34% to 66%, and from 67% to 100%, respectively. In Figures 40 and 41 we aggregated the countries according to the percentage of their gross national products spent on military expenditures.[5] The three categories of S, M, and L include countries with 0—2.5%, 2.6%—5.0%, and more than 5%, respectively. Finally, Figures 42 and 43 aggregate the countries according to their colonial past, that is, according to which country they used to be colonies of.

In all the Figures 34—43, not the absolute numbers of authors or authors per capita are shown, but the ratios with respect to the absolute numbers in 1971. The ordinate therefore is in percentage increase with respect to the 1971 values.

J. BLICKENSTAFF, M. J. MORAVCSIK: SCIENTIFIC OUTPUT

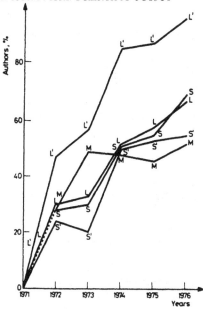

Fig. 36. The compound percentage change of scientific authors as a function of time for countries with small, medium, and large populations. Population figures are in millions, S = 0−5, M = 5−20, L = more than 20. Figures for large populations do not include Japan. Countries without complete data were omitted. Dashed lines represent shared values. S' = S − Israel; L' = L − India

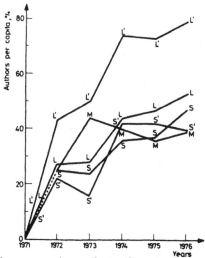

Fig. 37. The compound percentage change of scientific authors per capita as a function of time for countries with small, medium and large populations. For explanation see Fig. 36.

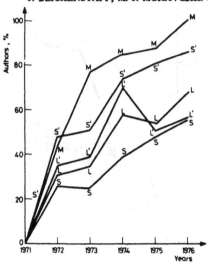

Fig. 38. The compound percentage change of scientific authors as a function of time for countries with small, medium, and large percentages of literacy about 25 years earlier. Countries without complete data were omitted. S = 0–33%, M = 34–66%, L = 67–100%. S' = S without India, L' = without Israel

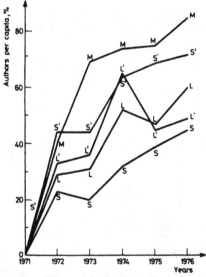

Fig. 39. The compound percentage change of scientific authors per capita as a function of time for countries with small, medium, and large percentages of literacy about 25 years earlier. For explanation see Fig. 38.

J. BLICKENSTAFF, M. J. MORAVCSIK: SCIENTIFIC OUTPUT

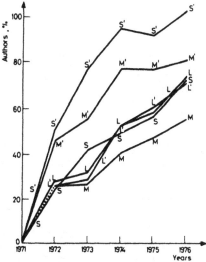

Fig. 40. The compound percentage change of scientific authors as a function of time for various percentages of GNP spent on military expenses. Countries without complete data were omitted. Dashed lines represent shared values. S = 0–2.5%, M = 2.6–5.0%, L = more than 5.0%, S' = S without Japan, M' = M without India, L' = L without Israel

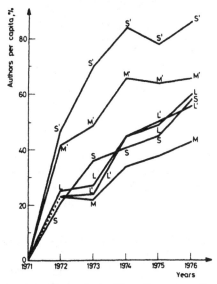

Fig. 41. The compound percentage change of scientific authors per capita as a function of time for various percentages of GNP spent on military expenditures. For explanation see Fig. 40.

J. BLICKENSTAFF, M. J. MORAVCSIK: SCIENTIFIC OUTPUT

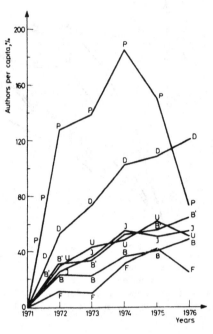

Fig. 42. The compound percentage change of scientific authors as a function of time for countries with various colonial backgrounds. Countries without complete data were omitted. B = British, D = Dutch, F = French, J = Japanese, P = Portuguese, U = American (United States), B' = British without India

Foci for interpretation

In drawing conclusions from this huge amount of information, it is useful to agree on certain interesting questions that one would want to answer. The ones listed below are certainly far from exhausing the array of such questions but they will nevertheless serve us as an initial guide to the interpretation of the data.

1. To what extent is science universal and independent of local factors?

Our data include countries which vary among themselves drastically in terms of the political, cultural, geographic, social, religious, and other parameters. In fact, the four major regions we are considering are different from each other also.

J. BLICKENSTAFF, M. J. MORAVCSIK: SCIENTIFIC OUTPUT

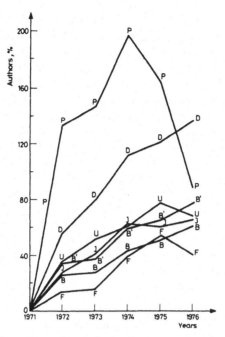

Fig. 43. The compound percentage change of scientific authors per capita as a function of time for countries with various colonial backgrounds. For explanation see Fig. 42.

Of all the developing regions of the world, Latin America is, relatively speaking, most homogeneous and coherent. Broadly speaking, countries in Latin America have similar cultural traditions, the same or similar language (with the exception of Guyana and Surinam), and, again relatively speaking, a remarkably peaceful recent history as far as wars between countries are concerned. As a result, co-operation and interaction among Latin American countries are significantly stronger than among countries in Africa or Asia. In the area of science and technology, not only the Organization of American States serves as vehicle for a very extensive regional cooperation, but in addition more informal organizations, like Centro Latinoamericano de Fisica, have been very active.

For these reasons it is interesting to present a unified set of statistical informa-tion on the scientific output of the Latin American countries, since it permits the study of how certain parameters (like size or educational standards) influence the output, without having to worry also about the influence of other parameters which vary so drastically from country to country in other parts of the world.

In contrast, Africa, as well as Asia, represent geographical areas which are extremely inhomogeneous and incoherent in terms of the personalities of the various countries. There is no common language, no common religion, no common race, no common social or political system, and in fact considerable animosities or even hostilities exist among the various countries of the region. The Middle East appears somewhat more homogeneous, thought still riddled with conflicts.

In view of this kaleidoscopic background of the countries, is the scientific output equally kaleidoscopic?

2. Is science, on the whole, on the upswing in the developing countries? If so, to what extent is this growth uniform in time and geography?

3. Even if science is on the upswing, is the growth comparable to the growth in the scientifically advanced countries? Is the gap (in absolute or percentage terms) increasing or decreasing with time?

4. To what extent do small countries have a chance? Is scientific output primarily a matter of the size of the country, or can a small country compensate for the difference in size by a considerable edge in scientists per capita?

5. It is often said that political instability has an adverse effect on the evolution of science in a country. Can this be documented from the data?

Interpretation

Let us first consider the figures for the authors *versus* time.

It is immediately evident that in every geographical area, there are countries where the total scientific output is represented by less than 20 authors. In such countries, the variation in the output curve from year to year is largely meaningless, since the statistical fluctuations in personnel and in the exact time of publication are likely to be responsible for much of that variation. One can conclude that in those countries all of science depends on a very few individuals and hence is in an extremely rudimentary and fragile state. Admittedly, such a minuscule output marks the beginning of science in every country, but beyond this general and vague statement, there is little one can further say about these countries at this time.

The second general observation is that in every geographical region, there is a huge variation among the outputs of various countries, so much so that this large range overwhelmingly masks any average differences that may exist between two regions. Roughly speaking, one can distinguish three groups of countries in each region. The first, already spoken about, has a negligible output of less than 20 authors. The middle group contains countries with an output between 20 and about 80 authors. Finally, in each region there are large producers, with more than 80 authors, ranging up to many hundreds of authors. Not counting Japan (which is plotted only for comparison anyway), India, and Israel, the outputs nowhere exceed about 1300 authors. The three above exceptions exhibit a number of authors in the thousands, up to 14 000 for Japan. In comparison, the figure for the United States in 1976 was 156 000.

This spatter of statistics gives us an idea of the relative orders of magnitude involved in the worldwide distribution of scientific authors.

The third observation pertains not to the absolute numbers of authors but to the growth during the period under consideration. Here also, we can distinguish three groups of countries. In the first, the number of authors remained approximately constant during the period or may have even declined. The second group of countries involves those where the growth during the period amounted to less than 60%. Finally, there are countries in which the growth has been larger than 60%. It should be emphasized again that in the countries with less than 20 authors such growth curves cannot be established so the following remarks pertain only to the countries in groups two and three of the previous classification. With respect to these two groups, it appears that in either of them we can find stagnant, slowly growing, and fast growing countries. Thus it appears that a country with a small scientific output is not necessarily doomed to stay with such a small output, but has about as good a chance to grow as countries with an already larger output.

One can also ask whether growth rates are equally large in the various geographical areas. Here there appears to be some difference among the various regions. Although there are fast growers in Latin America (notably Brasil, Costa Rica, and Mexico), they are relatively smaller in number than in Africa, in the Middle East, or in Asia. A glance at Figures 8, 9 and 10 as well as 12, 13, 15, 16 and 17 indicates a fairly large number of countries in these other regions with a quite fast growth rate, including a number with growth factors over 2 in this period (Lybia, Ethiopia, Ivory Coast, Zaire, Zambia, Tanzania, Sudan, Nigeria, South Africa, Iraq, Kuwait, Saudi Arabia, Iran, Indonesia, Taiwan and Thailand). We will return to this point below.

How do these growth rates compare with growth rates during the same period in the scientifically advanced countries? The numbers for those countries (not given in this article) suggest that the average growth factor during this period there was about 1.6, or a 60% growth. When we compare this with the rates indicated above for the developing countries, we conclude that the percentage gap between the advanced and the developing countries is narrowing for the fastest growing developing countries, but is widening for those which grow only moderately or which are fairly stagnant, and that this latter group (for which the gap is widening) is larger than the fast growing group. One must recall that this statement leaves out those numerous countries in which the growth rate cannot be derived yet because of the very tiny absolute numbers in output.

The next question to investigate is whether one can find some correlation between a low growth rate, or a fluctuating one, on the one hand, and political uncertainties and upheavals, on the other. In other words, can one explain somehow the less than adequate performances by some connection to political problems?

This questions cannot easily be answered in a thorough way. There are clearly examples where the correlation is obvious. Peru and Argentina in Latin America, Uganda in Africa, Lebanon in the Middle East, Bangladesh and Sri Lanka in Asia represent some such examples. In other cases correlations may also exist but a more detailed study of the history of the countries involved would be needed to trace them down. There are also examples for the same correlation in a favorable sense: Countries with relatively stable political systems tend to show a steady and fast growth curve. Mexico, Brasil, Costa Rica, South Africa, Kuwait, Saudi Arabia, Iran (during the period under consideration), Indonesia, and Malaysia offer some examples for that phenomenon. The diversity among these countries with regard to their political systems also shows that steady and fast growth in science is not a monopoly of any particular political system. One is tempted to hypothesize that as long as politics does not interfere with science and vice versa, and the regime is a progressive one with respect to the support of science and technology, science can grow.

There are, however, a number of large fluctuations and trend reversals which do not lend themselves to such an immediately evident interpretation but which would be interesting and important to trace down. When dealing with countries with an output of more than 50 authors, a sudden change in authorship by, say, 25% is not likely to have a purely stochastic explanation. The huge one-year peak in Venezuela, the large sudden decline in Senegal, the zig-zag patterns in Tanzania, the erratic output of Ghana, the steady decline in Pakistan and the erratic variation in Singapore are some examples of phenomena on which further research is warranted.

It should also be noted that although India is a very large producer, its growth rate is only moderate.

Now let us turn to the authors per capita figures. To start again with the absolute numbers, countries can be divided into three groups again. In the lowest category, the numbers are less than 2 authors per million population, in the middle category between 2 and 10, and in the highest category between 10 and 60. Even Japan does not go very far beyond this upper limit, reaching about 140 at the end of the period. The one exception is Israel where the number reaches about 1000, which is higher than the corresponding 750 or so for the United States.

It is immediately obvious that high absolute numbers of authors do not go hand in hand with high absolute numbers of authors per capita. For example Brasil, one of the top producers in the number of authors among the developing countries, is only in the medium group for authors per capita. To be sure, on the whole, the countries with the lowest authors per capita figures are also mainly in the category of very low producers, but there are exceptions to this rule, such as Ethiopia, Korea, Indonesia, and Bangladesh, which are medium producers. We see, therefore, that it is quite possible for a small country to make up for its disadvantage in size by having a relatively large author per capita figure, and, conversely, some countries which are large can show their impact even though their authors per capita figures are modest. This may be taken both as an encouragement for small countries and a warning for large ones.

It is perhaps needless to say that in the authors per capita figures, small numbers do not necessarily go with large fluctuations, as was the case for the number of authors, because the random fluctuations affect only small absolute numbers but not necessarily small ratios.

Since for most developing countries the populations show a steady growth during the period under investigation, the growth rates for the authors per capita figures tend to be more modest than for the number of authors, but the relative standing among countries does not change much.

Let us now turn to the aggregated graphs. Grouping by continents is shown in Figures 34 and 35. Both these figures show that the fastest growth occurred in Central and South America and in Africa, with somewhat less rapid growth shown by the Middle East (with or without Israel) and by Asia without India. India's growth rate was considerably lower than that of either of the above two categories. Nevertheless, it is clear that all parts of the Third World have grown quite substantially during the period under investigation.

Figures 36 and 37 give the growth rate according to the size of the country's population. We see that small and medium size countries have grown slower than the large ones, except for India which, as we learned also from Figures 34 and 35, has grown relatively slowly.

Figures 38 and 39 show the relationship to literacy rates. We see that the "medium literate" countries grew the fastest, the "little-literate" countries (excluding India) almost as fast, and the very literate countries the slowest. It would probably be erroneous to conclude, however, from this that too much reading hinders progress: The effect is more likely to be due to other parameters which happen to be correlated with the literacy rate. The important conclusion from

166

these figures is that a low literacy rate, *per se,* does not prevent a country from making good progress in building up its science.

Figures 40 and 41 show the relationship to the relative amount spent by the country on military expenditures. Conventional wisdom, at least in some quarters, would claim that science and the military live in an unholy symbiosis and thrive on each other. The actual results of these two figures, if taken at face value, would prove just the opposite. The less you spend on the military, the faster your science grows. Here again, however, other correlated parameters may be responsible also for causing the observed effect.

Finally, Figures 42 and 43 group the countries according to their colonial[6] past. The results for former Portuguese and Dutch colonies should not be taken to be too indicative since the number of such countries is very small. For the other groups, the growth rates appear to be roughly equal, with the possible exception of a lower growth rate for the former French colonies.

These conclusions and interpretations by no means exhaust the potential of these graphs, and the reader will undoubtedly stare at them some more and formulate his own interpretations of some of the variations. The above discussion nevertheless suggests that output figures plotted against time do represent a rich set of data for learning about the development of science in the Third World.

In conclusion we want to stress that ours is by no means the first presentation of scientific output data comparing the production of various countries. On the contrary, a number of papers exist in the literature[7] exhibiting such information in various contexts and with varying aims. Nevertheless we believe that by offering the time variation of outputs country-by-country and in aggregated sets, both in absolute numbers and per capita, our paper establishes certain patterns that would not have been easy to discern in earlier work.

*

We are indebted to J. Davidson *Frame* who suggested that we include some aggregated figures among our results.

References

1. *Who Is Publishing In Science,* Institute of Scientific Information, Vols 1972–1977 inclusive.
2. *United Nations Statistical Yearbook,* United Nations, Vol. 1977.
3. Private communications with Isabel Wertheimer, Institute of Scientific Information, Philadelphia, PA.
4. Data for literacy rates were taken from the U.N. Statistical Yearbook, 1957. United Nations, New York.

6

J. BLICKENSTAFF, M. J. MORAVCSIK: SCIENTIFIC OUTPUT

5. Data for the percentage military expenditure were taken from *Military Balance*, 1974 and 1977, International Institute for Strategic Studies, London, England. We are indebted to Mr. S. R. *Elliot* for providing us with the data.
6. The colonial backgrounds were determined on the basis of The Times Atlas&Gazetteer of the World, 1922, The Times, London.
7. See for example D. de S. PRICE, Measuring the Size of Science, *Proc. Israeli Acad. of Sci. and Human*, IV—6 (1969); I. SPIEGEL-RÖSING, *Sci. Stud.*, 2 (1972) 337; H. INHABER, K. PRZEDNOWEK, Geoforum 19 (1974) 45; H. INHABER, *Geoforum* 6 (1975) 231; J. D. FRAME, F. NARIN, M. P. CARPENTER, *Soc. Stud. of Sci.* 7 (1977) 501; J. VLACHY, *Czech. Journ. of Phys.* B29 (1979) 455; J. D. FRAME, *Soc. Stud. of Sci.*, 9 (1979) 233.

Appendix[3]

List of Scientific Journals Published in Developing Countries

Argentina	Acta Physiologica Latinoamericana, Anales de la Asociacion Quimica Argentina, Medicina-Buenos Aires, Phyton-International Journal of Experimental Botany
Brazil	Anais da Academia Brasileira de Ciencias, Arquivos da Escola de Veterinaria, Revista Brasileira de Pesquisas Medicas e Biologicas
Chile	Archivos de Biologia y Medicina Experimentales, Revista Medica de Chile
China	Acta Entomologica Sinica, Acta Zoologica Sinica, Botanical Bulletin of Academia Sinica, Chinese Medical Journal, Geochimica, Scientia Geologica Sinica, Scientia Sinica, Vertebrata Palasiatica
Costa Rica	Agronomia Costarricense, Turrialba
India	Acta Botanica Indica, Advances in Invertebrate Reproduction, Central Glass and Ceramic Research Institute Bulletin, Cheiron, Comparative Physiology and Ecology, Current Science, Entomon, Indian Journal of Agricultural Science, Indian Journal of Animal Sciences, Indian Journal of Biochemistry and Biophysics, Indian Journal of Chemistry Section A-Inorganic Physical Theoretical&Analytical, Indian Journal of Chemistry Section B-Organic Chemistry Including Medicinal Chemistry, Indian Journal of Experimental Biology, Indian Journal of Genetics and Plant Breeding, Indian Journal of Medical Research, Indian Journal of Nutrition and Dietetics, Indian Journal of Pure and Applied Physics, Indian Journal of Physics and Proceedings of the Indian Association for the Cultivation of Science-Part A, Indian Journal of Physics and Proceedings of the Indian Association for the Cultivation of Science-Part B, Indian Journal of Theoretical Physics, Indian Journal of Zoology, Indian Veterinary Journal, Journal of Feed Science and Technology-Mysore, Journal of the Geological Society of India, Journal of the Indian Institute of Science Section A-Engineering& Technology, Journal of the Indian Institute of Science Section B-Physical and Chemical Sciences, Journal of the Indian Institute of Science Section C-Biological Sciences, Journal of the Indian Chemical Society, Journal of Scientific and Industrial Research, Kavaka, National Academy of Sciences, Nucleus, Proceedings of the Indian Academy of Sciences Section A, Proceedings of the Indian Academy of Sciences Section B, Phytomorphology, Pramana, Psychological Studies

Kenya	East African Medical Journal
Mexico	Archivos de Investigacion Medica, Patologia-Mexico City, Revista de Investigacion Clinica
Rhodesia	Rhodesia Agriculture Journal, Rhodesian Journal of Agricultural Research
South Africa	Journal of the Entomological Society of Southern Africa, Journal of the South African Institute of Mining and Metallurgy, Ostrich, South African Journal of Chemistry, South African Journal of Physics, South African Journal of Science, South African Journal of Surgery, South African Journal of Zoology, South African Medical Journal, Transactions of the Royal Society of South Africa, Water S.A.
Taiwan	Bulletin of the Institute of Zoology Academia Sinica, Journal of the Chinese Chemical Society
Thailand	Journal of Ferrocement-Bangkok, Journal of the Science Society of Thailand
Venezuela	Acta Cientifica Venezolana, Interciencia

AN ANALYSIS OF CITATION PATTERNS IN INDIAN PHYSICS*

MICHAEL J. MORAVCSIK, POOVANALINGAM MURUGESAN AND EVELYN SHEARER

Institute of Theoretical Science, University of Oregon, Eugene, Oregon 97403, USA

THE study of the citation structure of scientific literature has, in the past decade, become an interesting and promising tool in the analysis of the methodology and sociology of scientific communities[1]. Simple citation counts have been used to evaluate individual scientists, to trace down the development of groups of scientists around a new discovery or around the person of an outstanding scientific leader, as well as for many other investigations in the "science of science"[2].

Citation counts are said to be preferable to the reliance on the number of scientific papers, because the former attempts to include a measure of quality also. In other words, much cited papers are thought to be more influential in the development of science and hence are given greater weight in citation counts.

At the same time, a mere counting of citations also have their shortcomings, because an article can be cited in another scientific article for numerous reasons, some of which have little to do with the ability of the cited article to bring progress to science. In order to refine the use of citations, therefore, a relatively simple classification scheme was developed[3,4] which sorts citations according to the context in which they are embedded in the citing article. This classification method can be shown to be able to produce some intriguing insights into the workings of the scientific community.

In particular, in one of the articles[4] on this method, comparisons were made between journals published in various geographical areas of the world, and some significant differences were found. Although the interpretation of the reasons for these differences is at this time not unambiguous, the results themselves will serve as a foundation for further sociological studies of scientific communities in various geographical areas of the world.

The articl just referred to covered the United States, Western Europe, Japan, and the USSR. In the present article the analysis is extended to some articles in theoretical physics published in Indian journals, and the results are compared to the results found in the other journals. In conclusion, some attempts are also made to interpret differences found between citation patterns, but it must be emphasized that while the differences themselves rest on solid foundations, the interpretation of such differences is purely conjectural, based on personal experience and intuitive generalization, but not on any in-depth investigation of the Indian physics community.

Definitions

The definitions of the classifications we are using have been given in our previous articles.[3,4] Since some of the journals where these articles were published are rather inaccessible to the Indian scientific community, we reproduce here the complete statement[4] of the definitions.

As a general remark about our classifications, it should be emphasized that their criterion is not what the actual format or content of the cited article is, but to what use it is put in the citing article.

*Work supported by the US National Science Foundation.

The detailed definitions are as follows :

(a) *Conceptual vs. Operational*

The purpose of this classification is to distinguish between citations which are used to develop a theoretical and conceptual basis for the citing paper from citations which are used as tools in the development of the citing paper.

More specifically, citations are considered to be conceptual when a concept or theory from the cited paper is used directly or indirectly in the citing paper in order to lay theoretical or conceptual foundations to build on or to contribute to the citing paper.

By "using directly" we mean that a concept is taken from the cited paper and the material of the cited paper form the basis of the citing paper (whether the author states this explicitly or not).

By "using indirectly" we mean the situation when in the course of the development of his ideas, the author of the citing paper finds it necessary to incorporate certain ideas or concepts which are not strictly necessary to formulate the basic ideas of his paper but may add more insight or help clarify certain key concepts of the citing paper.

In contrast, the definition of an operational citation involves the situation when a concept or theory is referred to as a tool to substantiate the author's claim (e.g. the author may compare his results based on his theory with results of another theory) or to indicate alternative approaches. In addition, a reference is also called operational when it borrows mathematical or physical techniques, results or conclusions from the cited paper.

(b) *Organic vs. Perfunctory*

The purpose of this classification is to distinguish necessary citations from dispensible ones.

More specifically, citations are considered organic when concepts or theories are taken from the cited paper which are necessary to lay the foundations of the citing paper, or when certain results (including numerical ones) are taken from the cited paper in order to develop the ideas, procedures, and results of the citing paper, or when the cited paper is used to clarify certain concepts or results in the citing paper.

In contrast, perfunctory citations are those which refer to descriptions of alternative approaches which are not utilized in the citing paper, or references which are used to indicate the fact that a certain method employed in the citing paper is routine in the literature. In short, perfunctory citations are not really necessary for the development of the citing paper, and often can be considered as existing only for ceremonial or decorative purposes, or for purposes of advancing personal relations with other scientists.

It might be mentioned that the proper use of this dichotomy, more than any of the other three, depends on the individual making the classification having sufficient amount of scientific knowledge and background, since scientific judgement must be exercised to decide whether the cited paper is really needed to understand the content of the citing paper, or whether the cited item is generally known anyway and hence would not need a specific reference. This difficulty is sometimes compounded by the author of the citing paper providing a reference for an item which is not the primary (original) source of the item. We have followed the convention that if the item is judged to need a reference at all, even such secondary references are considered organic if we have no information about the primary source.

(c) Evolutionary vs. Juxtapositional

The purpose of this classification is to distinguish material in the same line of development from material in parallel or divergent lines. In other words, evolutionary citations are references to material which is in the same line of logical development of the subject as the citing paper itself.

More specifically, a citation is evolutionary if it provides a concept or theory to build on in the citing paper, or a mathematical technique to use in the citing paper, or results of an analysis which is used in the development of the citing paper or in the notation of the citing paper.

In contrast, a citation is called juxtapositional if it refers to alternative approaches, or gives mere references to works using the same general approach but which have branched off so that they do not contribute to the development of the citing paper, or refers to other analyses used in the citing paper only to make comparisons, or refers to other works which may help to clarify some ideas but do not contribute to the development of the citing paper.

(d) Confirmative vs. Negational

The purpose of this classification is to distinguish material considered correct by the citing author from material disputed by the author.

More specifically, a citation is confirmative if the author of the citing paper considers the item referred to in the cited paper to be correct. Almost all evolutionary citations are also confirmative, since authors generally do not base their papers on other work they consider incorrect.

In contrast, a negational citation describes a situation when the author of the citing paper is not certain about the correctness of the cited paper and says so. There are two types of negational citations. In the first type the author of the citing paper claims that the cited paper is incorrect. Such a claim may be based on experimental evidence or on a contradiction with accepted and well-established theory. In the other type of negational citation the author of the citing paper disputes the cited paper but cannot come to a definite conclusion, because the issue is still being tested experimentally and theoretically. We have so far not made a distinction between these two types of negational citations.

If the citing author does not express his doubt about the correctness of the cited paper, the citation is in general considered confirmative. The only exception is made when, though the citing author does not say anything explicit about the correctness of the cited paper, the results of the citing paper are in clear contradiction with the results of the cited paper.

(e) Comparative remarks about the first four classifications

To use a simple language, the main goals in these four classifications are as follows:

In (a) : To distinguish *ideas used* in the citing paper from *tools used* in that paper.

In (b) : To distinguish *necessary* citations from *dispensible* ones.

In (c) : To distinguish material used in the *same* line of work from material in *parallel or divergent* lines.

In (d) : To distinguish material judged *good* from material judged *bad*, the judgements being done by the citing paper.

In general, any of the two categories of a given classification will contain pieces of both categories of any other classification, since these four classifications are cuts along four different dimensions.

The abbreviated definitions given in the simple language above serve mainly to get across the main intent of the classifications. For the operational purpose of actually performing the classification, they should be used in conjunction with the longer definitions given earlier.

(f) *Redundancy*

In case more than one paper is referred to in a given context just to show that more than one author has dealt with that topic, but the citing paper does not make specific use of the theories, mathematical techniques, or results of all those papers, then all those cited papers but one are classified as redundant.

On the other hand, when the author refers to several papers in a given context and then uses something specifically from each of those papers, then those cited papers are not judged to be redundant.

There is a difference between redundant and perfunctory citations. The latter could just stand by itself, and still be judged perfunctory because it does not contribute to the development of the citing paper. A redundant paper, on the other hand, could possibly contribute to the development of the citing paper except that it stands in a group with other papers cited, all of whom make the same contribution to the citing paper.

The purpose of redundant citations in papers is to be fair to all prior authors, and to show explicitly that the citing author has acquainted himself with the literature. From the strict point of view of the scientific development of the citing paper, however, all redundant references could just as well be omitted without affecting the reader of the citing paper. Hence the classification of redundancy.

The Extent of the Study

The investigation of citation patterns in our previous work included articles on theoretical physics, subdivided into various fields. In order to be able to make comparisons, we restricted ourselves to similar articles in Indian scientific journals. Furthermore, in order to have a sufficiently large sample for meaningful statistics, we confined our investigation to theoretical articles on solid state physics. Most other fields of physics are either very scarcely represented in the Indian journals we considered, or were not among those fields we studied in the non-Indian journals.

As in our previous work, we restricted our attention to citations to other theoretical articles. Thus, references to experimental articles, to books, or to non-published material were not included. In most cases this restriction is not a serious one from a quantitative point of view, since most references in theoretical papers are to other theoretical papers. At the same time, inclusion, for example, of references to experimental papers might have brought in some additional variables thus making the results more difficult to interpret.

The Indian articles were taken from two journals : the Indian Journal of Pure and Applied Physics, and the Indian Journal of Physics. A brief survey of other Indian journals suggested to us that the two we selected are likely to be fairly representative of most of the Indian journals and at the same time would contain a sufficiently large number of relevant aticles. The articles were taken randomly from the articles dealing with solid state physics from the period 1968-1972, the same period studied by us for non-Indian journals.

Altogether 813 citations were considered. Using our previous way of giving a rough estimate of random errors connected with such a sample, we can say that a dichotomous classification of 813 "measurements" would then result in a random error of $(1/2 \times 813)^{1/2}$ in each of the two classes, so that, roughly speaking, we can assign a random error of $\pm 3\%$ to the *percentages* given in our tables. As previously, this error estimate is not to be taken very literally, since the conditions for purely random deviations are not altogether satisfied in this case. At the same time, effects amounting to 2 or 3 standard deviations thus defined should probably be taken seriously.

Results

The results of our analysis are shown in Table 1. Also shown in that table are the corresponding results from our previous study of non-Indian journals in solid state physics.

The material in Table 1 can be summarized as follows :

1. There are no significant differences between the two Indian journals we studied in the percentage of conceptual or of confirmative citations. There are, however, significant differences in the percentages of the organic and of the evolutionary citations, both of which are higher for the Indian Journal of Pure and Applied Physics. Furthermore, the percentage of redundant citations is significantly lower for this journal than for the Indian Journal of Physics.

2. As to the comparison of the two Indian journals with the two non-Indian journals we had studied, the most striking difference is in the percentage of conceptual citations, which is much lower for the Indian journals than for the non-Indian ones. It will be recalled that in this category there was a significant difference even between Physical Review (solid state) and Soviet Physics Solid State, but the difference between the average of these two journals and the average of the two Indian journals is even larger. There is no significant difference between the Indian and non-Indian journals in the percentage of confirmative citations. Concerning the percentage of organic cita-

tions, there is no difference between the Indian and non-Indian *averages*, but the percentage of the Indian Journal of Pure and Applied Physics is close to that of Physical Review (solid state), while the percentage of the Indian Journal of Physics is close to that of Soviet Physics Solid State. There is a smaller but significant difference also between the percentage of evolutionary citations, the Indian *average* being lower than either of the two non-Indian journals, though the difference becomes marginal if one considers only the Indian Journal of Pure and Applied Physics.

As mentioned earlier, these differences between percentages can be considered as facts. Any explanation of the causes of these differences, on the other hand, are at this point only speculations, based mainly on anecdotal type of personal observations and intuitions. Entering, nevertheless, into this speculative domain, one can venture the interpretation that research which is routine and narrow in the scientific sense is likely to have few conceptual and mostly operational citations, because the concepts underpinning the limited objectives of a routine problem are few and are likely to be well known so as not to need citations. By "routine" research we mean an investigation which is not very different from previous ones, perhaps directed in a somewhat different direction, and which thus produces results which are perhaps new in the sense of never having been attained before, but are not novel and do little more than add a piece of information to a huge collection of similar information already accumulated. Perhaps another way of stating this is to say that many routine pieces of research are motivated mainly by the argument that it has never been done before, and not by the feeling that the result will be esthetically pleasing, or will contribute to our view of nature, or will be useful in technological applications.

TABLE 1. Comparison of citation patterns in Indian journals with patterns in other journals. All data refer to papers on theoretical solid state physics. The table shows the number of citations in each category and the percentage of the total citations for that journal. For the definition of the categories, see the text The abbreviations for the headings are as follows : TR : Total number of references ; TP : Total number of papers referred to ; ER : Extraneous references ; CC : Conceptual ; OP : Operational ; NI : Neither ; OR : Organic ; PF : Perfunctory ; EV : Evolutionary ; JU : Juxtapositional ; CF : Confirmative ; NE : Negational ; RD : Redundant. Since some citations can be used in several contexts, the percentages do not add up exactly to 100 %.

	TR	TP	ER	CC	OP	NI	OR	PF	NI	EV	JU	NI	CF	NE	NI	RD
Indian Journal of Pure and Applied Physics	801	573	284	60 11%	492 86%	21 3%	194 34%	374 65%	0 0%	212 37%	354 62%	3 0%	536 94%	41 7%	0 0%	192 34%
Indian Journal of Physics	313	240	90	16 7%	216 90%	8 3%	54 23%	186 78%	0 0%	60 25%	180 75%	0 0%	232 97%	8 3%	0 0%	114 48%
Total Indian	1114	813	374	76 9%	708 87%	29 3%	248 31%	560 69%	0 0%	272 33%	534 66%	0 0%	768 94%	49 6%	0 0%	306 38%
Physical Review (Solid State Physics section)	538	423	208	163 38%	195 46%	66 16%	164 39%	267 63%	5 1%	189 45%	226 53%	22 5%	410 97%	29 7%	4 1%	99 23%
Sov. Phys. Sol. State	542	438	134	99 23%	323 74%	20 5%	104 24%	346 79%	1 0%	189 43%	250 57%	0 0%	399 91%	61 14%	7 2%	153 35%
Total Non-Indian	1080	861	342	262 30%	518 60%	86 10%	268 31%	611 71%	6 1%	378 44%	476 55%	22 2%	809 94%	90 10%	11 1%	252 29%

The above interpretation appears to be confirmed by (and, to some extent, might owe its origin to) a subjective assessment by trained physicists of the articles used in our study. In fact, peer judgement has so far been the best (or perhaps even only) gauge of quality of scientific research. Our work in studying the nature and context of citations aims at trying to find more "objective" criteria for measuring quality. We cannot claim at this point that we have achieved this goal, but a demonstration of the correlation between the more "subjective" (or intuitive) peer judgement and the citation classification measures we have developed appears certainly to be a step in the right direction.

References

1 See, for instance, J. R. Cole and S. Cole, Social Stratification in Science, University of Chicago Press, (1973) and references therein.

2 For example, see D. de Solla Price, Little Science, Big Science, Columbia University Press, New York, (1963) or, for more recent results, H. Small and B. C. Griffith, The Structure of Scientific Literature 1 : Identifying and Graphing Specialties, Science Studies, 4, 17, 1974.

3 Michael J. Moravcsik and Poovanalingam Murugesan, Some Results on the Function and Quality of Citations, Social Studies of Science, 1, 5, 86-92, 1975.

4 Poovanalingam Murugesan and Michael J. Moravcsik, Variation of the Nature of Citation Measures with Journals and Scientific Specialties, Journal of the American Society for Information Science, (in press), (1975).

C. Communication

Reprinted by permission of *Management of Research and Development*, OECD, 1970.

COMMUNICATION AMONG SCIENTISTS AND ITS IMPLICATIONS TO DEVELOPING COUNTRIES

INTRODUCTION

The pursuit of research in the natural sciences is a very well structured activity, at least compared to most other human endeavors. Whereas there are definitely deviations on a small scale and in the short run, on the whole one can say that in science progress is made in a way a brick wall is laid. Individual contributions serve as a basis for others to make further advances, which in turn lead to even further progress. Even the so-called great break-throughs that occur from time to time are generally not as whimzical and ad hoc as they sometimes appear to a scientist who looks at it fifty years later but who is not an avid student of the history of science. This coherent structure of scientific progress is one of the main factors behind the very international nature of science. It stimulates a link among scientists everywhere, because every scientist knows that the step on which his next accomplishment might be based could just as well be generated in Tokyo, Moscow, London, or Berkeley as by his fellow worker in his own institution.

To be sure, not all approaches in science come to fruition, and sometimes false starts must be made before one finds the right approach. But as long as the outcome is in doubt, it is difficult to distinguish the "false" approach from the "correct" approach, and hence even the pursuit of the approach which later turns out to be false is a service to science.

The situation is quite different in other areas of intellectual activity. In fine arts or music, for example, there is no well defined direction to "progress", and development consists more of the exploration of alternate approaches to age old esthetic problems, with no clear value judgement as to what is "better" or "worse". There, consequently, progress does not necessarily build on previous accomplishments in any direct

way, and therefore there tends to be less of an international brotherhood of artists or composers than there is of scientists.

This difference has several consequences. For one thing, I believe
that a second rate scientists has more justification to feel that he contributes to science than a second rate artist with respect to art. This is so
because the work of the artist is either appreciated or not, and if the
latter is true, the only other justification that remains is a matter of
personal expression, completely unrelated to others and hence completely void of external recognition. In contrast, the scientist can always
console himself with the knowledge that even if his work is not epoch-
making, it does contribute to the overall progress of science.

A second and quite different consequence is that artists can work
quite independently from other artists, while the work of scientist
depends crucially on the interaction with others. It is therefore of utmost importance for most scientists to keep in constant touch with their
fellow scientists. This does not mean, of course, that they need actual
daily contact with thousands of other scientists around the globe. What
it does mean, however, is that they should have very prompt access to
any scientific development around the globe which is relevant to their
work, and that they could have access to a reasonable number of knowledgeable people with whom they can discuss their work, and with whom
they can exchange ideas for future research.

This is particularly essential to a scientist in a developing country,
because his work is made more difficult by his relative isolation compared to his colleagues in the more advanced research centers. Developments in science occur very fast these days, and the pace is constantly
accelerating. Since most of the scientific developments occur in the more
advanced countries, our scientist in a developing country is in greater
need of quick communications than his colleague in whose institution the
development might take place. Furthermore, in some areas of science,
team work is used to facilitate progress, and such team work is
often impossible in scattered institutions in the developing countries.
It is therefore of utmost importance for research and development
in the emerging countries to assure excellent channels of communication with the rest of the scientific world, so that a priori disad-
vantages due to geographic location can be minimized though communications.

Absence of such communications is the main contributor to the
feeling of scientific isolation,* which is one of the primary enemies

---

*   See for example A. Salam, _Minerva_ IV, 461 (1966).

of scientific morale in the developing countries. Indeed, it is a disheartening thought to realize that compared to other colleagues, one is handicapped from the outset by a lack of information and contact.

One possible reaction to such a state of affairs in some cases has been to select a research problem which is not pursued by anybody else, and which is far off the main branch of science. The trouble with this approach is that such a research problem is also likely to be one whose outcome nobody cares about, and which in effect is along one of those approaches that most other scientists would judge to be "false" by now. It is perfectly possible to continue research in such a direction, particularly in theoretical work, since if one dissociates oneself from the question of whether a certain mathematical construct has relevance to the way nature actually is or not, one can always find innumerable, very beautiful theories whose internal consistency one can take delight in, and whose development one can chalk up as a personal accomplishment. One principal element of science, namely the relevance to "reality" is of course gone, but such a sacrifice might be deemed necessary under the circumstances. To an outsider, however, this is hardly the solution to the problem of isolation.

The problem of isolation must be faced also because it is one of the main contributors of the brain drain. The purpose of this article is not to deal with the brain drain problem, and there is in fact already a large literature on the subject. * I would like, nevertheless, to digress for a moment to respond to a very recent article** on this subject which presents a somewhat unusual and quite optimistic view on this problem. In effect it claims that brain drain as such does not really exist, and that the migration of highly trained individuals from the developing to the developed countries is simply an overflow of personnel that the "effective demand" of the economy of the developing country at that point cannot accomodate, even though the "human needs" of the society would in principle need. The article therefore suggests that one should not worry too much about this problem, which "will largely take care of itself" if "both the developed and developing countries concentrate on economic growth". The article supports this conclusion by reminding us that a number of organisations, both international and American, have studied the question of brain drain but their reports did not recommend any drastic governmental action to remedy it.

Whereas I was impressed by the eloquence of the article and found much of its discussion illumating and educational, I would nevertheless like to take issue with its main conclusion. It seems to me that there

* For some earlier remarks of mine on the brain drain, see J.M. Moravcsik *Minerva* 4, 381 (1966).
** G.B. Baldwin, Foreign Affairs, 48, 358 (1970).

are at least two possible major effects that the article neglected to take into account, and which make the overall picture much less rosy than the article contends. First, whereas it might be true that on the face of it, all jobs for highly trained scientific personnel that the economy can offer at that particular time are filled, it is by no means obvious that they are occupied by the best men the country has to offer. The article we are discussing admits that there is no statistics comparing the quality of the personnel that leaves for abroad to the personnel that stays behind, and hence, it claims, "all we can do is to fall back on the assumption that migration of the critical elite would be roughly proportional to the total number of professional migrants". Although of course I cannot invalidate this assumption by statistics either, it is my personal experience that it is grossly incorrect, and that it is often exactly the most talented, and the most aggressive that leave the country. Neither is this very surprising since the international market in professionals is a competitive one, and naturally the best people get picked up first. Furthermore, the most aggressive and management-wise most adroit individuals also have a higher chance of success in acquiring for themselves a suitable position abroad. Thus I suspect that the fraction of the professional personnel that migrates abroad represents, from the point of view of future developments of science and technology, a rather cricial element in quality, even though perhaps not in quantity.

Secondly, I believe that this same group of people, as well as the others who migrate, would, if they stayed home, represent an additional force to accelerate the over-all development of the country in the direction so that in fact the gap between the "effective demand" and the "human needs" is narrowed. In other words, migration is also a valve for relieving a pressure, but this pressure should not be altogether unwelcome.

At least for these two reasons, I tend to continue to believe that the brain drain is a problem to worry about. As to what to do about it, I do not disagree very much with the various organisational reports that large governmental actions would probably be inadvisable. At the same time I do believe that a number of channels are available on a smaller and less conspicuous scale through which useful measures can be taken. Many of these pertain, in fact, to alleviating the problem of isolation, and more specifically, facilitate the communication of the scientist in the developing country with the rest of the world.

I.   COMMUNICATIONS IN PERSON

A.   Internal Communication

When looking for things to correct, one should always begin on home grounds. In this connection I would like to suggest that in some developing

countries communications among scientists <u>within</u> the country is far
from ideal. Sometimes the reason appears to be that rivalries and
negative competition are too strong to allow amicable relationships
among different institutions or individuals. In other situation, travel
funds for <u>internal</u> personal contact are scarce. This latter deficiency
is relatively easy to correct, since in many developing countries long
distance public transportation is subsidized anyway and is therefore
quite inexpensive, in addition to not requiring any foreign exchange, a
commodity hard to come by in many countries. In planning a country's
internal scientific expenditure, it would be advisable to give considerable
weight to the benefits stemming from the contact between the country's
own scientists stationed at different points.

In this connection I would also like to repeat the suggestion I made
earlier that a program should be experimented with under which a <u>group</u>
of young scientists with closely related interests are repatriated to the
same institution in their home country, perhaps even with partial outside
subsidy during the initial few years, so that they can form a research
group among themselves.

According to the suggestion, half of the normal domestic salaries
of these scientists would be paid by some external source, thus freeing
them of half of the customary teaching load so that they can maintain a
reasonable level of activity in research. The university where these
scientists are placed would receive an additional amount, namely 20%
of the half saleries of the scientists, as an overhead payment in recognition of its contribution to this cooperative program. The amount of
money needed for such a program is not large and can be spent in local
currency.

Such a program would greatly decrease the feeling of isolation and
improve the communication problem for these scientists. To make the
arrangement even more attractive, some funds could be made available
so that at least some of them could spend 2-3 months each year abroad
to further improve their contacts with the outside world. Such "package
deal" type of hiring of scientific personnel has also been used in the
United States by less developed institutions intent on acquiring rapid
excellence, and it has in fact worked in many cases.

In this connection let me also suggest that in spite of the proliferation of scientific journals and societies, there is much to be said, from
the point of view of the internal scientific life of a country, in favour of
local journals and local science associations. Even if the journal and
the meetings of the society are not utilized by many from abroad, these
activities help to develope an <u>esprit de corps</u> which can greatly contribute to the scientific morale of a country

B. Travel Abroad

The opportunity to travel abroad periodically is one of the most important facets of communication. At the same time, it is also one of the most delicate questions for a number of reasons. First, on a purely material level, travel abroad almost always involves foreign exchange expenditure, if for nothing else than for transportation. It might be worthwhile considering the possibility of flying appropriately selected scientific and technical personnel from the developing countries, on an if-space-available stand-by basis, for considerably reduced rates. I realize that such a suggestion reaches into the hornet's nest of international airline operations and fares, and that a worldwide agreement among airlines would be necessary to open up such a possibility. But perhaps it is worth exploring whether this could be done. In any case, the airline of the developing country involved (and what developing country does not have an airline nowadays?) should and perhaps would be in favor of such a proposal, and perhaps even could act on it unilaterally.

There are, however, other, more profound difficulties also. For one thing it is difficult to determine what the "proper" fraction of time is that a scientist should spend abroad. One extreme is to allow the scientist abroad only in say, five years, for a short conference. At the other end of the spectrum is an acquaintence of mine, originally from a developing country, now holding a medium level academic appointement at an American university, who wishes to spend 3-4 months every year in his country of origin, and demands rather royal arrangements in return. Both of these are inadequate: The first person would not be able to function adequately enough in the interim periods to take much advantage of the conference, while spending resources on the second person would be to a great extent a waste, because the infinitesimal amount he could contribute in widely spaced, brief intervals would not justify the great expenditures.

Whereas admittedly there is no absolute, objective way to point at the correct answer in this question, I would like to suggest that a paid sabbatical year every seven years, and a one year leave of absence without pay halfway between the sabbatical leaves should be an adequate arrangements for anybody. In addition, unpaid leaves during periods when the university is not in session should be allowed. The sabbatical salary would not have to be paid in foreign exchange, or certainly not fully in that. I have seen similar arrangements put into practice and prove successful. This custom would also conform roughly to the practice at American universities, without seriously disrupting the economic or the scientific life of the country in question.

A more difficult question is to assure that the scientist from the developing country has an adequate position abroad for his sabbatical leave. With the positions for scientists getting definitely less abundant in the developed countries than they used to be, this might constitute a real bottleneck. International agencies like IAEA and UNESCO should be very active in providing funds for such sabbaticals, but their budgets are not in a favorable position to do this on a grand scale.

A type of organisation specially geared to this problem is the International Centre of Theoretical Physics in Trieste. It has received its rather meager support from a number of sources, such as the IAEA, the UNESCO, the Ford Foundation, the Italian government, and the City of Trieste. Its main aim is to accommodate theoretical physicists from the developing countries for a temporary stay (lasting from a few months to a year) and assure for them an environment in which they can catch up on developments in their fields and communicate with their colleagues from the more developed countries. It has functioned very well indeed, and has contributed immensely to the strengthening of the scientific life of many developing countries. It is, however, too small and is geared to only one of the many scientific disciplines. Using its by now well proven structure as a model, more such institutions should be established.

At this point, I would also like to argue that if indeed some more of these institutes are formed, at least some of them should be made regional and placed geographically in the proximity of a group of developing countries. Even at the inception of the present institute which eventually was located in Trieste, I argued that it be placed in a developing area, and while I am very happy indeed with the success of the Trieste institute, I still believe that regional location has its merits.

This brings me to the discussion of another type of international cooperation, namely the regional one. While contact with the scientifically more developed countries is a necessity, much could be derived also from the regional cooperation among developing countries. There is now a European Physical Society, which points the way to a South Asian Physical Society, or a Latin American Physical Society. I realize that on the one hand political differences matter elsewhere much more than in Western Europe, and that some regional progress has in fact been made in some areas. But from what I have seen in practice, much more could be done with very little additional resources, although with considerably increased enthusiasm and aggressiveness.

C. Visitors from Abroad

Another, opposite source of communication is travel to the developing countries. Fortunately, in this respect the natural trend is to some extent in our favor. With travel becoming relatively cheaper, faster, and more comfortable all the time, the number of scientists from the developed countries who do travel abroad is increasing fast. I am now talking mainly about short term travel; longer stays will be discussed later. It nevertheless remains somewhat of a problem to take optimal advantage of such travel.

It would, for example, be helpful if in each area of science, there were a central registry where scientists who intend to travel abroad (particularly to the developing areas of the world) would indicate their intent and where educational and research institutions in the developing countries could register their desire to receive visitiors. It would then be easy to coordinate these two aspects of the potential exchange. At first sight such a coordination might seem unnecessary, because one would believe that most people have sufficient personal initiative and contact to make their own arrangements. It has been my own observation, however, that this is not at all so, and that such coordination might be of considerable value to a number of people and institutions. In any case, it is worth trying: it costs next to nothing, and can be discontinued at any time if it proves unnecessary.

If, on the other hand, it is successful, an extention of it might also be possible. Such an extention would consist of actually arranging say, a one month long lecture and discussion tour for a scientist from a developed country to a group of developing countries, which could then share the expense. Assuming that the scientist would be able to receive his salary from the organisation sponsoring his research (which in any cases is not an unrealistic assumtion), the cost of a round-the-world, one-month trip of that sort would be about $ 2000. If this is shared among ten institutions and considering that about half of the expense can be contributed in terms of local currency, the cost per institution might in fact be not entirely unmanageable? As to whether such an expenditure of perhaps $ 200 to pay for a two-day visit of a distinguished and active scientist from a developed country is worthwhile or not, examples I have seen would suggest that the answer is definitely in the affirmative.

There is, however, also need for long term visitors, who stay for an academic term, or a year or two. Such visitors can contribute in a different way, and their own education in affairs of science in a developing country will also be more profound? Interestingly enough, there are a number of programs that are designed to support such long-term visits, and it is often the lack of appropriate applicants that limits the outcome. Better publicity of the existing programs is certainly called for. In addition, however, thought should be given also to determine the optimal level of such programs, and then take measures to reach it if it differs drastically from the present level.

Particularly useful have been in this respect bilateral programs set up between an institution in a developed and an institution in a developing country. The feeling of personal responsibility and identification adds greatly to the effectiveness. Furthermore, such arrangements also have more flexibility to experiment than large international or intergovernmental arrangements. It would be very salutary indeed if more universities in Western Europe could organise such bilateral programs with some university of their own choosing.

In this connection I would like to resubmit a suggestion I made recently* in connection with national laboratories in the United States. I made the point that older scientific personnel in such laboratories, who is perhaps less eager to work full time on current research than he was fifteen years previously, has very few alternatives to rechannel his energy and experience, since national laboratories have no teaching facilities, and scientific administration is neither to everybody's taste, nor is everybody suited to engage in it. I therefore suggested that some of this older scientific personnel in national laboratories could very well be encouraged to take advantage of his accumulated experience and knowledge by spending a year or so in a developing country. I have known examples when older scientists of this type became engaged in activities in developing countries. In some cases at least it was very successful indeed, when the person in question acted not as a "flying advisor" taking occasional trips to developing countries, but rather was willing to put in a number of years in residence, learning about the problems and then improvising help as it was needed.

In this connection I would also like to reiterate a suggestion made by F.E. Dart that there is particular value in a recurring visitor, somebody who spends a fairly extended period of time in a country, then leaves, and returns some years later for another stay. He is then in an unique position to observe progress or the lack thereof, and suggest areas and methods that seem to have worked particularly well, or particularly badly. At the present, recurring visits occur more or less only by chance, and the point I want to make is that since there is special value in such repeated visits, more systematic efforts should be made to increase their frequency.

* M. J. Moravcsik, Reflections on National Laboratories Bulletin for the Atomic Scientists, 26, 2.11 (1970).

Finally, a word about the assets and the limitations of visitors from abroad. As to the advantages, I have listed some in a previous article of mine. * One of the most important, although perhaps least rational advantage of a foreign visitor is based on the proverb that nobody is a prophet in his own country (and on the corrollary that often even a fool becomes a prophet if he travels far enough). Thus sometimes advice which could very well have been solicited from local scientists is not so solicited or in any case not believed until a foreign visitor comes and repeats it. This peculiarity is by no means restricted to developing countries, and it is not entirely without a rational foundation either, because often an outsider does have a degree of detachment that is locally absent.

One should always remember, however, that visitors from abroad are <u>not</u> a substitute for the development of the indigenous scientific manpower resources. Visitors can be helpful, but their presence can also lull people into a certain amount of complacence, since things sometimes look fine as long as the visitor is present. Just as education, the scientific research and development structure of a country must also be fundamentally indigenous and on the whole self-supporting, at least in the long run. Self-supporting of course does not mean isolation. In fact the main point of the present article is that <u>lack</u> of isolation is a prerequisite for becoming self-supporting. What I mean, however, is that in the long run visitors must be regarded not as crutches but as colleagues of more or less equal standing who come to have contact as part of the natural exchange that is an essential ingredient of scientific work.

II. COMMUNICATIONS IN WRITING

Even with all the conferences and the considerable amount of traveling scientists do, written information has remained the most important mode of communication. It does not compete with, but rather supplements personal, face to face contacts. The latter is indispensible for extended discussions when instantaneous feed-back is essential, and also to cover research or ideas which have not reached the written stage yet. In contrast, written communication covers a much broader area, and it also covers work done in the past which might have remained unknown to the scientist in question. Written communication also offers a more leisurely time schedule, allowing therefore a more thorough digestion of new ideas and hence a more considered response.

Written communications can assume a broad range of formality. Perhaps the most formal such communications are books and journals. Next come reports and so-called prepints, and finally there are personal

---

*   M.J. Moravcsik, <u>Minerva</u> 2, 197 (1964).

letters from scientist to scientist. We will comment on each of these
in their turn.

A.   Journals and Books

The number of scientific articles published in journals is increasing
exponentially. In physics, for example, the largest but by no means only
journal, Physical Review, extends over several tens of thousands of
pages every year. It is now subdivided into many subsections, each
covering a rather narrow part of physics. My own subsection, elementary
particle physics, hits my desk once a month with some 700 pages. There
are a number of other journals around the world containing articles on
elementary particle physics, which are available to me in our university
science library. The journals are mainly of archival nature, that is, they
contain presumably carefully written and carefully refereed articles on
research which, at the time one receives the journal, is at least one
year old, and in many cases older.

In addition, in the last ten years the so-called letter journals have
become popular. They contain shorter reports on presumably important,
much more recent, research results. They are refereed much faster,
and are not printed but reproduced by a much faster process. One such
journal, Physical Review Letters, appears once a week, and contains
papers on all parts of physics.

The trouble with journals from the point of view of the developing
countries is that they are generally quite expensive and require foreign
exchange. I hasten to add that journals are not really expensive if one
considers the number of pages one receives for the money. This, however, is little consolation to the individual or instituation in a developing
country when it comes to spending, say, $ 1000 a year in hard currency
for even the scantiest of journal coverage in physics.

One possible remedy would be to make special arrangements in needy
cases so that journals could be paid for in local currency, which then
would be used by some international agency which has business in that
country anyway. In turn, that agency would reimburse the journals in
hard currency. Such arrangements have been considered from time to
time. Another possibility would be to make arrangements with the journals that for a reduced overall price, special copies are made of some
journals, to be distributed, under strict supervision, only in the developing countries. Such schemes would not be very much to the liking of some
journals which already face very serious financial problems.

Whatever the method, it is imperative that research centers in the
developing countries have an adequate coverage of journals. It is sufficient if several institutions, geographically immediately adjacent, get

together and share subscriptions. The overall coverage within each city, for example, should be adequate. By adequate I mean a subscription to, say, the 15-20 most important journals in physics. I would not like to venture a guess for the corresponding number in other sciences.

It is my feeling that books play a much less important part in physics research than articles do. This is of course not the case when it comes to physics education, but researchers use journals much more often than books. Thus, if a hard choice must be made about expenditure of a finite amount of foreign exchange, I would favor journals over books. The exceptions include review volumes (like Annual Reviews of Nuclear Science), and conference and advanced seminar proceedings. These are in their nature much closer to journals than to conventional monographs or textbooks, and hence should be lumped in with the former. It should be remembered, however, that the relative importance of books varies from field to field, and from discipline to discipline, and hence the scientists in question should always be consulted for their own preference.

B.   Reports and Preprints

Since scientific research is making progress at an ever increasing rate, the very slow turn-around time of the article journals, and the slight tardiness and scanty coverage of the letter journals leave no choice but to place considerable fraction of the responsibility for quick written communication on reports and so-called preprints. The latter are mainly version of articles to be published later, although sometimes they also include informal report which do not get published in that form ever. These prepints are duplicated by some relatively inexpensive and fast reproduction process, and then mailed to those around the world who are known to work in the field in which the topic of the preprint falls. The importance of preprints varies from field to field. In my own specialty, elementary particle physics, preprints are in fact the most important single form of communication. My "preprint library", which I share with some 6 other elementary particle physicists at my institution, receives several preprints a day, from all around the world. Naturally one can only browse through such a volume of literature, but browsing helps one to keep generally current about developments and also helps locating those preprints which are more directly related to one's own research.

In principle, preprints are sent free of charge to all who might be interested in it. In practice, preprints are generally sent to groups of people who can be expected to share a copy. Also in practice, many institutions around the world receive only a fraction of the relevant preprints, because each author has his own mailing list which is usually not at all complete. The fact that each author has to duplicate and

distribute preprints of his own papers makes the whole preprints communication scheme not only inequitable but also inefficient and uneco-
nomical. The US Atomic Energy Commission Technical Information
Division suggested a number of years ago that, as an experiment, one
should try to duplicate and distribute centrally preprints in elementary
particle physics.* This suggestion then was investigated in detail by
a special study conducted by the American Institute of Physics, with a
resulting recommendation** that such an experiment be carried out.
In spite of this, the American Institute of Physics itself decided not to
do so, but referred the matter to the Division of Particles and Fields
of the American Physical Society. This division is an organization with
no financial resources, no permanent staff, and only a few part time
officers who are distinguished researchers themselves and have very
little time for division matters. So the matter now rests on the shelves.

I believe, however, that the time has come to circumvent this
particular blind alley and assure in other ways that preprints reach
research centers in the developing countries. Perhaps one more general
campaign could be waged to get individual authors to place every interested institution on their mailing list. If this fails, one should consider organizing a central duplicating station (or perhaps several regional ones),
where preprints could be duplicated and distributed to institutions in the
developing countries only. This could be done selectively so as not to
overlap with those preprint communication channels which do exist between the developing and developed countries. With a very modest initial
investment, preprints can be duplicated for a fraction of a cent per page
per copy, and even this cost does not need to be paid for in hard currency.
In any case, it is very important that much more attention be paid to this
increasingly more essential problem concerning communications with the
developing countries. If preprint distribution could be made quite equitable around the world, it would reduce (though not eliminate) the reliance of scientists on letter journals to the extend that perhaps these
would not have to be delivered via air mail as they are now. Letter
journals, however, would still serve the important purpose of alerting
researchers to important developments in areas outside their own specialty. Since preprints are unrefereed, unindexed, and appear in a
bulky and perishable form, they cannot be considered substitutes for
journals which would therefore have to be subscribed to in any case.

---

* See J. M. Moravesik, Physics Today, 19: 6, 62 (1966) and S. Pasternack,
Physics Today 19: 6 (1966). See also M. J. Moravesik, Physics Today, 18 : 3, 23 (1965).
** Report of the American Institute of Physics: AID/SDD-1 (Rev.) Miles A. Libbey
and Gerald Zaltman, the Role and Distribution of written Informal Communication in Theoretical High Energy Physics, New York, 1967.

C. Personal Written Communications

When science was smaller and slower, personal letters between important scientists constituted a very important form of communication. These days, the importance of personal letters has greatly diminished. Nevertheless, such communication exists. It would be very helpful if more scientists would develop personal scientific penpalship with scientists in some developing countries who have similar interests, and would keep them systematically current of matters of concern to them. The initiative often has to be taken by the scientist from the developed country, since local social forms make their colleagues in the developing countries often too shy to initiate a personal correspondence with somebody they never met in person.

Within the United States, telephone service, even from one end of the country to another, has become cheap enough so that communication by telephone between researchers is getting increasingly common. The telephone supplies the instant feed-back and the speed of spoken word that is missing in written letters. We should look forward to the time, I hope in the not too distant future, when the satellite technique of communication will become sufficiently wide spread so as to lower trans-oceanic telephone rates to the level of about $ 1 per three minutes, at which time the advantages of telephone communication will be available to scientists in the developing countries also.

III.   CONCLUSION

Compared to salaries of scientific personnel, to the cost of equipment, and to development and production expenses, the cost of communication among scientists is small. It would therefore be foolish penny pinching and administrative shortsightedness to be damagingly stingy with money destined to develop these communication channels. As I tried to outline at the outset, communications can play a crucial role in the competence, relevance, and timeliness of the research and development in an emerging country, and can also affect in a major way the scientific morale, which in turn is a fundamental ingredient of a a successful program. The thoughts expressed in this article, however, are at best a tiny contribution to the many problems that still face us in this respect. Perhaps having considered some of these problems in the present article will stimulate some further efforts in this direction. After all, pure discussion is easy, and by itself does not represent progress. Instead, progress is made by people with energy and determination to convert some ideas into reality.

Reprinted by permission of *Scientometrics* 11 (1987) 53.

Scientometrics, Vol. 11. Nos 1–2 (1987) 53–57

IN THE BEHOLDER'S EYE: A POSSIBLE REINTERPRETATION OF VELHO'S RESULTS ON BRAZILIAN AGRICULTURAL RESEARCH

M. J. MORAVCSIK

Institute of Theoretical Science University of Oregon Eugene, Oregon 97403 (USA)

(Received January, 8, 1986)

Using the data recently presented by *Lea Velho* on the citation rates in and on Brazilian agricultural journal articles, it is suggested that a given such paper is cited by the non-Brazilian scientific literature at the same rate as a paper written anywhere else in the world would be, and that is cited by other Brazilian papers very much more than a paper elsewhere would be. These conclusions are surprizing in view of the prevailing conventional wisdom, and are also exactly opposite to the conclusions *Velho* herself derived from the same data.

In a recent article[1] *Lea Velho* analyzed the citation patterns in Brazilian agricultural research and inferred from them a set of general conclusions concerning the communication patterns to and from the scientific communities in the developing countries. Since scientometric studies of science in the Third World are still very scarce, and therefore science polycy efforts in that part of the world are based either on anecdotal experience or formalistic conceptual schemes, Velho's work is more than welcome. It can be considered a pivotal contribution on which much future research will be based.

Knowing this view of mine, what follows should be taken as a small step toward stimulating such research. In particular, I will concentrate on two conclusions drawn by *Velho* from her data and suggest that the same data can also give rise to radically different, in fact exactly opposite, conclusions. The choice between the two directions can be made by adding to the *Velho* data a small amount of additional information.

I want to stress that in my argument I will *not* question any of *Velho*'s methodology, such as the definition of the problem, the sampling, the data collection, the interview techniques, and alike. Indeed, I will use her data tables as presented in her article. My only deviation from her paper will be at the phase in which the data are interpreted and conclusions are drawn.

Elsevier, Amsterdam–Oxford–New York Akadémiai Kiadó, Budapest

M. J. MORAVCSIK: IN THE BEHOLDER'S EYE

The two conclusions I want to take issue with can be found on pages 77 and 80 of *Velho*'s article, although they are repeated in other parts of the article also. Here are the direct quotes:

1. "The data in Table 4 indicate that authors publishing in advanced country journals make negligible use of work emanating from Brazil..."

2. "There is evidence here that Brazilian agricultural scientists have very weak linkages with Brazilian colleagues working outside their own institutions..."

These two conclusions are based on the data given in Table 4 of *Velho*'s paper, in which the origin of the papers referred to by Brazilian agricultural scientists and by a comparable sample of advanced country papers is shown. The categories offered are self-citations, in-house citations, references to work produced by Brazilian scientists at other domestic institutions, citations to journals in other Latin American countries, citations to journals in other "peripheral" countries, and citations to journals published in "advanced" countries. In this table, I want to concentrate only on two numbers which are the ones revelant to the two above listed conclusions. These two numbers are as follows:

A) The percentage of citations made by Brazilian agricultural scientists to work produced by Brazilian scientists at other domestic institutions is 16.2%.

B) The percentage of citations made by the sample of advanced country papers to work produced by Brazilian scientists at other domestic institutions is 1.1%.

There is a footnote in *Velho*'s paper after the second of these numbers, remarking that that out of the 19 citations, 9 occur in a single paper which was authored by Brazilian scientists. So, *Velho* remarks, the actual number of references from *foreign* to Brazilian scientists is 10, or 0.59%.

Let me first discuss conclusion 1, which is based on percentage B quoted above.

There are at least two different meanings one can attach to conclusion 1 (*Velho* does not state which one she means). One is that the *absolute* fraction of citation in advanced country journals to Brazilian papers is very small. This is a trivial meaning in view of the fact that a very small fraction of the world literature in agricultural science is produced in Brazil. The other meaning of the statement is that *relative to the fraction of agricultural papers produced in Brazil* the citations of such papers in the literature produced in the advanced countries is low. To my mind, this would be the significant meaning of conclusion 1, and it would also be the one that would give rise to the set of explanations offered by *Velho* for the low fraction, in terms of language barriers, problems of access, etc.

As I said, the first meaning is well known to be true and is of little interest. Whether the second meaning is true or not cannot be decided without knowing the total number of agricultural papers produced worldwide and the number of agricultural papers produced in Brazil.

The first of these numbers could probably be ascertained without too much difficulty from international data bases such as that of ISI, subject to the assumption that such a data base covers adequately the agricultural papers produced in the advanced countries (which will make up the lion's share of the worldwide total). The second of these numbers, however, could be determined only from local Brazilian data bases. These should be available to *Velho*, and hence she is urged to present that number and then ascertain whether the second of the above meanings is true or not.

In the absence of these "hard" numbers, we can make a tentative and somewhat dubious approximate calculation, by simply taking the percentage of scientific papers from Brazil according to the ISI database. This is a questionable approximation for at least two reasons:

a) The ratio may be different in agricultural science from that for all sciences taken together,

b) The ISI database may be biased against applied papers in agricultural science and also against Brazil vis a vis the advanced countries.

This second question has been discussed frequently, most recently at the Philadelphia meeting,[2] but a definite answer, which is not available anyway, depends on a number of qualifiers, for instance on how to define the quality of papers which should be included in publication counts.

Inspite of these problems with using this approximation, let me use it anyway for a rough estimate. Information on the percentage contribution of Brazil to the worldwide scientific literature can, for instance, be found in a paper submitted[3] to the Philadelphia meeting just referred to, in which this figure is given as 0.37%.

Comparing this figure with the 1.1% cited as "number B" earlier, or even with the 0.59% which was the 1.1% corrected to include only citations by non-Brazilian authors, we see that there is no basis whatever to claim that a given Brazilian paper is undercited compared to a non-Brazilian paper. On the contrary, if one takes statistics at its face value, one might even conclude that Brazilian papers receive more attention than one would expect on a purely statistical basis, but because the absolute numbers in Table 4 of *Velho*'s paper are quite small for the entries that provide the 1.1% and 0.59% figures, such a conclusion would not be on a very firm ground. Nevertheless, we can say that if the ISI data base underestimates the Brazilian contribution to worldwide science by a factor of two, even then there is no basis for claiming that a given Brazilian paper is undercited compared to a non-Brazilian paper.

It should be noted that the type of underciting that *Velho* claims for the Brazilian papers does not exist in her Table 4 with regard to the totality of papers coming from Third World countries either. The share of the Third World in the ISI database is 4–5%, which is about the same as the 3.9% of the citations received by Third

World countries in advanced country papers. It would therefore appear that in fact a given paper written in the Third World is as likely to be cited by advanced country papers as a given paper written in an advanced country.

Let me now turn to conclusion 2, which is based on number A quoted earlier. There our situation is much easier. Whatever the fraction of agricultural science papers produced in Brazil is compared to the worldwide number of agricultural papers may be, it must be very much lower than 16%. Therefore the conclusion is that the amount of attention paid by Brazilian agricultural papers to a given Brazilian agricultural paper is very much more than the attention paid by Brazilian agricultural papers to a given agricultural paper published elsewhere in the world. One would therefore conclude from this that the communication links within the Brazilian agricultural research community are quite strong, perhaps surprizingly so. This is a conclusion which is diametrically opposite to *Velho*'s conclusion 2 quoted earlier.

In fact, even the *absolute* percentage of 16% is quite large (as compared, in *Velho*'s Table 4 with 59% which indicates the percentage of references in the samples of advanced country papers to articles in journals published in advanced country journals). In my view, even the absolute percentage does not warrant *Velho*'s conclusion 2, quoted earlier.

Lest I be accused of trying to interpret data according to my own preconceived notions, let me say that the invalidity of *Velho*'s two conclusions comes as a considerable surprize to me, since, *a priori*, I would have agreed with *Velho* in holding the two *Velho* conclusions as true. I would therefore venture to suggest that in fact the two most important, unexpected, and far reaching results of *Velho*'s work are the invalidity of her two conclusions 1 and 2. Interestingly, *Velho* herself offers descriptions of characteristics of the Brazilian agricultural community which could easily serve as explanations for the two altered conclusions, particularly for the second. The fact that a large fraction of Brazilian scientific literature in agricultural science in published in Portuguese may suggest the community has a considerable degree of cohesion, national consciousness, and solidarity which, if true, would facilitate scientific interation among the members of the community. My own personal experience with other countries in the Third World would suggest that Brazil is an exception in this respect, though research on other countries along *Velho*'s lines may very well prove me wrong again.

Incidentally, according to Table 4 of *Velho*, papers written in other Third World countries are cited in a total of 4% of the citations which agrees with 4% of the worldwide papers coming from the Third World. Hence what *Welho* shows is that a paper published in the Third World is as likely to be cited by a Brazilian paper as a paper written in an advanced country. This is also a surprizing conclusion, since conventional wisdom (including my own) would have it that the scientific interaction between two Third World countries is much weaker than the interaction between a Third World country and an advanced country. This is sometimes described in terms of the picture of a center (the advanced countries) with strong spokes going out to the peripheries (the Third World countries) but with non-existent links between different parts of the periphery. I still believe that this picture is correct with respect to many or even most aspects of scientific interaction, but, at least in this example, it does not appear to be true for citations.

In summary, I suggest that *Velho*'s data (subject to confirmation from a small amount of additional data) tell us that a given paper written in a developing country has as good a chance to be cited by the worldwide scientific literature as any other scientific paper, and that a Brazilian paper has a much better chance to be cited by another Brazilian paper than a paper written elsewhere would have. In this case, therefore, articles from the Third World are not discriminated against, and the ties within the Brazilian scientific community appear quite strong. As mentioned earlier, these conclusions are exactly opposite to those of *Velho*'s which were derived from the same data. Indeed, it is the eyes of the beholder. . .

References

1. L. VELHO, The "Meaning" of citations in the context of a scientifically peripheral country, *Scientometrics, 9 (1986) 71.*
2. M. J. MORAVCSIK (Ed.), *The Bibliometrics of the Third World's Contribution to Science,* Deliberations, Conclusions, *and Initiatives of an ad hoc* International Task Force for Assessing the Scientific Output of the Third World. (To be published in early 1l86).
3. T. BRAUN, W. GLÄNZEL, A. SCHUBERT, A. TELCS, Facts and Figures on the Publication Output and Citation Impact of 107 Countries as Reflected in ISI's SCI Database, paper submitted to the International Workshop to Assess the Scientific Output of the Third World, Philadelphia, 1985.

D. Managing Science

LOCAL INSTITUTIONS FOR THE BUILDING OF SCIENCE AND TECHNOLOGY IN DEVELOPING COUNTRIES

(A brief assessment of motivations, roles, operations, and limitations)

Written as background for the U.S. delegation to the United Nations Conference
of Science and Technology for Development, Vienna, August 1979.

Prepared for a workshop, sponsored by the Department of State through the
American Association for the Advancement of Science, Washington, April 1979.

Michael J. Moravcsik
Institute of Theoretical Science
University of Oregon
Eugene, Oregon 97403
USA

SUMMARY
I.   Introduction
II. A characterization of the institutions
     a) Sectors
         Education: Schools, colleges, universities, technical training schools, public communication media
         Research and Development: Universities, governmental research and development centers, industrial
             laboratories, international centers
         Organization: Planners, managers, supporting functions (information, repair and maintenance, etc.)
             professional societies
     b) Functions
         Aspirational
         Operational
         Social and Political
     c) Personnel
         Bureaucrats
         Hierarchophiles
         Accomplishers
     d) The background of the personnel
         Science and technology
         Civil Service
         Politics
         Economics
III. Problems
     1. Overinstitutionalization
     2. Overformalization
     3. Overbureaucratization
     4. Civil Service
     5. Fragmentation and strife
     6. The few and overextended activists
     7. Separation from the S&T community
     8. Quantity vs. quality
     9. Lack of evaluation
IV. Suggestions for action at UNCSTD
     Parenthetical numbers refer to subsections in III.
     A. New organizations (1)
     B. Generalities vs. specifics (2)
     C. Person-to-person programs (3,4)
     D. Local interaction (5)
     E. Manpower, quality (6,8)
     F. S&T community (7)
     G. Evaluation (9)

I. INTRODUCTION

        The purpose of this note is to offer a very brief summary of some aspects of institutions in develop-
ing countries, connected with the building of science and technology. The aim is to assist the U.S. delegation
to UNCSTD in formulating effective action at that conference.
        This note will be concerned with institutions participating in the creation of science and technology.
Institutions directly involved in technological products (patent offices, ministries of industry, trade centers,
etc.) will be covered by other summaries presented at this workshop.

II. A CHARACTERIZATION OF THE INSTITUTIONS

        The institutions will be described along four dimensions: The sector in which they belong, the func-
tions they strive to fill, the motivations of their personnel, and the educational and professional backgrounds
of their personnel.

     a) Sectors

        On the whole, institutions pertaining to the creation of science and technology in developing countries
can be sorted into three sectors: education, research and development, and organizational. A given institution

might very well belong to more than one sector.

i) Education

In this category we find, first of all, elementary and secondary schools which are supposed to offer the rudiments of science and technology and should also arouse interest, enthusiasm, and commitment toward science and technology on the part of at least some students.

On a more advanced level, we find colleges and universities, technological institutes, and technical, vocational schools, the primary functions of which should be the creation of specialists in science and technology, and the providing of the necessary scientific and technological ingredient in the education of students in science-related or technology-related professions (e.g. nurses, positions in transportation and communication, officials in ministries of commerce, of health, of agriculture, etc.)

Finally, this category should also include a much-neglected aspect of education in science and technology, namely the mass communication media which are to bring elements and aspects of science and technology to the masses. Newspapers, radio, television, community centers, and others are part of this. Of the three groups mentioned above, this one requires the most improvement and enhancement in developing countries.

ii) Research and development

With occasional exceptions, research and development takes place in developing countries in four types of institutions:

α) Universities and institutes of technology. Compared to the situation in the more advanced countries, universities and institutes in most developing countries engage in little research and development. The aims of these institutions are generally <u>not</u> formulated in terms of research being a primary function, but they are instead considered as almost exclusively teaching entities. Correspondingly, the practice of generous government research grants awarded to university personnel is not at all common in developing countries. In the sciences, the little university research there is tends to be "basic" in motivation, and the technological development activity also tends to be somewhat abstract and remote from the users. There are some exceptions, especially in agriculture, when a research institute is associated with certain university departments (e.g. the relationship between Ahmadu Bello University in Zaria, Nigeria and the adjacent Institute of Agricultural Research.)

β) Governmental research laboratories. In many developing countries these are highly visible, at least in an organizational sense, and consume the lion's share of the research and development expenditure of the country. Their professed purpose is to engage in applied scientific research and technological development. In many cases, however, applied scientific research (i.e. the creation of new scientific knowledge with techno-logical applications in mind) is minimal, and the laboratories are almost exclusively engaged in low-level techno-logical manipulations based much more on trial and error than on the systematic utilization of scientific know-ledge. There are exceptions to this state of affairs (e.g. KIST in S. Korea, the Leather Institute in India, the Rubber Research Institute in Malaysia), but they are relatively few in number.

γ) Industrial research laboratories. These are very rare in developing countries, compared to prac-tices in the more advanced countries. Industries owned locally (by the government or by private capital) have not developed the tradition for such research and development, and often utilize derivative technology for which, they think, research and development is no longer needed. On the other hand, industries which are subsidiaries of foreign or multinational companies almost always rely on research and development performed at the headquarters of the mother company in a different country.

δ) International research centers located in the developing country. These are relatively few in number. In addition, they are, not infrequently, too preoccupied with the research activities to build extensive and functional ties with the scientific and technological community of the country in which they happen to be located. Some examples of these centers are the International Institute for Tropical Agriculture in Ibadan, Nigeria, the International Institute for Rice Research in Los Banos, the Philippines, and the Asian Institute of Technology outside Bangkok, Thailand, the latter being also an educational institution. The quality and level of research activity in such centers tends to be quite good, sometimes in stark contrast with the national insti-tutions located in the same country.

iii) Organization

These are institutions which are concerned with the planning, the management, the support, and the logistics of scientific and technological activities. Among them are ministries, councils, and other foci for science and technology, commissions of atomic energy, of agricultural research, of industrial research, etc. and supporting organizations like centers for scientific and technological information, workshops, stockrooms, and maintenance facilities for research and development equipment, etc. Finally, there are the professional societies, that is, organizations which at least nominally belong to the local scientific and technological communities them-selves, and are concerned with meetings, journals, and other communication channels.

b) <u>Functions</u>

The functions perceived by these organizations follow from the motivations and justifications for under-taking science and technology, as viewed by the developing countries, and from the fabric of the social and political context in which these motivations must be realized. Although the functions are spread over a wide range, they can be summarized under three headings.

The first, and perhaps strongest function is what might be called aspirational. The state of under-development of a country (or of any groups of people or even of an individual) is considered detrimental, in-tolerable, humiliating by the country (or the group, or the individual) not only (and in fact not even primarily) because it reflects material poverty, but because it closes options and choices, because it implies a de facto colonial status in which everything that happens in the country (culturally, socially, economically, artistically, scientifically, etc.) is determined from abroad, and because the feeling of accomplishment, the pride of achieve-ments, the exhilaration of being a pioneer and leader, are frustrated. In other words, the country is void of purpose, aim, and aspirations, since the country's fate is determined by factors and forces external to it.

This aspirational motivation of development does not, of course, directly appear in national development plans, because it is not part of economics, and such plans are, at best, purely economic documents. It does appear, however, in political speeches, in the behavior of and conversations with officials in the country in the United Nations and other world bodies, and it is also reflected in the de facto way science and technology are organized, performed, and utilized in developing countries.

In this context, engaging in science and technology becomes symbolic of the country having come of age, having joined the community of nations on equal footing. From this point of view, what matters is the act of participating in science and technology, the activity of organizing, the process of putting up buildings, the involvement in making plans and in setting up institutions, the opportunity to award degrees, bestow titles, and prepare policy documents. The actual output of science and technology is secondary in the framework of this aspirational motivation.

The second function of scientific and technological institutions in developing countries is operational, that is, the direct promotion of the production of scientific and technological results. This is the function that is primarily in the minds of the international and bilateral assistance agencies, and this is also the one which appears, on the surface, in the national development plans. It involves the promotion of scientific and technological education, the mainten-ance and stimulation of scientific and technological manpower, the providing of the supporting environment in which scien-tific research and technological development work can be undertaken productively, and the fostering of interaction between science and technology, between technology and the country's productive sector, and between scientists and technologists on the one hand, and the population as a whole, on the other. One might say that this is the pragmatic function of such insti-tutions, the one that is designed to produce relatively tangible, concrete, and to a large extent material results.

Finally, the third role of scientific and technological institutions is social and political, in the sense that they have to serve as spokesmen, advocates, and propagators of the interest, of the needs, and of the products of the scientific and technological infrastructure of the country. They must create an understanding of what science and technol-ogy are and what they are not, generate public support (both moral and material) for their activities, and then transfer the products of science and technology to the rest of the country.

It is not difficult to see that these three types of functions are quite different and in fact potentially contra-dictory. Action which appears advantageous for one function might very well be judged detrimental from another one. This conflict of aims, functions, and directions is more severe in a developing country, where science and technology have not become so firmly embedded in the fabric of the country, than in the more advanced countries which have lived with the scien-tific and technological revolution for several centuries.

c) Personnel

Institutions are primarily manifestations of the motivations, aspirations, qualifications, and devotions of the people in them. The success of any institution is determined mainly by the manpower within it. Formally similar institu-tions will perform in drastically different ways depending on the characteristics of the human ingredients in them. Thus we must survey these characteristics.

Human motivations are generally complex, and it is virtually impossible to pinpoint one single motivation behind a person behaving, acting, and thinking in a given way. This is true also in the case of the manpower of institutions in science and technology in the developing countries. Yet, we can summarize the types of motivations in terms of three headings, granting at the same time that a given person generally has an admixture of all 3 motivations, with varying weights attached to each. I will call these three motivational types bureaucrats, hierarchophiles, and accomplishers.

The bureaucrat is motivated by being part of an orderly system. He takes pride in his participation in such a system, especially if it is large and has high visibility. His primary consideration will be to live up to the rules of the system, which in fact are transfigured, in his eyes, into goals rather than means. In many developing countries the civil service system is well developed and constantly increasing, and the members of this system tend to have strong strains of the bureaucrat in them. At his best, the bureaucrat provides reliable support for the transaction of routine matters. On the other hand, the bureaucrat finds it difficult to deal with unusual occurrences, with new types of activi-ties, with matters which, by their very nature, are not easily amenable to the application of formalized rules.

One generally thinks of bureaucrats in the context of administrative offices, but the type can just as easily appear in educational institutions, as a teacher of science who presents well set, formalistic lectures, unchanged over the years, and demands the smooth, well prescribed, and accurate reproduction of those lectures on examinations. Or, our bureaucrat can also take shape in the person of a researcher in a governmental research laboratory, who contributes in an orderly way to work on research problems that were selected on the basis of being next on the list of problems that have not been solved before.

The second type I would like to call a hierarchophile, that is, a person for whom life consists of a long ascend-ing staircase, with the aim being to rise through the grades of this hierarchy, and success being measured by the number of rungs one has passed on the way. To the hierarchophile, the nature of the staircase is of secondary importance, and he is likely to accept the obstacle course as it was laid out for him by others, by tradition, or by other factors. Colonial regimes in the 19th and the first half of the 20th century were particularly successful in creating such hierarchies, and their shadows continue to be cast on the aspirational structure of people in the developing countries even today. But indigenous social structures, especially when old, static, and emphasizing stability, are equally good in setting up such aspirational staircases.

The third type of person is the rarest, and at the same time the most crucial for development. Let me call him the accomplisher. His aim is to create new things, to accomplish something, to be able to see something that before his intervention was not there. I do not mean to imply that the accomplisher is always an altruistic and selfeffacing soul, and his creations always significant, functional, and operationally effective. On the contrary, at his worst, he might be a grandiose self-adulator who produces white elephants with no utility other than serving as a monument for himself. But at his best, the accomplisher is the person on whom practically all development hinges. To be sure, he, by himself, cannot change the world. But he can supply the ingenuity, the talent, the devotion, the inspiration, and the enthusiasm to carry others with him to bring about changes in the direction judged productive and significant by the hindsight of subsequent generations. Virtually everybody who has had extended contact with science and technology in developing countries can pinpoint individuals in various countries who are such accomplishers, and on whose character much of the development in that country depend. Accomplishers come on various levels, some having a large influence over a whole country, others affecting only their own institutions, disciplines, or immediate environments. Among them, one can mention as a few of the many examples, Bhabha of India, Usmani of Pakistan, Sabato of Argentian, Nayudamma of India, Riazuddin of Pakistan, Saavedra of Chile, Ndili of Nigeria, Odhiambo of Kenya, Munroe of Guyana, or Wijesekera of Sri Lanka.

d) The background of the personnel

Having tried to describe the various motivations of the personnel in the scientific and technological institutions of developing countries, let me now summarize their professional and educational backgrounds. Here again, mixtures of back-grounds are most likely to occur. The backgrounds can be classed into four groups.

First, some of the personnel has background in science and technology. This background may be just formal, that is, a degree in science or technology without having ever "practiced" either science or technology. In other cases, the background may be more extensive, involving even distinguished accomplishments in scientific research or technological development work. In the developing countries it is even more difficult than in the more advanced countries to distinguish these two types from the other, since in the former the holding of a degree counts more, seniority is respected much more, and prominence in scientific bureaucracy is equated to being a distinguished scientist or engineer. For somebody who is active in science and technology it is rather easy, even after just a few minutes of personal acquaintance with the person, to tell whether he is an experienced scientist or engineer, or just a degree holder who afterwards switched into administration. To the uninitiated, however, such a distinction may not be immediately clear.

The second type of background is that of civil service. As mentioned earlier, civil service is a very prevalent way of life in many developing countries, and can be a very desirable career option for even the marginally educated. Typically, the background of a civil servant would be that which is called a "generalist" nowadays in the United States (i.e. a person who knows nothing about many things). Specifically, in the sciences and in technological matters, the background of a civil servant would be especially weak, matching a similar situation in the more advanced countries which, in part, prompted C.P. Snow to write his Two Cultures. Even more specifically, a civil servant would be unaccustomed to quantitative thinking, would be mystified by the process of creating results of science and technology, and would not be favorably inclined toward claims by scientists and technologists that creative activity in these areas has its own rules and needs that may not match those of the more common occupations in the country.

A particularly detrimental feature of civil servants, again paralleling a similar situation in the more advanced countries, is that they will feel insecure and unsure vis a vis technically educated persons, and that scientists and technologists, on the other hand, a priori and based on what might be called "class prejudice", are likely to regard civil servants with suspicion and as inferiors.

The third type of person important in institutions of science and technology in developing countries is the politician. Needless to say, much is to be gained for science and technology through the presence of politicians favoring and understanding science and technology, and specific examples in a number of countries could be named to support this assertion. These examples would always point at great politicians, with statesmanship, vision, skill, and breadth of horizon. On the other hand, the domination of science and technology by small, partisan, and fumbling politicians implies great dangers, not only in diverting science and technology from their legitimate aims to serving as footballs in political squabbles, but also by dragging science into partisan political conflicts and thereby depriving the development of science and technology from the continuity and stability that they need and which can only be secured by an apolitical composure in the midst of political instability and of the alternation of extreme political regimes. A good example for such a detrimental effect of politics on science and technology can be found in the majority of the Latin American countries, at least at one time or another. In most of these cases both politicians and scientists are to be blamed, the former for barging into science, and the latter for giving higher priority to their partisan political activities than to their service to the building of science and technology.

Finally, the fourth type of background found among the personnel of science and technology in developing countries is in economics. This background is most prevalent in institutions of organization, and particularly in planning institutions. It is a curious although understandable feature of the national development plans of developing countries that while they nominally include an extremely broad spectrum of elements in science, technology, art, culture, sociology, philosophy, religion, psychology, economics, and many other aspects of life, they are almost invariably composed by economists and are viewed in an almost exclusively economic light. As a result, economists have a considerable influence on institutions of science and technology. At the same time, the treatment of technology, and especially of science, in economic theory is still rather rudimentary, in that the short term connection between scientific and technological activities and economic production remains unarticulated. This gives rise to a number of problems and distortions in the treatment of science and technology in developing countries.

III. PROBLEMS

In this section, an extremely brief list is given of some of the most important deficiencies of the institutional structure of developing countries in science and technology.

1. Overinstitutionalization. Developing countries tend to have a very much heavier institutional structure than it is warranted by the size of their scientific and technological manpower and efforts. Extensive planning bodies, research institutes set up regardless of the availability of manpower to do such research, universities founded in the absence of qualified faculty for instruction, grandiose central documentation centers organized when even the simplest scientific and technical journals are missing from the ordinary libraries, are some of the hallmarks of this situation. It is unfortunate that some bilateral and international assistance agencies have greatly encouraged such a proliferation of premature and stillborn institutionalization. Such overinstitutionalization is a result of the aspirational motivation for science and technology, but it hinders the operational function.

2. Overformalization. The institutional structure of science and technology tends to be overly formal, concerned with appearance and structure rather than essence and output. For this reason, much of the emphasis is on planning, much less on decisions making, and very little on implementation. Universities tend to regard science education as a formalized ceremony aimed at the awarding of a degree. This trait is also a consequence of the aspirational function of these institutions but it hinders the operational function.

3. Overbureaucratization. Institutions tend to be overburdened by rules and procedures. This bureaucratization hinders the functional practice of science and technology and also robs science and technology of some of its talented practicioners who are forced to spend much of their time in the quagmire of these rules and procedures.

4. Civil Service. In many developing countries scientists and technologists are placed within the civil service. The result of this is a lack of recognition of the talented, the outstanding, and the energetic, and science and technology are straightjacketed into a system which is rigid, pedestrian, and unresponsive to the special needs and structures of the different types of activities it encompasses.

5. Fragmentation and strife. Institutions of science and technology in the developing countries are more fragmented, and have more strife within and among them than is the case in the more advanced countries. Beside possibly traditional reasons, this is likely to be due to the law of scale: In a country where science and technology is small, personal enemies have no way to find a course other than the collision course, and differing opinions, aspirations, and directions cannot be easily given different but equally satisfactory outlets. Partly as a result, internal communication within the scientific and technological communities of a given developing country is astoundingly weak, thus enhancing the isolation of the individual researchers or developer.

6. The few and overextended activists. As mentioned earlier, there are relatively few accomplishers in any group of people. Where this number is expecially low, the lone accomplisher or activist tends to work by himself, oblivious of

the necessity of finding allies on an equal footing and of grooming successors for future times. Thus many great accomplishers, once their time is up, find their work aborted and disrupted. A national or institutional tradition in scientific research or technological development work is well known to play a crucial role in the excellence of such activity, and such a tradition is difficult to establish through the overextended lone wolves who struggle valiantly to build science and technology in developing countries.

7. Separation of the scientific and technological community from the institutional management of science and technology. As mentioned earlier, the active members of the scientific and technological communities in developing countries are not well involved in the decision making over institutions of science and technology, whether these institutions are educational, research, or organizational. Conversely, those in charge of such institutions are often not part of the scientific and technological communities and sometimes are even antagonistic or at least incomprehending of the goals and needs of the latter.

8. Quantity versus quality. Most institutions in developing countries stress quantity over quality. Educational institutions are geared toward maximizing the number of diplomas awarded, and the idea of small, quality institutions are often denounced as elitist. Research institutions have funding and personnel policies wich favor sheer size at the detriment of quality. This is again understandable in the light of the aspirational function of science and technology in developing countries, but it hinders the operational function. National development plans and other organizational bodies all stress numbers and are generally silent on quality.

9. Lack of evaluation. To a casual observer of the science and technology scene in developing countries, perhaps the most startling feature of this scene is the almost complete absence of an evaluation procedure of research completed, or of development work performed. In fact, a similar absence is also noted in the assessment of research of development proposed, or in the granting of funds for such research or development. Correspondingly, national development plans are virtually always formulated in terms of input into science and technology (buildings, manpower, money, organizations) and practically never in terms of output. Similarly, no attempt is made to assess whether previous development plans have been met in terms of the achievements of science and technology. Not only is evaluation absent, but there is no methodology for such an evaluation either. Again, from the point of view of the aspirational function of science and technology, this is not surprising, since from that standpoint the results of science and technology are unimportant compared to the process of being involved in science and technology. From an operational point of view, however, the lack of evaluation is quite disturbing and detrimental.

IV. SUGGESTIONS FOR ACTION AT UNCSTD

The suggestions included in this section follow directly from the list of problems offered in the previous section. As a result, the parenthetical numbers following the capital letters denoting the subsections below refer to the subsections in the previous section.

A.(1) UNCSTD should avoid urging the establishment of, or actually setting up, new organizations and institutions, national or international, unless these are radically different from the existing ones both in purpose and in structure. Organizational manipulations are highly unlikely to contribute to the amelioration of problems of science and technology in developing countries, and can very well result in a worsening of the situation by siphoning away needed manpower and funds and by diverting attention from truly effective remedies.

B.(2) UNCSTD should avoid making declaration of generalities, since they are already readily available in previous UN conference proceedings as well as in countless UN documents and reports. Instead, UNCSTD must propose specific action to remedy particular problems. Such action proposals must consist of seven parts:
  a) Specific identification of the problem
  b) Summary of previous attempts at remedies and of the reasons why these have failed,
  c) Detailed outline of the new idea for action, including reasons why it is expected to work better than its predecessors
  d) Specification of what will be done
  e) Specification of how much will be done
  f) Specification of who will be doing it and how
  g) Specification of how it can be determined, ex post facto, whether the new program is successful or not.

C.(3,4) UNCSTD should stress programs involving person-to-person interaction, as contrasted with large, bureaucratically conceived, impersonal projects. The former is more effective, less costly, and its participants feel more responsibility toward making them into successes.

D.(5) UNCSTD should find ways of supporting specific programs which enhance the local interaction within the scientists and technologists of a given developing country, as well as the local interaction between scientists and engineers, between engineers and the productive sector, between scientists and science policy personnel, and between scientists and engineers on the one hand, and the population at large, on the other.

E.(6,8) Manpower is the key to a creative scientific and technological infrastructure, and quality in manpower development is crucial. Hence UNCSTD should emphasize the support of specific programs which stimulate the creation of high quality indigenous manpower in science and technology.

F.(7) At the Vienna conference, try to seek out those, possible few, participants who are members of the active scientific and technological community in the developing countries, and solicit their views on what specific programs would be most helpful to those communities in their attempt to engage in creative science and technology. This interaction with scientists and technologists from the developing countries is best accomplished through active scientists and technologists who are members of the US delegation and who therefore already have professional ties with their counterparts in the developing countries.

G.(9) UNCSTD should see to it that international assistance is freely available to the developing countries for the purpose of making scientific and technical evaluations of particular research projects. Such an international network of referees, consultants, and assessors is already well established, in an informal way, within the scientists and engineers in the more advanced countries, and this network is very widely, in fact routinely, used in these countries for the assessment of past research, for decisions to support proposed research or development, and for the promotion and appointment of scientists and engineers. At the moment, the developing countries are, to a large extent, excluded from benefiting from such a network, and this unsatisfactory situation must be altered.

References

The list below offers a short and quite incomplete collection of writings that might serve as a good background for the content of this paper.

General books

Jones, Graham, The Role of Science and Technology in Developing Countries. Oxford University Press, Oxford, 1971.

Moravcsik, Michael J., Science Development - The Building of Science in Less Developed Countries. PASITAM, Indiana University, Bloomington, Indiana. Second Printing, 1976.

Spaey, Jaques, et al., Science for Development. UNESCO, Paris, 1971.

Reports written specifically for UNCSTD

Kidd, Charles V., Manpower Policies for the Use of Science and Technology in Development. Graduate Program in Science, Technology, and Public Policy, George Washington University, Washington, DC, 1978 (Supported by the National Science Foundation).

Moravcsik, Michael J., Science and the Developing Countries. Institute of Theoretical Science, University of Oregon, Eugene, Oregon, 1977 (Supported by the National Science Foundation; copies available from Andrew Pettifor, Room 1229, National Science Foundation)

Morgan, Robert P. et al, The Role of U.S. Universities in Science and Technology for Development: Mechanisms and Policy Options. Department of Technology and Human Affairs, Washington University, St. Louis, MO, 1978 (Supported by the National Science Foundation).

National Research Council, U.S. Science and Technology for Development: A Contribution to the 1979 UN Conference. Department of State, Washington, 1978.

Articles, reports, and more specialized books

In this category there are thousands of items. The list below contains some "classics" and/or some pieces which are especially relevant to the problems we have discussed.

Basalla, George, The Spread of Western Science. Science 156, 611 (1967).

Bhabha, H. J., Indian Science--Two Methods of Development. Science and Culture 32, 333 (1966).

Blackett, P.M.S., Science and Technology in an Unequal World. Science and Culture 34, 16 (1968).

Cernuschi, Felix, Educacion, Ciencia, Tecnica y Desarrollo. Universidad de la Republica, Montevidea, Uruguay, 1971.

Dart, Francis, and Pradhan, Panna Lal, Crosscultural Teaching of Science. Science 155, 649 (1967).

DeHemptinne, Yvan, The Science Policy of States in Course of Independent Development. Impact 13, 233 (1963).

Gasparini, Olga, La Investigacion en Venezuela. Publicaciones IVIC, Caracas, Venezuela, 1969.

Glyde, Henry R., Institutional Links in Science and Technology: The United Kingdom and Thailand. International Development Review Focus 15, 7 (1973/4).

Herrera, Amilcare (Ed.), America Latina - Ciencia y Tecnologia en el Desarrollo de la Sociedad. Editorial Universitaria, Santiago de Chile, 1970.

Lomnitz, Larissa, Organizational Structure of a Research Institute. (report 1975, to be published).

Nayudamma, Y., Promoting the Industrial Application of Research in an Underdeveloped Country. Minerva 5, 323 (1967).

Price, Derek deS., The Relations between Science and Technology and their Implications for Policy Formation. Report by the Research Institute of National Defence, Stockholm, Sweden, 1972.

Roche, Marcel, Descubriendo a Prometeo. Monte Avila, Caracas, Venezuela, 1975.

Roche, Marcel, Early History of Science in Spanish America. Science 194, 806 (1976).

Saavedra, Igor, The Problem of Scientific Development in Chile and in Latin America. In Chile: A Critical Survey , Institute of General Studies, Santiago, Chile, 1972.

Sabato, Jorge, Quantity versus Quality in Scientific Research (I): The Special Case of Developing Countries. Impact 20, 183 (1970).

Sabato, Jorge, Atomic Energy in Argentina: A Case History. World Development 1, #8:23, (1973).

Schlie, Theodore W., The College of Tropical Agriculture at the University of Hawaii: A Case Study in the U.S. Application of Science and Technology to Development in Developing Countries. Report, Division of Science Resource Studies, National Science Foundation, Washington, DC (1978).

Szmant, Harry H., Foreign Aid Support of Science and Economic Growth. Science 199, 1173 (1978).

Wijesekera, R.O.B., Scientific Research in a Small Developing Nation--Sri Lanka. Scientific World, #1, 6 (1976).

Zahlan, A. B., Science in the Arab Middle East. Minerva 8, 8 (1970).

Preprint, published in *Leite Lopez Festchrift —*
A Pioneer Scientist in the Third World, eds.
S. Joffily *et al.*

SCIENTISTS AS SCIENCE MANAGERS

Michael J. Moravcsik

Institute of Theoretical Science

University of Oregon

Eugene, Oregon 97403

USA

Abstract

Having in mind especially the situation in countries of the Third World where science is at the initial stages of establishment, the view is advanced and supported with arguments that people with significant personal experience in doing scientific research have a crucial role to play in the management of science, because they have acquired a point of view of an "internalist." This is illustrated in many facets of science. In research, for example, the internalist view is important in defining the proper nature of scientific investigation, in motivational aspects, in the selection of research problems, in the choice of personnel, and in the evaluation and assessment of research. Such experience is also important in forging a creative link between science and technology, and between technology and production, and in bringing the outlook of science to the population of the country as a whole. Some qualifications are suggested that a scientist should have in order to be successful in science management, and the problem of keeping a science manager current in science and of safeguarding his ability to make de facto choices are briefly discussed.

I. Introduction

As science becomes an undertaking in all countries of the world, there is an increasing tendency and necessity in these countries to develop science management also. In fact, in many countries the concern with science management is more pronounced than the concern with establishing science itself. This precipitous rush to evolve conspicuous science management has its origin in several factors. For one thing, the management of science can be exhibited in a conspicuous way, and therefore satisfies the motivation of showing to the outside world that the country is indeed involved in science. Second, there is considerable pressure from various international agencies to induce the countries to set up formal apparatus for science management. Seminars and workshops are held in which the formal aspects of such management are discussed, and the representatives of these organizations visit countries and urge the local leadership to establish such formal science management organizations. There is, of course, also the legitimate need for some framework to manage the incipient science that is beginning to establish itself in various countries.

The people who are chosen to be the personnel in such science management organizations generally come from several backgrounds. In most cases, they come from the civil service establishment, since most of the science management organizations are in fact governmental. Sometimes the people come from what the civil service considers the scientific establishment but which from a scientific point of view may not be that. I am thinking of people with some science degrees who, for reasons which may very well be outside their control, never had an opportunity to actually engage in extended scientific research, and hence, although they are nominally scientists, have not had personal experience with actually doing science. There also is a tendency more recently to send some younger people abroad to be trained as science managers. These people would again have just a general background on a university level, perhaps in the sciences, perhaps in other areas, but before they have an opportunity to engage in scientific research they are earmarked for science management and sent abroad to learn the tricks of the trade.

As mentioned above, the size of the science management establishment and the size of the personnel connected with it can be quite huge compared to the size of the scientific community in the country. Furthermore, since the working scientists in the developing countries are often not very skilled in articulating their needs and in taking into their own hands the management of their own scientific work, and since international organizations are often confined to dealing with these official science management organizations instead of with the working scientists themselves, the impact of these science management organizations and their personnel on a country's scientific development can be considerable. It is, therefore, extremely important to make sure that these science management organizations function in such a way as to aid and promote actual scientific activities in the country.

The purpose of the present discussion, therefore, is to analyze the background that such science managers need to have. In particular, I would like to advance the thesis, and support it with arguments, that good science managers on intermediate and top levels must be people with a considerable amount of personal scientific experience, that is, a considerable amount of personal involvement in scientific research activities. The point that is stressed here is that science is a rather unique and complex human activity, and there are certain elements of this activity that can be understood and judged almost exclusively only from the inside, that is, by people who have had personal experience doing this activity.

This debate on the "internalist vs. externalist" point of view is not limited to the discussions of science management. In the general and broad area of the science of science, that is, the study of science as a human activity, the same dichotomy can be observed. In that field, the majority of the researchers are in fact people who have not had personal experience with scientific research. Instead, their backgrounds are in philosophy, in psychology, in history, in sociology, in administration and management, and so on. It is claimed by them that in order to achieve the objectivity needed, one has to look at something from the outside using methodology that is well developed for such an objective analysis. In response, those who favor the internalist point of view would say that looking from the outside misses certain important factors which perhaps are not very conspicuous to an outsider and which are somewhat difficult to explain or unearth through the usual methodology used to study something from the outside.

Undoubtedly, the truth is somewhere in the middle. Both points of view should be pursued and integrated in some way. When it comes to science management, however, the point of view of the outsider will be represented anyway by the governmental and other societal elements with which science management is interlinked, interfaced, and interconnected. Therefore, it is my claim that the science managers themselves should come from the scientific community itself for reasons which I now want to enumerate.

II. Research

There are many elements connected with the pursuit of scientific research which, it seems to me, can be best judged from the internalist point of view of a scientist. First, perhaps, is the very nature of scientific activity. To many people from the outside, particularly if educated as a civil servant and as an administrator of other types of activities, science appears as a set of facts, and scientific research then turns out to be the process whereby these pieces of facts are collected. To a scientist who has had personal experience with research, science is a quite different undertaking. It is a method to ask questions and then, by a certain combination of thinking and experimentation, to find some approximate answers to these questions. In other words, science becomes a problem solving activity. The management implications for these two views of science are, of course, quite different, and provisions which might be useful for one may not be appropriate for the other one at all.

Somewhat connected with the first element is the question of motivations for doing science. There are societal motivations and the motivations of the individual scientist, and, although there is a considerable overlap in broad terms between these two types of motivations, there are also some differences in relative weights among the various parts of societal motivations and the various parts of individual motivations. It is extremely important for a science manager to understand the way of thinking of scientists on the issue of motivation, since without a motivated body of scientists a science establishment will fail.

This leads us to the selection of scientific personnel. There are various characteristics which are representative of creative scientists, characteristics which are much easier to judge from the inside than from the outside. Indeed, the scene of many developing countries is often open to the appearance of fake scientists, of people who are able to convince their non-scientist superiors that they should be given important resources and considerable power in the local scientific community, but who in actuality have very weak scientific qualifications. It is my claim that part of the reason why such people can be successful in many countries is the lack of strong representation of working scientists in the managerial system who would expose the superficial qualifications of these people.

Yet another element in scientific research is the selection of research problems. This is particularly an aspect of research which can be best judged from inside the scientific establishment. This is not to say that problems are selected entirely on the basis of scientific criteria. On the contrary, in problem selection, criteria external to science quite often play an important role and properly so. However, there is the important element of the scientific assessment of the feasibility and appropriateness of the selection of that particular problem, and that is an element which cannot be brought about without experienced scientists participating in the management process. Because this is not always the case in practice, very valuable resources of many countries are spent on scientific research and technological development work in which the problem is poorly conceived and ill-defined, and hence, success is highly unlikely.

Finally, in connection with research, I want to emphasize the extreme importance of assessment. Scientific research must be assessed prior to its inception, and also must be evaluated after the completion of the particular piece of research. Such assessment and evaluation activities are practically completely absent in many countries around the world. Research proposals are submitted which are then judged by non-scientists merely on whether their descriptions contain the "right words," and at the end of the research activity the closing report is read and evaluated by non-scientists in ministries and national science councils, again mainly on the basis of formal considerations. The result of this absence of evaluation is that scientific resources are often distributed in an egalitarian way or in ways which are determined by political considerations and personal influence rather than by the merit of the work.

In summary, therefore, I claim that it is extremely dif-ficult to establish a creative and efficient research program in a country without the involvement of scientists in the management of this research itself.

III. Science and Technology

One of the three important general motivations for pursuing science is that in modern times science is inseparably connected with technological development, which in turn is inseparably con-nected with the economic development of the country. Although the existence of these links can hardly be questioned, the nature of these links, and their complexity, is a very sophisticated matter which requires experience and analytical skill to decipher correctly.

To start with, there is the question of basic vs. applied scientific research. I have written on this subject previously and therefore will not repeat some of the substantive aspects of this problem, but let me just mention that all too often the nature of applied scientific research is misunderstood and confused with technological development work. Indeed, if one asks 100 science managers picked randomly around the world to explain the differences among basic research, applied research, and technological development work, and also the difference between theoretical and experimental scientific research, most likely well over 90 would flunk the test.

It is not surprising, therefore, that the link between science and technology in many countries around the world is extremely weak. Similarly weak is the link between technology and production in the country. This is regrettable, not only because it places an obstacle in the country's development, but also because it tends to create a negative image for science, an image that is not deserved by science itself but is brought about by the unreasonable expectations due to the misunderstanding of the relationships among science, technology, and production.

It is again my claim that this rather intricate relationship can be best understood by scientists who have in fact partici-pated in scientific research and particularly by scientists who have worked in what is referred to as applied scientific research. As a result, it will be these people who will be in the best position to initiate mechanisms for improving and strengthening the link between science and technology and between technology and production. In countries where recognized scientists have been in important positions in science policy matters, one can quite often see encouraging results in this respect. An example may be the history of South Korea in the last 15 or 20 years.

IV. Science to the People

By the above slogan, I mean not only formal science education in public schools, but other methods and channels also,
through which the average citizen of a country will be exposed to
the influence of science. Science, in fact, has been of enormous
importance in the scientifically advanced countries in the form
of a cultural force, an influence which molded the thinking of
the people and had an impact on their philosophy and on their
view of the world. This cultural impact of science has not yet
come to people in most countries around the world, even though
they might use transistor radios, might look at television, and
might even use pocket calculators in their everyday lives.

The means of getting science to the public include radio,
television, newspapers, museums and exhibits, continued education
opportunities, and many others. If one views these activities in
many countries around the world, one finds that not only are
they small in number and tiny in extent, but in addition to that
they are also deficient in their form and structure. Science is
represented in these efforts mainly as a mysterious activity
which produces interesting facts. Descriptions of science present results of scientific research. They tell the latest laws
of science discovered by scientists. They concentrate on form
rather than content and on results rather than process.

Now, it is of course to some extent helpful if people in
general have some idea of the main features of the scientific
laws that are being discovered, since understanding these laws
helps them to deal, for instance, with technological gadgets that
are based on these laws. But much more important in my view is
the process of science, the attitude of science, the method of
science which, suitably transformed, can be an important factor
in the life of anybody even if he himself is not involved in
scientific activities. Science has a certain belief system,
science has a certain methodology, science encourages a certain
attitude, the utility of which is much broader than just in
scientific activities themselves. The fact that science claims
knowledge to be open and describes a process whereby through
experimentation we can learn something is an epistemological element which has an impact in general on our everyday lives.
Indeed, it seems to me that the concept of development, almost
regardless of how it is defined, involves many attitudes and many
elements which are in common with those of scientific
investigation.

I therefore believe that this effort to bring science to the
public must be one in which scientists play a major role. For
the same reason, I also believe that the presence of scientists
among managers and policy makers has a similarly important
effect, namely it injects into the deliberations a certain
attitude, a certain methodology, a certain belief system, which
can be extremely helpful. It has been shown in the United States
that those trained in physics, in particular, can be found in an
extremely broad variety of different positions throughout our
society, positions which often are very remote from that of a
professional physicist. Nevertheless, it is found that quite
often these people, trained in physics and imbued with the

scientific approach, can make important contributions in other
fields very remote from the actual practice of science.

V. Qualifications

Whereas in the previous sections I have tried to argue that
it is important in scientific management to involve people with
personal experience in scientific research, in this section I
will now describe some of the qualifications which such scientists should have in order to have a reasonable chance of being
successful as science managers. It is by no means true that all
scientists, or even the majority of scientists, are suitable for
science managerial activities.
Perhaps the most important qualification of a scientist for
science managerial positions is his personal satisfaction with
his own scientific research career. I do not claim that this
scientist has to be one of the greatest of the century. No, he
can instead be somebody who has done recognized and solid
scientific research that has contributed to development in a
given field. The importance is not so much in the absolute level
of his scientific accomplishments but instead in his personal
view of these scientific achievements. It is interesting to note
that personal satisfaction with one's scientific contributions
and personal serenity with respect to the extent to which one can
contribute to science does not seem to be correlated at all with
the absolute importance of the person's scientific achievements.
I have known many outstanding scientists who have made very
important contributions toward science as judged by their
colleagues and by people from the outside who, nevertheless,
continued to be dissatisfied with their own contributions. They
continued to be insecure in their personal attitudes, and they
continued to act as if they had strong inferiority complexes.
They simply did not manage to achieve personal serenity vis a vis
the goals and achievements of a life of an individual. In
contrast, I have known scientists who have made much more modest
contributions as judged by objective standards who, nevertheless,
appeared to be satisfied with their contributions and as a result
of it acted without feelings of inferiority, without feelings of
insecurity, and with a sincere desire to make another contribution indirectly to science by helping other scientists to
make scientific contributions. This last point is extremely
important. In the actions of a science manager there must not be
any trace of envy, of feelings of jealousy, of feelings of
competition with the scientist that he is managing.
This element is also very important because scientists are
very sensitive to the nature of the person who is superimposed on
them as a manager. This is one of the most important reasons why
I believe scientists must be involved in science management.
Should the science manager be without a scientific background,
whom the scientist cannot look to as a scientific colleague, the
relationship between that science manager and the scientist is
likely to be tenuous, tense, and in many ways negative. One
might philosophize that this is regrettable and unfair, and that
scientists are narrow-minded in this respect, but such

philosophical statements would hardly change the reality, namely that scientists get along best in a managerial situation where the manager himself is "one of them."

A considerable problem in drawing managers from the ranks of scientists is the fact that if scientists go entirely into science management, after a relatively short period of time they will have lost touch with the field of science in which they had been doing research, and if this situation continues they will in fact lose their touch with science in general. It is, therefore, important to set up a mechanism whereby scientists who are at the present time working in science management have an opportunity to continue to educate themselves in developments in science and, in fact, preferably have an opportunity to continue their scientific research, at least part-time and perhaps with the help of associates. Indeed, I would select scientists to be in science managerial positions only for a limited amount of time, after which I would make a serious effort to return them to their previous scientific positions, where they would continue their research activities. There may be an exception to this toward the end of a person's career, when his ability or interest in participation in scientific research itself may have diminished and therefore there is no incentive or time to return him to research after the completion of his assignment.

Having a scientist with research background in a science managerial position also has the advantage that such a scientist, if he is good at all, is likely to continue to be attracted to scientific research and therefore will more easily withstand some of the temptations that surround someone in a science managerial situation. Particularly in Third World countries, becoming a manager can be a springboard for domestic political ambitions, for domestic bureaucratic aspirations, and, most importantly, for international bureaucratic goals, particularly within the United Nations and its special agencies. It is unfortunate that a substantial number of people from the Third World who would have had an opportunity to contribute to their own countries through scientific research in a local context were attracted to bureaucratic positions in international organizations, positions the utility of which is certainly much more remote and indirect, and perhaps not even as well established as the utility of doing good scientific research in the home country. The point I am making is that if the science administrator has an alternative option to return to his scientific work, then when he is faced with thinking about his future career, or when, for instance, due to a domestic political turn of events, he is asked to leave his science managerial position, he is not constrained entirely to thinking in terms of continuing on the administrative road he has already entered, but he has the option of returning to his former scientific career. Generally, people who, when having to make a decision, have hardly any options to choose from will end up in a situation which is less creative and less advantageous for everybody involved than people who have genuine choices to make.

VI. Conclusion

As explained at the beginning, in this discussion I tried to argue for the necessity of scientists being well-represented among the managers of science, particularly in countries where science is a relatively new undertaking and where infrastructure-building is in progress. I have tried to demonstrate that, of many points of view which are essential to the creation of a healthy science, the internal point of view, the point of view and intuition of a person who has had personal experience in pursuing scientific research, is an invaluable resource which cannot be replaced by the views of outsiders, no matter how well they may be formally trained.

Since I am a scientist myself, I may be accused of having a somewhat distorted and biased view toward this matter. I don't think this is the case. You certainly would not want to learn tennis from a person who has never held a tennis racquet in his hand but instead sat in the stands at the Wimbledon tennis matches for 20 years <u>watching</u> the best people in the field play tennis. Neither would you like to have as a bus driver a person who just got his degree in automotive engineering but, in fact, has never sat behind the wheel. Or would you want to place your private investment decisions in the hands of a fresh Ph.D. in economics who has never had any experience with the stock market? Science management is not a science, or at least not yet a science, and, therefore, success in it is based more on empirical experience than on theoretical brilliance. Especially at the beginning of one's career as a manager, such empirical experience comes from having been a working scientist and having had some contact with the way science works. What I am urging is the incorporation of this practical experience with doing science into the structure of scientific management.

Preprint, published by permission of *Interciencia*
11 (1986) 90.

Priorities and Decision Making for Science in Developing Countries

Michael J. Moravcsik
Institute of Theoretical Science
University of Oregon
Eugene, Oregon 97403
USA

Abstract

    Some suggestions are offered to facilitate the work of a maker
of decisions concerning scientific or technological projects in developing
countries. It is claimed that "picking the best" is an ill-defined and
impracticable objective which should be replace by "picking a good one".
Practical consensus can be created even in the absence of resolving all
conceptual and ideological disputes. Proposals for projects should be
formulated in terms of the three broad objectives of science, should
specify certain contextual and structural criteria, and should document
the availability of appropriate manpower. Techniques like cross-
critique, outside critique, and requiring a specification of evaluation
methods of the result help decision making.

I. Introduction

    Managers of science in national governmental and private agencies,
in international organizations dealing with scientific assistance, in international agencies handling "foreign aid", or in regional organizations
active in fostering scientific collaboration in the region are constantly
faced with making choices among objectives, among proposals, among
projects, among institutions, and among individuals. Such decisions may
be absolute ("Is this worth doing at all?") or relative among competing
alternatives. The factors that such a decision maker is forced by the
circumstances to take into account are very numerous ranging from rather
abstract scientific considerations to elements of local politics within
the agency of the decision maker. As a result of this and of the usual
time pressure, the selection of priorities and the making of decisions
often appear quite haphazard and unsatisfactory to many thoughtful participants of this process.

    The purpose of the present note is not to define a simple, quick,
objective, and infallible method for making decisions -- there is no such
method. The aim is very much more modest: It is to offer some observations and suggestions that may make such decisions somewhat easier,
somewhat more effective, somewhat more orderly. The medical profession
can be called successful even if it cannot altogether eliminate death,
pain, suffering, and loss. Similarly, science management can be helped
also by measures which are fragmentary and only very partially effective.

II. The nature of the task

a) Do not aim for the "best"

The aim of making choices among alternatives is sometimes construed as picking out the best of the available options. I want to argue that this statement defines the task incorrectly, for at least two reasons.

First, the justification for a choice is almost always multidimensional, that is, there are many arguments in favor (and against) a choice, coming from different "dimensions" of the situation. In such a multidimensional situation it is impossible to define a superlative ("best", "largest", "most beautiful", "most effective", etc) in an unambiguous way. The definition will depend on the relative weightings of the various dimensions of the situation, which can at best be done by a subjective value judgement. Thus, even if one could operationally decide which alternative is the best, this judgement would be subjective.

Second, choices are made of course a priori, and such a priori judgements are very shaky at best. All the needed information is usually not available, and even if it were, extrapolating from it entails large uncertainties. Furthermore, the time needed to take into account all the factors and to extrapolate from them is generally much longer than the time at one's disposal.

Thus, the task should be defined as selecting a good one out of the available alternatives. The practical consequence of this redefinition is that the decision maker will not spend time with the impossible task of trying to define what "best" is, and will not spend long hours in painstaking attempts to intercompare alternatives that may be, in principle and in practice, impossible to intercompare, at least given our present understanding of the workings of science. Instead, after a respectable amount of effort on studying each alternative in detail and on its own merit, and afterwards intercomparing them in a rough way, the decision should involve a certain element of intuition and even experimentation.

b) Practical Consensus

Exactly because of the multidimensional nature of the justifications of the various alternatives, it is not necessary to bring about theoretical agreement among a group of people when one wants them to agree on a choice among alternatives. The theoretical foundations of science and technology as human activities are very unclear at the stage of our present understanding, and abstract and ideological debates are frequent, presenting apparently diametrically opposite points of ivew. Indeed, if the settling of all theoretical disputes were a prerequisite for any action at all, nothing would ever be done. In reality, however, people with quite different theoretical conceptions can very well agree on the desireibility of a particular practical project, even though different people will favor the project for quite different reasons.

This realization suggests appropriate procedures for engineering agreement among a group. It involves a delicate balance between analytical and empirical ways of making a decision. On the one hand, as I will argue more in detail later, articulating the justification for a project in terms of a well defined analytical framework is very useful in order to test the maker of the proposal, to pinpoint the source of possible disagreements, to acquire further background information about a given project, and to defend a decision once made. On the other hand, such a method, should not destroy any practical consensus that may exist by thrusting the issue from a practical arena into a theoretical debating rink.

III. The framework

As just mentioned, I want to argue that projecting the situation at hand into a structured and multidimensional framework of well defined arguments and justifications is likely to be helpful. Indeed, it is helpful to request the originators of proposals to follow such a framework when writing their proposals. Such a procedure will direct the proposal writer to think through the project in a broader sense, and will also make it easier to argue about the relative merits of different alternatives.

Such a framework can be constructed in three simultaneous directions.

a) The three broad objectives of science.

Justifications of science tend to be classed into three main headings:

1) Science is intimately related to technology which in turn is strongly linked nowadays to the improvement of the material standard of life.

2) Science is a human aspiration in the 20th century, and hence is closely tied to feelings of self-sufficiency, of being on an equal footing with others, of prestige, of morale, of national pride.

3) Science has a major influence on Man's view of the world and on Man's perception of what Man's role in the world is. In other words, science is an important cultural factor.

These headings have been discussed in detail previously, and hence will not be elaborated on here. Each heading has innumerable subheadings, that is, ways in which these three main influences of science manifest themselves in a given situation.

b) Contextual and structural criteria

These have been discussed previously in some detail, for example in the seminal article by Alvin Weinberg. In order to engage in a project, we must assure that the results would in fact be of interest or value to others beside the participants in the project. If we talk about a research project, for example, we must make sure that the results will be of interest to other scientists, or to technologists in their development work. If we talk about an implementation of a scientific application, we must assure that the project brings about the needed change and thus affects those needing it. The project must also be ripe for exploitation, that is, doable in terms of present day knowledge and preparation. Such criteria serve to assure that the various effects of science enumerated in the previous subsection will actually occur as a result of the project under consideration.

c) Manpower

In some discussions this is lumped together with other contextual
and structural criteria, but it is so important that it deserves special
emphasis. Most projects that a decision maker of science or of assistance
in science comes to consider are limited by manpower considerations, and
in fact more by the quality of the manpower than by the quantity. We
must assure that a high quality and motivated scientific manpower exists
for the project to be undertaken. If this is not the case at the outset,
we must make allowance for the fact that acquiring such manpower is a long
and very unpredictable process, which may postpone the actual onset of
the project by a larger number of years than one may at first estimate.

In order to bring some of these abstract considerations down to a
plane of practicality, let us consider the example of a proposal to
establish an Institute of Electronics in some medium-level developing
country. In accordance with the above scheme, one might ask the following
questions:

In what way will the institute contribute to the level of technology
of the country?
In what way will the institute contribute to the national morale and
self-confidence?
In what way will the institute affect the understanding and usage of
science and technology by the average citizen of the country?
Is the proposed program of such an institute feasible in terms of
the present state of electronics?
Is there sufficient manpower of an appropriately high scientific
and technical quality to carry out the projected work in such an institute?

Each of these questions will result in numerous subsidiary questions,
thus orienting the type of information that will be needed for decision
making.

The aspects discussed so far by no means comprise all the relevant
factors to be taken into account. As mentioned earlier, there may be
many others bearing no relationship to the scientific, technical, or
developmental components of the situation. For example, an international
agency to some extent feels compelled to distribute its resources in a
geographically not too skewed way. In another situation, the influence
of a very powerful and assertive personality, with special values,
priorities, and interests, may distort the purely "objective" part of the
decision making. Nevertheless, it is very important first to lay out
the framework of decision making in terms of the more objective elements
as discussed above, and make an assessment in terms of it. The result
may not be final but may be modified by the other, more extraneous
factors, but then the weight which these played will be known. This is
very important also for the learning process in decision making: If a
certain mix of objective _versus_ subjective factors produce too many
decisions judged bad _a posteriori_, a good argument can be made for
changing the mix in the future.

IV. The practice of decision making

In this last section I want to list a few practical tools that will tend to make the decision making process easier and more reliable.

1) Structured justification of projects

As mentioned earlier, in addition to a freely flowing exposition of the proposed project in the customary way, the proposer should also be required to justify the project along a preset structure of questions, as illustrated earlier. This can be implemented either in written proposals or in conversations or other oral presentations.

2) Cross-critique

In situations in which several different points of view compete and conflict, the evaluation of proposals should be done through peer reviews of not only advocates of a given viewpoint but also of opponents of it. In other words, each proposal should be followed by a mini-debate which uncovers both the strengths and the weaknesses of the proposal. In order to achieve this, a strenuous and conscious effort must be made to find such a spectrum of reviewers. Even in situations in which there is no extensive peer review, and the decision is made primarily by one person in an administrative capacity, the presentation and discussion of the proposals should be done in terms of such a cross-critiquing manner.

3) Outside critique

It is often helpful to include in the roster of reviewers at least one person who is <u>not</u> from the specialty that the proposal deals with but from a neighboring discipline, but who has a broad enough view of scientific research and technological development work to be interested in and able to judge projects not strictly in his specialty. Such an outside critic often has a more balanced view and can intercompare projects of different types better than narrow specialists even if the latter are very competent.

4) Methods of evaluating results

The last practical tool I want to mention is the requirement that a proposal should state ways in which proposed project can be evaluated and assessed at a later stage or at the conclusion of the project. This relatively inocuous requirement can add enormously to the soundness of the proposal, since it forces the proposer to view the proposed project, so to speak, <u>ex post facto</u>, and to articulate ahead of time what he would consider success to be in the particular circumstances.

V. Epilogue

The observations and suggestions in this note intend to form a practical tool toward making the task of a decision maker easier. Only practical experience with the implementation of some of these suggestions will decide whether they contributed to the objective. The recounting of such practical experience would, therefore, be valuable to those wanting to experiment with such suggestions, and would certainly be appreciated by the author.

Reprinted by permission of *Approtech* 4 (1981) 1.

Special Obstacles in the Organization of Science In Developing Countries

by Michael J. Moravcsik

Perhaps the most far reaching hallmark of science and technology in the developing countries is their novelty. Unlike the scientifically more advanced countries, where science as a cultural force has been in evidence for several centuries (or, in Japan, for over 100 years), and where modern technology as a national productive activity has existed for a long time also, in developing countries the practice of science in the modern sense of the word and the use of modern technology are at best a few decades old, and have to a large extent remained a yet unintegrated graft onto the traditional culture and way of life. The aim of this note is to list some consequences of this state of affairs for the organization of science in the developing countries.

The purpose of such a listing is *not* to intimate that the task of evolving science in the developing countries is *a priori* hopeless. I believe that just the contrary is true: In a few decades there will be a large scale change of the guard among the ranks of the countries in the world and the scientifically and technologically leading countries will not be the present leaders but will arise from among the developing countries of today. The purpose instead, is to build an awareness of some of the obstacles which result from the lack of a long history of science in such countries so that correcting action can be focused on the elimination of these obstacles. They are by no means insurmountable, if they are not swept under the rug but faced squarely.

There are, in particular, seven circumstances that I want to list which impede the rapid development of science in the Third World. The first three pertain to the general milieu in which the development of science is embedded. The next two are related to the manpower involved in science development and the organizational aspects of this manpower. Finally, the last two have to do with the special educational background that is demanded by the special conditions in developing countries but which are generally not available to the scientist in those countries.

1. The Scientific Tradition

In the developed countries, science began as a small undertaking, practiced by a few, and the traditions for scientific work grew out of this modest beginning. The organizational superstructure came much later, as science became more visible, and as a clear need developed for such a structure due to the increasing needs and requirements of the scientific infrastructure itself.

In contrast, in many developing countries the decision to get involved in science and technology came much before there was

Michael J. Moravcsik is with the Institute of Theoretical Science, University of Oregon, Eugene, Oregon 97403, USA.

any substantial scientific or technological infrastructure in the country. Thus the bureaucracy of science preceded science itself. One sees ponderous and pretentious science councils, ministries of science and technology, or similar bodies arising from nothing, often goaded on by international "science policy" bodies. They erect impressive buildings marked as institutes for this and that, draw up development plans for scientific and technological activities, but all this in the absence of any substantial body of competent scientists or engineers in sight, and without setting up any criteria for evaluating the results of all these activities. The result gives the appearance of all the trimmings of science but without the soul and, even more importantly, the output.

2. Non-Adiabatic Development

The world "adiabatic" is used to describe the evolution of a system in which the rate of change of the system as a whole is much slower than the characteristic time scale of its components. For example, the growing up of a human being is an adiabatic process compared to the normal bodily functions of humans, such as heart beat, breathing, metabolic exchange, sleeping and being awake, etc.

The evolution of the scientific infrastructure in a developed country is generally adiabatic compared to the metabolic rate of its institutions and manpower. A given scientific organization has an opportunity to evolve and function before it must be modified in view of the overall development of science in the country. Similarly, a scientist can utilize his education and skills in an experienced way before they have to be strongly modified because of new situations and requirements in the scientific community he is a member of. Also, the generation of a given scientist is not very markedly different from the preceding generation of scientists, so that a rapport and cooperation can exist between the two generations.

In developing countries the situation is diametrically the opposite. The evolution of science is strongly non-adiabatic. An organization is hardly formed before it becomes obsolete in view of the new demands and new stage of development science reached in the country. The demands exerted on a scientist in a developing country 20 years after his Ph.D. are totally different from what they were at the time of his Ph.D., and so are the opportunities for science education and for scientific research. As a result, the elderly members of a former scientific generation, less fortunate in having had access to favorable opportunities for a functional education in science, are looked down upon by the succeeding generation, better educated and finding strength in numbers. Such a catastrophic generation gap is very much in evidence in developing countries, and is accentuated by the strong tradition of chronological seniority which places older and less accomplished scientists over the younger and better equipped ones in the organizational hierarchy and power structure.

3. The Rationale for Science

Understanding and articulating *why* science should be practiced is clearly a prerequisite for a purposeful and productive scientific activity. To do so, however, to establish a philosophical, cultural, and conceptual basis for being engaged in science as a human activity takes a long time. It is a process that money cannot buy, foreign assistance agencies cannot implant, and imitative manipulations of foreign examples cannot internalize. Even in countries where science has been in evidence for centuries, the justification of science and the motivations for doing science constitute a recurring, often controversial subject of public discussion.

In the developing countries this turmoil is even more bothersome. National development plans, political speeches, the personal statements of scientists, the silent view of the population as a whole all deal with this subject, often each in a one-dimensional way, and all in different dimensions. Furthermore, words point one way, while deeds in another. The usual rationale ascribed to science in the developed countries cannot automatically be taken over in the developing countries, where the cultural motivation may not yet be present, the industrial production process too weak to link up with technological innovation in a self evident way, the impact of science on the national Weltanschauung still too rudimentary to attract notice, etc. Long term motivations for science are laid aside in favor of short term ones, only to be remembered when a crisis stage is reached due to the neglect. Because of the perpetual crisis management, no time is found to explore these teleological problems of science and hence both the managers and the practitioners of science suffer of uncertainty, of a lack of self-assurance.

4. The All-Pervasive Government

In many developing countries the traditional economical and political sector is highly decentralized and rests in the hands of many small groups of people or of many individuals. In contrast, in the same countries the modern part of the economy and of the administrative machinery is in the hands of a centralized government. There are, of course, developing countries, ruled by totalitarian regimes, where even the traditional sector is under the thumb of centralized power, but in any case, the modern sector is virtually always highly concentrated in the hands of the government.

However regrettable this may be, it is easy to understand how this came about. The rapidity by which the modern influences descends on most of these countries, the conservatism and low educational status of much of the population, together with occasional ideological overtones favoring centralization lead almost inevitably to this trend.

On the whole, the effect of this centralization in most countries has been disastrous. The development of a widespread system of mindless civil servants by colonial powers in the old days burgeoned, after independence, into a horrendous quagmire of bureaucracy which consumes large sums and

suffocates innovation, incentive, and flexibility. Effectiveness and efficiency are not among the criteria for such bureaucracy, and often formal promotion is the only yardstick by which an employee of such organizations evaluates his work.

The detrimental effect of this state of affairs is much more evident in technology than in science. To be sure, in the sciences also, the disregard of merit in appointments and in the distribution of funds, the indifference toward a functional assessment of past work, and the rigidity in clinging to rules rather than seizing novel opportunities do much harm. When it comes to technological development and the linkage of this to the agricultural and industrial production of the country, however, the situation is even more grave, since large amounts of resources are at stake, and the detrimental results are also much more visible in the short run.

5. The Waste of Exceptional Individuals

In almost all situations that I have ever encountered when there was an unusually strong, creative, and productive upsurge in some aspect of the progress of a country, the upsurge could be traced to the presence, active involvement, and dynamic leadership of one or a few exceptional individuals. In the scientific areas, and on various levels of impact, names like Bhabha, Usmani, Choi, Baiquni, Saavedra, Sabato, and many others are quite familiar to observers of the corresponding developing countries. The same is also true in the advanced countries, except that there, one single person usually does not have as great a leverage as his counterpart may have in a developing country. This is simply a matter of the absolute size of the community. When the scientific manpower is large and well educated, the leadership tends to be distributed among a sizeable number of exceptional people.

In view of exceptional individuals playing such a crucial role in the progress of developing countries, it is especially sad that the environment in which they work usually tries to make their efforts as difficult as possible. To some extent they themselves are at fault: they are often too aggressive and even arrogant, they try to go too fast because they realize how little time they have at their disposal, and they fail, in their zeal, to provide for successors to themselves so when they vanish from the scene, there is no continuity. But much of the blame must be laid on the others in the country, who view such exceptional people with incomprehension, envy, jealousy, and suspicion. This results, not infrequently, in the downfall of the exceptional individuals through the concerted machinations of their peers and colleagues. The destruction of Usmani in Pakistan through ludicrous spy charges is just one of the many examples of this.

6. The Double Burden on Scientists

Scientists in developed countries, for the most, need to be concerned only with their direct professional responsibilities in scientific research or education. The infrastructure and tradition for doing science are already in place, and the management of the scientific establishment can be carried out by a relatively small number of scientists with particular interest in and aptitude for such activities.

In sharp contrast, in a developing country even the youngest scientist, fresh out of school, will immediately be confronted with a double responsibility of discharging his direct professional duties *and* at the same time also contributing to the creation of an infrastructure, of an environment in which such professional duties can in fact be carried out. Such an infrastructure does not exist or is in a rudimentary state, the environment is indifferent or even hostile, and unless the scientist himself tends to these problems, bureaucracy will take over and scientific work will become impossible.

Scientists in the developing countries are ill prepared to assume this double duty. Their education even in the best universities of the developed world did nothing to prepare them for this task, and so they are not only inexperienced but also shocked when they confront the realities upon their return to the homeland. The reaction can be in diverse directions. The easiest one is simply to braindrain out of the country with the claim that science cannot be done there. The equally common one is to stay but to effectively give up doing science and to join the bureaucracy itself which promises security and social status, although little challenge or internal satisfaction. The most difficult alternative is to stay, to try to continue being an active scientist, and at the same time try to form the infrastructure also. That only relatively few people have the strength and skill to opt for the third alternative means that even within the relatively small scientific manpower that most developing countries have, an unusually large fraction has a stunted scientific career.

7. The Neglect of the Context of Science

In an environment where science is not yet established and not yet part of the local culture, one would think that the education of the local scientists in the broader setting of science (the history, philosophy, psychology, sociology, economics, methodology, and management of science) would be considered a high priority goal, so that these scientists could eloquently explain their work and could transmit to the population as a whole those elements in science which are of help to an everyday person in everyday situations. Nothing could, however, be farther from the truth. These contextual aspects of science are hardly ever, if at all, included in the curriculum of schools, colleges, and universities, and the number of local specialists in these areas is also practically nil. Not only the background is lacking, however, but also the tradition works against scientists assuming such a role of an advocate for science. Public speaking, debating, and extracurricular group activities are not often the lot of a secondary school student in a developing country, and class, political, or geographical separations may also prevent a scientist from exerting a broad influence on his environment. The result is that the few scientists in the country do not generally function as carriers of the new scientific ingredient in local culture, and in fact they may not be able to defend even their own profession from the onslaught of bureaucrats, traditional conservatives, or political functions.

Epilogue

One way of quickly summarizing all this is to say that the building of science in a developing country is a very much broader, more multidimensional, and complex task than the mere maintenance or slow advancing of science in a scientifically already well developed country. This remark is perhaps particularly topical at the beginning of the 1980's when, at last, there appears to be a quite widespread recognition that the building of scientific infrastructures in the developing countries must be a top priority of international assistance, at least on an equal footing with the short term crisis management of famines, epidemics, and population explosions that have been in existence for decades. In this decade, therefore, when, hopefully, considerable attention, skill, and resources will be devoted to such infrastructure building, it is essential that the efforts be set in a sufficiently broad framework to assure that the assistance does not wither as soon as a program expires, but takes root in a context and environment that was carefully prepared and improved for this purpose.

From International Development Review,
1976, No.3

The Missing Dialogue—An Obstacle
in Science Development

There is a lack of communication between scientists and decisionmakers on the role of science in development because most of the decisionmakers have insufficient knowledge of science, on one hand, and because the scientific community is inexperienced in articulating their position, on the other.

Michael J. Moravcsik
Institute of Theoretical Science
University of Oregon

☐ It is well known that the building of indigenous science in the less developed countries faces many difficulties. Perhaps the most crucial problems are related to the absence of adequately trained scientific manpower, while others hinge on the lack of material resources, or on the isolation of the local scientific community from a sufficiently productive interaction with the mainstreams of the scientific world.

In this essay, however, I want to analyze a situation which is not directly related to any of the above difficulties, but which nevertheless is a very serious obstacle in the path of science development. It pertains to two groups of people who are essential in the science development process, and to the lack of interaction between these two groups. As I hope to demonstrate, the problem is multidimensional: both of the participant groups lack the qualities which the demands of science development would call for, and in addition, the foundations of the dialogue are also absent so that even if the participants themselves rose to the occasion, interaction and mutual understanding would still be hard to achieve.[1]

The Decisionmakers

One of the two groups in question consists of those in the country who are in the position of making the important decisions regarding development. We will not be interested here in the qualities, characteristics, knowl-

edge, and attitudes of the decisionmakers in any other way except as they pertain to science. In this respec., the picture is dismal. The decisionmakers have practically no background as far as science is concerned. Their contact with science throughout their educational process has been either nonexistent, or rudimentary, sporadic, formal, and misleading. It is not at all uncommon to encounter quite high officials in ministries of science or education, or similar bodies, who cannot tell the difference between science and technology, and consequently do not understand the relationship between the two. They do not realize that the dichotomy of basic vs. applied science is not the same as the dichotomy of experimental vs. theoretical science, and so on.

A similar confusion reigns with respect to the role of science in the development of a country. The prevailing concept one encounters is that the only justification for

[1]It will be noted that the ideas in this article are discussed in a general way without giving reference to specific instances or examples. This is intentional, and stems from the rather delicate subject of the article, dealing with human qualities, characteristics, personalities, and interpersonal relations. In fact, such topics have been generally taboo in written contributions to science development, a tradition that I intend to violate by offering this article for public discussion. Anybody even moderately familiar with the development scene around the world can easily think of his own examples and illustrations for the various points made in the article.

pumping large sums of money into local science is the amount that will emerge the day after in terms of factory output. That longer range and more complex relationships exist between the presence of indigenous science and a country's material production is ignored. Furthermore, that the presence of science contributes to development in a broader sense, in terms of shaping attitudes, fostering and aiding the fulfillment of aspirations, bolstering national morale and consciousness, and forming the new cultural milieu of a country, is altogether obscured.

Indeed, it is often pathetic to visit a country with a long standing and rich history of former greatness along many dimensions, a flourishing culture in arts, and a broad and profound philosophical and religious tradition, and then meet decisionmakers whose horizon along these lines is very much narrower than that of men from the villages.

Perhaps the most tragic element in this picture is that in their singleminded obsession with immediate material development, these decisionmakers do incalculable damage to the very cause they aim to pursue. As a result of their misunderstanding of the role of science, of the nature of scientific activity, and of the process of fertilization of technology by science, their actions often virtually guarantee that science and technology in their country will not rise beyond the meager stage of low-grade, imitative technology kept alive only by the availability of pitifully cheap labor.

It might be countered that construing the above remarks as being universally valid would be unjust. In fact, in virtually every country, one meets some exceptions. There are everywhere isolated figures among the decisionmakers who have a sufficient understanding of the problems, a sufficiently broad view of the situation, and a sufficiently realistic set of expectations to be able to act constructively in the science development process. Whenever some progress is evident in a country in the building of its science, such progress can almost invariably be traced to one or several such key individuals who happened to be in the right position at the right time with the right qualifications. Nevertheless, one can conclude that on the whole, decisionmakers in less developed countries tend to lack the knowledge, the background, the outlook, and the understanding minimally required for a national policy toward science development.

It could also be countered that decisionmakers in advanced countries are often no better equipped in scientific matters than those in the developing countries. Among the more than 600 members of the United States Congress, for example, perhaps a half dozen have a formal background in the sciences equal to, or exceeding, a Master's degree. In spite of the general intelligence and the best of intentions shown by most congressmen, if the providing for science in the United States had to rely on Congress's scientific expertise in science or in science policy, the situation would also be quite dismal.

There are, however, two elements in the United States which contribute to a much more favorable context of science policy. First, science as an activity is already established, and the infrastructure on the whole is already there. Thus, what is needed is "only" the maintenance and further development of this activity and infrastructure. This, although by no means trivial or automatic, is a less demanding task than the creation of science where there was none previously.

Second, while Congress itself has no significant competence in scientific matters, and neither does the topmost echelon of the executive branch of the US Government, the latter branch has, at somewhat lower levels, an appreciable corps of scientifically trained people who have some influence in decisionmaking. Furthermore, and perhaps even more importantly, the executive branch has a long tradition of rather active seeking of scientific advice through consultants, committees and other study groups, thus linking the scientific community directly with formulation of decisions.

As an epilogue to this section of the discussion, I would like to add that the above remarks are not at all meant to be derogatory to the decisionmakers of the less developed countries. It is not a matter of their inferiority to anybody else. It is simply the fact that there is a disparity between, on the one hand, the almost superhuman demands that the development of a less developed country places on the individuals who play a leading role in it, and, on the other hand, the quality and quantity of opportunities these individuals have to prepare themselves for this role. The demands of development call for persons with a broad and prescient view of the world, with a rare capability of assimilating the known and at the same time creating the new, and with a talent for organization and human leadership that can stand up to the turbulence of the development process. It should not be surprising that only in rare cases can individuals even approach these requirements in terms of their personal abilities and background, regardless of their nationality, race, religion, ideology, or philosophy. This analysis therefore does not aim to blame, but rather to establish the factual framework which exists, perhaps through nobody's "fault," but in which we have to work at the present.

The Scientists

The second group consists of those directly participating in the scientific or parascientific activities of the country. Let us call them the scientific community. It is not my purpose here to assess the purely technical

MICHAEL J. MORAVCSIK is professor of theoretical physics at the University of Oregon, and an active researcher in nuclear and particle physics. During the past 15 years he has also been involved, all over the world, in the study of science development problems and in action to remedy those problems. He is the author of many articles and a book on this subject.

competence of this group in science proper. Instead, I will single out that particular segment of the group which in fact is irreproachable in terms of technical scientific competence and achievement. Thus it will be clear that the critique of characteristics to follow does not arise from any lack of scientific standards in a purely technical sense.

However admirable this group may be in a purely scientific sense, it tends to have a number of serious shortcomings when it comes to its role in science development. I would like to mention two of these which are particularly pertinent to their interaction with the decisionmakers.

The first is an almost complete absence of an articulate position on the context of science in the life of the country and on the requirements which the establishment of science implies. If the decisionmakers deserve blame for not knowing about the difference, for example, between science and technology, the scientific community must take the blame for being inarticulate on this subject. In fact, the scientist in a less developed country tends to be curiously and almost frighteningly timid and apologetic about his pursuit of science. More often than not, he succumbs to the illusory and misleading view of science that generally prevails in the country and tries to prove how his scientific efforts will somehow result in increased agricultural production in the immediate future. The argument, of course, usually turns out to be quite unconvincing even to the local decisionmakers. It could hardly be otherwise, since today's scientific research activity in any country is highly unlikely to have any direct effect on production in such a short time.

The second great weakness of the scientific communities in the less developed countries is the lack of initiative, the absence of a drive toward creating a local context for scientific activities, and consequently a total dependence on the government or other decisionmakers in matters of action. When discussing problems of science development, one invariably receives a response from a scientist to a particular suggestion which runs like this: "Yes, that is a very good idea. I will write a memo to the appropriate governmental office requesting that they follow it up." The realization that a very large part of the science development process can be, and in fact must be, done by the local scientific community itself, within its own rank and independently of (though not contrary to) the decisionmakers, has hardly percolated yet into the scientific communities.

One might again counter these remarks by pointing at conspicuous figures in the scientific communities of less developed countries who have a long and admirable list of accomplishments in science development to their credit. True enough, but again they are the rare exceptions, and thus the impact of such personalities is fragile and diluted. Added to this, the highly accentuated infighting within the scientific communities (due in part to the small size of these communities) makes those active persons particularly vulnerable to envy and a consequent lack of co-operation.

These shortcomings are certainly experienced in the more advanced countries as well, but, again, the existence of a certain tradition helps. Moreover, in the larger scientific communities of the advanced countries, members are likely to have enjoyed a broader general education and some practice in the public expression of ideas. It is therefore easier to find someone who can assume the role of an articulate spokesman.

In summary, it can be stated that scientific communities in less developed countries tend to be inexperienced in articulating the conceptual and practical elements of the context of science and also tend to show very little initiative in taking direct action in science development matters.

The Basis for a Dialogue

Let us imagine for a moment that somehow a consensus could be achieved between decisionmakers and scientists about what scientists are supposed to do and what they need in order to perform. The dialogue would still be futile because of the lack of another crucial element—namely that of any attempt to evaluate the output of science in less developed countries.

It is granted that since the output of science is knowledge, it is not at all easy to measure.[2] This is not the place to enumerate the difficulties in detail, or the often ludicrous ways economists and others from outside the sciences attempt to deal with this problem. It is sufficient to state that there are now increasingly varied and reliable ways, using indicators of scientific communication media and assessments through peer judgment from the scientific community itself, to evaluate the creativity and production of individual scientists, of scientific institutions, of scientific fields, and of the scientific production of whole countries.[3] In fact, these methods have been in use for some time in the worldwide scientific community and also in the scientifically more advanced countries.

This is hardly surprising, since the quality of scientific work is of paramount importance. Work which formally

---

[2] The problem of measuring scientific output has recently acquired a rather large literature. For some examples, see Jonathan and Stephen Cole, *Social Stratification in Science* (Chicago: University of Chicago, 1973); Stephen and Jonathan Cole, "Scientific Output and Recognition: A Study of the Operation of the Reward System in Science," *American Sociological Review* 32, 377 (1967); Maurice B. Line and A. Sandison, "Progress in Documentation," *Journal of Documentation* 30, 283 (1974); Henry Small and Belver Griffith, "The Structure of Scientific Literature I: Identifying and Graphing Specialities," *Science Studies* 4, 17 (1974); M.J. Moravcsik, "Measures of Scientific Growth," *Research Policy* 2, 3, 266 (1973); M.J. Moravcsik and P. Murugesan, *An Analysis of Citation Patterns in Indian Physics, Science and Culture* (in press, 1976); C. Freeman, *Measurement of Output of Research and Experimental Development: A Review Paper* (UNESCO Statistical Reports and Studies No. 16). Paris, 1969. These references also cite many additional sources of interest. It should be stressed that these assessment measures can be applied to *both* "pure" and "applied" scientific research.

[3] For a most interesting application of one of these communication-based indicators, see Derek de Solla Price, *Proceedings of the Israel Academy of Science and Humanities* IV, 6, 1969, pp. 98-111.

can be called scientifc, but which lacks quality, is no contribution to knowledge, no influence on technology, and no factor in shaping man's general view of the world, and hence is a waste of manpower, time, and resources. In this sense science is a fundamentally elitist undertaking. It can be pursued only by a relatively small group of people, and even among them, those who produce higher quality have an inordinately greater impact on, and hence influence in, science than those of lower quality.

In light of this, it is shocking to realize that scientific output in developing countries is practically never subject to evaluation. The process of supporting scientists, inasmuch as it exists at all, stops when a grant is given to the scientist. Very seldom does anybody worry about trying to assess whether the work was successful.

Not only decisionmakers shirk the responsibility of evaluating science when they offer support to it, but the scientists themselves also shy away from measuring output. In fact, critique within such scientific communities is very rare except for generally disparaging remarks uttered usually behind people's backs. Promotions, posts within the scientific communities, or honorary awards are offered primarily by seniority, and recognition of all sorts is given without objective reference to a person's scientific accomplishments in the real, functional sense of the word.

This is all the more distressing because in a small scientific community experts in a given specialty are usually very few in number, and hence a given person finds it very difficult to assess his own work by simply talking to the few colleagues who happen to exist in the same country. Furthermore, scientists in less developed countries are often very much isolated from regular and substantial interaction with fellow specialists in other parts of the world because they have minimal access to the scientific communication channels.[4] Thus, an evaluation by the usual international methods would be of even greater value there.

It is sometimes said that these methods could not be applied fairly to those countries because the conditions under which science is practiced are different enough so that a straightforward application of the usual methods would put the local scientists at a disadvantage. I am confident, nevertheless, that if the evaluation is conducted by scientists from the worldwide scientific community who have some understanding of the different conditions in those countries, this problem could be easily overcome.

The use of an international method of evaluation could considerably bolster the position of the local scientific community in the eyes of the local decisionmakers. Moreover, it could serve as the basis of a meaningful

dialogue concerning the quality of science in that country and the support science needs, without which science development, if not completely impossible, is very much more difficult.

What Can We Do?

Once the above diagnosis is accepted, the remedies are rather evident. In particular, there are measures one can take which are long term in nature and which will show results in 5 to 10 years. There are also steps which, though in an overall sense not as effective, might bring relief almost immediately.

Let us first turn to the measures which would affect the decisionmakers. The long-term solution would be to provide them with a much broader and deeper educational background. Educational systems in less developed countries tend to be oriented along very narrow channels determined by the conception of what will be "relevant" (some 10 to 20 years hence) to the strictly professional pursuits of a given person. Therefore, often painstaking care is exercised to prevent students from having contact with anything "useless" and to make sure that they learn techniques rather than ideas, since the former are judged to be what the country needs.

Specifically, students who aim at nonscientific professions are carefully shielded from exposure to science. University courses on science for the nonscientists are extremely rare in less developed countries, and since secondary education certainly does not fill this gap either, C. P. Snow's two cultures are intentionally created and artificially enhanced.

This clearly must stop. The less developed countries must either begin to acquire the spirit, and not only the form, of the educational systems they emulate or, even better, develop their own brand of education which truly responds to the demands of development in the most general sense—and that means education with a much broader scope than has existed in the past. Such a reform will by no means be an easy task, but with determination and with the help of the international intellectual community, it should be possible.

The shorter-term remedy with respect to decisionmakers might either consist of steps toward reorienting those already in decisionmaking positions. For example, a simple and brief brochure, mass-produced and mass-distributed to decisionmakers in all countries, which explains the most elementary ideas, concepts, results, and methods surrounding "the science of science", scientific methodology, science policy, and related areas would certainly be useful. Similarly, brief but intense meetings aimed at decisionmakers, in which these concepts are discussed, could have an impact.

There have been past efforts in terms of publications and conferences by various organizations, international and otherwise, but all these efforts were fundamentally misplaced because they dealt with the context of science on a very much more sophisticated level than the situation calls for. As a result, one encounters decision-

---

[4]For discussions of the communication problem in general, see M.J. Moravcsik, "Communication in the Worldwide Scientific Community," *Science and Culture* 41, No. 1, 10 (1975), and M.J. Moravcsik, *Science Development—The Building of Science in Less Developed Countries* (Bloomington, Indiana: International Development Research Center, 1975), Chapter 4.

makers who are ignorant of even the most fundamental concepts in these areas, but can carry out a formally sophisticated conversation on the relevance of Delphi methods to science policy decisionmaking. As usual in such discussions a great affinity is shown toward acquiring techniques, tricks, and formal structures, without sufficiently penetrating the ideas, conceptual problems, and fundamental spirit of the subject under investigation. What is really needed is interaction with decisionmakers on the most elementary level, starting right from the beginning and building up a solid structure of ideas and knowledge concerning the context of science in less developed countries.

In an international meeting arranged by high-level governmental bodies and attended by formally selected senior decisionmakers, the atmosphere is by necessity filled with ceremonious orations and platitudes aimed at impressing fellow delegates as well as the readers of the sumptuously printed proceedings. Instead, the need is for small intimate meetings without fanfare and with the minimum of set programming, at which a real collaboration in search for new ideas and understanding can take place. There are precedents for such meetings, and even though they cannot instantly solve all problems, their effectiveness is many times that of the ceremonial science policy conferences, and at a fraction of the cost.

Now what about measures aimed at the scientific communities? The long–term ones again pertain to education. Both in the local educational system and in training abroad at universities in the more developed countries, special emphasis must be placed on exposing scientists to a far-ranging discussion on the context of science in less developed countries. That this is not done in universities in those countries themselves is perhaps not so surprising, since there is at the moment neither the realization of the need for, nor the manpower to teach, such a curriculum. But the fact that such education is not offered to these students in universities in the more advanced countries is much less excusable. I have commented on this previously[5] and also suggested special mechanisms to experiment with such programs, but so far no funds have been able to be raised even for an experimental exploration of this area.

The shorter-term measures have to do with providing the local scientific communities with written material,

personal contacts, and opportunities for discussion of the context of science in the less developed countries, and of the need for personal initiative. Examples of exceptional people in less developed countries who have blazed a trail in science development against heavy odds and in a discouraging environment should be documented, summarized, and widely distributed to stimulate others. These exceptional people should also be given opportunities to travel in other less developed countries to talk about their experience and spread their credo. There have been isolated occasions when this was done, and the impression created was quite significant.

Finally, let me turn to the action needed to create a foundation for the dialogue between scientists and decisionmakers, that is, a system of evaluation for the output of science in less developed countries. As I discussed in a recent article,[6] the two types of tools used in such an evaluation are publication and citation data, and peer judgment. The former is relatively easy to administer, though undoubtedly a certain care must be exercised to adjust its absolute standards to conform with what is possible under the conditions that generally prevail in less developed countries. The peer assessment system, on the other hand, is somewhat more elaborate, but its long-standing use in many facets of worldwide scientific interaction (in connection with journal articles, research proposals, etc.) should offer precedents that greatly facilitate the task. It is in this area that international cooperation would be most effective—not through formal international bodies but through the professional organs of scientists themselves, who could offer direct assistance in providing referees and in creating contacts.

Modest Funds, Minimal Bureaucracy

In all these measures, the main source of initiative must come from the indigenous scientific communities, who, rising out of their lethargy, have to realize that without concerted efforts in upgrading the two confronting groups of scientists and decisionmakers and establishing a functional dialogue between them, the hopes for creative science in the less developed countries are dim. At the same time, cooperation from the worldwide scientific community is crucial, since many of the suggestions outlined above imply enhanced interaction between scientists around the world and the participants in the science development process in the less developed countries.

Nowadays appeals such as these are likely to be dismissed as unrealistic because science has fallen on hard times and resources are thought to be scarce even in the more advanced countries. It is therefore important to stress that most of these measures require very few resources, extremely modest funds, and very minimal bureaucracy. The enrichment of the curriculum of "foreign" graduate students in the sciences who are being educated in universities of advanced countries can be done for practically nothing if the understanding and

[5] I discussed the problems pertaining to the education of science students from less developed countries at universities in the more advanced countries in M.J. Moravcsik, "Foreign Students in the Natural Sciences: A Growing Challenge," *International Educational and Cultural Exchange* 9, No. 1, 45 (1973).

[6] M.J. Moravcsik, "The Context of Creative Science," *Interciencia* 1, No. 2 (July-August 1976).

[7] Moravcsik, "Foreign Students in the Natural Sciences: A Growing Challenge," loc. cit.

[8] There are now projects like the SEED program of the U.S. National Science Foundation which are aimed in this direction. In addition, many such contacts can be made as piggyback detours on regular professional travel by scientists. In coordinating such detours, programs like the Visitors Registry of the U.S. physics community can be of considerable help.

will exist.[7] Simple brochures dealing with the fundamentals of the context of science could be prepared, duplicated, and distributed on a small budget and with no new organization or overhead expenses. The sending of articulate and experienced spokesmen on matters pertaining to the context of science to interact with the appropriate people in the less developed countries involves a minimal expenditure and very little new organization.[8] These and other examples demonstrate that, if the worldwide scientific community had the awareness and concern, much could be done for sums that would hardly eat at all into the existing resources available for scientific activities in the more advanced countries.

□ □ □

The Incomplete Alliance

Michael J. Moravcsik

Institute of Theoretical Science
University of Oregon
Eugene, Oregon 97403
USA

Science is probably the most universal and most collective human undertaking. In the work of a scientist previous contributions by other scientists form an indispensible basis and his work in turn must be communicated to other scientists and technologists in order to make an impact on the overall edifice of science and technology. This can be readily done since science can be propagated relatively easily across national, racial, political, religious, and cultural boundaries.

Because of this universality and collectivity, the scientific development of various parts of the world has always been through intense interaction with other parts already more advanced in the sciences. This "scientific alliance" benefits both sides: As new countries and parts of the world are increasingly included in science-making, science makes faster progress to the benefit of all. The prominence in science of countries like Japan or the United States is founded on the learning process these countries underwent, 50 or 100 years ago, by interacting with the then leading countries of science, mainly in Western Europe. Today, the Third World countries, intent also to become part of the worldwide scientific community, interact with the "developed" countries in order to build their science.

The most crucial element in establishing a scientific infrastructure in a country or region is scientific manpower. Thus science education is a pivot on which the rest of development hinges. In this brief article I want to comment on the extent to which the "scientific alliance" between Third World countries and the developed countries has been productive in strengthening science education for the Third World countries.

If at all possible, science education should be indigenous to a country or region. At the beginning of the development process, however, science education, particularly at an advanced level, is not available locally, and hence must be obtained abroad. Thus there are two parts of the "science education alliance" between the Third World countries and the developed countries. The first pertains to the education of Third World students in the educational institutions of the developed countries, while the second includes all aspects of the indigenous science education in the Third World.

It is my contention that the first of these has been very extensive and successful, even though improvements in it could be made. On the other hand, the "alliance" with respect to the second of these aspects has been much less productive and successful, and needs considerable additional development.

Let me first consider the first aspect, namely the education of Third World students in the developed countries. There are, at the present, well over 300,000 "foreign students" in educational institutions of the United States, and similarly impressive, though smaller numbers of students are located in other developed countries. Of this number, a considerable majority comes from Third World countries, and a goodly part of those study some part of science or technology. Not only are these numbers huge, but they also grow very rapidly, so that an increasingly larger percentage of students in US educational institutions come from Third World countries. Since in most US educational institutions education is heavily subsidized from public funds even for those students who pay the "full" tuition, the total contribution, in financial terms, of the United States to the education of Third World students is enormous, probably exceeding $1 billion a year. This sum never appears in the statistics of international assistance, and yet is probably the largest single item in such assistance.

As I mentioned earlier, this large contribution to education could nevertheless be improved. Methods of determining which applicants from Third World countries should be admitted and given financial aid by universities in the developed countries could be improved. Once the student arrived, methods of advising him in professional and educational matters need to be much more efficient. Also, the Third World student should obtain, as part of his education abroad, some exposure to those contextual problems of science (machine shops, libraries, science funding, national science policy, how to assess scientific work, how to order from a catalog, etc. etc.) which will become crucial to him upon his return to his home country. There have been some developments in these areas, but much more needs to be done.

The second aspect of science education, namely that which takes place, in various forms, in the Third World country itself, is in a much less fortunate state as far as the "alliance for science" is concerned. It is of course true that the responsibility for this aspect of science education rests primarily with the Third World countries themselves. Yet in many respects international cooperation is sorely needed. To begin at the end, once a student returned from abroad with his advanced degree, his success in scientifc work and his opportunities for continued education throughout his career depend very much on such cooperation. Isolation is the main killer of productive scientists in the Third World, and such isolation can be relieved only with the cooperation of the international scientific community.

But the "alliance" must be effective in other areas also. To give an example from the other end of the spectrum, the scientific education of the population as a whole is usually very much neglected in Third World countries. Newspapers seldom if ever feature science columns, and even if they do, often they are bad translations of dispatches by news agencies from abroad. Radio and television also lack science features, and museums, exhibits, and fairs dealing with science and technology are also in great need of development. Experience with these public media in the developed countries should be made available in the Third World to a much greater extent, and cooperative ventures in exploring such media should be instituted.

In between these two extreme aspects of indigenous science education there are many others pertaining to teaching science in elementary and secondary schools, and in colleges. Although there have been a number of cooperative projects oriented toward these stages of science education, it is still too common in Third World countries to see science education which stresses memorizing instead of understanding, and which specializes the student much too early, thus greatly narrowing his expertise and utility.

Who is to be "blamed" for the alliance not being effective in these aspects of education? In part, the blame is on the developed countries for not recognizing to a greater extent the crucial role such education plays in the evolution, stability, and compatibility of Third World countries and hence in the bright future of the whole world. But I would also place the blame on scientists and science managers in the Third World countries for being far too timid and vague in placing their needs before the worldwide scientific community. The literature dealing with problems of science in developing countries is being written primarily by people from the developed countries. This should not be so. Proposals for fruitful international scientific cooperative programs emerge very rarely from the scientific communities in the Third World. I would therefore strongly urge ASPEN to establish a group of its interested and knowledgeable members for the purpose of creating a set of very concrete proposals for international cooperative projects in science education. By proposals I do not mean lofty sentiments and moral preaching combined with vague and non-quantitative expressions of needs. Instead, what is needed is a quantitative, concise, and factual set of concrete proposals, consisting of a) description of a specific need or problem, b) an explanation of why past efforts have been absent, weak, or futile, c) an exposition of the new program proposed, with indications of why this should work, d) a delineation of who will do what in the project, how much of it, and when, e) a quantitative analysis of the financial needs of the project, f) a methodology for how the project can be evaluated after it has been in place for a while. Such proposals then need to be submitted to appropriate governmental, international, private, and other organizations for consideration and eventual funding. The important point to be noted here is that generating new projects in science and science education requires much homework, and that if the scientific communities in the Third World do not do this homework, others will certainly not do it for them.

For most scientists in the developed countries the infrastructure is readily provided and they have to be concerned "only" with doing science. The scientists in the Third World countries are in a different position. There the infrastructure is not yet in place, and each scientist find himself occupied with two tasks simultaneously: To do science and to create the circumstances under which science can be done. What I tried to indicate in this brief review is that the international dimensions of local, indigenous science education form an important part of this infrastructure that require much more energy, work, and devotion.

LAS RESPONSABILIDADES DEL HOMBRE DE CIENCIA EN CALIDAD DE ESTADISTA CIENTIFICO [1]

por **Michael J. Moravcsik**

Institute of Theoretical Science
University of Oregon
Eugene, Oregon 97403
USA

INTRODUCCION

Falta de conciencia

La mayoría de nosotros, los investigadores científicos, nos imbuimos en la ciencia porque ella nos fascinaba y, por lo tanto, queríamos tomar parte de la investigación científica. Cuando éramos jóvenes leímos libros científicos, tal vez alcanzáramos la buena suerte de haber tenido un profesor fervoroso de ciencia en la escuela, la oportunidad de visitar museos de ciencia que nos emocionaran, tal vez hayamos visto programas de televisión que trataban de la ciencia, y después de tal iniciación concluimos que simplemente teníamos que llegar a ser uno de esos hombres de ciencia que se encuentran a la vanguardia de todas esas cosas maravillosas.

En ese momento la mayoría de nosotros (y esto es algo que puedo decir definitivamente de mí mismo) ni siquiera nos detuvimos a considerar la manera en que esas atractivas oportunidades de practicar la ciencia se habían manifestado, la forma en que se· sustentan, cómo se enlazan con lo que los demás hacen. Sin saber cómo, aceptamos que todo lo que tenemos que hacer es consagrar nuestro talento, nuestro entusiasmo, y nuestra perseverancia a la creación de nuevos conocimientos científicos y que la estructura para hacerlo va a estar allí naturalmente, esperándonos.

Esta actitud persiste aun después de los primeros estudios. La mayoría de los estudiantes graduados de ciencia en los Estados Unidos, y creo que en otras partes también, le prestan poca atención a los problemas infraestructurales de la ciencia y a la forma en que la ciencia encaja en la sociedad en que existe. No hay duda de que una vez que están por obtener su grado definitivo tienen que enfrentarse al asunto de cómo conseguir el primer trabajo, pero eso lo consideran ellos

---

1. Título original de la ponencia: «La responsabilidad del científico como hombre de Estado».

un problema personal, lo mismo que sonsacarse una «ganga» al comprar un automóvil.

No sólo eso, sino que la actitud hasta persiste después del doctorado. En la mayoría de las comunidades científicas que yo conozco, la mayor parte de los miembros se interesan sola o casi exclusivamente en «desempeñar la ciencia», es decir, en la investigación científica y tal vez en la enseñanza. Esto, hasta cierto punto, tiene razón de ser. La investigación es el objetivo más importante de las actividades científicas, pero también es verdad que no podemos permitirnos la exclusión total de otra exigencia de ser hombre de ciencia, me refiero a la de ser estadista científico.

Hay dos aspectos de tal calidad de estadista: El primero pertenece a la estructura interna de la ciencia, la infraestructura, la misma comunidad científica, la «política dentro de la ciencia». El segundo aspecto atañe a las relaciones de la ciencia con otras actividades e intereses humanos, los efectos que la ciencia produce en el mundo y los efectos que el mundo produce en la ciencia. Le podemos dar el nombre de «política con ciencia».

La necesidad de la calidad de estadista

¿Cuáles son las razones que requieren la consideración de las funciones de la calidad de estadista? Cuando la ciencia era una actividad menor, esotérica y despaciosa, poco evidente en todos los países, es decir, por ejemplo, en la década de 1930 a 1940, a muchas de las actividades científicas se les podía distinguir de las del resto del mundo e individualmente de tal forma que la estructura interna de la comunidad científica no requería gran atención. Es claro que aun en ese tiempo algo de dinero se necesitaba para la ciencia, que había que establecer *cierta* comunicación entre los investigadores, que los filósofos a veces trataban de ver lo que estaban haciendo los hombres de ciencia, etc. Pero, en general, la preocupación por esos asuntos se mantenía relativamente sin complicaciones y era, en grado sumo, discrecional.

Todo eso ha cambiado hoy día. En parte debido a la naturaleza de la misma indagación científica, y en parte debido a que el hombre depende de la ciencia crecientemente para su adelanto material y cambios de actitud, la ciencia ha llegado a ser considerable, conspicua, y sumamente entrelazada con el resto de nuestras solicitudes y actividades. Esto impone su efecto tanto en los aspectos internos como en los externos de la estructura de la ciencia. Todos ustedes han oído decir que un 90 % de los investigadores científicos que han vivido están vivos hoy día. El tamaño de la comunidad científica mundial ha aumentado enormemente en comparación con la de hace cincuenta años. Por lo tanto, la sociología de la comunidad científica, que anteriormente constaba más de la psicología de unos pocos individuos, es ahora un tema complejo e importante, tanto en la teoría como en la práctica. Grandes proyectos científicos que exploran fenómenos naturales muy alejados de nuestras experiencias cotidianas y que por consiguiente requieren instrumentos complejos, la participación de equipos de diversos especialistas, y considerable preparación logística, ocupan gran parte de la ciencia hoy día.

Al mismo tiempo los aspectos externos de la ciencia también han cambiado. Por un lado la ciencia, más que nunca, tiene que contar con el apoyo de la sociedad para procurar sus recursos materiales. Por otra parte, los frutos directos e indirectos de la ciencia se destacan definitivamente en la conciencia del público, ya se les entienda o no, ya se les alabe o condene. Además, el «punto de vista científico», bien o mal interpretado, ha llegado a ejercer una influencia poderosa en nuestra forma de ver el mundo y nuestra posición en él, ya seamos hombres de ciencia o no.

Es, por lo tanto, necesario ahora que por lo menos *algunos* de nosotros, los investigadores científicos, dediquemos una parte *considerable* de nuestro esfuerzo hacia estos aspectos del contexto de la ciencia. No sólo eso, sino que también es necesario que *todos* nosotros consagremos una *pequeña* (con tal que no sea cero) parte de nuestra atención y nuestro esfuerzo en esa dirección.

A este tema, que se discute más y más en nuestros días, se le da a veces el nombre de «la responsabilidad social del hombre de ciencia». Yo no estoy seguro de que este título sea en realidad apropiado para aquello de que quiero hablar hoy. Este título en realidad abarca un campo mucho más amplio y es muy complejo, como ya lo indiqué en un artículo mío publicado hace poco en *Interciencia*. Hoy quiero limitarme a hablar de un tema más específico, sin alejarme de aquellos aspectos de la vida del investigador científico que verdaderamente corresponden a su calidad de estadista científico.

Discusión en dos partes

Voy a dividir la discusión en dos partes: Los problemas dentro de la ciencia, y aquellos que enlazan a la ciencia con actividades externas. Como toda dicotomía, ésta también es imperfecta y en realidad los dos aspectos son inseparables en algunas formas. Sin embargo, en la práctica, es sorprendente que muchos de los problemas en este sentido se pueden discutir dentro de los términos de esta división.

Antes de comenzar quiero explicar algo de la terminología. Una gran parte de nuestra discusión se referirá a todos los países del mundo, ya tengan una infraestructura científica bien desarrollada y establecida por algún tiempo, o ya se encuentren en el punto de partida hacia su desarrollo en la ciencia con sólo unos cuantos investigadores. Sin embargo, habrá diferencias cuantitativas entre los países de las dos categorías mencionadas. Para lograr brevedad voy a llamar estas dos categorías los países avanzados y los países en desarrollo. Tiene que entenderse que mi uso de esta terminología se refiere estrictamente al desarrollo científico del país, y que en realidad el espectro es continuo, así que esta terminología dicótoma tiene un tanto de simplificación exagerada. Venezuela, por ejemplo, se encuentra en un lugar difícil de definir entre los extremos de tener una ciencia crecida y bien desarrollada o de tener una ciencia pequeñita y muy recién nacida.

LA CALIDAD DE ESTADISTA DENTRO DE LA CIENCIA

Educación hacia la calidad de estadista

La participación de la creación interna, de la organización y del gobierno de la infraestructura científica en los países avanzados, es en general evidente entre los hombres de ciencia de «mayor» edad, es decir, aquellos que llevan más de diez años de practicar la ciencia. Esto es así porque en esos países una gran parte de la infraestructura institucional y de otras, ya está establecida, y por eso los investigadores científicos pueden dejar que esta infraestructura los absorba sin tener que preocuparse por su gobierno. En los países en desarrollo esto ocurre muy rara vez. Allí, a la mayoría de los investigadores jóvenes, el día siguiente al de recibir sus grados más avanzados, se les empuja hacia el medio de actividades febriles que intentan llevar a cabo la investigación científica y al mismo tiempo crear las circunstancias que permitan la investigación científica. Como lo dijo Jorge Sabato con tanta percepción, la crisis es el estado normal de las cosas en los países en desarrollo.

No obstante, ni en los países avanzados ni en los países en desarrollo, están los investigadores jóvenes en ninguna forma entrenados, preparados, ni equipados para arreglárselas con esta situación. El caso es que una preparación totalmente teorética y académica para el papel de estadista científico no es suficiente en su totalidad. Pero la introducción al conocimiento, teórico y práctico, que se ha ido acumulando alrededor de la ciencia en su contexto de actividad humana, sería enormemente útil para los investigadores jóvenes y les ayudaría a encararse con el choque que se produce al hacer el cambio del ambiente muy artificial de la facultad en la universidad y del laboratorio de investigación al campo actual en que se practica la ciencia. La nueva disciplina de la ciencia que se está desarrollando muy rápidamente y que está compuesta por un conjunto de filósofos, economistas, psicólogos, hombres de ciencia, sociólogos, gerentes científicos, etc., ya ha llegado a producir muchas cosas útiles e interesantes en ese sentido. También hay una gran cantidad de conocimiento puramente empírico acerca de los aspectos más mundanos de la ciencia, como por ejemplo la administración de un taller de maquinaria, cómo hacer pedidos por catálogo, cómo dirigir un departamento de física, etc.

Puesto que soy de los Estados Unidos y por lo tanto me interesa particularmente la educación de jóvenes investigadores científicos (y en especial la de los estudiantes extranjeros de ciencia en las universidades de los Estados Unidos), he estado tratando de organizar por algún tiempo un seminario que dure 3 o 4 semanas durante el verano, para los estudiantes de los países en desarrollo que hayan ido a estudiar en las escuelas graduadas de los Estados Unidos. Durante este seminario los estudiantes penetrarían a estos aspectos de la ciencia. Hasta ahora los esfuerzos han sido en vano, puesto que no he logrado que las entidades gubernamentales o privadas que pudieran proveer los fondos se interesen en la organización de tal experimento.

Parece más probable que los mismos países en desarrollo se dieran cuenta de la importancia de esa clase de preparación y que la incorporaran a sus propios sistemas de educación científica. Desafortunadamente no es ése el caso. La ciencia de la ciencia como parte de la educación del investigador científico ha sido tan menospreciada en los países en desarrollo como lo ha sido en los Estados Unidos. Yo soy de la opinión que una de las responsabilidades que los hombres de ciencia tienen en su calidad de estadistas científicos es asegurarse de que las generaciones por venir reciban una educación mejor en estos aspectos de la ciencia que la que ellos mismos recibieron.

El mantenimiento de una comunidad científica

Un segundo aspecto de la calidad de estadista científico dentro de la ciencia que quiere discutir, es con respecto a la comunidad científica del país. Ciertamente, en principio, la ciencia es universal e internacional, y, en efecto, hablaré más tarde con mayores detalles acerca de esta dimensión. Pero, en la práctica, por varias razones en gran parte ajenas a la misma ciencia, la unidad de «ciencia nacional» es una de mayor importancia, aunque se encuentre incluida dentro de una mucho más grande «ciencia mundial». Es precisamente en este nivel nacional que la mayoría de las conferencias científicas se podrían llevar a cabo. También es en términos de la comunidad nacional (o tal vez regional) que los problemas científicos que le atañen específicamente al país se pueden discutir. Esto quiere decir que hay que encontrar editores para las gacetas, establecer sistemas de referencia entre ellas, crear cadenas de seminarios en los que se presente a visitantes de otras instituciones del país, discutir asuntos de política de ciencia y publicar sus resoluciones, etc.

En mi calidad de observador, quiero expresar que este aspecto de la calidad de estadista científico parece irse desarrollando bien aquí en Venezuela. El hecho de que estemos reunidos aquí lo confirma. La existencia de instituciones científicas de importancia como el IVIC, también sirve de ejemplo, y otras instituciones que se están organizando también. El hecho de que Caracas sea el foco editorial de la gaceta regional *Interciencia* es otro ejemplo.

Las relaciones con la comunidad científica mundial

Para presentarles el tercer ejemplo de otro aspecto de la calidad de estadista científico dentro de la ciencia, permítanme adentrarme un poco al concepto de la relación de los investigadores científicos de un país con la comunidad científica mundial. Yo no quiero ponerle énfasis aquí a los contactos individuales que un hombre de ciencia establece con sus colegas en el extranjero dentro de su especialidad profesional. Ese aspecto de la situación es generalmente bastante favorable. La mayoría de los investigadores competentes en su especialidad hacen todo lo posible para establecer esa clase de contacto. No hay duda de que el sistema de comunicación interna de la comunidad científica mundial pone al investigador que vive en un país en desarrollo en una situación desventajosa en comparación con su colega en un país avanzado. Para rectificar esto, sin embargo, toda la comunidad científica mundial tiene que cooperar y el investigador individual en un país en desarrollo, tiene que hacer todo lo posible por el momento, dentro de ese sistema desequilibrado, para establecer sus contactos. En general él lo trata de hacer en el mayor grado que le sea posible.

También hay, sin embargo, otro elemento en la relación con la comunidad científica mundial. Es precisamente este elemento que còn tiempo puede llegar a lograr que el sistema de comunicación en la escala mundial sea más equitativo. Lo que se requiere es una participación mayor de parte de los investigadores de los países en desarrollo para hacer sus problemas y las deficiencias de la estructura científica mundial públicos, para proponer programas específicos que alivien esas deficiencias y, en general, para crear una conciencia mundial entre hombres de ciencia que demuestre la necesidad de hacer ajustes en el sistema científico para reducir las desventajas de los investigadores en los países en desarrollo. Es un hecho deplorable que una gran parte de la literatura que trata de los problemas del investigador científico en los países en desarrollo, haya sido producida por gente *fuera* de los países en desarrollo. No hay duda de que la gente de los países en desarrollo ha contribuido algo, y en particular la gente de Latinoamérica, pero si uno se fija en las grandes convenciones internacionales, en las grandes gacetas que los investigadores leen, en las conferencias y discusiones a que los hombres de ciencia asisten, a uno le sorprende la extrema escasez de representantes entre los investigadores de los países en desarrollo. Yo creo que ésta es una de las responsabilidades más urgentes de los investigadores científicos de los países en desarrollo en cuanto a la calidad de estadista científico se refiere.

Estos tres ejemplos de ninguna manera completan los varios aspectos de la calidad de estadista científico dentro de la ciencia, pero ahora necesitamos pasar a ver el enlace de la ciencia con el resto del mundo.

LA CALIDAD DE ESTADISTA EN LOS ASUNTOS QUE ENLAZAN A LA CIENCIA CON SU MEDIO AMBIENTE

Aquí también quiero limitarme a discutir sólo cuatro aspectos, aunque haya muchos otros de interés e importancia. Los cuatro aspectos son: la ciencia y el público, la ciencia y los que hacen las decisiones de un país, la ciencia y la tecnología, y la ciencia y las agencias cooperativas internacionales.

La ciencia y el público

Comenzamos con la ciencia y el público. Uno de los tres motivos principales que animan la práctica de la ciencia se encuentra en el efecto que la ciencia ha ejercido en la manera en que el hombre ve el mundo y su posición y función en él. Otro motivo de importancia es la realización de las aspiraciones humanas, es decir, que la ciencia sea parte de aquello que haga que el individuo, la comunidad, el país y la humanidad en conjunto, se sientan estimulados, curiosos, orgullosos, satisfechos y felices. No tenemos tiempo en esta ocasión para extendernos a discutir estos dos puntos, y en todo caso yo ya lo he hecho en un artículo publicado en *Interciencia*. Para cumplir con el propósito de esta discusión es suficiente decir que para llevar a cabo estos dos motivos, se requiere la comunicación de la ciencia, de sus resultados, de su método, de su punto de vista y de su influencia a las amplias esferas de los que no son hombres de ciencia. No sólo logrará esto llevarles la ciencia a los amigos de ustedes, sino que también le traerá amigos a la ciencia, amigos que le pueden ser necesarios en una época en la que la investigación científica tiene que contar con individuos y organizaciones fuera de la ciencia para obtener sus recursos.

Bajo este respecto me gustaría especialmente subrayar la necesidad de que el hombre lego no sólo vea los resultados de la investigación científica, sino que entienda el «porqué» y el «cómo» de practicar la ciencia. Muchos de ellos que hayan leído libros populares acerca del estudio del cáncer, de «black hole» o de «quark» por varios años y que hayan visto programas de ciencia en la televisión y exhibiciones en los museos por mucho tiempo, bien pueden seguir muy desconcertados en cuanto a las razones y maneras que los hombres de ciencia tienen para hacer lo que hacen. Por consiguiente, ellos no se aprovecharían de aquellos aspectos del método y de la actitud de un investigador científico que pueden ser de provecho en la vida cotidiana de alguien, aunque no sea profesional científico. La actitud hacia la forma de resolver problemas, el tener fe en la existencia de soluciones para cada problema, el uso de pensamiento cuantitativo hasta cierto punto, la confianza en demostrar la teoría a base de experimentos, y la forma de guiar el experimento a base de conceptos teóricos, todos son elementos que pueden ser de provecho para todo el mundo en su vida diaria. En otro artículo[3] yo presenté la idea de que el concepto del desarrollo, en su calidad de actitud, coincide en parte con las cualidades que son importantes para la práctica de la ciencia. Por lo tanto, yo quiero encarecerles a los investigadores científicos que consideren la revelación de la metodología y de la filosofía científica al público como parte de sus responsabilidades en calidad de estadistas científicos. No hace mucho que yo hice un esfuerzo para contribuir algo en esta dirección[4], pero en total hay muy pocos esfuerzos de esa clase, especialmente en los países en desarrollo.

La ciencia y los que hacen las decisiones

El segundo aspecto de la calidad de estadista hacia otras actividades es el de la influencia de la ciencia y de los que hacen las decisiones del país. Los oficiales del gobierno, los ciudadanos de distinción e influencia en la industria, los dirigentes intelectuales del país, etc. Al hablar de esto, yo sé muy bien que, especialmente en Latinoamérica, estoy enfocando un asunto muy controversial y explosivo, es decir, las relaciones sobre la ciencia y la política.

Es lamentable que en la mayoría de las lenguas, con excepción del inglés, las palabras «policy» y «politics» son idénticas. La semántica tiene mucha influencia en nuestro modo de pensar y no es nada sorprendente que la gente encuentre difícil hacer la distinción entre los dos conceptos, ya que la misma palabra se usa para describirlos a ambos, «política». De todos modos, yo quisiera que en la traducción de esta discusión al español la distinción se pudiera preservar de alguna forma.

La política tiene significados múltiples. En el sentido más amplio y menos útil se refiere a la forma en que cualquier cosa se logra entre cualquier grupo de personas. El uso que yo quiero hacer de la palabra, sin embargo, es el que se refiere a la esfera en la que las varias facciones que mantienen ideas y enfoques opuestos en un país o en todo el mundo ponen en práctica sus encuentros, sus negociaciones y sus batallas. Este es, de hecho, el significado cotidiano de la política.

Usando esa terminología, yo creo que el hombre de ciencia, en su calidad de estadista, tiene la responsabilidad de establecer un contacto recíproco y vigoroso con los que hacen las decisiones del país, en tal forma que se le dé a la ciencia mayor importancia que a la política y que se mantenga separada de ella. Hay muchos otros aspectos de la sociedad que se mantienen alejados de la política, así que mi sugerencia no es sin precedente. La discusión de si Caracas debe tener un depósito de agua pura o no, no es un asunto acerca del cual los varios partidos políticos del país disentirían mucho.

La ciencia, en realidad, se encuentra en una buena posición para que se le maneje fuera de la política. No importa qué imagen tenga cualquier persona de lo que debe ser Venezuela dentro de veinte años. Sea lo que sea, el desarrollo científico será una parte absolutamente necesaria de ella. De esta manera virtualmente todos los elementos que se encuentran en la política de la ciencia y que incluyan una acción recíproca de la ciencia con los que hacen las decisiones del país van a tener esa cualidad objetiva, imparcial y duradera. Claro que siempre habrá excepciones, como por ejemplo en ciertas direcciones bien marcadas de la ciencia aplicada o en asuntos tecnológicos a los que la investigación científica les dé su apoyo. Esas excepciones, sin embargo, no impiden que la dirección de la política de la ciencia se mantenga en gran parte separada e independiente de la política nacional.

En Latinoamérica no es necesario hacer notar el daño que se les puede hacer tanto a la ciencia como al país cuando la política y la ciencia se entremezclan. Venezuela, especialmente, ha sido el refugio de muchos investigadores científicos latinoamericanos que vinieron de los países en los que los hombres de ciencia siempre tomaban parte de la política y en los que los gobiernos constantemente consideraban a la ciencia en calidad de tema político, que requería purga y reglamentación para mantenerla bajo su control.

Al mismo tiempo también hay, afortunadamente, casos en los que se mantuvo un *statu quo* en el que la ciencia no se entremezcló con la política ni el gobierno con la ciencia, y en los que, por lo tanto, el desarrollo científico pudo seguir su curso de una manera constante aún durante los trastornos políticos. También quiero añadir que en este sentido Venezuela presenta un ejemplo luminoso para los demás países latinoamericanos, puesto que en este país la calidad de estadista científico de gente como Marcel Roche y Raimundo Villegas, que tratan de desarrollar la ciencia en tal forma que se mantenga más allá de la política, puede declararse un gran éxito. Es importante que muchos investigadores científicos de la comunidad hagan esfuerzos para crear una acción recíproca entre ellos y los que hacen las decisiones, en varios niveles de generalidad y con individuos de varias categorías.

La ciencia y la tecnología

El aspecto de la calidad de estadista científico fuera de la ciencia que quiero discutir brevemente es el que trata de la ciencia y la tecnología. Además de los dos motivos que conducen a la práctica de la ciencia a los que me referí antes, hay un tercero que es bien conocido: La ciencia en el mundo moderno es la base de casi todo el progreso material. Para utilizar completamente este aspecto de la ciencia, la larga y complicada cadena que une la investigación científica básica, la investigación científica aplicada, el desarrollo tecnológico y la producción (una cadena que se mueve en ambas direcciones) tiene que estar en su sitio y tiene que funcionar sin estorbos. La mayor responsabilidad del investigador científico en este sentido es la de crear el vínculo entre los resultados de la investigación y los peritos en la tecnología. En ese caso también, el movimiento es en ambas direcciones: Lo que es de interés para la tecnología tiene que producir su efecto en las actividades de los hombres de ciencia, y hay que comunicarles los frutos de estas actividades a los peritos en la tecnología eficazmente.

En este sentido es de gran importancia recalcar un punto. En el caso de Venezuela, por ejemplo, este país produce alrededor del 0,1% de la nueva ciencia que se crea en todo el mundo. Al mismo tiempo, las necesidades tecnológicas de Venezuela quieren utilizar el total de la base de conocimientos científicos. Es evidente, por lo tanto, que la investigación personal de un solo investigador venezolano, o aun de todo el grupo de investigadores venezolanos, va a resultar en tan sólo una pequeña fracción de la ciencia que la tecnología venezolana requiere. De esto se infiere que la contribución mayor de un investigador científico venezolano desde el punto de vista de la tecnología del país, no es directamente basada en su investigación científica personal, sino más bien que le sirva como un embudo por el cual la información científica de todo el mundo llegue a Venezuela y sea digerida allí y convertida a una forma que la tecnología local pueda utilizar.

Esto indica, por lo tanto, una responsabilidad muy importante del hombre de ciencia de un país en desarrollo: El tiene que tener un amplio campo de intereses dentro de su especialidad científica y tiene que mantenerse al día en los desarrollos de importancia en esta especialidad. Un requisito para esto es que él mismo desempeñe su propia investigación en la especialidad, puesto que indudablemente la mejor forma de mantenerse al corriente en un campo de la ciencia es la participación en la investigación dentro de ese campo. Tal actividad investigativa, de su propia cuenta, no es, sin embargo, suficiente para servir en calidad de embudo para el conocimiento: Además de mantener un alto nivel de excelencia en su investigación personal, el investigador científico también tiene que hacer un esfuerzo consciente para mantener su amplitud y para digerir los nuevos desarrollos dentro de todo el campo del que su investigación personal representa sólo un pequeño segmento.

La ciencia y las agencias internacionales

Por fin quiero enfocarme hacia el cuarto aspecto de la calidad de estadista científico fuera de la ciencia, el de las relaciones de la ciencia al nivel local con las agencias cooperativas internacionales de la ciencia. Para usar algunos ejemplos, voy a concentrar la atención en las agencias de las Naciones Unidas como la UNESCO, FAO, WHO y LAEA.

Hay muchas razones por las que estas agencias están muy aisladas de las comunidades científicas trabajadoras de los países miembros. El personal de las agencias de la ONU incluye una fracción de esos investigadores activos, la cual es mucho menor de lo ideal. Lo que se necesitaría más que todo sería el asignarle estos investigadores activos a las agencias por períodos de tres años, sin posibilidades de renominación y con la condición de que el cargo de esos investigadores se les reserve en su país para su vuelta. También las agencias de la ONU se han limitado en sus comunicaciones casi completamente a un enlace con los oficiales del gobierno de los países miembros, y ellos también están frecuentemente muy fuera de contacto con sus propias comunidades científicas. Para continuar la lista, los programas y actividades de estas agencias en los países miembros, muy rara vez, si es que se hace del todo, se someten a una evaluación significativa que incluya el avalúo y las sugerencias de los investigadores locales que son los que están más al tanto de los efectos que estos programas producen. El resultado es que es muy imaginable que cientos de millones de dólares se desperdicien cada año en tales programas de la ONU, recursos que se podrían utilizar muy efectivamente hacia la creación verdaderamente eficaz de comunidades científicas en los países en desarrollo.

¿Qué hay que hacer entonces? Hay dos formas en las que el estadista científico puede contribuir. Primero, los investigadores locales deben insistirle a los gobiernos locales que se aseguren de que los investigadores activos obtengan una mayor parte de las posiciones administrativas y organizadoras (también las de «expertos») en las agencias internacionales, y que se ponga énfasis en que los períodos de servicio sean cortos, de tal forma que el investigador no sienta que ha abandonado a la ciencia para siempre para convertirse en «burócrata».

Segundo, la comunidad científica local debe intentar, de su propia cuenta y dentro de la estructura de las asociaciones profesionales y otros grupos, la evaluación de los numerosos proyectos que las agencias internacionales han patrocinado u operado, y que hasta ahora no han sido evaluados. Esta faena requeriría primero que todo la creación de una metodología evaluativa, por sí mismo un objetivo cautivador y muy educativo. Una vez que exista un acuerdo en cuanto al método, varios grupos de investigadores se podrían concentrar en la evaluación de proyectos dentro de su área de competencia y experiencia. Los resultados de las evaluaciones se les remitirían a las agencias gubernamentales locales tanto como a las agencias apropiadas de la ONU. La responsabilidad de organizar esa clase de estudios evaluativos es también algo que yo quiero recomendarles con mucho ahínco a los investigadores científicos que quieran desempeñar el cargo de estadistas.

Epílogo

Quiero concluir con una reflexión. Aunque se tratara solamente de los pocos aspectos de la calidad de estadista científico que se han discutido aquí como ejemplos, si se tomaran todos juntos, serían demasiados para que un solo investigador los acometiera. Es claro que cada investigador tal vez quiera seleccionar una o dos de estas áreas de actividad para participar en ellas personalmente. Una actividad, llevada a término de una forma competente y a fondo, es más útil que la participación superficial en una docena de causas. De todos modos, es de importancia que la estructura recompensativa para investigadores científicos en las universidades, en los laboratorios de investigación y en las otras instituciones, reconozcan que la calidad de estadista constituye una contribución estimable y que, por lo tanto, ofrezcan su agradecimiento por tales actividades. Si esa clase de participación sigue siendo puramente «suplementaria», su alcance e intensidad seguirán siendo firmemente sofocados. Hay muchas formas de contribuir a la ciencia y a pesar de que la investigación siga siendo, y con razón, la prioridad de mayor importancia, hay otros aspectos de las funciones del investigador científico que se deben incluir en los criterios que se utilizan para apreciar su contribución total.

Bibliografía

1. M. J. Moravcsik, *The Scientist in a Developing Country and Social Responsability. Interciencia 4*, 326 (1979).
2. M. J. Moravcsik, *The Ccontext of Creative Science. Interciencia 1*, 71 (1976).
3. M. J. Moravcsik, *Scientists and Development. Journal of the Science Society of Thailand 1*, 89 (1975).
4. M. J. Moravcsik, *How to Grow Science*, Universe Books, New York, 1980.

III. ACTION

A. Directions

Preprint, published in *Science Policy* 2:6 (1973) 179.

NEW DIRECTIONS IN SCIENTIFIC AND TECHNOLOGICAL ASSISTANCE

Michael J. Moravcsik
Institute of Theoretical Science
and
Department of Physics
University of Oregon
Eugene, Oregon 97403 USA

I.

The development of scientific and technical assistance programs in the United States has reached an exciting and challenging juncture. Following the conclusions and recommendations of the so-called Peterson report[1], President Nixon sent a message[2] to Congress on September 15, 1970, in which he announced a radical restructuring of the foreign assistance program. For the present discussion, the most pertinent part of this message is the separation of the technological and scientific functions of the foreign assistance program from the economic, commercial, and investmental functions. The latter, which have been and will remain financially the bulk of the program, will be handled by the U.S. International Development Bank and the Overseas Private Investment Corporation, while the scientific and technological assistance will come under a new U.S. International Development Institute(IDI).

The operation of the IDI was not spelled out in detail in the presidential message. The only specification was a general one, as follows:

"[We will establish] a U.S. International Development Institute to bring the genius of U.S. science and technology to bear on the problems of development, to help build research and training competence in the lower income countries themselves, and to offer cooperation in international efforts dealing with such problems as population and employment."

During the coming months, while the new IDI is in the formative stage, there is therefore a particularly good opportunity to fill the above general outline with detailed content.[3] It is the intention of the present article to make a contribution in this direction, and to outline some of the main directions in which the activities of IDI must aim. It is, however, not my purpose to propose a complete and fixed program. On the contrary, one of the main emphases will be placed on the requirements of adaptability and flexibility for IDI, so that as new ideas arise, they can be experimented with in the framework of the organization.

II.

It might be good to start our discussion with the basic dichotomy of the aim of technical and scientific assistance. On the one hand, we want to assist the less developed countries (LDC) in specific scientific and technological research and innovation directed at present problems. This is the immediate, short term objective of assistance, akin to the administration of a painkiller and the treatment of symptoms in the case of an ill person. The other, long range objective of assistance is to help the LDC to create its own, indigenous capability to deal with future problems as they arise. This can be likened to the treatment of the causes of an illness and to the general strengthening of the patient's body to resist future infections. It is needless to say that these two requirements must go hand in hand, so that the patient neither dies of the symptoms before the causes can be cured, nor does he become a chronic returnee to the hospital on account of his inability to deal with any new forms of illness.

The basic difference between these two aims is in their time scale. Working on specific projects has a natural time scale of 5-10 years. It can be reasonably expected that after 2-3 years of initial inertia the project shows marked results, visible even to those not trained in science or technology, and every year a quantitative, definite, and tangible "progress report" can be written. Such projects also have a more or less definite end, when the stated objectives have been attained.

In sharp contrast, the assistance to achieve general scientific and technological capability has a much longer time span, perhaps 3-5 decades. Furthermore, its time table is much more difficult to predict, and the progress made along the way is also much more subtle. For this reason, people trained in science and technology are needed to evaluate the success or failure of these projects.

It is therefore not surprising that scientific and technological assistance efforts in the past have been heavily slanted in favor of the short term components. AID's own in-house capability as well as the type of consultants it chose to have easy access to could deal with the evaluation of the specific, short term projects, but these capabilities were mismatched with the requirements to manage and evaluate long term projects. As a result, some short term projects have been built on the quicksand of indigenous personnel deficient in competence in the basic sciences on which those projects were based, and the training received by the indigenous personnel has sometimes proven irrelevant and out of date as soon as the particular project was completed, since the training did not contain an education in the basic sciences which provides the versatility to apply one's knowledge to new problems.

III.

In order to achieve a balanced mixture of short and long range objectives, scientific and technological assistance must forge ahead along three fronts.

First, assistance must be provided to carry out work to attack specific, immediate problems of considerable local importance. Desalinization in Pakistan, or malaria eradication in many tropical areas are examples. This phase of the work will often involve the use of known technology, so that the main task becomes the transfer of such technology to the local scene, involving minor adaptations of methods, and the major assignment of training adequate indigenous manpower on a large enough scale and in a short enough time. Purely from this point of view, it might seem plausible to train local manpower along very specialized and narrow lines, with an expressly utilitarian outlook. Past activities along these lines have often been successful.

Second, assistance must be provided to indigenous research and development pertaining to problems specific to the particular LDC or geographical region. Some of these problems might be presently pressing, others only on the horizon. The applied research in science and technology for these problems might not exist in and might not be forthcoming from the more developed countries. Thus these research activities should be viewed as a needed support for the work done on specific problems that was mentioned above. Our assistance program has also been fairly compatible with these types of activities in applied research and development.

The third area toward which assistance must be directed, and which has been somewhat neglected in our past programs, is the creation of a capability to carry out research and development through the strengthening of the scientific, technological, and educational infrastructure of the LDC's. This phase is a prerequisite to a successful applied research program mentioned above, which in turn supports the attack on specific problems. Thus a creative and viable indigenous scientific and technological understructure is the basis of success in all phases of an assistance program.

This third area has been neglected in the past not only because of the less-than-adequate weight attached to it, but also because the nature of the process of strengthening of the infrastructure is not well understood by anybody, and in particular by those primarily responsible for decisions in technical and scientific assistance. In a situation like this, in which the road to progress is not clearly outlined, the wise course is to explore several parallel avenues, suggested by various people with some claim to relative competence in the matter. This certainly has not been done in the past, but must be incorporated into the future.

IV.

We will now return to the proposed IDI and outline its activities in view of the general considerations discussed above. These activities can be classified into three groups: research on development, strengthening of the infrastructure, and assistance with specific projects.

In the first of these, research on development, an enormous amount remains to be done. In fact, our knowledge of how to formulate a systematic scientific and technological development program in any country, advanced or less developed, is extremely rudimentary. This hardly needs documentation nowadays when the role of science in the United States itself is undergoing the convulsions that sooner or later accompany any issue about which our ignorance is profound. The IDI therefore must serve to bring together past and present efforts to study "the science of science", particularly as it pertains to development, and to foster increased efforts in this direction. One can give a long list of immediate questions that deserve study. Is it possible to define an optimal fraction of the gross national product (GNP) of an LDC that should be spent on research and development, and if so, what is it? How can one establish the interface between science and technology on the one hand, and industry on the other? How can one train scientific and technological administrators and managers? What are the optimal methods of technology transfer? What is the role of the attitude of the population as a whole toward science and technology? What are the best evaluative indices in the growth of science and technology? These and many other questions must be answered with the help of detailed case studies, and this effort must be a joint one between qualified personnel in the United States as well as in the LDC's. Past attention to research on development by AID and other agencies has been almost negligible, so this aspect of IDI will be a novel one.

Second, IDI must devote some of its resources to the strengthening of the scientific and technological infrastructure of the LDC's. As I indicated, this has also been underplayed in past efforts, and must receive much more careful attention. It is a difficult task since it involves many components which are interrelated.

On the most elementary level, it involves assistance for the education of the population as a whole about the outlook of science and technology. Contrary to frequent charges, such an education need not result in the eradication of traditional outlooks and values, just as the scientific view lives in "coexistence" with "non-scientific" philosphies and values in the advanced countries also. Nevertheless, it is important to bring about a mass awareness of the fact that man does have some control over his environment, and that change can become the rule rather than the exception in human development.

On the next level, such strengthening of the infrastructure must involve assistance to improve the school systems. The most pressing need in this area is to make sure that science is taught not as a collection of set rules to be memorized, but as a method of solving unsolved problems and answering unanswered questions. An ancillary requirement is to teach science as an experimental discipline, involving laboratory work with a meaningful content.

Next, one must also make sure that institutions where future scientists and technologists receive professional training are up to this task. These institutions are most often universities, although there are examples of countries where the university structure is poorly developed so that, at least temporarily, such education must take place in research institutes.

What are the crucial characteristics of such an educational institution? Above all, it must contain teaching personnel that is up to date on technical matters, alive in terms of actual scientific activity, broad in its view of science, and willing to transmit science and technology as a live discipline.

In order to develop and maintain such an attitude, the university personnel must be given assistance in maintaining its own research activity at the university. As long as such research consumes only a small fraction of the resources available for science and technology, good university personnel should be thus supported regardless of what area the personal research is carried out in. When it comes to the teaching, say, of an exciting and relevant course in mechanics, electricity, or modern physics, on an undergraduate level, it is much preferable to have a first class teacher whose personal research is in an esoteric subject than a second rate person who works on something apparently more down-to-earth. Of course, one might wish for a first rate person with an interest in something down-to-earth, but such matters are almost always beyond the control of the LDC, particularly at the beginning stages of development, and hence a realistic choice must be made. Since quite often first rate people develop an interest in one of the esoteric, pioneer areas of science because they find it more stimulating intellectually, the realistic choice is between taking the person as he is, or losing him to the brain drain. In such instances, the choice should be clear.

It is important to dwell on this point at some length since it remains a controversial one. Some claim that there is no room whatever in a LDC for research on apparently non-applicable areas of science. This view, beside explicitly relegating the LDC's to second class citizenship in the world of science, which is untenable from a political and philosophical point of view and produces low morale, also ignores a whole string of secondary benefits that accrue from the presence of an excellent scientist, regardless of his personal research interests. Since I have discussed this question in detail elsewhere[4], I will not reiterate the arguments. It suffices to say that to expend, say, 0.03% of the GNP (or 1/30 of the research and development funds) of an LDC for fundamental scientific research by first class people, no matter in what area, is a very defensible economic proposition with benefits much in excess of the investment.

Beside scientific and technological education, the LDC's overall scientific institutions should also be assisted in their development. To build up a viable scientific or technological tradition in a country, one must have lively professional societies, means for technical communications of all sorts, a cadre of science managers and policy makers, auxiliary personnel and facilities, and the like. In all this, quality must be constantly emphasized, if for no other reason than because the easier evaluability of quantity (as compared to quality) makes it a more ready target for developers.

Finally, and perhaps most importantly, all these activities must bring about a high indigenous scientific morale, a purposefulness, a faith in what is being done and a belief that it can be done. The presence or absence of this somewhat intangible ingredient will make the difference between success and failure, even if all other factors are favorable[5]. The realization of this crucial element and the resulting sensitivity in any type of scientific or technical assistance must be a hallmark of IDI to an extent that exceeds that of its predecessors.

The third area in which IDI activities must be concentrated is assistance for specific applied projects. This type of assistance involves, among others, applied research with immediate objectives, the improvement in the interface of completed research and its field application, progress in quick and effective technology transfer and import, and advancement in project definitions, administration, and evaluation. Whereas there is room for innovation in these areas also, they are somewhat closer to the more traditional ones carried out by AID and other agencies.

Some general comments will conclude this part of the discussion.

It is important to emphasize that in all these activities, assistance must work with indigenous plans for development. There must be a fine balance between injecting new ideas and proposals on the one hand, and avoiding the semblance of interference or imposition on the other. Some means of accomplishing this objective will be discussed later.

A second general comment should be directed at the way in which existing resources should be distributed among the three objectives of research on development, strengthening of the infrastructure, and specific projects. A hard and fast general rule is perhaps not appropriate, but one should warn against the frequent tendency to neglect longer term objectives compared to short ones, because the former appear less urgent. The specific applied projects usually take the lion share of the funds in any case, and perhaps justly so, since they often involve operations on a larger scale than the other two objectives. It is therefore important to guard the integrity of the budget for these other two objectives, perhaps only 20-30% of the total, so that they are not eliminated to bring about a rather marginal improvement in the specific, applied projects.

It is also important for scientific and technical assistance to foster regional cooperation in a given geographical area, whenever possible. It is often easier to bring about such collaboration by an outside "neutral" party, and the resulting increase in regional activity will assure an earlier attainment of scientific and technological self-sufficiency.

V.

After outlining the activities IDI should undertake, one should turn to delineating the structure of the new organization which is best suited to carry out these activities.

There are two separate but related points in which the new IDI should be a departure from existing practices: a) It should be multipronged, flexible, and experimentally minded to accommodate a variety of new approaches, including some on a small scale; and b) It should draw collaboration from a wide segment of the American scientific-technological community, including governmental development personnel, universities, national laboratories, industry, scientific and professional societies, and very importantly, individual scientists and other technical people.

I would like to elaborate in particular on the necessity of incorporating into the structure some possibilities for individual scientists to contribute in their own ways. Their speciality might well turn out to provide the cutting edge of experimentation by devising and trying out new ideas on a small scale. But the value of individual scientists goes far beyond this. The indigenous scientific and technological community in an LDC is, like its counterpart in an advanced country, a priori distrustful of and condescending toward "beaurocrats". Whether such a sentiment is justified or not is irrelevant: the existence of this attitude means that the best way to "sell" a project is to work through scientists.

Furthermore, an individual scientist, through his personal acquaintances in the LDC's, can often achieve objectives easier than large projects can which, by necessity, have to work through governmental channels. To give an example, it might take months to procure a small and inexpensive spare part for an expensive piece of equipment if the official channels of large aid projects are used. But if the operator of the equipment in the LDC happens to have a friend in a more advanced country who had received a "petty parts" grant of $100 a year from IDI to cover such eventualities, the part might reach its destination in a matter of a week or two.

The utility of a scientist in such contacts ceases when he ceases to be an active scientist and turns into a full time administrator of a project. Hence provisions must be made so that the scientist can undertake such projects at his usual place of residence, whether a university or laboratory, with only temporary interruptions in his scientific career and temporary absences from his institution.

Such an outlet for activities pertaining to science development in the Third World would interest a sizeable number of people in the American scientific and technological community, who at the present find it difficult to convert their interest and know-how into activities because of the absence of small amounts of external "seed money". Thus the support by IDI of such activities would represent a large de facto increase in its personnel without much additional expenditure.

Even in a more general context, IDI might find it advantageous to contract out a considerable amount of its work to other organizations. Some American governmental entities, such as the National Science Foundation, the Atomic Energy Commission, the National Academy of Sciences, as well as some private foundations and professional societies have had varying degrees of practical experience in international activities in science and technology and it would be a waste to leave their expertise unexploited. Thus IDI could act as a coordinating center and wise catalyst for a large variety of programs without developing a huge in-house administrative machinery.

As to the matter of flexibility and spirit of experimentation, this is essential because, as suggested earlier, nobody at the present time has the final answer to how the scientific and technological resources of a country can be developed. Thus an undogmatic point of view is indispensible, especially in a situation when much of the resources for scientific and technological development are under the control of one single organization.

VI.

I will now summarize the above discussion by listing the major ways in which the new IDI should represent a departure from previous practices.

a) IDI should undertake research aimed at the understanding of the process of scientific and technological development and at the exploration of better methods to foster such development.

b) IDI should devote a relatively not very large but in absolute terms quite significant fraction of its resources to the strengthening of the infrastructure of science and technology in the LDC's. In doing so, IDI should use a comprehensive interpretation of what constitutes a strengthening of the infrastructure. In particular, IDI should be ready, on a small scale, to support the research activity of outstanding individuals in the LDC's, even if such activities are not applicable to present, specific development projects.

c) IDI should exhibit great versatility and flexibility in the type of projects it pursues. In particular, it should be ready to explore several alternative and sometimes quite different approaches to the same problem, and should be able to accommodate small scale projects in addition to the more traditional large projects.

d) IDI should structure itself so as to accommodate a variety of extramural programs, contracted through other governmental agencies, private foundations, national laboratories, industry, universities, professional societies, and individual scientists and technical people. Particular effort should be made to make the interest and competence of individual members of the scientific and technological community compatible with IDI opportunities.

Finally, we should remind ourselves of the justification of IDI activities in general. As it is the case for many worthwhile and successful projects, this justification can be made along a number of lines, among which at least some (though perhaps different ones in each case) will be found compelling by people of greatly varying views and philosophies.

To the humanist and altruist, the sharing of knowledge and competence should appeal. The scientific and technological revolution is perhaps the most significant development in the last two centuries, and spreading its benefits from a small group of advanced countries to mankind as a whole is a noble task.

To the international politician, the spread of science and technology will represent a reduction of economic and ideological differences between nations and a concomittant reduction of tensions.

To the world federalist, the scientific and technological ties among countries should represent a prelude to a more general unification of the many now divergent groups of people.

To the strategist, the scientific and technological strengthening of the LDC's represent an important opportunity for these countries to stand on their own against hostile external forces. This in turn lessens the burden of the United States in her responsibility for world stability.

To the anti-imperialist, the growing scientific, technological, and economic strength of the emerging countries should herald the decline of the domination of a few superpowers over the rest of the world.

To the penny-conscious or the isolationist, scientific and technological assistance promises the eventual discontinuation of the billions of foreign aid that is now distributed by the United States all over the world.

To the economist and businessman, the enrichment of the LDC's through science and technology represents vast new markets for international trade as well as new sources of valuable goods.

To the scientist and technologist, the emergence of the LDC's holds the promise of a vast acceleration of scientific and technological progress, so that more excitement and satisfaction in terms of new discoveries and inventions can be packed into one human lifetime.

To other scientists and technologists, international scientific and technological cooperation should offer an opportunity to convince the emerging countries to avoid the abuses of science and technology that the now advanced countries have perpetrated.

To end on a somewhat quizzical note, let me add one more reason for favoring scientific and technical assistance. As mentioned earlier, high morale, purposefulness, a belief in one's activities are crucial and indispensible ingredients in a scientific and technological development process. Recently there have been signs of a lowering of morale among some scientists in some advanced countries. Should this become wide spread, it is possible[6] that the scientific and technological development of the now advanced countries will come to a halt and then the inevitable decline will set in. It might then be reassuring to some that before such a decline hit us an orderly effort was made to pass on the torch of progress to others destined to become the leading countries of the next century.

References

1. U.S. Foreign Assistance in The 1970s: A New Approach (Report To The President From The Task Force On International Development, March 4, 1970. U.S. Government Printing Office, 1970-0-402-378).

2. Foreign Assistance for the Seventies, President Nixon's Message to the Congress, Obtainable from the Agency for International Development, Washington, D.C. 20523.

3. The Peterson report itself devotes two of its 38 pages to a very cursary discussion of the IDI, mainly in terms of some general statements and financial considerations. Some organizational matters are also mentioned. This very brief discussion agrees, as far as it goes, with everything said in the present article. In particular, the Peterson report also suggests the importance of supporting indigenous institutions, the importance of working through private channels rather than permanent agency personnel, and the importance of flexibility in organizational structure. Incidentally, the relative neglect afforded to technical and scientific assistance in the past, compared with economic matters, is demonstrated not only by the small space given to the former in the Peterson report but also by the composition of the Peterson task force, which contained among its sixteen members not a single representative of the active scientific or technological community.

4. See M. J. Moravcsik, Technical Assistance and Fundamental Scientific Research in Underdeveloped Countries, Minerva, 2, 197 (1964) (Reprinted in Edward Shils (Ed.): Criteria for Scientific Development: Public Policy and National Goals, MIT Press, 1968, and in A. B. Shah (Ed.): Education, Scientific Policy, and Developing Societies, Manaktalas, 1967). See also M. J. Moravcsik, Physics Today 17, 1:21 (1964) and International Atomic Energy Agency Bulletin 6, 2:8 (1964). A more comprehensive discussion will be published in the Proceedings of the Research and Development Seminar, organized by the Scientific and Technical Research Council of Turkey in cooperation with the Technical Assistance Program of the Organizations for Economic Cooperation and Development, held in Istanbul in May 1970.

5. M. J. Moravcsik: A Chance To Close The Gap? (to be published, 1971).

6. I elaborated on this theme at greater length in Reference 5.

Preprint, published in *Science and Public Policy*
8 (1981) 40.

SCIENTIFIC AND TECHNOLOGICAL ASSISTANCE AND THE CHALLENGE OF THE 80'S

Michael J. Moravcsik
Institute of Theoretical Science
University of Oregon
Eugene, Oregon 97403

Abstract

In view of the continuing efforts to establish a special organization in the United States for scientific and technological cooperation with and assistance to the developing countries, two new trends in development circles are stressed: The emphasis on the building of indigenous infrastructures, and the broadening of the concept of development to include the psychological and aspirational aspects which underline the varied role of science and technology far beyond short term material aspects. It is suggested therefore, that a new organization aimed at such infrastructure building is politically less controversial, will economize in financial outlay, will utilize the resources of the United States optimally, and will de-emphasize bureaucracies. To achieve this, it is urged that the new organization should think "vertically," be non-country-specific, and should operate catalytically on a person-to-person level. Such a mode of operation is likely to harmonize also with the style of the new administration.

Michael J. Moravcsik is a theoretical physicist with two decades of varied involvement in the building of science in developing countries. Beside originating and being involved in a number of "action" programs in science development, he is the author of over 200 articles in theoretical physics, science policy, and the science of science, including two books: "Science Development" (PASITAM, 1975) and "How to Grow Science" (Universe Books, 1980).

The scientific and technological assistance of the United States to developing countries will undergo significant changes in the next decade. Some of the apparent causes for such changes reside in specific recent events, while others are due to slower but steadier long term shifts in attitudes and perceptions regarding scientific and technological development.

One specific event that might be thought to herald such a change is the United Nations Conference on Science and Technology for Development (UNCSTD) which occupied the last week of the past August in Vienna. This is, however, more an appearance than a real cause: The conference ended in a mere gesture, consisting of a relatively small increase of funds poured into an already existing international channel with a doubtful past record of efficiency and effectiveness.

The other specific event is the recent attempt in the United States to establish a new semi-federal organization, the Institute for Scientific and Technological Cooperation (ISTC), aimed specifically at scientific and technological cooperation with developing countries. Although the first such attempt failed in Congress, support for sustaining such efforts remains strong. One of my main aims in this discussion is to outline how such an organization (which I will call ISTC) could be coordinated with the new trend in the development of science and technology in the Third World.

The direct causes of this new trend will therefore not come so much from sharp, conspicuous institutional, organizational, or policy initiatives, but instead from the long term attitudinal changes, from the slow but definite shifts in perception concerning the nature of the problem we face when dealing with the evolution of science and technology in the developing countries. These shifts are becoming increasingly evident in the underbrush of implied statements within political and policy pronouncements in the developing countries, as well as in discussions and analyses of development problems by professionals. There were definite examples of the "new look" even among the sterile maze of documents, speeches, and pronouncements surrounding UNCSTD. There are, in particular, two aspects of development which have been beneficiaries of such a reassessment.

The first and much more evident one is the realization that without a strong, long term, and concentrated effort of creating an overall indigenous scientific and technological infrastructure in the developing countries, no significant and lasting improvement can take place in those countries. In particular, it is becoming increasingly evident that the massive projects of the past, consisting of "international welfare" (that is, the short-term oriented injection of resources and foreign manpower, supplemented by local personnel with a "quickie" training for a specific purpose), are little more than band-aids in the absence of a local infrastructure that can utilize such shots-in-the-arm to achieve the capability of preventing similar future crises.

In particular, the prevailing philosophy in the past has been to tie any support of local scientific and technological efforts to specific short term projects, and thus, at best, create a scientific and technological community with competence, orientation, and capabilities trimmed to the fashionable crisis of the day. It was said that such a community must not be supported per se, but must be justified only in terms of specific material problems already in evidence in the country in question.

The results of such an approach are now beginning to be in evidence. Most developing countries continue to stumble from crisis to crisis, still lacking an indigenous body of scientists and technologists who could act in such crises, who could play a significant part in policy formulation, and who could play a leading part in the appropriate education of future generations. Some of the most promising individuals in these communities leave their countries, disillusioned by the lack of exciting and gratifying opportunities to exercise their talents, and the countries remain under the perhaps unintended but quite complete scientific and technological domination of the few leading powers of the world.

This has now become sufficiently evident so that there is a quite widespread and still growing emphasis, at least in words, on the creation of broad indigenous scientific and technological communities and infrastructures in the Third World. At UNCSTD, for example, "infrastructure building" was not only an "in" word (after having been in exile for decades within the international "development community"), but in fact it was one of the central themes in various speeches at that conference.

The second change in the attitude toward and perception of the task of scientific and technological development is at the present perhaps more subtle and less far progressed, and yet potentially it has an even larger impact on practical efforts than the emphasis on infrastructure building. This second trend pertains to the concept of development itself.

Although it was not stated in quite as explicit terms, in the past the prevailing conception about development in the Third World held that the main (in fact exclusive) task of such development was to bring about short term material (economic) changes. It was thought that the primary burden of being "underdeveloped" was in having a GNP/capita 20 or 40 times smaller than that of the leading countries, and therefore, it was said, the developing countries should devote their entire efforts toward closing this gap. Only when this task was accomplished would such countried earn the "right" to be concerned with aspects of development other than economic or material.

In contrast, it is now increasingly realized that development is a much more complex process, and that "being underdeveloped" places many burdens on the carrier of this label which are much more bothersome, frustrating, humiliating, and crying for removal than mere material poverty. In conjunction with this, the view is also gaining weight that if the developing countries are to have an indigenous infrastructure in science and technology which is to remove the country from the international welfare list, the impact of science and technology must be felt in the country in ways other than the fixing of purely material problems.

The point here is that the main handicap of being "underdeveloped" is the lack of the ability of the country or of the individual to make his own decisions and to govern its own fate. In all areas (economic, cultural, political, social, spiritual), developing countries find themselves unintended slaves of innovations, influences, forces, and trends originating abroad, and in the absence of indigenous know-how, of indigenous leadership potential in any of the areas of international concern, there is nothing these countries can do to avoid such a frustrating state. Much of the often mercurial, and apparently "illogical" political behavior of many developing countries can be interpreted as the venting of such frustrations.

It is not too surprizing that such non-material considerations can play such an important role. History has been full of pieces of evidence that non-material aspirations and factors have a dominant influence in human affairs. For example, the analysis of the various wars and near-wars that are taking place in the world today also indicates the predominance of nationalistic, ideological, religious and other non-material forces. Indeed, if only material factors governed human affairs, the world would be a much easier arena in which to adjudicate disputes.

In particular, science and technology, beside having a very pivotal effect on the material well being, influence us in two other broad ways. First, excelling in them constitutes an important non-material aspiration in the 20th century. Second, their aims, outlook, and methodology have a significant impact on Man's view of the world and of his place in the world.

These cultural, attitudinal, and teleological influences of science and technology have been active in the "Western" civilization for several centuries and hence they are tacitly acknowledged and in fact not even noticed any longer. In the civilizations which prevail in most developing counties, however, science and technology as such a cultural force have been present only for a few decades, and even then only in traces. It is no wonder, therefore, that these countries have not become yet environments in which the evolution of science and technology proceeds with ease and in which therefore the material consequences of science and technology could be realized effectively.

For example, the exposure of the population as a whole to the conceptual, atti-tudinal, and practical aspects of science and technology in developing countries is usually extremely spotty and rudimentary. Public communication media devote little attention to this problem, schools are woefully defective in dealing with it, and, on the whole, no "tradition" has developed yet to integrate science and technology with the elements of local culture.

The task is, of course, one of long range. Even though the developing countries will undoubtedly be able to indiginize the scientific revolution much faster than the "Western" countries have been able to do, the process will take quite a few more decades. Thus one has to adopt a longer range view of development, and a much broader one, which, simultaneously with short term "emergency" programs, must be tended to with vigor and energy.

We have thus seen that two important new concepts are gaining acceptance: That infrastructure building is one of the most important tasks of developmental aid, and that development must be viewed from a broad and long range vantage point in order to satisfy all needs, material and otherwise, that exist in the developing countries. In the remainder of this discussion, I will outline the practical advantages of this new framework and then suggest ways in which this framework can be a strong foundation for the activities of the new ISTC.

In particular, I first want to list four advantages of the new approach toward scientific and technological development over the past views and practices.

First, infrastructure building and the recognition of the broad nature of develop-ment will result in less political controversy than the aspects of development on which past emphasis was placed. For example, the explosive issues of short term technology transfer, patent rights, or the operation of multinational corporations would recede from the limelights in an atmosphere where both the "donors" and the "recipients" focus their attention on the creation of the indigenous capabilities of developing countries to generate their own technology, invent their own patents, and operate their own corporations. The issue of whether foreign aid should be focused on aiding the "poorest of the poor" among the countries or within a country would also shed some of its present high profile, since in the longer term efforts to create indigenous infrastructures and capabilities the distinction between what goes to the "rich" and what to the "poor" loses much of its meaning.

Second, the new orientation in developmental activities is likely to result eventually in considerable financial savings on the part of the United States. In contrast to short-term band-aid measures, where the limitations of programs are often purely financial, the bottle-neck in infrastructure building is most likely to be found in the shortage of appropriately educated indigenous scientific and technological manpower. Thus the instant application of billions to such problems is unnecessary, in fact, counterproductive, and in some cases outright impossible.

On the other hand, and, so-to-speak, in exchange for financial savings, the new approach will much more strongly rely on something that the United States has in abundance (in fact, according to some, in overabundance), namely human resources in science and technology. The new approach will involve a much greater and better utilization of individual American scientists and technologists in person-to-person interaction with the nascent scientific and technological communities of the Third World, and will make use of many of the latent resources and interest that exists within the American scientific and technological infrastructure. Considering that the American educational system at the moment is straining under the impact of a shortage of domestic students on a college and university level, and also considering that full employment in science and technology has not been achieved in recent years, the new international orientation could contribute in a most salutary way to the creative solution of some American domestic problems also.

Fourthly, the new approach to scientific and technical assistance to and cooperation with developing countries will operate in modes which deemphasize large bureaucracies. This is in contrast with some of the past activities in which huge projects of "technology import" (or imports of products of technology, such as the building of a dam) required large scale organization and hence resulted in large scale bureaucracies. With the present ubiquitous revolt in the United States against big government, against huge and even not so huge bureaucracies, against centralization, and against the impersonalization of human activities, the changed mode of international projects will be highly welcome. In infrastructure building within a broad concept of development, the organizational structure, by the nature of the task, is best kept flexible, decentralized, and attached to existing educational, research, and developmental institutions.

These four features, if adopted and utilized by ISTC, would greatly add to its chance of being established by Congress and of its operating successfully. It is well to recall in this respect the various elements of controversy and difficulty that surrounded past discussions of ISTC. At the risk of somewhat oversimplifying a complex situation we might say that the attitude of members of Congress toward an institution like ISRC could be grouped into three categories.

First, there were those who are inclined toward international cooperation in general, and who, in addition, realized the benefits that an organization with a longer range objective and a more flexible structure could bring about in international scientific and technological cooperation. Such congressmen were instrumental in bringing about the existence of ISTC, and it must be made sure that their trust, confidence, and hopes are not disillusioned by the actual practical performance of the new agency.

Second, there were congressmen who, though generally internationally inclined, were more advocates of quick aid than of deeper and more thorough assistance. There are, beside genuine differences in philosophy, very practical reasons in the career of congressmen why they would want to prefer rapid and visible manifestations of United States presence in developing countries. If, however, ISTC adopted the new approach and made a significant impact in terms of infrastructure building within a broad front of developmental tasks, visible signs of such work would also appear, even though they may differ from the traces of the past activities. It is most likely, therefore, that at least some of these congressmen could be won over by the success of the new type of activities, but only if ISTC boldly charts its own course and does not simply become a miniature carbon copy of AID.

Finally, the third group of congressmen represents various versions of isolationism. To them American activities and expenditures abroad, apparently resulting in little "gratitude" on the part of the recipients or little "influence" over them, is an undesirable situation. But ISTC could win the acquiescence of this faction also by pointing out that huge programs of foreign aid, in existence for some years, must utilize a "weaning mechanism" before they can be discontinued, and that ISTC serves as at least one such "withdrawal organ" through its efforts to enable the developing countries to help themselves. Here again, the argument is convincing only if the programs and methodology of ISTC are sharply different from past practices in developmental aid.

It might be worth mentioning at this point that the practical question is not whether to fully adopt the "new" approach toward scientific and technological assistance or to stay completely with the old one. There is no doubt that AID with its budget reaching into billions will be with us for some time, and not only because of bureaucratic inertia, but also because some of its short-term oriented activities fill a real need in the handling of economic crises in developing countries. Without such an instant-aid, the amount of human suffering in the world would increase considerably. In contrast, the size of the efforts along the "new line" are tiny (roughly one-hundredth of that of AID in financial terms). Furthermore, one would hope that this apparent disparity remains so in the future, in the sense that ISTC must direct its attention primarily toward the flexible, low-key, resource-coordinating programs which catalyze the existing latent potential in the American scientific and technological community, and it must not turn into a money-gobbling empire of its own. What is important, however, is that ISTC should exist, should be allowed to pursue its own novel and often experimental course, and that it be supported adequately. An initial yearly budget of, say, $20 million would well suffice.

In conclusion, I would like to list three general guidelines for ISTC which would help assuring its exploitation of the "new approach" to development.

The first of these, couched in what might be called "UN-language", says that ISTC should emphasize vertical classifications of developmental tasks and problems as compared to horizontal ones. The latter is used to mean divisions along specific problem areas, such as food, health, industry, transportation, etc. In contrast, vertical divisions cut across such problem areas and emphasize general infrastructural tasks, such as education, manpower development, research capability, etc. The two classifications constitute the rows and columns of a matrix and hence are complementary ways of looking at the whole field. It is therefore clear that both views must be given leeway. In past activities, however, both in American agencies and in international organizations, the horizontal classification was predominantly used. This resulted in narrow ways of looking at problems, and in a neglect of the development of a general scientific and technological infrastructure which supports activities in all problem areas but is not directly chargable to any particular one. The horizontal classification also neglected the non-material aspects and tasks of science and technology since this classification, in practice, had only material subheadings.

In order to balance this predominance of horizontal thinking, ISTC must think in vertical terms. This will also be much better suited to its longer range role of a broad infrastructure building.

The second guideline for ISTC is that it should mainly undertake programs and activities which are not country-specific. There are a huge number of reforms needed, for example, in the communication structure of the worldwide scientific and technological community. The patterns at the moment disadvantage scientists and technologists in the developing counties. Changes in these patterns would, by their nature, be universally beneficial to developing countries, and hence their initiation might be difficult in an organization which aims at helping only countries X and Y, but not Z. The salutary feature of such universal efforts is that the more countries are included in them, the more each country is helped. This again follows from the very nature of some of the infrastructure problems in science and technology and is entirely opposite to the situation in quick-aid measures where, for example, a given amount of grain, if divided between two countries, results in less grain going to each of them. Avoiding being country-specific also lessens the political controversy surrounding programs. It might be thought that it may also lessen the amount of "gratitude" earned by the program. It is my strong feeling, however, that the desire of "earning gratitude" through foreign aid programs has been a flop in the past anyway, and so the less of this unrealistic expectation there exists the better off we are.

It is often said that country-specificity is needed because the needs of different countries are different, depending on their political, social, cultural, geographical, ideological, and other characteristics. It would appear, however, that this statement is more true in principle than in practice. It is sometimes astounding to see how people with diametrically opposite ideologies, locked in acrimonious debates, can be made to agree on the necessity of certain specific practical measures. Besides, needs and measures in science and technology tend to be even more universal than in some other areas of development.

Thirdly, and finally, ISTC must operate almost exclusively through the person-to-person interactions among the scientific communities in the United States and the developing countries. This direct, debureaucratized, individually-motivated format has been sorely missing from past developmental activities which operated, even at best, in terms of huge "bilateral" links between a complex of universities at each end, connected on both sides by governmental bureaucracies. Scientists and technologists in the United States could be greatly revitalized in their attitude and willingness to interact with developing countries if opportunities were given to do so without red tape, proposal writing, trips to Washington, fights with bureaucrats abroad, report writing, and alike. Scientists and technologists in the "advanced" countries have already their own system of interaction and communication, and the main task is simply to assure that the colleagues in far-away and evolving countries have an equitable place in this spontaneous communal system. The credibility of ISTC in the eyes of scientists and technologists in the United States and abroad will crucially depend on this point.

It is encouraging for future discussions that ISTC's mandate, as laid down by the wording of the bill originally introduced in Congress, and by the legislative history of that bill as reflected in the hearings in Congress, is a very broad one which encourages ISTC to be novel, experimental, and imaginative. There is nothing in the "new look" outlined above which is in any way in discrepancy with what most congressmen regarded as desirable during the discussions of the original bill. The failure of the first attempt was due not to a lack of support in principle, but a number of practical, logistic, and organizational details of the bill and to the way it was handled during the discussions. In fact, the articulated ideology of the new administration coincides to a large extent with the philosophy of ISTC as outlined above: An aversion to large bureaucracies, an emphasis on the catalytic and coordinating role of government, a stress on individual initiative and on person-to-person interaction, a preference for offering opportunities for self-help rather than for welfare, and enhanced attention toward the national image of the United States throughout the world which can be well represented by an international utilization of something in which the United States is paramount, namely science and technology. There is no reason, therefore, why ISTC could not begin its intriguing and potentially very pivotal activities in the not too distant future.

14. The Role and Function of a Scientific and Technological Infrastructure in the Context of Development Policy

MICHAEL J. MORAVCSIK

INTRODUCTION

This discussion was prepared for an international symposium, the aim of which was to generate ideas, proposals and programmes for the consideration of the United Nations Conference on Science and Technology for Development (UNCSTD) to take place in Vienna in August 1979. Since that conference as all United Nations evens of similar character, is in the danger of being merely a string of oratories with no tangible effect on the actual state of science and technology in the developing countries, it is important to orient any input into such conferences toward a realistic and frank analysis of the situation followed by a set of specific recommendations.

Since the problems of science and technology in developing countries have by now a large literature (perhaps too large compared with the amount of action generated by it), I will limit myself to a relatively brief exposition of the analysis and of the recommendations without giving arguments or examples to the extent that I would like to if I had unlimited time and space. (1) The basic concepts will be reviewed so as to avoid semantic ambiguites. Then the roles, functions and the importance of a scientific and technological infrastructure in a developing country will be listed. This will be followed by an account of some of the most common deficiencies in development policy pertaining to scientific and technological infrastructure. The discussion will end with specific recommendations to UNCSTD with respect to an enhanced and improved support in the building of scientific and technological infra-structures in developing countries.

CONCEPTS

Science and technology (2)

Science is a human activity resulting in knowledge about nature, utilizing a parti-cular methodology called the scientific method. This method, based primarily on observational verifications of ideas about and explanations of natural phenomena, is characterized by the existence of criteria which can resolt disputes and build up an "objective" set of knowledge and understanding. "Objective" here means merely that the members of the scientific community form a consensus in accepting that method as the arbitrator of disputes.

Technology is a human activity resulting in procedures, processes, prototypes, gadgets, oriented toward the ability of doing certain things or making certain things. In the old days technology was primarily based on trial-and-error type of tinkering with out immediate environment but in the last hundred years or so tech-nology has become increasingly science-based, that is, technology now utilizes the knowledge created by science to replace the mere trial-and-error approach prevalent earlier.

It should be noted that the difference between science and technology is in terms of their results, their products. The product of science is new knowledge, the product of technology is a new process or gadget.

People are motivated to do science by a number of aspirations. Among these are the delight in expanding the realm of scientific knowledge per se, and the urge to generate new knowledge because of its possible applicability in other human activities. According to these two motivations, scientific research can be characterized as "basic" or "applied" research. The two labels are not very distinct ones, however. The two may be present simultaneously in the same person. One may be predominant in the researcher but the other in the provider of research funds. All good scientific research, even if it appears exclusively "basic" at the time of its creation, becomes "applied" sooner or later, so that the division between "basic" and "applied" is also fuzzy in time. In spite of this lack of a sharp division between the two categories, in the management of science and technology it is useful to set up a somewhat arbitrary boundary between them. For example, one might call "applied" research some activity in which the dominant expectation is knowledge that is likely to become applicable in an a priori specified area within the next 8-10 years.

Development (3)

Development is a set of actions leading to the increased realization of aspirations. This concept holds both for individuals and for groups. Note that development is action, not plans, intentions, or preparations.

Aspirations of individuals and of groups of people are varied. Some aspirations are material, pertaining to either the basic necessities of life, such as food, shelter, health, or to more luxurious items like TV sets, cars, or swimming pools. Others are non-material, such as religious or spiritual fulfillment, intellectual growth, artistic achievements, achievements in sports or in exploration, ideological domination, wide-ranging political power, etc. Throughout human history, non-material aspirations of people have played a much more dominant role than the material ones. (4) A quick analysis of the small and not so small wars and conflicts taking place in the world today will show the extent to which these non-material aspirations dictate the events on the international scene. Similarly, the behaviour of few individuals, if any, can be analyzed successfully in terms of only material factors. Indeed, the world would be a much easier arena to manage if only material motivations propelled its inhabitants. In fact, however, the non-material factors enormously complicate the situation while, at the same time, also make life infinitely more interesting and absorbing.

Policy

Policy is a set of actions working in the direction of the realization of some goal. Sometimes policy is conscious, enunciated in advance and coordinated, while at other times policy is de facto, developing en passant and consisting of a mosaic of seemingly disjointed elements. Whether centralization and planning constitute the more effective policy is open to debate and may depend on the nature of the goal and the environment in which the realization of the goal is to be effected. For example, happiness is a policy goal for every individual, and it is also a policy goal of society as a whole, but whether this goal can be achieved better by centralization and planning has been an often debated question in human history.

Explicitly or implicitly, policy consists of four stages:

(i) Planning, that is, the projection of action to be taken;

(ii) decision making, that is, the choice among alternatives in the course of the realization of the goal;

(iii) implementation, that is, the actual efforts converting intention into reality;

(iv) evaluation, that is, the assessment of the success or the failure of the previous three stages, as compared with the original goal or aspiration.

THE IMPORTANCE OF A SCIENTIFIC AND TECHNOLOGICAL INFRASTRUCTURE IN A DEVELOPING COUNTRY

An analysis of this importance can be neatly arranged in terms of the three main motivations (5) of science and technology. Each of these may include several specific roles.

Infrastructure as a generator and as a carrier of science into technology and technology into production

This is, for the most part, the material aspect of science and technology. In this respect the infrastructure has the following three essential functions:

(i) Receptor and channeller of scientific and technological information

knowledge and know-how from abroad into the country

All countries in the world are net importers of science and technology. Even the United States produces only about 30% of the world's science, and hence imports 70% of it. In the case of a developing country, this ratio may be 0·1% versus 99·9% or even more extreme. Thus such a country will be dependent to an overwhelming extent on science and technology generated elsewhere. Only an indigenous scientific and technological manpower, kept au courant through personal research and development activity, can serve as an effective recipient and channel of such worldwide scientific and technological knowledge. Centralized and computerized information systems, tapes, libraries, journals, reports, seminars and other tools are of no use whatever in the absence of such a high quality local group of scientists and technologists.

(ii) Selector and adaptor of science and technology

The amount of raw information on scientific and technological developments around the world is of a staggering magnitude. To be able to use that information in any sense, a broadly educated, highly competent and imaginative scientific and technological manpower is needed to make selections among the flood of information available and to adapt the selected information for the specific use in question. Such an adaptation is an essential step both in scientific research and in technological development work. In science no two research problems are identical, and hence previously generated scientific knowledge must either be brought down from a general plane to the specific problem in question or must be transmuted from its application to a very different problem to the application to the problem on hand. In technology, the particular environment of a developing country, both physically and in terms of the human, cultural and economic aspirational aspects is likely to be different from the environment where the technology was invented, and hence adaptation is necessary.

(iii) Generator of new science and technology

Although, as we saw, in the case of a developing country, this function of the scientist and technologist may, on average, be a small fraction of his role in terms of the previous two functions, the ability to generate new science and technology is nevertheless of crucial importance both in the short term and in the long term. Quite apart from the non-material motivations to be discussed below, new science and technology also play prominent roles in material considerations. Imitative science and technology are likely to result in products of relatively low specific value, not optimally suited for the domestic market and not highly competitive on the international market, unless propped up by the availability of cheap and yet sufficiently skilled labour. To be sure, this last combination is often quite effective and can bring a temporary boost to the country's economy, as we have seen in the examples of Japan, Hong Kong, Korea, Singapore, and other countries, at one time or another. Sooner or later, however, the time comes when the rising standard of living in the country eliminates this mode of operation and the country can progress further only if it also has the discovering, innovating and investing capability at the forefronts of science and technology. Japan passed this stage some time back, Korea and Singapore are at it and many countries are fast approaching it. To prepare for having such a capability at the proper time, the scientific and technological infrastructure must be carefully nurtured for decades prior to that time since it takes a long time to bring it up to a significant level.

Infrastructure as a foundation for the country's dignity and independence in science and technology

(i) Science and technology as human aspirations

At any given time in history, people and peoples have many and different aspirations. In the 20th century, one of the many highly valued human activities and achievements is in science and technology, and scientific discoveries and technological inventions are matters of individual and collective pride thus forming an important part of the raison d'etre of an individual or a country, far transcending the purely material tools of human survival. Forefront achievements in science and technology can raise the country's morale the same way as achievements on the soccer field, in art, in mountaineering or in entertainment can.

(ii) Science and technology bring independence

Perhaps the foremost burden of a developing country, even transcending that of material poverty, is a feeling of complete dependence on forces external to the country and the feeling of having no choices and no options. In a world in which knowledge and know-how are so pivotal for power in all walks of life, the relatively few countries in the world which are advanced in terms of such knowledge and know-how automatically dictate the life patterns for the whole human race, whether

152   Michael J. Moravcsik

they wish to be thus "imperialistic" or not. Specifically, without a scientific
and technological infrastructure in a country, everything in these areas will be
determined by organizations, individuals and groups from abroad. Slavery in a
pedestrian sense was presumably abolished some time ago but the status of being a
slave (with intentional or unintentional masters) very much continues to be in existence and will remain so until knowledge and know-how are more equally distributed around the world. Such a distribution is impossible without strong indigenous scientific and technological infrastructures in existence in every country.
Only such an infrastructure will permit the country to begin to be its own master,
to develop several options and to choose among them and, in general, feel the degree of independence and dignity that befits an equal partner rather than a subordinate.

The feeling of frustration and inferior status brought about by a lack of capability and independence is very much in evidence today around the world. It shows up
in a tour given by a visitor through the scientific and technological facilities of
a country, during which the visitor is often left with the impression that "deep
down" the demonstration of the country's participation in a worldwide scientific
and technological activity takes precedent over concerns about the specific, short
term, and material results of the research and development that is done, even
though that concern is stressed in formal documents. It manifests itself in the
international political arena, where many hot issues are directly attributable to
a desire to assert dignity and pride in the face of the realities of a helpless and
inferior status in de facto self-determination. For this reason alone, the development of the country's capability in science and technology is an urgent and highly
desirable task.

(iii) Infrastructure as an influence on the country's world view

While science and technology do not constitute a whole culture, they can be powerful forces within any culture. The basic assumptions implied in scientific or
technological work, such as the emphasis on change, the openness of human knowledge
and capability, the ability of Man to influence his fate, the existence of a con-
sentaneously objective method to resolve disagreements and thus facilitate and define progress, and many others, are ideas and attitudes without which any definition of "development" or "progress" is hard to imagine, and which therefore can have
a crucial influence in determining whether the country will "develop" or not.

It is, therefore, advantageous for any country to have the opportunity to incorporate into its own traditional culture some of the view and attitudes science and
technology provide. Such an assimilation process can take, and has taken, place
without altogether killing the traditional value systems. In fact, throughout
history, the successful and influential countries and civilizations have excelled
in amalgamating the new with the old, the borrowed with the traditional, the foreign
with the local.

The message of science and technology concerning Man's view of the world must be
transmitted to the country by natives to that country who are themselves personally
involved in doing science and technology. Talking about it is not enough. Few
people will cherish a book on sex written by a virgin with twenty years of voyeur-
ship.

The creation of a scientific and technological manpower that can thus radiate the
ideas and attitudes of science and technology to the population as a whole is a
special challenge, since by no means all scientists and engineers will be suitable
for this function. This is one more reason why the establishment of such an infrastructure must begin very early in the country's overall development, so that one
can count on statistical fluctuations over a long time to produce publicly influential leaders of science and technology. The particular examples one can find in
the history of the developing countries in the last 3-4 decades suggest that such
leaders can have a monumental impact on their country's attitudes and morale.

COMMON DEFICIENCIES IN SCIENCE AND TECHNOLOGY POLICIES

A complete list would be a long one indeed. It will simply select a few representative problems, some of which might be particularly appropriate for attention by
UNCSTD.

(i) The false equation: Development = Economics

As was evident from the previous discussion, the development process is a highly
complex conglomeration of cultural, philosophical, economic, social, political,
attitudinal, geographic and other elements, both in motivation and in evolution.
It is, therefore, shocking to see the widely pervading practice of equating development with economics, that is, considering only short-term material elements. That
single-purpose, simplistic organizations like the Agency of International Development of the United States would exhibit such as attitude is perhaps not surprising.

But the malady is much more widespread. Formalistic and vacuous science and technology councils, ministries, committees and directorates in many countries, as well as many United Nations agencies, fall into the same trap. It "policy" is formulated on the basis of such narrow conceptions, it is not surprising that most of them fail even in the narrow area in which they are intended to be applied.

(ii) The false time scale - how the crisis is perpetuated

There is an overwhelming tendency in de facto development activities pertaining to science and technology to concentrate on the tomorrow and ignore long-term considerations. Countries order their young people to be narrowly trained in problems of momentary prominence, forgetting that by the time these young students reach maturity in terms of scientific and technological activity, different crises will be confronting them in which their narrow education is useless. Virtually all efforts are directed toward the existing problems thus enhancing the seriousness of problems arising in the future.

This attitude tends to upset the desired balance between scientific and technological activities in favour of low-grade technology, the tangible results of which are expected to manifest themselves immediately. The proper balance between basic and applied scientific research is also upset, disfavouring the former and hence de facto handicapping both.

This obsession with the immediate future may appear to contradict the other obsession of many countries, to be discussed next, namely that of planning. The resolution of this apparent paradox is simple and will be explained in the next subsection.

(iii) The false emphasis: "Planning, planning über alles"*

When a visitor is given a summary of the country's scientific and technological capability, the briefing always begins, and often continues and ends, with the exhibition of national science and technology development plans. Indeed, much of the time and effort of the national science and technology policy apparatus appears to be spent on planning. Marvellous organizational charts showing vertical and horizontal relations, time evolution charts, computerized planning facilities, etc., adorn the walls of the offices, and the training of personnel is concentrated in the areas of planning and forecasting.

In comparison, if and when the visitor has an opportunity to acquaint himself with specific research or development projects, he finds that very little if any attention is paid to the development of skills and capabilities in decision making and implementation. Decisions are often made haphazardly, by unqualified people, in the absence of pertinent information, and hence the actual activity resulting from such a procedure is feeble. There are, of course, exceptions and they readily stand out among the crowd: those projects which are led by imaginative, skillful, experienced, enthusiastic, and highly competent scientists and engineers can be easily identified merely by the exciting, lively and energetic atmosphere that prevails among the people in the project as well as by the results of the project itself. Yet, on the whole, the appearance of the overall system suggests that many consider planning 90% of the whole job, with a meagre 10% assigned to all the rest.

But not only is planning so prevalent, it is also poor in quality. A functional model of planning involves some goals, an input, some processes and an output. Planning for science and technology is particularly deficient in working with output. As we saw earlier, the output of science is knowledge and the output of technology is a process, a method, a prototype. Neither of these is easy to quantify. When we talk about the output of a tomato factory things are infinitely simpler: the number of cans of tomatoes, with some minimal attention to quality, might do the job. But neither science nor technology is directly connected to such easily quantifiable production. Furthermore, the quality of the output in science and technology is extremely crucial, something that is completely ignored in most planning processes. In fact, national plans often simply ignore output altogether and substitute input for it, thus resulting in such grotesque statements as "the goal in science for Country X in the next five-year plan is to produce 500 more scientists", as if the purpose of science were to produce scientists.

It is not surprising that planning for science and technology turns out to be so poor and pointless. Apart from the conceptual question of limits on the extent to which scientific and technological activity can be planned at all, such planning is done mainly by people who are not scientists or engineers, know very little about the de facto workings of scientific research or technological development and thus the whole operation turns into an isolated academic exercise in economics.

The result of this is that the ensuing plans have very little relationship to what really happens in the country in science and technology. This is the clue to the apparent paradox mentioned earlier, between the predilection for planning (which presumably takes into account the future) and the almost exclusive emphasis on short-term considerations. Whatever the plans may say, the actual decisions made by a medium-level bureaucrat in the ministry or council are conceived in the light of a very narrow and short-term view of science and technology.

(iv) The false climax: No evaluation

One of the most amazing aspects of science and technology in most developing countries is the virtual absence of an evaluative procedure which could determine the success or failure of work in science and technology, either prior to committing support to it, or after the work has been completed. Infrastructure for such evaluation is non-existent. To be sure, nominal attention is given to research proposals to ascertain, usually from the title of the project, whether it "fits" the national "policy" or not, and completed projects are perfunctorily monitored in terms of whether the "progress report" written by the researcher himself satisfies certain formalistic requirements. But a substantive assessment, in terms of peer review or other well-established methods, is indeed very rarely used.

There are a number of factors contributing to this state of affairs, which have been discussed previously in the literature (6) and will not be repeated here. In this light it is however not surprising when, as mentioned earlier, planning is non-functional, decisions are divorced from plans and are frequently made badly and implementation is lagging, and then all this is left unchanged. Remedy is not forthcoming because in the absence of evaluation these discrepancies are not even discovered.

(v) The false classification: Applied scientific research versus technology

As we saw earlier, applied scientific research and technology are two different activities. The former results in new knowledge about natural phenomena, generated with the motivation that such knowledge may be applicable in some process or gadget in connection with some material problem. Technological development work, in contrast, is the utilization of existing knowledge (whether common empirical knowledge or knowledge generated by science) in investing processes, building gadgets or constructing prototypes which then can be utilized in production.

When a country is faced with a practical problem of a material nature, its solution generally involves both applied scientific research and technological development. The two types of activity will be carried out by two different groups of people, with different education, training, skills, motivations and capabilities, sometimes working together in a coordinated fashion but sometimes working separately and even at different times.

Nominally, a considerably fraction of the financial resources of most developing countries, earmakred for research and development, are spent on applied scientific research. Huge and expensive governmental institutes and laboratories are established, carrying names like Council of Scientific and Industrial Research, Federal Institute of Forestry, Applied Science Research Corporation, and the like. It turns out, however (7), that much of the activity in many of these organizations is low-level technological manipulation based on trial-and-error. There is simply no tradition of applied scientific research, that is the reaction of new knowledge which may be useful in applications.

One of the many reasons for the absence of such applied research activity is in the overly narrow scientific and technological education that often prevails in these countries. (8) Citing, curiously enough, the needs of applied research, the education of much of the scientific and technological personnel is narrowed and constrained at a much too early stage to deal only with matters that are judged "relevant", usually by people who themselves come from an overly constricted educational background. What is ignored by them is the fact that the more applied the research is, the greater need there is for a broad educational background, because while topics for basic research arise in single disciplines and can be determined by the internal criteria of science only (and by the researcher himself), problems in applied science are defined by external circumstances and hence are likely to be inter- and multi-disciplinary. Hence education for applied research demands a much greater breadth than education for basic research.

(vi) The false community: Lack of internal communication

Communication within a given country between scientists and others is often appallingly weak. (9) By "others" I mean other scientists, or technologists, or governmental figures, or the population as a whole. The same can be said for technologists who are isolated from other technologists, technology policy makers, from representatives of the productive sector of the country and from the public.

Such an internal isolation is in part a matter of a lack of communicative tradition, partly a matter of questionable personal relations within a small group but also partly a matter of negligent science policy. Stimulation of such internal interaction, in terms of administrative procedures, catalytic financial incentives and other relatively easy means is generally absent and in fact the problem itself is not well appreciated by the bureaucratic machinery handling the day-to-day matters of science and technology.

(vii) The false linkage: External isolation of the scientific and technological community

One of the most debilitating handicaps of doing science and technology in developing countries is the isolation from the worldwide scientific and technological communities. (10) This isolation manifests itself in all respects: scientific and technological journals (current issues and back volumes), scientific and technical reports, scientific and technological meetings and conferences, opportunities for "hands-on" technology transfer, short and long term visitors from abroad, apportunities for visits to institutions, laboratories, or development centres abroad, etc.

Analyses of the "communication gap" have also appeared frequently in the literature (10) and will not be repeated here. Although full remedy cannot be effected without the cooperation of the international scientific and technological community, some helpful programmes can be instituted even by local science policy organizations to benefit the scientists and engineers in that particular country. In reality, however, I know of no such programmes.

(viii) The false superstructure: The separation of science policy from the active scientists and technologists

In most of the countries where science and technology has a fairly long-standing tradition, the mechanism for making policy for science and technology incorporates representatives of the active scientific and technological community. For example, in the United States the President's Science Advisor is generally a reputable scientist with contemporaneous or recent active research involvement, and the innumerable advising committees of the various governmental departments consist of working scientists. The same is generally true for the top leadership of the important governmental agencies such as the National Science Foundation. Although there are always some grumbles, it would be quite unfair to charge that a gap exists between the active scientific and technological community and the science policy mechanism of the United States. A similar situation exists in Great Britain, Germany, France and the other scientifically significant countries.

In contrast, in many developing countries policy for science and technology is made exclusively by people with no present and, for the most part, no past credentials in scientific research or technological development. Civil servants with little if any scientific background, economists and politicians join with some people with scientific degrees but either no or no recent research involvement. Since seniority per se is often a strong force in the tradition of these countries, preference for older people is strong and such people either had no opportunity to participate in significant research (because the state of the development of science and technology in their youth was too primitive) or have given up research some years prior.

A particularly "dangerous" type often in evidence in developing countries is the young social scientist who received "training" in "science policy" and then is appointed to work in a "science policy organization". Such a person in "dangerous" for a number of reasons. First of all, his view of science and technology will be that of the voyeur already mentioned earlier. This view is likely to turn out to be a formalistic one, because for an outsider form is easier to grasp than substance. Second, he is in for trouble in his relationship with working scientists and engineers who, justly and unjustly, will consider him as a "mere bureaucrat" and will find substantive communication with him rather difficult. It would be an exaggeration to say that in my encounters with science and technology in developing countries I never met a person of the above type who was at least moderately successful in his activities. It is, however, an accurate statement that such successes are rare exceptions.

This concludes the brief listing of eight important deficiencies of policies for science and technology in developing countries. The next section will propose some action to be taken by UNCSTD to compensate for some of these deficiencies.

SPECIFIC RECOMMENDATIONS FOR UNCSTD (11)

Concentration on infrastructure building

Our discussion tried to indicate that the rapid and effective development of the scientific and technological infrastructure in a developing country, and in particular the energetic creation of a high quality and broadly educated scientific and technological manpower is the key to the future of the developing countries in

156   Michael J. Moravcsik

science and technology. It is therefore recommended that UNCSTD concentrate its
attention mainly or even exclusively to this problem area, in which international
assistance and cooperation are very essential. Help in infrastructure building has
lasting consequences and is not a band-aid type of measure. Also, the building of
infrastructures is a politically rather non-controversial problem area. One of the
big dangers of UNCSTD is that it will develop into a sterile shouting match between
political demagogues of various persuasions, in which accusations are exchanged and
labels thrown at each other while the actual issues affecting science and technology in the developing countries remain untouched. This danger is enhanced by the
fact that a goodly portion of the delegates sent to UNCSTD by the various countries
appears to be politicians and others ignorant about scientific and technological
policy issues who therefore, because of this ignorance, will be constrained to
engage in political rhetorics since they do not know anything else. It is, in my
opinion, the duty of the professional community around the world to warn UNCSTD
ahead of time and try to expropriate at least one or two days at UNCSTD during
which something of substance will be discussed.

Breadth of education

As is evident from our discussion, the narrowness of education in science and technology is a significant cause of the weak position of many developing countries in
these areas. There are a number of specific measures, both in local education and
in the education abroad of students from developing countries that can be taken to
remedy this defect. Many of these have been outlined in the literature (11) and
hence will not be repeated here. UNCSTD should direct its attention to these proposed programmes and find ways of implementing them.

Reorientation of policy activities

The education of science policy makers, as well as the actual science policy activities, need urgen reforms. In particular:

    (i) The overbearing emphasis on planning should be replaced by a much
         greater concern with decision making and, above all, with implementa-
         tion.

    (ii) The personnel involved in science policy making should be revamped to
         attain a new balance: many fewer economists, civil servants, bureau-
         crats, politicians and pro-forma scientists and technologists, and a
         much greater representation of working scientists and engineers.

These two objectives should be reflected both in the local structures in the various
countries and in the science policy activities of international organizations, for
example UNESCO. To achieve this, some drastic personnel changes may be needed both
locally and internationally.

UNCSTD might find it difficult to be very effective in this in the short run. If,
however, a consensus is reached in this respect and mechanisms are established in
the various bilateral cooperation programmes between countries to reflect this
consensus, eventually relief will be achieved.

Establishment of evaluation

This is a large task. It consists, first, of establishing a generally acceptable,
realistic and functional evaluation methodology for assessing the operation of
scientific and technological institutions and of science policy bodies in the developing countries, as well as the operation of international agencies and organizations. Methods for assessing have in fact been developed in the past (12) and some
are successfully used in some of the advanced countries so the task, while challenging, is far from impossible.

Second, mechanisms should be created so that organizations internationally and
locally are able to use these procedures. This may involve schemes of utilizing
international peer review networks, computerized "objective" indicators or whatever
method appears most suitable for the purpose. UNCSTD would have a good opportunity
to be effective in this area since it could enter this field without being tied to
any particular country or group and its contributions to this field would therefore
be less construed as a sign of weakness on the part of the Third World. Specific
programmes have already been mentioned in the literature.

Boost for applied scientific research

As mentioned earlier, in many developing countries, apart from a tiny amount of
basic scientific research, the resources for research and development are predominantly spent on low grade, trial-and-error type technical manipulations and applied
scientific research is on the whole missing from the spectrum.

UNCSTD should direct its attention to this void and find ways to boost applied scientific research in developing countries through various means of communication, cooperative research, education, etc. Some of these have also been mentioned in the literature (7), and hence need not be repeated here.

Scientific and technological communication

This is a prime area for UNCSTD's attention since many of the defects are caused by the distorted patterns of communication within the worldwide scientific and technological community. A specific list of projects has been suggested to UNCSTD elsewhere in the literature. (13) It should also be stressed that at least some of the very helpful programmes in this area are financially extremely economical and hence are easily within the capabilities of the leading countries even in times of financial hardship.

Mechanisms to bring active scientists and engineers into science policy making

There are various vehicles to accomplish this objective. Among them are:

   (i) Seminars and practical working experience in science policy and the science for practising scientists and engineers. These seminars should be conducted by scientists and engineers with a long-standing background in active work in science and technology and practical experience in decision making, implementation, evaluation and general management of scientific and technical projects.

   (ii) Auxiliary education of students from developing countries who are being educated in a more advanced country, to expose such students to elements of practical science policy ranging from operating a workshop, a library, or a university, and performing in the electronic or glass blowing shop to elements of national science policy and the fundamentals of the science of science. (14) Such auxiliary education could be carried out during summer vacations and should be conducted by a staff with qualifications similar to those mentioned under. (1)

   (iii) Special financial and other incentives for scientists and technologists in developing countries to be able to carry the double load and double duty of continuing with research and development work and being involved in science policy activities. In many advanced countries where university teaching duties are lighter, laboratory assignments more flexible, and organizational, logistic and technical support for research as well as for policy are better developed, such double activity is somewhat easier to bear. Similar conditions might be at least approximated also in developing countires by some relatively minor special arrangements for deserving personalities.

Footnote

* A variation on "Deutschland, Deutschland über alles" (Germany, Germany above all), the supernationalistic German hymn between the two World Wars.

References

(1) I will, however, give references to other papers of mine where some of these points are discussed in greater detail. In fact, most of the references will consist of such "selfcitations". The many interesting writings by others will not be referred to here explicitly but can be found as references in the other papers of mine that I am referring to

(2) See for example M. J. Moravcsik, "What is Science", Science Centre Bulletin (Singapore), 4:1, 6 (1977), and M. J. Moravcsik, How to Grow Science (book now in manuscript form, to be published), Chaper 1

(3) See for example M. J. Moravcsik, "Scientists and Development", Journal of the Science Society of Thailand 1, 89 (1975)

(4) See for example Kenneth Clark, Civilizations, Harper and Row, N.Y. (1969)

(5) See for example M. J. Moravcsik, "The Context of Creative Science*, Interciencia 1, 71 (1976) [reprinted in Everyman's Science (India) 11, 3, 97 (1976)], and M. J. Moravcsik, How to Grow Science (book now in manuscript form, to be published), Chapter 2

(6) See for example M. J. Moravcsik, "Developing Countries and the Fruits of Science", Leonardo 11, 214 (1978). See also Reference 11

158   Michael J. Moravcsik

(7) See for example M. J. Moravcsik, "Applied Scientific Research and the Develop-
     ing Countries" Science and Public Policy 5, 82 (1978)

(8) See for example M. J. Moravcsik, Science Development - The Building of Science
     in Less Developed Countries. PASITAM, Bloomington, Ind. (1976) (Second
     Printing), Chapter 2

(9) See for example Reference 8, Chapter 4

(10) See for example Abdus Salam, "The Isolation of the Scientist in Developing
     Countries", Minerva 4, 461 (1966)

(11) See for example M. J. Moravcsik, "Science and the Developing Countries", posi-
     tion paper toward the United States national paper for the United
     National Conference of Science and Technology for Development (1977).
     Copies obtainable from Andrew Pettifor, Room 1229, National Science
     Foundation, Washington 20550, USA

(12) See for example M. J. Moravcsik, "A Progress Report on the Quantification of
     Science", Journal of Scientific and Industrial Research (India) 36, 195
     (1977)

(13) See for example M. J. Moravcsik, "Something Concrete for Vienna", International
     Development Review (to appear in the April 1979 issue)

(14) See for example M. J. Moravcsik, "Foreign Students in the Natural Sciences: A
     Growing Challenge", International Educational and cultural EXCHANGE 9:1,
     45 (1973). The summer seminar described there is yet to be implemented,
     in spite of various attempts to attract funds for such a programme

Bull. Sci. Tech. Soc., Vol. 1, pp. 355-377, 1981. Printed in the USA.
0270-4676/81/040355-23$02.00/0 Pergamon Press, Ltd.

MOBILIZING SCIENCE AND TECHNOLOGY FOR INCREASING THE INDIGENOUS CAPABILITY IN DEVELOPING COUNTRIES

Michael J. Moravcsik

Institute of Theoretical Science
University of Oregon
Eugene, Oregon 97403 USA

CONTENTS

PREFACE

This article was originally commissioned by the United Nations Office of Science and Technology for the ACAST-sponsored meeting during the week preceeding the 1979 United Nations Conference on Science and Technology for Development (UNCSTD) in Vienna. Its language and contents having been found markedly different from what usually appears in United Nations documents, the article was not publicized by the United Nations in its original form. Instead, together with another commissioned article, it was amalgamated and purified into a background document for the ACAST-sponsored conference. Hence this is the first time that the article

appears in its original form. Only a few specific and by now irrelevant
references to UNCSTD were deleted in the text. Although some two years
have passed since UNCSTD, the contents of the article are just as timely
as they were then. In the present context the purpose of the article is
to pinpoint some of the major issues of science development. For the
reader who is close to a novice in this problem area, the article is mainly
a map of the territory to be explored in detail. Further sources can be
located by perusing the references given in the bibliography and by branch-
ing off from there by locating the many references quoted in these references.

I. INDIGENOUS CAPABILITY AS A SELF-GENERATING SYSTEM

If we ship coal to the proverbial Newcastle at the rate of 100 tons
a day, the increase in coal at Newcastle will be 100 tons after one day,
200 tons after two days, 3,000 tons after 30 days, etc.

If, on the other hand we dump algae into an appropriately prepared
pond at the rate of 10 tons a day, the increase in algae in the pond will
not be 300 tons after 30 days, but much more, and in fact we can soon stop
dumping algae altogether and still the amount of algae will continue to
increase by leaps and bounds.

This, in a nutshell, is the essence of indigenous capability. A system with an indigenous capability is not linear but, under appropriate
circumstances, exponentially increasing. At the beginning, it might have
to be fed externally, until a "critical mass" for self-generation is
attained, but thereafter, it can generate further development without
external force-feeding. To be sure, the system still needs to maintain
contact with the outside world (in the case of the algae, to absorb energy
and chemicals) but now it has the ability to convert the raw ingredients
into the growth of itself.

Most "dead" systems in nature are linear in this sense, while "live"
systems are self-generating, because life implies an indigenous capability
to survive and develop. We might therefore also call a self-generating
system a live system.

A country is live if it has such indigenous capability for self-
generation. Countries perpetually living on foreign handouts and surviving
through never-changing, traditional, and primitive efforts could justifiably
be called dead. Dead countries are a burden on the world, and are doomed
sooner or later to die, since international charity has its bounds.

A self-generating country is also live in a different sense. The
knowledge of having the indigenous capacity for creative work reflects in
pride, self-confidence, high morale, and a purposefulness of existence that
contrasts sharply with the deadly indifference and fatalism encountered in
societies which are stagnant or on their way down. Such existentialistic
elements in people's thinking and feelings give meaning to the life of
individuals and of communities.

International assistance and cooperation has long been guilty of
merely feeding needy countries instead of focusing on the more ambitious,
difficult, and long-range task of helping those countries toward the
establishment of an indigenous capability and hence turning them into a

self-generating system. Short-range aims blotted out the long-range
vision, and the mechanical and unsophisticated motions of simply delivering goods and the accompanying instructions of "how-to-use" took prece-
dence over an attempt to catalyze an indigenous development toward self-
propulsion. The highly misleading example of the recovery of postwar
Europe stimulated by the Marshall Plan obscured the need for first
creating an "appropriately prepared pond" before the dumping of algae can
begin.

There are three dimensions in which indigenous capability for self-
generation is created in developing counties by science and technology.

First, there is the economic capability. It is the rule of the last
two centuries that with a few rare exceptions of countries possessing
large amounts of very valuable raw materials, economic size is directly
related to scientific and technological size. The correlation is established beyond doubt, and the causal character of this relationship can be
demonstrated on specific examples. The causal link goes both ways to
form a self-generating loop: Scientific knowledge and the ability in
science-based technology produce economic wealth which in turn nurtures
further acquisitions in scientific and technological capabilities, provided that some other factors (such as, for example, willpower) are also
present.

Fortunately by now the demonstration of this link is not restricted
to the so-called developed countries. In the last 3-4 decades a number
of countries, formerly considered toward the bottom of the economic
hierarchy of countries, have risen rapidly through the bootstrap process
of science-technology-economy-science-etc. and have in fact attained
close parity with the formerly "developed" countries. This gives hope to
the other countries still behind, that given the appropriately conducive
indigenous environment, they may be able to make significant economic
strides also within the span of, say, 5 decades, a time period which is
very short compared to the length of the similar historical evolution of
the now developed countries.

Second, and perhaps even more important than the purely economic
aspects, indigenous capability in science and technology gives the countries
a self-generating capability toward realizing their non-economic aspirations. Individuals and society thus gain an opportunity to convert talent
and potential to accomplishment, the country acquires an individual and
collective pride in its accomplishments, sheds its humiliating total
dependence on other countries in all aspects of life (economic, artistic,
political, social, cultural, etc.) and thus enters the phase when it can
make a mark in history. Great civilizations in the past, the ones that
burst with energy, optimism, action, and ingenuity were all leaders of
their times and not sad followers and dependents. Science and technology
both represent a strong non-material aspiration of the 20th century, and
serve as tools to realize such aspirations in other walks of life.

Finally, science and technology increase the indigenous capability of
the country in its outlook on the world, in its world-view. Science and
technology do not constitute, by themselves, a culture, but they are significant ingredients in Man's view of the world. Furthermore, the ingredients
that science and technology provide are exactly those which are indispensible for any concept of development. The belief that knowledge is open,
and that by a constant effort toward gaining new knowledge we can keep

increasing our understanding of the world around us is indeed a crucial foundation of development. Coupled with this is the belief that with the knowledge thus gained, and with ingenuity in applying it, we can gain an ever increasing control over Man's fate which, after all, is not the fragile boat tossed around by unknown forces on a rough sea as it was once thought. That change is a natural element of the world is also a basic tenet of science and technology, and this can also be a powerful addition to many traditional systems of belief in which the world is determined once and for all and never changes significantly. Since the "developed" countries have lived with the scientific revolution as a cultural force for at least three centuries, these attitudinal, philosophical, epistemological, and psychological influences of science and technology have become unnoticed and taken for granted. Yet, in the countries which are just in the process of absorbing the scientific revolution on a large scale, these influences are likely to play a crucial role in development.

II. INFRASTRUCTURE FOR INSURING INDIGENOUS CAPABILITY

In the illustrative analogy used earlier, I emphasized that algae will reproduce and form a self-generating system only if they find themselves in an appropriately prepared pond. In this section, therefore, we must try to describe what an appropriately prepared pond for growing science and technology is like. A complete specification of the conducive conditions would hardly fit into the confines of the present brief discussion, but some of the major areas of concern will be listed and analyzed briefly.

A. Education and Training

Perhaps the major difference between the "developed" and the developing countries as far as science and technology are concerned is that in the former, science and technology are well entrenched, in existence for a long time, with a functioning infrastructure and a size large enough to ensure a considerable amount of continuity and stability. In contrast, in the developing countries science and technology are often in a rudimentary state, with relatively few practitioners, with a very tenuous network of connections with the decision makers and the public, and with a fragile infrastructure.

This state of affairs poses a special challenge for the educational system in these countries, a challenge much greater than in the "developed" countries. And yet, the educational system itself is much less well developed and hence much less able to rise to this challenge. What are these special features needed in the developing countries?

First, as the scientific and technological infrastructure struggles to establish and strengthen itself, there is need for a great amount of breadth in the education of scientists and technologists, since the luxury of specialization in a well defined small subdiscipline cannot be afforded in a situation where the manpower is small, and the problems varied and large. It is one of the most tragic features of the actual educational systems in many developing countries that instead of stressing this needed breadth and wide range in the education of individuals, they veer in the exactly opposite direction and cripple their scientists and technologists right at the start by premature specialization and a shortsighted and distorted focus in education. Especially damaging is this in applied science and technology where problems tend to be "interdisciplinary" and hence requiring great breadth.

Second, the approach to education in science and technology must be that of problem solving. Whether the "problem" is "basic" or "applied" is of secondary importance at this point and in fact is in part a matter of semantics. The important stress is on viewing science as a method of solving problems rather than a memorization of a rigid set of "laws," and on considering technology as the art of inventiveness and ingenuity rather than the mechanical application of tables and formulae found in handbooks. In this respect also, educational systems in developing countries often fail.

Third, education in science and technology must generate flexibility and readiness for change. In a small scientific and technological community, and with enormous developmental changes expected during one human lifetime, scientists and technologists must be ready to deal with a variety of very different problems, to change their "field" a number of times during their career, and in fact to contribute in several directions at the same time.

Fourth, scientists and technologists in developing countries must be prepared to pursue science and technology and at the same time contribute to the building of the infrastructure that permits the pursuance of science and technology. Thus they must be educated not only in the technical aspects of science and technology, but also with respect to the elements of the context in which science and technology are practiced. This latter includes the operations of machineshops, stockrooms, libraries, universities, and research laboratories, as well as features of science policy, the science of science, and related topics. Such an education is not only absent today in the developing countries, but unfortunately is also absent in the universities of the developed countries where many of the scientists-to-be of the developing countries study.

The features enumerated so far primarily relate to the professional scientists and engineers. If, however, science and technology are to make an attitudinal impact on the country as a whole, mass education of and communication with the population as a whole in science and technology are also tasks of the highest priority. This is a very neglected dimension of education in most developing countries. Newspaper, radio, and television coverage of science and technology is rudimentary and often derivative of the "developed" countries, and even where it exists, it covers only specific results of science and technology and not the attitude, the method, the outlook, or the approach of science and technology that should impact on the everyday lives of the common listeners.

Related to this last problem, there is also little done in most developing countries to educate, in science and technology, the decision makers of the country who, like their counterparts in the "developed" countries, seldom have a significant background in these areas. This split between the "two cultures" results in less damage in the "developed" countries where science and technology are better established and have their constituents and traditional ties, but in the developing countries the gap often results in misshapen "science policies" and a scientific and technological community crippled in its efforts by an uncomprehending bureaucracy.

B. Supporting Structures

An appropriately functional institutionalization of the scientific and technological infrastructure is an integral part of developing the indigenous capability. In this respect the developing countries often find themselves in a curiously ambiguous state of having simultaneously both a surfeit of inappropriate supporting structures and a lack of appropriate ones.

In the educational field, one sometimes encounters huge universities with indiscriminate admission standards and hence an enormous but non-functional student body. At the same time, quality institutions are rare, and advanced education continues to be available only abroad. Universit-ies are also highly bureaucratized, with strict formal requirements, centralized "external" examinations, set syllabi and inflexible and static curricula, which make educational experimentation almost impossible.

The sole aim of universities and technical institutions is often con-strued to be teaching, and hence the well-tested and highly productive symbiosis of teaching and research (i.e., the transmission and generation of knowledge) traditional in some other parts of the world is absent. This automatically creates a gap between universities and the large governmental research institutions which undertake the lion's share of research in developing countries (at least, judged by the size of the research funds they receive).

On the other hand, the link between these research institutions and the productive sector of the country is also often tenuous. There are some understandable and natural reasons for this. The scientific and technological capability of a country must be built up slowly and care-fully over a long period of time so that it is available at the crucial moment when the country needs large scale science-based technology. In the interim, while the rest of the country still primarily subsists on the production and sale of raw materials and items of traditional (empirical) technology, the scientific and technological work performed in the country has less than a full opportunity to make direct and contemporaneous con-tact with the infant beginnings of science-based technological production. But in addition to this natural and probably unavoidable mismatch, the gap is further widened by a lack of communication between the personnel in the research institutions and the people in the productive sector, and by the frequent lack of a sound evaluation procedure for scientific or technolo-gical work proposed or performed. This last area is a particularly ripe one for increased international cooperation.

A serious bottleneck in the structures supporting science and tech-nology in the developing countries is in equipment repair and maintenance. Technicians are few in number and not well enough educated or sharply enough trained. There is usually also an internal brain drain within each country of technicians moving from universities and government institu-tions to the better paying and less bureaucratic private industry, and while this movement represents no net loss and in fact has some advantageous effects, it does handicap research and development work in the universities and governmental research laboratories.

Spare parts for equipment are often hard to get, partly because of a lack of hard currency and the red tape surrounding such purchases, and partly because of the time consuming logistics involving the shipment of

such parts. The ensuing delays not only slow down work, but also seriously degrade the morale of the workers in research and development, thus
further deteriorating the efficiency and effectiveness of science and
technology in the country.

An area where the double spectre of the excess of the inappropriate
and the shortage of the appropriate is very evident is in bodies managing
and administrating science and technology. Many developing countries
abound in organizations for "science policy," particularly those concerned
with "planning," but effective organs for actual decision making and, even
more so, for implementation are frequently lacking. In creating this
skewed situation, some of the international scientific and technological
assistance agencies are accomplices through their encouragement of mere
organization formation without a parallel scrutiny of the effectiveness
and functional nature of these newly created bodies.

C. Sustainability

The second law of thermodynamics says that an isolated system, if left
to its own devices, will go from a state of greater order to a state of
greater disorder. To prevent this, one must pump energy into the system.
Applied to the process of development, this means that even the maintenance
of gains scored in the development process requires constant work and
devotion, and if the momentum of the development is to be kept up, even
greater efforts and greater amount of devotion are needed.

The scientific and technological infrastructure of a country is in a
particularly acute danger of relapse or collapse when it is still small
and rudimentary, and the loss of a few people or of one or two institutions
represents a potentially fatal blow.

Of the many factors which contribute to sustainability under such
conditions, I will mention four.

The first is sustained will and purposefulness, both on a collective
and on an individual level. Jorge Sabato, in his classic analysis of this
problem, wrote (after discussing the great initial enthusiasm of governments in developing countries to build flashy buildings for scientific and
technological institutes, inaugurate them ceremoniously, purchase for them
showy equipment, and staff them with a string of administrators):

"And it is now, at this very moment, that the government loses its
enthusiasm for the centre, and casts it aside. The honeymoon is over:
funds to meet operating costs, which are not spectacular and do not
allow of inauguration ceremonies, are more difficult to obtain;
salaries, particularly for the auxiliary staff--'Why is a glass-blower
so important?' is a typical remark often heard--are not increased; and
bureaucratic difficulties become positive nightmares. The 'operational phase' slows down and research becomes intermittent, taking
on a stop-go tempo; then discouragement and frustration spread and the
brain drain sets in. The centre which seemed so promising now becomes a veritable graveyard for expensive equipment, where only the
second-rate staff members remain, continuing to draw their salaries
without doing any creative work."

But the sustenance of willpower is needed not only on the part of the
government, but also in the individual scientists and technologists, in

spite of the often difficult conditions surrounding research and development, and in the face of a constant struggle with bureaucrats. It is crucial, for this reason, that the international scientific and technological
community give at least moral, and preferably also material support to
colleagues in the developing countries so as to strengthen the local will
power to continue work.

Second, steady, continued, and predictable financial support is
needed. The characteristic time span of individual research and development projects is a number of years, and the span of the effort to build
up a productive scientific and technological infrastructure is decades.
There must be some assurance that financial support for such efforts will
be forthcoming, even if on a modest level, for some time into the future.
For governments used to yearly budgeting and short term activities, such
a pattern of funding is frequently difficult. Fortunately the mere
inertia of every large bureaucracy to some extent shields scientific and
technological projects from too abrupt financial fluctuations.

Third, one needs a substantial isolation from the vagaries of everyday politics. In many developing countries political life is turbulant,
governments change frequently and drastically, and an unfortunate tradition exists which considers political non-involvement amoral. As a
result, science students at the universities spend more time demonstrating
in the streets or sitting in jails than in the classrooms and laboratories,
and often every change of government empties the country of half of its
scientific and technological personnel because such personnel, having been
deeply involved in partisan politics, is purged or sent into exile by the
government equally thoughtlessly devoted to placing politics above science
and technology.

The fourth requirement for sustainability of the scientific and technological infrastructure is the ability to reproduce and expand scientific
and technological manpower as older figures vanish from the scene. This
reproduction and expansion must stress quality and not only quantity. A
number of developing countries ignored this last point, and ended up with
large manpower pools of predominantly low quality, thus creating a class
in society with high expectations and low productivity. Some recent
studies showed that if quality is not emphasized, scientific and technological production can actually decrease with time in spite of an exponentially increasing number of scientists and technologists. From this
point of view, it is particularly important to cater to the relatively
few exceptionally outstanding people in the scientific and technological
community. Such an apparently "elitist" guideline is frequently ignored
in countries with a leaning toward mindless egalitarianism.

D. Information Systems

Science is, to a high degree, objective, universal, collective, and
cumulative, and technology also has these properties, although perhaps to
a somewhat lesser extent. Thus information on old and new developments
in science and technology are of great importance to the worker in research
and development. As a result, the marked isolation of the scientist and
the technologist in developing countries represents a considerable handicap in his ability to pursue his objectives.

Indeed, on the whole, the worldwide patterns of dispersal of scientific and technological information is based on the desire to maximize the

overall scientific and technological output in the very near future.
Thus those performing best in the recent past receive the most preferential
treatment in information systems. This is, to a large extent, a matter of
policy by the worldwide scientific and technological community itself,
since "information systems" organized by non-scientific organizations notwithstanding, the structure of the worldwide scientific and technological
information dispersal is predominantly determined by the scientific and
technological community itself. Thus, in this respect, this community
could make enormous contributions to the cause of science and technology
in developing countries by modifying this policy.

Information transmittal requires three elements: An information donor
or source, a system closely linking the source to the user, and a recipient
with sufficient knowledge, skill, and motivation to use the information.
It is unfortunate that much of the official effort by national and international bodies is directed only to a small part of the second of these
elements, resulting in mechanisms to bring taped, computerized, or other-
wide "processed" information to large and centralized information centers
in the developing countries, which may or may not have effective links
with the individual scientists or technologists who can actually use the
information.

Similarly, "official" information systems often greatly neglect
person-to-person communication modes in favor of the impersonal, written
or computerized modes. And yet, in spite of the huge "information explosion," the most effective and frequently used method of communication
between scientists and technologists is through personal contacts. In
a recent survey in a country with a well organized central technological
information system, it was found that the industrial organizations in the
country received their information from that center only 3% of the time,
while the remaining 97% came through personal contacts, visits, letters,
telephone calls, and other decentralized modes.

In the area of technology especially, the "hands-on" method of
learning about new developments domestically or abroad is highly preferable to written and formalized channels of communication.

The relative deprivation of scientists and technologists in the
developing countries in areas of scientific and technological information
is evident in many ways. Scientific and technical journals rarely and
only tardily reach them because of the high cost to be paid in scarce
foreign currency. They have difficulties participating in scientific and
technological conferences which are mostly held in the "developed"
countries many thousands of kilometers away. Scientific and technical
visitors are quite infrequent, compared to the flood of exchanges and
guest speakers common in institutions in the "developed" countries. Conversely, they have hardly any opportunity to visit other institutions, such
as the ones leading in their field which are likely to be located in the
"developed" countries. Catalogues, technical reports, exhibitions of new
hardware, listing of patents, or manuals of technological products relatively rarely reach the developing countries.

But not everything is the fault of the international scientific and
technological community. The internal communication system within a given
developing country is often also very weak. Exchanges and visits between
the few institutions and organizations in the country are rare, infighting
among the few scientists and technologists is not infrequent, and the

contact between the scientific, technological, and industrial sectors is minimal.

III. SOCIAL RESPONSIBILITY OF SCIENTISTS AND TECHNOLOGISTS

When the activities of a scientist or technologist coincide with those envisaged by the motivations and aspirations of the society around him, he is said to be a socially responsible person. This does not necessarily mean that the scientist's motivations and aspirations are also the same as those of societies. Many activities have a multitude of different justifications in terms of various motivations and aspirations.

Furthermore, motivations or aspirations, of individuals and societies, are themselves pluralistic. In a developing country, societal aspirations include national pride, a rise in the material standard of living, independence, a desire to be on an equal footing with the "developed" countries, spiritual and cultural fulfillment, and many others. Thus the perception of what these aspirations are and which of them are the dominant ones may vary from one observer to another.

Individual motivations are partly internal and partly external. Among scientists and technologists internal motivations often play a decisive role in determining success or failure. Scientists intensely curious about natural phenomena, technologists fascinated with inventiveness and the creation of new products will be likely to have the determination, drive, energy, and persistence that is very much needed in an environment where external motivation and support cannot necessarily be counted on.

The basic motivations of scientists and of technologists hardly differ from a "developed" country to a developing country. Yet the optimal way in which they can convert this motivation to accomplishments will be different in part. For substantial effectiveness in the context of a developing country and for a stamp of maximal social responsibility, the scientist and technologist must be a very forceful, radiating and energetic person with considerable leadership qualities.

An aspect of social responsibility that is often discussed is the degree to which a scientist or technologist in a developing country is oriented toward the immediate material goals of his own country, as compared to the broad general goals of the worldwide scientific and technological community and through them humanity as a whole. For example, a brain drained scientist is often called socially irresponsible because he left his own country, whereas he might try to defend his behavior by claiming that his talents can serve humanity better if they have the opportunity of being realized in the more conducive environment of the "developed" countries. It is extremely difficult to make univocal and definitive judgments in such matters which may depend on the circumstances. It is certain, however, that without some of the outstanding scientists and technologists being willing to operate in their native countries, these countries will never attain scientific and technological capabilities, no matter how much international assistance is showered upon them.

In some respects, an outstanding scientist or technologist who could very well find acceptance in any country of the world but who decides to stay in his own country, makes certain personal sacrifices. At the same

time, however, he might also gain benefits which would be unattainable
for him in a "developed" country. Being among the top three scientists
in a country with a small scientific community has some definite advantages over being No. 7,594 in a country with tens of thousands of scientists. Also, being able to see directly the fruits of one's effort in
creating science and technology where there was none previously is a
rewarding experience, in contrast to activities in a huge scientific community where no single person can possibly have a very large impact. Thus
what may be judged an exhibition of social responsibility by some onlookers may very well have less altruistic and more self-centered roots.
In the end, what matters is that both the scientist and the society he
lives in benefit, in one way or other, from his activities, something that
can be definitely ascertained only in retrospect.

IV. INTERNATIONAL MECHANISMS FOR COOPERATION

As is evident from the foregoing, indigenous capability, willpower,
and human and material resources are indispensible in the creation of a
scientific and technological infrastructure in developing countries.
Thus one of the main limitations of international cooperation in science
and technology is the necessity of linking up with such an indigenous
potential within the country. If such potential is absent, the cooperation will fail even in the face of massive external resources.

This point is of particular importance in the area of technology
transfer. It is certainly true that some technology in the "developed"
countries is deliberately kept from the developing countries out of fear
of competition, and some more is available only at a considerable cost in
royalties and patent fees. On the other hand, an enormous amount of
technical information is freely available in technical journals, expired
patents, governmental technical reports and publications, and other
sources. This wealth of information and know-how remains unexplored and
unutilized because the knowledge and expertise of how to use this technology is lacking in the developing countries. As an example, a gigantic
amount of crucially useful information generated by natural resource
survey satellites placed in orbit by the "developed" countries and covering the developing countries is freely available from the governments of
the "developed" countries, but most developing countries lack the know-
how to use it.

Although international cooperation in science and technology can
operate through many channels, these can loosely be classified into four
groups: International agencies, national governmental agencies, private
organizations and companies, and the scientific and technological communities.

Starting with the international agencies, we find among them some
within the United Nations family, some regional ones like the Organization
of American States, and some specialized ones like the International Council
of Scientific Unions.

Most of the scientific and technological efforts in the United Nations
are channeled through the special agencies like the International Atomic
Energy Agency (IAEA), the World Health Organization (WHO), the United Nations
Educational Scientific and Cultural Organization (UNESCO), and others.
Although there has never been an independent, systematic, and substantial

evaluation of the results of the activities of these organizations (and in
fact even the methodology for such an evaluation remains to be constructed),
there is no doubt that these organizations have scored some successes and
can exhibit some programs which have had a noticeable impact. Yet, in
the judgment of many scientists and technologists in the developing countries, these UN agencies also suffer from a number of very serious ills
which greatly impede their effectiveness. In an attempt to stay apolitical, their internal operations become intensely political, with personnel
selected on a regional (and hence political) basis instead of on scientific, technological, or managerial merit. Their bureaucracies are horrendous, and in fact a very major proportion of the funds of many of
these agencies is spent on administrative personnel at the headquarters
rather than on actual projects in the field. Because of the perceived
necessity not to "interfere" in the internal affairs of countries, the
agencies deal most often with governmental bureaucrats and hence are
greatly isolated from the working scientist and technologist. The
agencies are also quite timid, conservative, and formalistic in the formulation of their projects, partly due to a multiple and politically
diverse screening mechanism of committees which passes mainly the innocu-
ous, non-controversial, and hence often routine and unimaginative programs.

Turning to the national agencies in the "developed" countries which
are concerned with scientific and technological cooperation with the
developing countries, experience seems to indicate that the huge multi-
purpose "foreign aid" agencies (like AID in the United States or CIDA in
Canada) are poorly suited to administer the subtle task of international
cooperation in science and technology. In contrast, much smaller, light-
footed, and specialized agencies (like IDRC in Canada, SARAC in Sweden)
appear much more successful. Recent attempts to establish a similar
organization in the United States (named ISTC) have failed, but efforts
will continue in this direction.

Private agencies and companies can assume a great diversity of forms
and modes. The private foundations such as the Ford and Rockefeller
Foundations in the United States, the von Humboldt and Volkswagen Foundation in West Germany, and a large number of others in many countries constitute one such mode. Although operating with modest resources compared
to national agencies or even to UN agencies, they have a low profile (and
hence produce little political controversy) and, in principle, have more
flexibility and spirit of experimentation than their governmental or UN
counterparts. Research and development leading to the Green Revolution
is one example of what such private organizations can do.

A much more controversial private agent for international activity in
science and technology is the multinational firm. Its record contains conflicting claims from various sides of great contributions to the educational and training infrastructure of developing countries, a neglect of
the evolution of local research and development in favor of such work being
done in the mother country, an exploitation of the natural resources of
developing country, a great role in developing such resources and in
transferring practically oriented knowledge and know-how into the developing country, etc. The abstract debate on these features is indeed intense.
At the same time, most countries managed to work out a practical modus
vivendi with such multinational firms in which both parties receive benefits in an operation of the firm regulated by local governments. It is
better to avoid the theoretical debates on this topic and instead propose

specific ways in which developing countries can benefit more from the presence of such multinational firms. For example, multinational firms could greatly contribute to the realistic orientation of indigenous research and development of a developing country by establishing small research and development laboratories of their own on the soil of these developing countries.

It is intriguing to note that in recent years the number of multinational companies owned and operated by some developing countries (like South Korea, Hong Kong, Brazil) has been growing precipitously. To be sure, the total assets of such companies is still only at most a few percent of the assets of the multinationals owned and operated by "developed" countries, but the trend favors a rapid increase of this percentage in the next few years. It is likely that such a development will help to defuse the political oratory surrounding the existence of multinational firms, and will help to concentrate attention on the practical measures that can make such firms more effective in the evolution of the indigenous capabilities of developing countries.

Finally, there is the worldwide scientific and technological community. Much remains to be done to turn this community into an effective instrument for the assistance of developing countries in their creation of an indigenous capability. As already mentioned, communication patterns within this community need restructuring to cease to disadvantage the colleagues in developing countries. Professional societies in the "developed" countries, most of which show little if any concern for these colleagues, must be reoriented. The argument for such a change need not be given in altruistic terms: On purely scientific grounds alone, such action is warranted. If the worldwide scientific community arose from the whole population of the earth rather than, as it does at present, from only one-quarter of the population (that living in the "developed" countries) science and technology would make much more rapid progress, thus benefitting all countries and delighting scientists everywhere.

Among the many areas in which scientists and technologists in "developed" countries can play a pivotal role is the education of students from developing countries at institutions in the "developed" countries. While in the long run indigenous education is preferable to education abroad, in many developing countries educational facilities are not extensive, advanced, and broad enough to serve all needs of the expanding scientific and technological manpower. Thus education in the "developed" countries remains a crucial source of such manpower. Scientists and technologists in the "developed" countries must see to it that in addition to the generally high level of education such students receive, their experience is supplemented by some special features geared toward the needs and the context of science and technology in their home countries. Specific suggestions for such auxiliary education exist in the literature.

V. NATIONAL MECHANISMS

On the whole, the indigenous domestic mechanisms operating in developing countries among scientists and technologists and between them and the rest of the country are small and weak. This is in sharp contrast with the bureaucratic structures surrounding science, technology, and production, which are huge and rigid. Although the strengthening of these tenuous mechanisms is primarily the task of the indigenous scientific and technological communities, international cooperation by colleagues in other countries can be very helpful.

Scientific societies exist in many countries but are often not very active. Their manifestations can be found frequently in social events, in formal business meetings involving the election of officers, and in similar ceremonious occasions. What is needed is a much greater emphasis on action programs in such societies, such as the holding of scientifically substantial meetings and conferences, the representation of the scientific communities in public issues involving the operation of science and technology in the country, and the improvement of the already mentioned weak internal scientific communication systems. There are also important tasks to be done in connection with local journals.

Editors of the large international scientific journals generally frown upon the existence of the many tiny and very local scientific and technical journals in the developing countries, which they consider inefficient, ineffective in communication, of poor quality, and in general detrimental to the worldwide scientific literature. I believe this is a onesided view. Local journals, if properly organized, can serve very important functions by providing quick, inexpensive, and undiluted ways for the researchers in a country or region to keep up with scientific and technological developments in the region. Such journals also boost the local morale, and serve as educational tools for learning about science policy and the management of science. In many cases they also help to bridge the language gap and contribute to the development of scientific and technological terminology in the local language. The functions of local journals are sufficiently different from those of the large international ones that in many cases double publishing should be accepted as a useful method of fulfilling the objectives of both channels of communication.

The problem, however, is that local journals are often not operated optimally. Perhaps the most difficult problem is the refereeing of the submitted articles, which cannot be done in the tiny local scientific community, but at the same time the editors of the journals are reluctant to use the customary international refereeing system. In this area much can be done by the international scientific and technical community to overcome these problems.

As mentioned earlier, even more tenuous is the communication between the local scientists and technologists and the public at large. "Popular" literature on science and technology is rare, programs on radio and television dealing with science and technology are few in number and often not very effective, and whatever programs exist deal more with facts than with ideas, attitudes, and the way of thinking that characterize science and technology. Here also, cooperation with the worldwide scientific community could be beneficial.

Mechanisms such as consulting agreements, joint seminars, mutual visits between the scientific and technological community and the industrial sector of the country are seldom in existence, and hence industry becomes accustomed to lean on foreign science and technology, while the local scientists and engineers become isolated from those problems which arise in the context of their countries.

In utilizing technology, originating abroad or in the local laboratories, financial institutions are indispensible to offer the necessary credit for the conversion of this technology into production. The taking of risk in business is often not a local tradition in developing countries, and hence financial incentives may be necessary to induce the entrepreneur to choose the risky but potentially more lucrative alternative over the traditional, low-technology, but relatively safe one. For this, financial institutions themselves need a spirit of bold entrepreneurship, combined with considerable scientific and technological expertise in order to tell the swindler from the pioneer, the crackpot from the expert, and the risky but possible from the far-out, harebrained schemes. These qualifications are not always present.

Finally, there is a burning need in many developing countries to reform the bureaucratic structures pertaining to science and technology, so that they can involve a much larger number of active scientists and technologists. The gap between administrative fiction and scientific and technical reality can be bridged only if there are a large number of people who are simultaneously involved in scientific research or technological development and in decision-making in science policy. The sharp division between the working scientific and technical manpower on the one hand, and the science policy makers no longer personally in contact with science and technology on the other hand, is highly detrimental not only for pragmatic but also for psychological reasons. Provisions must be made to enable outstanding scientists and technologists to play a dual role of researcher and scientific statesman. Here also, the international scientific community can be very helpful by offering opportunities to those in the scientific and technological leadership of developing countries to periodically spend an extended visit at an active scientific or technical institution in order not to become overly separated from the spirit and results of up-to-date research and development.

This enumeration of national mechanisms is very scanty indeed. Many other aspects are equally important, pertaining to education in science and technology, to the manufacture of research and demonstration equipment, to repair and maintenance facilities, to libraries and other modes of information transmittal, to extension services to link technology with the non-technologist user, and many other activities. Some of these were mentioned in previous sections, others will have to be content with just being listed.

There is, however, a general international mechanism that can have a great effect on the improvement of such national mechanisms. This is the bilateral link, in which a small group of scientists or technologists in a developing country join forces with an equally small group in a "developed" country, with a common objective and a large body of overlapping interests. In strengthening the activities of professional societies, in the editing of journals, in the enhancement of interaction with the local public, in the creating of technological information sources, and in most other areas also, such informal, inexpensive, personal, and well-focused collaborative schemes can work wonders. It has been shown that the smaller the linked groups are, the more productive the link is. The initiation of such small links should and generally does come from the potential participants, which ensures liveliness, the absence of undue red tape, and a feeling of personal responsibility toward the project, in contrast to the "I-just-work-here" attitude that one can often find in huge, centralized collaborative programs.

VI. SPECIFIC RECOMMENDATIONS

In this section, I will first list items on which proposals should
be made. I will also offer, in an appendix, a set of nine specific proposals dealing with certain aspects of the development of science and
technology. These proposals are quantitative, with the amounts having been
determined from the data and results provided by science policy and the
science of science, keeping also in mind that the financial willingness
of the "developed" countries to carry the lion's share of large programs
is rather limited nowadays. The proposals are structured in such a way
that their implementation can be performed country by country, and hence
a partial participation of countries will result in partial benefits, in
contrast to centralized schemes which usually fail to work unless everybody cooperates. I want to stress that these proposals, while having some
merit of their own, are cited here mainly as examples of the type of
specificity and functionality that is likely to result in tangible improvements.

The list of items given below has been selected with international
cooperation primarily in mind. It would appear that an international conference should more properly emphasize those aspects of the building of
science and technology which are not exclusively the concern of a developing county within its own confines.

A. Conceptual

1. The problems of developing countries in building their science
and technology must be publicized broadly to the worldwide scientific and
technological community to raise awareness and catalyze cooperation.

2. Objective research, preferably using specific case studies, must
be enhanced on various aspects of science and technology in the developing
countries so that we can learn from past experience.

B. Education

3. In spite of the bad economic times now prevailing in the
"developed" countries, the opportunities for students from developing
countries to obtain advanced education in the sciences and technology in
the "developed" countries must at least be maintained.

4. The selection of students from developing countries to receive
education abroad should be improved so that the deserving is chosen and
the undeserving not selected.

5. Revolving funds should be established for the travel of students
from developing countries to and from the educational institution abroad
where they receive advanced education.

6. The education of students from developing countries at institutions in the "developed" countries should be supplemented by aspects of
science and technology which are of particular importance in the context
of the developing countries.

7. The advising of students from developing countries at institutions
in the "developed" countries should be improved to conform with the special
needs of those students.

8. Students from developing countries studying science or technology at institutions in the "developed" countries should be given opportunities for practical employment in scientific or industrial institutions during the vacation periods.

9. Cooperative programs should be established with the public mass communication media in the developing countries to strengthen their ability to bring science and technology to the population as a whole.

10. Scientific and technological institutions in the "developed" countries should cooperate with their counterpart institutions in the developing countries to produce positive incentives for stemming and reversing the "brain drain."

C. Communications

11. Unused back issues of scientific and technological journals should be channeled to libraries in the developing countries.

12. Current issues of scientific journals should be made de facto available to libraries in the developing countries.

13. Publication charges in international scientific and technical journals should be adjusted to fit the financial circumstances in the country of the potential author.

14. International cooperation should be organized to assist scientific and technological journals published in the developing countries in the refereeing of the submitted articles.

15. Microfilm, microfiche, and other techniques should be explored to provide the developing countries with less expensive and more rapid communication channels for scientific and technical journals, reports, catalogs, manuals, and abstracts.

16. The attendance of scientific and technical meetings should be made more equitable for deserving scientists and technologists from the developing countries.

17. Periodic opportunities should be created for scientists and technologists for short or longer visits to active scientific or technical institutions, laboratories, or production facilities in the "developed" countries.

18. The flow of scientists and technologists from "developed" countries to institutions in the developing countries for the purpose of visits and collaborative research and development should be increased.

19. International cooperation should be arranged to help developing countries who wish to avail themselves, at least temporarily, of the services of scientists and technologists from the "developed" countries.

20. International cooperation should be arranged to find means of enhancing the internal communication between scientists and technologists in a given developing country.

21. Contributions of various "developed" countries to the United Nation's "associate expert" scheme should be strengthened.

22. More scientific attaches of "developed" countries, in the persons of active scientists or technologists, should be stationed in the developing countries in order to facilitate the contact between the scientific communities through governmental channels.

23. A communication satellite (or part of one) should be earmarked for communication among scientists and technologists the world over.

24. International cooperation should be enhanced to strengthen regional interaction and communication among scientists and technologists in developing countries.

D. Research and Development

25. Multinational companies should be persuaded to establish research laboratories in all the developing countries where they operate.

26. Bilateral links between groups of scientists or technologists in developing countries and their counterparts in "developed" countries should be increased in number and variety.

27. Governmental regulations in many "developed" countries should be modified so that joint research projects between scientists or technologists from that country and their counterparts in a developing country should be easier to support.

28. International cooperation should be arranged to help in maintaining and providing spare parts for research and development equipment in the developing countries.

29. Existing international organizations offering support, on an individual basis, to scientists and technologists in the developing countries (e.g., International Foundation for Science, Fund for Overseas Research Grants and Education, etc.) should be supported on a more extensive scale.

30. International cooperation should be arranged for a better utilization by the developing countries of freely available technological information, procedures, expired patents and techniques.

31. International patent practices should be reviewed to facilitate the utilization of such patents by the developing countries.

32. The "developed" countries should refrain, even in economically adverse times, from creating trade barriers against the technological products of the fledgling industries in the developing countries.

E. Organizations and Management

33. Literature (books, journals, reports, studies, statistical data, etc.) on the science of science, on the organization and management of science and technology, on technology assessment, on the history, philosophy, psychology and economics of science and technology, and related matters should be made available to scientists and technologists in the developing countries in order to encourage the development and absorption of these areas of inquiry for utilization in the policy for science and technology.

34. Assistance should be offered to developing countries in their efforts to evaluate and assess research and development being proposed or being completed.

35. International assistance should be organized to aid scientists and technologists in developing countries who wish to maintain their active involvement in research and development while also participating in the policy making, organization, and management of science in their own countries.

36. Suitable methodology should be created and then applied to the evaluation of the effectiveness of international and national organizations working in international assistance and cooperation in science and technology.

APPENDIX. An example of a set of specific resolutions aimed at assisting the increase of the indigenous capacity of developing countries in science and technology. These proposals first appeared in International Development Review 21:2, 16 (1979).

WHEREAS

it is generally recognized that the existence of a functional scientific and technological infrastructure in each country is an indispensible ingredient of the New World Economic Order, and

RECALLING THAT

this country paper repeatedly stressed the crucial importance of international cooperation in the creation of such an infra-structure, particularly in those countries where such an infrastructure is at the present not developed,

BE IT RESOLVED THAT

a NEW STRUCTURE FOR WORLD SCIENCE AND TECHNOLOGY be formed, aiming specifically to assist each country in establishing its own capacity in science and technology through its own infrastructure.

An an important first step in the implementation of this NEW STRUCTURE FOR WORLD SCIENCE AND TECHNOLOGY,

BE IT RESOLVED THAT

the countries with an already developed scientific and technological infrastructure (henceforth designated as countries A) and the countries where such infrastructures are still in a rudimentary state (henceforth designated as countries B) jointly commit themselves to take the following concrete steps to accelerate the creation of such infrastructures in countries B:

1. Each country A assures that by 1990 key libraries in all countries B are supplied with back issues of those major scientific and technological

journals (from 1960 on) which were published in that country A. The number of key libraries in countries B should be one per country B or one per 10 million population of country B, whichever higher. Criteria for selecting major journals should be such that the worldwide total number of journals thus reaching each country B is approximately 1,000. Correspondingly, each country B benefitting from this program commits itself to providing orderly space for the incoming journals, including cataloging and shelfing, and to assuring easy access to such journals by the scientific and technological community.

2. Each country A supplies each key library in each country B with current issues of those major scientific and technological journals which are published in that country A. The transaction of journals should be in exchange for 1/4 of the regular individual subscription rate of that journal and the payment should be made in the currency of country B. The key libraries and the number of journals are to be defined as in Section 1 above. Correspondingly, participating countries B assume obligations similar to those listed in Section 1 above.

3. Each country A assures that for any international scientific and technological conference held in that country, resources are made available to make possible the participation of scientists and/or technologists from countries B, in numbers up to one such participant from all countries B together for each 30 overall participants at that conference, provided that such participants from countries B are available and measure up to the usual scientific and/or technological criteria used to select participants for that conference. Correspondingly, participating countries B pledge to facilitate the participation of their scientists and/or technologists through all administrative means.

4. Each country A assures that resources are available to support scientists and technologists from that country on short term visits to counterpart institutions or individuals in countries B. These short term visits, with durations up to a month or two, might include several countries B and the number of such visits per year should be one for every 1,000 research scientists and technologists in country A, provided that a sufficient amount of mutual interest exists between the participants of such a visit. The participating countries B correspondingly pledge, during each such visit agreed on, a sum equivalent to US $10 per day, in the currency of the country B in question, toward a partial defrayal of the local living expenses of the visitor.

5. Each country A assures that resources are available to support longer visits by scientists and technologists from countries B to counterpart institutions in country A. These longer term visits are to be up to one year in duration, with occasional renewals for a second year but never beyond that, and the number of these long term visits is to be one for every 300 research scientists and technologists in country A, provided that a sufficient number of qualified applicants are available as judged by the customary standards prevailing in the worldwide scientific and technological community. Correspondingly, participating countries B pledge that, (a) the customary local salary of the scientist or technologist continues to be paid in the currency of that country B to locally maintain the family of the scientist or technologist while he is abroad, and (b) the transportation of the scientist or technologist will be provided by country B on its national airline, if there is such an airline, to a point closest to the institution to be visited that is a regular stop of that national airline.

6. Each country A assures that resources are made available for the placing into scientifically or technologically meaningful practical positions, during the longer academic recesses, students of science or technology from countries B who are being educated toward a degree at an institution of higher learning in country A. The number of such positions is to be up to one for every 500 research scientists and technologists in country A, provided that qualified applicants for such positions are available. The participating countries B, correspondingly, pledge that the time spent in such practical positions will be counted as time spent in the service of country B when matters of seniority and promotion arise during the career of this student.

7. Each country A assures that resources are available for the formation of bilateral links between a group of scientists or technologists in a country B on the one hand, and a counterpart group in country A, on the other. The nature of the link is to be determined by the two parties involved, but the resources should also allow for the possibility of some collaborative research between the two groups. The number of such bilateral links is to be one for every 2,000 research scientists and technologists in country A, provided that qualified groups are available, as judged by the usual standards of the worldwide scientific and technological community. Correspondingly, the participating country B pledges to contribute, in its own currency, to each such linked group in country B, a sum amounting to 10% of the amount contributed by country A to that bilateral link.

8. Each country A assures that resources are available for the support of technicians from country A stationed in a country B or covering a group of countries B. Such a technician is to be experienced in the repair and maintenance of scientific and technological research equipment, have some ability to offer on-the-job training to local technicians, and he should also have access to minor spare parts from a stockroom which is to be replenished through the eventual arrival of parts ordered through the usual purchase procedures of country B. The number of such technicians is to be one for every 10,000 research scientists and technologists in country A, and the distribution of such technicians among the participating countries B should be such that no country should have less than one-half such technicians from among all the technicians contributed by all countries A together. Correspondingly, each participating country B pledges that (a) it expedites to the greatest possible extent the order and delivery of the spare parts borrowed from the revolving stock room, free of custom duties, and (b) it contributes, in its local currency, and equivalent of US $10 for each day the technician spends in country B, in order to partially defray his living expense.

9. Each country A assures that it will make a contribution annually into an overall fund for home fellowships, to be administered by a designated agency of the United Nations or some other international entity so designated. Home fellowships are to be used for the salary and research expenses of outstanding and promising scientists and technologists from countries B for work by them to be carried out in countries B. The support of such researchers should be comparable to the level customary in most countries A. Each country A should provide funds sufficient for the support of one scientist or technologists for every 10,000 research scientists and technologists in that country A. The fellowships should be awarded on the recommendation of a committee of scientists and technologists representing the participating countries A, the selection being made

from among individual applications received by this committee from candidates from countries B.

The provisions of this resolution are to be in effect until 1990, at which time a review is to be undertaken to ascertain the effectiveness of these provisions and the advisability of continuing them in the original or a modified form.

BIBLIOGRAPHY

The literature on problems of science and technology in developing countries is enormous. What is given below is a somewhat arbitrary selection of items which appear to be particularly relevant to the content of this report. These references, in turn contain bibliographies of hundreds of other writings, which can serve for further reference.

Cooper, C. (Ed.), Science, Technology, and Development. Frank Cass & Co., London (1973).

Djerassi, C., A Modest Proposal for Increased North-South Interaction Among Scientists. The Bulletin of the Atomic Scientists 32, 56 (Feb. 1976).

Eisemon, T., Emerging Scientific Communities: What Role does Counterpart Training Play? International Development Review: Focus 19, 14 (1977).

Frame, J.D., et al., The Distribution of World Science. Social Studies of Science 7, 501 (1977).

Frame, J.D., Mainstream Research in Latin America and the Caribbeans. Interciencia 2, No. 3, 143 (1977).

Glyde, H.R., Institutional Links in Science and Technology: The United Kingdom and Thailand. International Development Review: Focus 15, 7 (1973/4).

Heenan, D., and W.J. Keegan, The Rise of Third World Multinationals. Harvard Business Review, Jan.-Feb. 1979, p. 101.

Hentges, H., The Korean Institute of Science and Technology: A Case Study in Repatriation. International Development Review: Focus 16, 27 (1974/5).

Herrera, A., Social Determinants of Science Policy in Latin America: Explicit Science Policy and Implicit Science Policy. Journal of Development Studies 9, 19 (1972).

Kidd, C.V., Manpower Policies for the Use of Science and Technology in Development. Pergamon Press, Elmsford, NY (1980), 183 pages.

Layton, E., Conditions of Technological Development. In I. Spiegel-Rösing and D. DeS. Price (Eds.), Science, Technology, and Society: A Cross-Disciplinary Perspective. Sage, London and Beverly Hills (1977).

Lomnitz, L., Hierarchy and Peripherality: The Organization of a Mexican Research Institute. Minerva 17, 527 (1979).

Moravcsik, M.J., Science and the Developing Countries. A Contribution to
    the U.S. Country Paper for the U.N. Conference on Science and Tech-
    nology for Development, National Science Foundation (1977) (unpub-
    lished: copies can be obtained from Dr. A. Pettifor, Room 1229,
    National Science Foundation, Washington, DC 20550, USA).

Moravcsik, M.J., Science Development--The Building of Science in Develop-
    ing Countries. PASITAM, Indiana University, Bloomington, Indiana
    (2nd Edition, 1976).

Morgan, R.P., et al., Science and Technology for Development, Pergamon
    Policy Studies No. 38, Pergamon Press, New York (1979).

Roche, M., Early History of Science in Spanish America, Science 194, 806
    (1976).

Rosenberg, N., Science, Invention, and Economic Growth, The Economic
    Journal 90 (March 1974).

Sabato, J., Quantity versus Quality in Scientific Research. (I): The
    Special Case of Developing Countries, Impact 20, 183 (1970).

Sagasti, F.B., Science Policies in Developing Countries, Science and
    Public Policy 2, 56 (1975).

Skolnikoff, E., Science, Technology, and the International System. In
    I. Spiegel-Rösing and D. DeS. Price (Eds.): Science, Technology
    and Society: A Cross-Disciplinary Perspective. Sage, London and
    Beverly Hills. (1977).

Wijesekera, R.O.B., Scientific Research in a Small Developing Nation--Sri
    Lanka. Scientific World 1:6 (1976).

Wionczek, M., Prospects for the UNCSTD-Three Major Underlying Issues.
    El Colegio de Mexico (1978).

Zahlan, A.B., Science and Science Policy in the Arab World. St. Martin's
    Press, New York (1980).

* *

A MAP OF SCIENCE DEVELOPMENT

Michael J. Moravcsik
Institute of Theoretical Science
University of Oregon
Eugene, Oregon 97403, U.S.A.

The purpose of this brief note is to suggest a schematic description of
the various steps in the development of science and tehcnology,
particularly in the Third World. Like all such schemes, this one
suffers from being somewhat oversimplified, from describing continua in
terms of discrete categories, etc. Yet such descriptions can help in
orienting activities in desirable directions and in avoiding false
impressions of accomplishments.

In the development of science and technology, in science policy, or in
all events surrounding the building of science and technology, we may
distinguish five types of occurrences:

1. <u>Formation and clarification of concepts</u> (C). One might call this
the theoretical aspect of science development, which discusses the
differences and connections between science and technology, the
characteristics of underdevelopment, the methods of evaluating
scientific research and technical applications, etc. Much
conceptualization has taken place in past decades, and the work is by
no means completed. This aspect is important for science development;
but even on the basis of the already existing body of concepts, broad
agreement can be found with respect to a very large number of specific
steps necessary for the development of science and technology. Hence,
while conceptualization will continue, its incompleteness is no serious
impediment to the other steps listed below.

2. <u>Proposals</u> (P). In order to carry out development activities in
science and technology, concrete proposals need to be formulated. A
proposal seeks to answer the following questions: What is the problem?

Why were previous efforts unsatisfactory? What is the new plan and why should this be superior to previous efforts? Who will do what, when, and how? How much will the proposed activity cost? What are the anticipated results?

3. Tools (T). Activities in science development are carried out by organizational and other tools. They may be institutions, organizations, societies, ad hoc groups of scientists, agencies, programs projects, financing, etc. There are a great multitude of already existing tools for science development, and although for special tasks new ones may be needed, the creation of tools in itself is not of very high priority in science development.

4. Auxiliary activities (A) We have now progressed from the preparations (conceptualization, proposal making, tool forming) to the actual science development activities. The ones we can classify as auxiliary pertain to the creation of a capability to do science. For example, an activity which improves the advising of Third World students studying in universities in the United States is such an axiliary activity, since it assists in the creation of the scientific manpower which in turn will be directly engaged in the development of science and technology in the Third World.

5. Basic activities (B) This is the fundamental purpose, the central objective of this process: increasing and improving scientific knowledge and its application to technological and other activities.

Why such a classification? Because every event surrounding science and technology development can be tagged in terms of any of these five stages. For example, the organization of a meeting on science policy (such as the Global Seminar), can be tagged (C,P,T). At worst meetings have no tags, although most meetings earn at least a (T) tag in that they usually strengthen some personal contacts among the participants even if they produce no new proposals. Note that no meeting can ever include A or B in its tag, although the appropriate follow-up of a meeting can result in A or B.

The main point I would like to make is that the map of science development abounds in tags in containing C,P,T, or combinations of these, but is very poor in tags with A or B in them. As if the world were divided into two non-overlapping parts: elaborate talk and complex preparations on a large scale, on the one hand, and a mere whimper of implementation, on the other. Huge sums of money are consumed in the former, but it threatens to remain sterile, self-contained, repetitively circuitous.

How can we avoid continuing this pattern? It is my hope that the introduction of the tagging suggested here might help. For example, in the midst of the follow-on planning for the global seminar, the aim should be to produce programs which can confidently be tagged A or B. Indeed, such tagging should be an indispensible criterion for even considering follow-on suggestions. Perhaps such an orientation can finally help breaking through the wall between the CPT-land and the AB-land, and thereby release our resources and energies for the real enhancement of science and technology in the Third World.

*   *

B. Latent Opportunities

Reprinted Papers

Reprinted by permission of *Minerva* 4, No. 3
(1966) 381.

SOME PRACTICAL SUGGESTIONS FOR THE IMPROVEMENT OF SCIENCE IN DEVELOPING COUNTRIES

Michael J. Moravcsik

In a previous article,[1] I discussed some of the reasons for my belief in the importance of a newly developing country creating a firm foundation not only in applied but also in basic research in the natural sciences. Following that general discussion, I would now like to turn to the listing of some specific and concrete steps that can be taken to achieve the general goal.

It might be well to begin my discussion of a controversial problem with a note of caution. My own personal experience with the problems described in this article has been gained mainly (although not exclusively) during the academic year 1962–63, which I spent in Pakistan as a temporary employee of the International Atomic Energy Agency, assigned to the Pakistani Atomic Energy Commission's Atomic Energy Centre in Lahore. It is possible that this experience does not allow an immediate generalisation to other developing countries with completely different historical, cultural and political backgrounds, such as Nigeria or Uruguay, and thus my conclusions might be valid mainly for the Indian subcontinent.

When a Western " expert " visits a university in a developing country, he often finds himself requested to review the local syllabus or curriculum and to suggest changes. He also often hears complaints about the absence of textbooks and other instructional aids. Indeed, there is much to be done to improve the situation in these respects. But he is also likely to come to the conclusion that the main trouble with science education at these universities lies in a different domain which is more difficult to convey to the local staff and even more difficult to remedy. The problem is not with *what* is being taught but with *how* it is taught. There is an overwhelming tendency on the part of the staff and the students to consider science as a collection of facts to be memorised, an abstract discipline consisting of general laws of nature, the faultless recitation of which makes a well-qualified scientist. That the best way to absorb science is to work out problems is very little appreciated and hardly ever practised. I have encountered undoubtedly very bright young men in Pakistan who represented the top of their class at the best Pakistani universities, and who

---

[1] " Technical Assistance and Fundamental Research in Underdeveloped Countries ", *Minerva*, II, 2 (Winter, 1964), pp. 197–209.

could rattle off the second law of thermodynamics much better than I can, but who, when confronted with the question of what happens to an insulated room when a working refrigerator with an open door is placed in it, not only did not know the answer, but failed to realise that this " silly " question had anything to do with thermodynamics.

What can be done to foster the transition from learning by rote to the perception and selection of problems among those who have concluded an undergraduate course in physics and who would like to go on to post-graduate work abroad? The Atomic Energy Centre where I was stationed in Lahore has a training programme for its young employees (most of whom have received an M.Sc. degree at one of the Pakistani universities) prior to sending them abroad for further training. This programme now includes three courses, in mechanics, electrodynamics and modern physics, respec-tively, based on American textbooks I recommended; the books and the courses lay heavy emphasis on problem-solving. Courses like these might make the transition to the style of work of a Western graduate school less painful. Not less important would be the experience which local teachers would gain in teaching in a more problem-oriented manner.

Postgraduate Education

Although it is very desirable in the long run for an underdeveloped country to be able to train its scientists at home on the postgraduate level too, this is likely to come only at a relatively advanced stage. In the meantime, such training must be obtained at Western graduate schools. This raises many problems.

The very first one has to do with obtaining admission and financial support at a Western graduate school. The pressure for places has been increasing every year and the better schools sometimes have 10 times as many applicants as they are able to admit. In such a situation an applicant from an underdeveloped country has many handicaps from the very beginning. Quite often the record of his predecessors in that department is not too distinguished. Furthermore, his application is supported by academic records which tell very little and by glowing letters of recommendation from completely unknown teachers, who often consider it a feather in the cap of their own prestige if another student is admitted to a Western institution. Since such an applicant represents so many question marks, he is most likely to be turned down by the graduate school which cannot afford to take chances with its scarce places.

There are several possible remedies for this problem. An admittedly stop-gap measure is the one I have tried with trainees of the Pakistan Atomic Energy Commission. I have made arrangements with about a dozen good American graduate physics departments, under which I help to assure them that the Pakistani students I recommend to them are in fact prepared for Western postgraduate physics education. In return the schools are willing to consider the applications of these students in the light of this additional information, although, of course, this does not mean automatic admission and financial support. My own recommendation, in turn, is arrived at in the following fashion. First, there are some Pakistani colleagues whose judgement I have come to rely on and who can compare for me the new applicants with previous applicants, some of whom might already be successfully engaged in postgraduate work in the United States. Second, I have prepared a written examination (consisting mainly of problems), based on the A.E.C. training programme I mentioned in connection with undergraduate education. This examination is then given by a colleague in Pakistan and is forwarded to me for review.

This method is admittedly awkward, piecemeal, and cannot be expected to work forever, as my personal contact with the Atomic Energy Centre becomes attenuated. One would think that some universal written examination could be worked out, similar to the Graduate Record Examination, which could serve as a fair indication of the candidate's ability and preparedness. Although several organisations are working on the construction of such examinations, I doubt very much that they will be successful. Because of the great discrepancy between the educational methods of the different parts of the world and because of the specific shortcomings I discussed above in connection with undergraduate education, I believe that such an examination would indicate very little.

My suggestion for a more permanent and more efficient solution of the problem of selecting students for postgraduate education in the West is the establishment of *ad hoc* interviewing committees which would visit the various universities and interview, for an hour or so, each candidate who wishes to apply to a Western graduate school. A committee of three physicists, travelling for one month, could probably take care of all applicants for physics postgraduate education in all the Asian countries, at a total cost of less than $15,000. It has been my experience that an hour-long interview offers an excellent opportunity to make a fairly reliable judgement of the applicant's accomplishments, capacities and weaknesses. This assessment, should he be admitted to a foreign graduate school, would be valuable information for his supervisor there. The information obtained by this committee would, of course, be made available to any university which might be interested in it.

During the past two years I have been trying to persuade a number of American foundations and government agencies to support this scheme and, while everybody seems to be eager to see it carried out, actual financial support has not yet materialised.

The next question that arises in connection with graduate education is related to the branch of science in which a student should be trained. In order to be able to offer the student, at the completion of his education, a position at home in the proper field, he must be trained in a field which is in fact being worked on at home. At the same time, practice has shown that a simple order prescribing the field of study for a student never works, and perhaps rightly so. The rather obvious solution of this problem is to place the student in a Western university where the field of science which his home university or laboratory has in mind dominates the department and where the student is likely of his own volition to choose that field. This arrangement should be followed even if a particular university is perhaps not as excellent in an overall sense as another in which the particular field is less strongly represented. Since the worship of " great names " among universities is even more pronounced in developing countries than in advanced ones, this procedure requires a certain amount of self-restraint. It is, however, indispensable because it is extremely difficult to change the field of interest of a student once he obtains a Ph.D. in that field.

The Effectiveness of Postgraduate Study

Now let us assume that the student has been able to find an appropriate university, obtained admission there and received financial support. What can be done at this stage to make his transition to his new environment easier and more effective?

First, it must be realised that in most cases such a student will need a lot of individual attention from his adviser and that this attention and cooperation must often be initiated by the adviser since the student is too shy or confused to do so. One of the most important functions of the adviser is to ensure that the student does not take too heavy a load of postgraduate lectures and seminars if his undergraduate background is not absolutely sound. More often than not, such a student will in fact need one or two terms to take some advanced undergraduate subjects to catch up with his Western contemporaries. He will consider the taking of such courses not only a personal affront (being used to being the much-admired top student of his class) but might even consider it a tragedy since it might delay the completion of his doctorate by a year. The emphasis on the formal aspects of education in the developing countries manifests itself not only through the formalistic attitude toward science teaching and learning but also in the excessive concern with degrees and titles. Even if the student's financial assistance is assured throughout his course of postgraduate study, he will want to obtain his degree in the shortest possible time and is sometimes willing to disregard all other considerations, including the quality of his education, to achieve this. Nevertheless, the adviser should make certain that the student has a very solid working knowledge of the basic subjects before he permits him to enter upon the advanced phases. By now, many of the larger physics departments have " experts " on the staff who have dealt with many foreign graduate students, but even then, dealing with some of these cases might require an inordinate amount of time which the adviser, himself a working research scientist, is not willing to give. A judicious allocation of such foreign graduate students among those in the department who realise the problems involved might in some cases " save " students who, if left without special supervision, would fall by the wayside.

Once the student has completed his work towards an advanced degree and received his diploma, he is confronted with the question of whether to return home or not. This is one of the thorniest problems, which will be solved only when postgraduate education on a high standard is established in the developing country itself. In the meantime, there are various measures that should be taken to encourage the young Ph.D. to return.[2]

The most important requirement is that a young scientist must have a good opportunity for research at a respectable salary in his home country. It cannot be expected, and in fact it is not even necessary, however, that these facilities or the salary match those that exist in some places in the West. For one thing, most such scientists would not be at the leading institutions in the West even if they did not return home, and to match the facilities of second- or third-class institutions in the West is not an insurmountable problem. As far as salaries are concerned, what seems to matter is not the absolute standard of living but the standard with respect to the rest of the population. Thus a salary of Rs. 1,000 in Pakistan is a very respectable one, although its dollar equivalent in the United States would be considered quite poor. I believe it is generally true that a young man, who lived in his native country up to the age of 21 or so and then has spent four years in getting a graduate degree in a Western country, prefers, other things being equal, to return to live in his own country. By " other

[2] One of these, among others, is the education on this matter of American public opinion which at the present sees the issue from a very one-sided point of view. It is not entirely uncommon for a well-qualified young scientist from a developing country, upon receiving his degree in the United States, to seek the help of newspapers and other information media to obtain permission to continue to stay in the country. The case is usually described in terms of a brilliant young scientist whose skills are greatly needed by the United States, whose training would be completely wasted in his homeland, and who might even be persecuted upon his return to his native country which is described as ruled by a non-democratic government. It is claimed that the United States Immigration and Naturalization Service, because of its bureaucratic outlook and cumbersome procedures, forces the poor student to leave the United States. Sometimes, for local political reasons, a congressman enters the case and the deportation is delayed, often long enough for the student to marry a young American woman, in which case the matter is settled. I believe that if the other side of this coin were explained to the American public in greater detail such incidents would be less frequent.

things being equal" I simply mean adequate research facilities and adequate financial support in the above sense.

Often the student from the underdeveloped country who wishes to stay abroad is not "brilliant" enough to be really in great demand in an advanced country, but his competence could be very valuable in his own country where persons with his kind, quality and amount of training are few and where his own prospects are, therefore, greater. One of the really exciting aspects of life in a developing country is that there a single person with good training and much energy and determination (and, of course, some support from the local leadership) can have a tremendous impact on the country as a whole. In the West, where well-trained, highly intelligent people appear in great numbers, most of them have to be contented with making themselves felt only on a rather microscopic scale. It is therefore in the best interest of everybody (including the scientist in question) if he is given an opportunity somewhere where his contributions can be maximised.

It has sometimes been a practice of governments of developing countries to require a bond from the student just about to be sent abroad, guaranteeing his return after he has received his degree. I do not think this is a good practice. For one thing, it simply does not work, since the bond is small enough (*e.g.*, $2,000) so that if the student did stay in the West, he could pay it off within a short time without prohibitive sacrifices. Secondly, such a bond establishes a rather strange relationship between the student and his home country, in which the former's obligations are expressed in purely financial terms. An alternative method would be for the Western scientist who arranges the admission or the head of the department to which the student is admitted to ask the student for his word of honour that he will return to his country for at least two years after getting his degree. This obligation could still be changed if really unusual circumstances arose but it would make the whole matter a part of the personal relationship between the student and a Western scientist, something that would undoubtedly appeal to many students.

Overcoming Intellectual Isolation

Some of the problems which face working research scientists in developing countries were discussed in my first article. The most important handicap of a scientist in a developing country is his relative isolation from contact with other scientists. The inferiority of physical facilities, although an important factor, is definitely less crucial and also much easier to remedy. I would like, therefore, to suggest five ways in which the isolation of these scientists can be relieved.

The first of these might be to grant rather frequent sabbatical years at

Western research centres. It is not unreasonable to expect that, in order to function near his capacity, a scientist in a developing country should spend one out of every three years in the West. The problem here is twofold, first to find positions for these scientists and second, in view of the market for scientific skills in Western countries, to dissuade institutions in those countries from trying to lure these scientists away from their own countries on a permanent basis. With respect to the former, one of the difficulties in finding a temporary position for such scientists is the same that arises with gaining admission for students at graduate schools: the lack of reliable references. As in the case of the students, the best type of reference is the one established by personal contact. This in turn can be brought about by increased travel of Western scientists to research establishments in the developing countries and by increased participation of scientists from the developing countries at international conferences, summer schools and other research meetings. Many universities in the United States have by now developed a considerable tradition and interest in offering temporary appointments to scientists from developing countries, but much more needs to be done, possibly with financial assistance from the United States Government. A significant step in the right direction has been the recent establishment of the International Centre of Theoretical Physics in Trieste, Italy, which is heavily slanted towards the temporary accommodation of scientists from developing countries.[3]

A second specific step could be the establishment of foreign-financed regional research centres with a significant number of foreign scientists on the staff. Although participation in such international centres might be hampered by local political considerations, the equitable distribution and impartial direction of such centres could be arranged under UN or other auspices. For example, two such centres in physics (one in Pakistan, and one in India, one specialising in solid state physics, the other in atomic and nuclear physics) would benefit not only those two countries but several others in that part of the world who, at present, are too undeveloped to do much on their own in scientific research. Such a centre would have all the advantages of being *in situ*, thus catalysing the scientific life of the region considerably beyond the primary effect it might have on the few scientists on its staff.

The third improvement involves the visits of foreign scientists to the research establishments in developing countries. (Since I dealt with this point at some length in my first article, I will not discuss it here in more detail.)

The fourth field where help can be provided has to do with written contact with the rest of the scientific world. The matter of books and

[3] *Cf. Minerva*, III, 4 (Summer, 1965), pp. 533–536.

journals was discussed in my first article. But in many branches of science (including the one I happen to be working in, namely elementary particle physics) books and journals are more and more relegated to being a depository of completed research for the purposes of later review or for access by later generations. The current research results, the "break-through", the "hot arguments" are propagated by conferences, personal letters and, above all, "preprints", which are rapidly duplicated copies of research papers, just completed, which might appear in journals six or 10 months hence. The distribution of these "preprints" is often done rather haphazardly and there is a tendency to flood well-established persons at leading research centres while omitting altogether little known scientists in less famous scientific establishments (who often need the preprint most). A new programme, soon to be put in operation, will centralise the duplication and distribution of such preprints, at least in high energy theoretical physics, and will send them to any group of scientists working in this field anywhere in the world, free of charge. Such a programme in other branches of science would be of great value in stimulating research in the developing countries. It is difficult for anyone not engaged in research in the pioneer fields of science to appreciate the psychological uplift and increase in research effectiveness brought about by the knowledge that vital information reaches the research worker simultaneously with those in the advanced countries.

Finally, cooperative research between Western and other scientists might also stimulate the scientific life in underdeveloped countries. A young scientist, returning home after recently obtaining his degree, often lacks the perspective to choose interesting problems to work on. But even for more experienced scientists, a congenial colleague interested in the same field is often lacking. In such situations, cooperation with a Western scientist on a certain research project might be very useful. Communication is possible by mail, although, of course, this is not at all as effective as personal contact. Such cooperation is clearly more feasible in theoretical research than in experimental work. It requires a certain amount of time on the part of the foreign half of the team, since writing out everything on paper is often time-consuming, but the rewards are often gratifying.

As I mentioned at the beginning of this section, the problem of physical equipment is also an important one. I would simply like to repeat here the suggestion made in my earlier article that a very significant improve-ment could be made in the efficient use of already existing equipment in the developing countries by establishing a programme of roving Western technicians, well equipped with spare parts, who would make sure that a $10,000 piece of apparatus does not lie idle for six months for the want of a 10 cent part and the know-how for its replacement.

Conclusion

Some of the above suggestions, if implemented, could contribute substantially to the stimulation of scientific life in the developing countries. In conclusion, however, two general points should be emphasised.

The first is that all the measures recommended above require patience and perseverance. It seems to be true that the less developed a country is, the more conservative its people are and the less amenable they are to changes in their lives and habits even if, from an " objective " point of view, such change is clearly " to their advantage ". The most important barrier to cross therefore when trying to contribute to any aspect of the development of an emerging country is to conquer the apathy, the inertia and the lack of urgency that generally prevails. Usually a few talented, visionary and energetic local figures will work hard at bringing about such changes but the results will be sometimes uncertain and almost always slow to emerge. That such slowness is in the nature of things is important to realise in order to avoid the disillusion and demoralisation that I have often seen among Westerners working in the developing countries and even sometimes among the more enlightened local leaders. But coercive methods can do nothing, and least of all in scientific research. A middle road can be found which accepts and seeks slow but steady change, fast enough to make considerable progress in the long run but not so fast as to make whole generations of people permanently miserable and insecure.

Finally, it must be stressed that the most important general factor in the success or failure of the scientific life of a developing country is the morale of its individual scientists. I have seen over and over again scientists from those countries, who worked hard and produced interesting results while staying in a Western country, falter and fail when returning home to research conditions which were by no means worse than those they had in the West. Partly, it might be the influence of the environment, the sight of too many people sitting around in the streets, doing nothing. Partly, it might be a secret conviction that it is just impossible to do research in an underdeveloped country. In part, it might also be due to their private life in a society whose standards, mores and values are different from a modern industrial society. In any case, it is primarily a psychological problem and, as such, might often be contrary to the norms of a " rational " analysis. At the same time, small and objectively speaking insignificant factors can often cause great improvements in the morale of a scientist in a developing country. An invitation to a conference, a Western visitor for a week, a joint paper with a Western scientist, being placed on the list of those who receive preprints, or even repeated reference in the literature to work done by him all contribute towards dispelling the feeling that he is excluded from the community of scientists, that he is cast out into the darkness where the handicaps are insurmountable. These are all steps that are easy to take but are often not thought of because their significance is not appreciated. And yet, in the last analysis, it is the enthusiasm and high morale of the leading individual scientists that will determine the rate of scientific progress in the emerging countries and hence will decide whether the gap between the advanced and emerging countries will continue to grow or whether it will begin to close.

Reprinted by permission of *Minerva* 9, No. 1
(1971) 55.

Science in Developing Countries 55

MICHAEL MORAVCSIK

Some Modest Proposals

THE problem of how to bring about the development of science and technology in the new states of Africa and Asia or in Latin America is not one of the great issues being debated in the advanced countries, or even in significant minorities within them. The so-called great issues change quite often, but this one is not likely to be among them. Even the scientific communities within the United States would rather spend their energy on matters, some rather far removed from their special competence (such as poverty, peace and pollution), than on scientific development of the developing countries, even though the latter, in the long run, might contribute much more towards the equalisation of income, the stabilisation of international relations, or the balanced utilisation of natural resources, than many of the present activities aimed directly at these problems.

This is not to say, of course, that nothing is being said or done at all about scientific development in the emerging countries.[1]

Much could be done if more competent manpower were available, and if

---

[1] As far as public discussion is concerned, readers of *Minerva* hardly need to be reminded of the contributions in this journal of Dedijer, Stevan, " Underdeveloped Science in Underdeveloped Countries ", *Minerva*, II, 1 (Autumn, 1963), pp. 61–81; Hyslop, J. M., " The University of East Africa ", *Minerva*, II, 3 (Spring, 1964), pp. 286–302; Crawford, Malcolm, " Thoughts on Chemical Research and Teaching in East Africa ", *Minerva*, IV, 2 (Winter, 1966), pp. 170–185; Salam, Abdus, " The Isolation of the Scientist in Developing Countries ", *Minerva*, IV, 4 (Summer, 1966), pp. 461–465; Nayudamma, Y., " Promoting the Industrial Application of Research in an Underdeveloped Country ", *Minerva*, V, 3 (Spring, 1967), pp. 323–339; Thomas, Brinley, " The International Circulation of Human Capital ", *Minerva*, V, 4 (Summer, 1967), pp. 479–506; Grubel, Herbert G., " The Reduction of the Brain Drain: Problems and Policies ", *Minerva*, VI, 4 (Summer, 1968), pp. 541–558; Zahlan, A. B., " Science in the Arab Middle East ", *Minerva*, VIII, 1 (January, 1970), pp. 8–35. I have also concerned myself with these problems on a number of occasions. For example Moravcsik, Michael J., " Technical Assistance and Fundamental Research in Underdeveloped Countries ", *Minerva*, II, 2 (Winter, 1964), pp. 197–209. As to organisations actively concerned with carrying out some of the programmes, one may mention as examples UNESCO and the International Atomic Energy Agency with their involvement in the International Centre of Theoretical Physics in Trieste, or the United States National Academy of Science and its connection with the chemistry programme in Brazil.

the problem could attain the status of a publicly declared national or international commitment, and thus enjoy reasonable financial support also. Yet even short of official commitment, it is possible to initiate some new projects which do not require large financial resources. Such projects, to which this paper is devoted, would instead be based mainly on a careful evaluation of the particular circumstances which handicap the emerging countries and on a skilful combination of the existing resources to remedy these handicaps.[2] Those mentioned in what follows are only examples of what might be done.

Physics Intervewing Committee

I would like to start with a suggestion that has already been tried out, and is thus now in the experimental stage. In many developing countries higher education in the sciences is not well enough established for most students to count on obtaining Ph.D.s locally. Thus they have to receive their advanced education in a developed country. In this connection a major problem is to evaluate the capacity of such students for successful study in a foreign institution of higher learning. At the moment, such an evaluation is based on formal school records which are often difficult to interpret, on a list of courses taken, of which the contents and method of teaching are unknown, and on letters of recommendation from local professors whose standards are often not known and who in turn are often not directly familiar with the requirements and expectations of the particular foreign institution.

To improve the information available about such students, four American physics departments combined their resources and sent a committee of two physicists on a month-long trip through Korea, Hong Kong, Thailand, Singapore, Malaysia, Pakistan and India.[3] The committee interviewed about 150 students who intended to apply for postgraduate education in physics in the United States. Each interview lasted about 45 minutes; at the end a brief report was written about the candidate, and this was later made available to any university in the United States in which the student was interested, or which was interested in the student. It is planned that those students who received admission to a university department of physics on the basis of this programme will be evaluated in their respective departments in January 1971, so that it can be ascertained whether the new selection procedure is in fact superior to the old one or not. If it does turn out to be superior, an annual interviewing tour will follow. It is hoped

---

[2] I do not want to claim originality for all of the examples I give. Some are my own, others arose recently during a small discussion meeting held at the University of Maryland, but all of them might have been thought of previously by a number of others. The emphasis is not on claims of priority but on a public presentation and on the hope that at least some of these ideas will soon be realised, at least on an experimental basis.

[3] This remedy to this problem was first suggested by me in 1966 (see Moravcsik, Michael J., *op. cit.*), and was then tried in 1969.

that in that case the programme will be extended to include also other geographical areas and other scientific disciplines.

The cost of the work of the first physics interviewing committee was $4,000.

Summer Seminar for Foreign Students in the Sciences

Another suggestion relating to the education of scientists for the developing countries is prompted by the observation that foreign students coming from developing countries and studying in the United States learn very little about how science works, what the scientific community is like, how science is organised and managed, and how science can be integrated into other activities.[4] Even American students gather little information on these topics, but somehow they are certainly more likely to do so than the foreign students who are often very preoccupied with their studies in science proper, with problems of adjustment, and with language difficulties. And yet, when the students from developing countries return home, they are often called upon to function not only in a purely scientific capacity, but also to take a part in the development of scientific institutions and scientific traditions. For this they are poorly prepared, with the result that their performance often does not reach the standard that their scientific training would have led an observer to expect.

The suggestion therefore is to organise summer seminars two or three weeks long for about 50 foreign science students from the developing countries who are engaged in advanced scientific studies at American universities. The seminar would deal with problems of science organisation and management, particularly as they pertain to the developing countries. The seminar leaders would be mainly American scientists with personal experience and knowledge in these problems.

Cost estimate: travel: 50×100 plus $4 \times 300 = \$6,200$; maintenance: $50 \times 20 \times 10 = \$10,000$; seminar leaders' compensation: $4 \times \$1,500$ and $4 \times 20 \times 15$ (maintenance) $= \$7,200$, or a grand total of $\$23,400$.

Bilateral Arrangements

Bilateral exchange arrangements between universities in the advanced and developing countries already exist on a small scale. They consist of the exchange of staff and students, and of coordination of the use of equipment and curriculum. A university or university department (A) in an advanced country selects a university (B) in a developing country and makes private, bilateral arrangements with it for such a collaboration. Members of the teaching staff from A would regularly visit B, for periods of a year or so, and *vice versa*. Students from B would have an easy opportunity to receive advanced education in particular subjects at A, while some students from A might be interested in spending some time during their studies partici-pating in the development of B, in conjunction with a member of the

---

[4] *Cf.* Zahlan, A. B., *op. cit.*

academic staff of A. Improvements in the pattern of courses of studies transmitted from A to B, and research activities of the members of the teaching staff would also be coordinated to encourage close cooperation between the two universities.

As a first step, a survey would be needed to ascertain which departments of universities in the advanced countries would be interested in participating in such bilateral programmes, which departments have already had experience in such programmes, which have staff members with experience and connections in certain developing countries, and what the main interests of these departments are in the various sub-fields of physics. On the other side, one would have to assess which universities in the developing countries have the interest and capacity to participate in such a programme, and what they could offer from their side. Only if there were complementarity of interests could relationships be established.

Some experience with such programmes already exists, and hence could be used towards further planning. For example, the African-American University Program, with the help of the Ford Foundation, organised a bilateral university exchange programme between American and African universities. Some 52 American and 23 African universities were involved, in various fields, and some 90 African scholars visited the United States under the arrangements. Whether such a massive but diffuse programme is to be preferred to a smaller project going to greater depth within a much smaller number of institutions and concentrated within a few fields of science can be decided only on the basis of experience.

The cost of such a bilateral programme is difficult to estimate, because it would depend on the extent of the programme. Some funds would be necessary mainly for travel, for auxiliary support of staff members of A while at B, and for support of students. A yearly amount of $30,000 would permit a fair amount of activity on an experimental basis.

A clearing-house for coordination of information about possible participants in physics has been established by a group of physicists at various West Coast universities in the United States, with headquarters at the University of California at Irvine.

Distribution of Preprints

One of the major handicaps of scientists in the developing countries is scientific isolation.[5] An important factor in this isolation is the lack of up-to-date information about the latest developments in the sciences. In many rapidly developing fields, communication within the scientific community about the latest developments is carried out mainly through " preprints ", *i.e.*, through locally reproduced research reports which are not yet published. These can be produced quickly and relatively cheaply, and then be sent to those interested in their contents. Most of these are

---

[5] See Salam, Abdus, *op. cit.*

later published in journals, but by that time they have mainly archival value. In order to carry out research in a rapidly developing field, one must have access to information more speedily than journals publish it.

At the moment " preprints " reach scientists in the developing countries only haphazardly and generally very slowly, partly because their active research interests are not very well known to scientists in the advanced countries who produce preprints, partly because preprints are made in limited numbers and are generally sent to the most active places, and partly because foreign air mail rates are often deemed too high.

A proposal was made several years ago to try an experimental scheme in which all preprints in high-energy physics would be reproduced and distributed from a central place, so that the distribution is fast, efficient and equitable. The proposal would have covered both the developed and the developing countries. The proposal was not however realised.[6]

The present suggestion would try such a scheme, in high-energy physics, covering only the developing countries. A copy of every preprint produced in high-energy physics would be sent to a centre where it would be reproduced and sent to all centres in the developing countries where there is activity in high-energy physics. One copy per centre would suffice.

Cost estimate: counting six preprints per day, 20 pages each, a reproduction cost of 0·2 cents per page per copy, and 100 centres on the address list, one would spend for reproduction alone $6 \times 365 \times 20 \times 100 \times 0·2$ cents a year, or $10,000 per year. Postage, via printed matter air mail, on the average of 15 cents per ounce, 10 pages per ounce, or about $70,000 a year. Thus the total yearly cost would be of the order of $80,000. The cost could be substantially reduced by the use of microfiche techniques.

Group Repatriation Scheme

The reluctance of a student with a freshly earned Ph.D. from an institution in an advanced country to return to his own country to take a university position is one of the main factors in the brain drain. Other factors are apprehension about the loss of contact with others working in the same field, and an excessive teaching load.

---

[6] The proposal to assure easy and fast access to preprints for scientists in the developing countries has on occasion been opposed in the past on the grounds that preprints are a poor form of scientific communication which should not be encouraged. Against these objections it should be said that at present, with communication patterns as they are in some areas of physics, such as elementary particles, it is virtually impossible to carry out " competitive " research without having access to preprints on a regular and extensive basis. If, therefore, one wishes to assure a minimally stimulating working environment for physicists in the developing countries who happen to do their research in these fields, one must make provision for them to have preprints. If, at some future time, communication patterns are reformed in such a way that preprints become superfluous, the proposed project will also become superfluous and can be discontinued. Since no elaborate administrative structure or substantial capital investment is needed for this project, it can be discontinued at any future date without difficulty. Some have also claimed that scientists in the developing countries cannot and should not engage in the most pioneering and most competitive areas of science. I believe that an acquaintance with some scientists in, say, Pakistan, Chile, Argentina and Ghana and with their profound effect on the local scientific scene would tend to dispel these claims.

As a means of counteracting the brain drain, I suggest that a group of young Ph.D.s, whose specialities are the same, and who come from the same country, be repatriated to a home university as a group, under the following conditions: they should be employed by that one university together; half of their local salary should be paid by the programme, so that they are assigned only half of the regular teaching load; their teachers in the developed country should maintain contact with them and advise them in their research; and the home university in question should receive an overhead payment of, let us say, 30 per cent. of the programme cost, in return for its agreement to the scheme.

The advantages are that the young scientists will have each other as intellectual company; they will have some time for research; and their former teachers will wish to keep in touch with them because they are serious scientists.

The programme is more feasible in the theoretical than in the experimental sciences. Local opposition can at first be expected but this will probably subside later. The payment of an overhead might help to decrease opposition, because it would assure that as well as the direct participants the departments would also benefit from the programme.

Cost estimate: for an example, consider four Indian physicists in such a scheme. Half their salary is about Rs. 7,000 or $1,000 per person. Thus 4 × $1,000 plus 30 per cent. as payment to the department comes to a grand total of $5,200.

Visiting Appointments

Scientists working at institutions in the developing countries must have frequent contact with their colleagues in the developed countries, so that their knowledge and expertise can be kept up to date. An active scientist in a developing country should, if it could be arranged, spend two years abroad after each three years of service at home. Such liberal arrangements might make it possible to reconcile a considerable number of first-class scientists to working in institutions in the developing countries.

The developing countries themselves are too poor and have insufficient foreign exchange to sponsor such long and frequent *séjours* abroad for their own staff members. Hence financial support of such scientists must come from the host country.

Most universities in the developed countries have some provision for visiting appointments, under which a scientist from another institution spends a year or two at the university in question. Such appointments are often used to attract outstanding visitors from one advanced country to another.

I suggest that all universities give serious thought to using some of their funds for visiting professorships for scientists from developing countries. Such a decision need not involve a compromise with standards of quality.

Some invitations might also act as catalysts for later bilateral arrangements between the two institutions in question.

Temporary positions like the one suggested are offered to theoretical physicists from the developing countries by the International Centre of Theoretical Physics in Trieste. That programme, which has been very successful indeed, is, however, limited in subject-matter and numbers because of the specialised nature and the limited financial resources of that institution. A much broader application of the same idea on the part of many universities would therefore be a very important contribution. Moreover, it would not require that universities go beyond the confines of their regular budgets to pay for it.

Dual Appointments

Dual appointments involve the collaboration of two universities, one in a developed country, and one in a developing country, under which a scientist of considerable stature, originally from the developing country, would receive a joint appointment by the two universities. The universities would share his time equally.

According to the account of one holder of such a dual appointment, great care must be taken to select persons who can stand the strain of periodically and frequently changing his place of work and residence. Once, however, the right people are found, the scheme works out to everyone's benefit. The graduate students of the dual professor follow him wherever he happens to be, and the host institution offers them support. The graduate students also seem to benefit from such a variety of experience. Travel, which, compared to the salaries involved, is a small amount, has either been paid by one of the universities or by the dual appointee from his own resources. Two-year periods seem to be optimal in many respects. To be sure, the family of the dual appointee must become bilingual, but such instances are rather numerous in other international professions.

It should be emphasised that dual appointments between an advanced and a developing country are in many ways easier to handle than dual appointments between two developed countries (for which there are also instances), and hence adverse experience with the latter should not necessarily preclude further experimentation with the former.

The cost of this scheme is almost nothing. Some currently unusual administrative arrangements would be needed, however. For example, provisions would have to be made to fill the position of the scientist when he is at the other institution. For this reason, such appointments might be made in pairs, appropriately phased to fit into each other.

The scheme would encourage outstanding scientists originally from a developing country to devote more effort to their countries of origin without losing in any sense the security, prestige and research stimulation which is connected with a prominent institution in an advanced country. It would also establish a link between universities which again might develop into a bilateral arrangement of the type already discussed.

Registry and Coordination of Travelling Lecturers

With travel becoming less expensive and more rapid, an increasing number of scientists from the developed countries travel in the less developed parts of the world. Scientific conferences, summer seminars, cultural missions as well as private reasons are the grounds for many such voyages. They could be much better utilised for the purposes of contact with science in the developing countries if there were a certain amount of coordination. The traveller often has no extensive contacts among scientists in the developing countries, while the universities of the latter have no way of knowing about the traveller being near at hand. Sometimes, even if the information exists, there is reluctance to ask the traveller to make a detour to visit an institution. Quite often only a negligible amount of local currency is needed (as well as the coordination) to bring about the needed detour. Even a two-day visit by such a touring scientist can have benefits for all parties involved, helping in the exchange of information, in the improvement of morale, broadening of horizons, and establishing personal ties which might lead to more extensive future contacts.

A central registry should therefore be established in each field of science, where those scientists from the developed countries who plan to travel in the less developed parts of the world would register their intentions, and where institutions from the developing countries would register their willingness to invite passing scientists.

The cost would be slight. The registry could easily be handled by any physics department or research laboratory with its existing administrative personnel, and publicity could be given through the semi-professional journals.

The organisation of lecture tours for scientists with brief stays at a number of universities and research institutes in the developing countries would be a reasonable extension of this scheme. This would involve some expenditure of foreign exchange. The costs would vary but they would not be high. For example, a scientist, whose salary is paid by his own university during his travels, and who takes a one-month round-the-world trip, could visit 10 institutions for two or three days each. The cost per institution would be $200, of which about $140 would have to be paid in foreign currency.

Joint Research Contracts

Young scientists from developing countries, who have just returned home having completed their advanced education in an advanced country, are sometimes handicapped in their work not only because of the lack of time and financial support, but also because of the lack of professional advice and direction by senior colleagues. It would be helpful to them if they could continue to collaborate with experienced scientists in an advanced country with whom they have previously worked.

I suggest therefore that bodies which support research in advanced countries should consider supporting jointly a senior scientist in an advanced country and one junior to him in a developing country. The grant would stipulate and provide for contact between them through collaboration on a common problem, correspondence and even periodic visits.

The scheme would not only directly help the scientist in the developing country, but would also contribute to the formation of a sense of responsibility on the part of a senior scientist in an advanced country for a particular project in a developing country. This in its turn might generate a much more far-reaching future collaboration between the parties involved and their institutions. The experience of a senior scientist in an advanced country thus might help to establish a " school " or " scientific tradition " in the developing country, without interfering with the performance of his ordinary responsibilities or taking him away from his accustomed locale.

The cost of this proposal would depend on the particular project. A grant with an annual budget of $20,000 would be more than adequate to experiment with one such joint research project.

Registry of Appointments in the Developing Countries

This proposal is similar to the one about registering travelling lecturers, in that it also serves to utilise existing resources more fully. There are today many bodies which support scientists from advanced countries who wish to spend a short period of time on work in a developing country. Among such bodies are governments, foundations, international scientific organisations and universities in the developing countries themselves.

I use as examples the National Academy of Sciences, the National Science Foundation, the Agency for International Development and the Fulbright programme in the United States. Private foundations in the United States which have supported or otherwise helped American scientists to work in developing countries include the Ford Foundation, Education and World Affairs and the Rockefeller Foundation. There are similar organisations in Great Britain, France and Sweden. Among the international organisations there are the International Atomic Energy Agency, UNESCO and the World Health Organisation. In the developing countries there are several physics departments in West African universities which are seeking staff members among scientists from the advanced countries, as is at least one physics department in East Africa.

There are also numerous scientists in advanced countries who are interested in undertaking a " mission " for a limited time, for example, during a sabbatical year. There is much to be done to coordinate the potential employers and potential donors. One particular matter which deserves attention and which would greatly enhance the willingness of younger scientists to engage in such activities is the problem of employment of the scientist in the home country on his return from an assignment. But unless the scientist in the advanced country has a permanent appointment from which he is on leave, he would have to negotiate a new appointment from a distance, which would put him at a considerable disadvantage. This arrangement affects younger scientists particularly. My present proposal therefore is the creation of a central registry in each field of science where the requests of potential employers and specifications of the posts they wished filled on a temporary basis could be gathered and coordinated, and which would also help those just returning from a temporary appointment abroad to find a new post on their return.

The cost of this scheme would probably be slight because it could be handled by the existing clerical staff of a university research institution. At most, a small budget of $500 a year would be needed to cover postage and minor administrative expenses.

Industrial Research in Local Branches on International Firms

At present, international firms with headquarters in an advanced country perform almost all of their research and development at their branches in these countries; they use the branches in the developing countries mainly for manufacture. In the long run, it would be in the interest of such firms to transfer some of this research and development to their branches in the developing countries. This would provide more specialised research adapted to regional needs, an increased sense of local identification with the firm and a more favourable political image, as well as the utilisation of local talent at a cost which is lower than the cost of similar talent in the advanced countries. In the developing country such arrangements would increase demand for well-trained scientific manpower (thus counter-ing the brain drain), and would also bring " academic " science closer to technological application; they might also open the possibility of consultant-ships in industry for local university staff. This would go far towards injecting a new element into the scientific life of a developing country. I propose therefore that such international firms consider the possibility of sponsoring a certain amount of research in their branches in the developing countries. Governments or influential scientists in the latter should explore these possibilities with firms operating in their countries and cooperate in the organisation of such arrangements.

The cost depends on the extent of the programme. However, even a yearly expenditure of $20,000 could begin such a scheme. In some countries tax incentives might be offered to the firms to encourage them to undertake such a scheme.

Promotion of Regional Collaboration

Scientific collaboration among neighbouring developing countries could contribute to the improvement of their scientific performance. Political reasons often prevent such schemes, and also the prevailing orientation towards advanced countries often makes it easier to seek a joint venture with an advanced country thousands of miles away than with an equally poor neighbour.

There are, as we know, some examples of successful regional cooperation in advanced countries. CERN constitutes a very successful collaboration among a large number of European countries. Its success is in part attributable to the felicitous circumvention of political obstacles and to the placing of the responsibility for the operation into the hands of an international scientific organisation. Another good example is regional science collaboration in Latin America. This consists, among other things, of regional meetings, seminars and summer schools, and of bilateral collaborative schemes between laboratories and universities of the region. In this case common language and common tradition played an important part in facilitating such cooperation. In Asia where strong tensions exist, or in Africa where language and tradition are not at all homogeneous, such regional collaboration might be much more difficult, though not impossible, to achieve.

I propose that foundations and scientific institutions should attempt to promote the regional cooperation of neighbouring countries. Sometimes an outside influence, particularly if accompanied by some financial support, can help to bring about such collaboration. Joint research facilities, regional institutes, regional seminars and conferences, exchange arrangements of scientific personnel as well as scientific knowledge could all be part of such a scheme. The advantages of such a scheme in enlarging the scientific communities in countries where these communities are too small to be effective under existing circumstances are obvious.

The cost of such a scheme would vary, but an initial appropriation of $10,000 would enable it to get under way.

Conclusions

The foregoing proposals are examples of inexpensive projects in science and technology in the developing countries; they are feasible even in a time of shrinking budgets. Although they are based on actually perceived needs, none of them is a certainty. They are no more than efforts to open the path of exploration, put forward with the intention of arousing the imagination and will of scientists and officials, governmental and academic.

We still know little about the institutional workings of science. There is no body of systematically studied experience available to us in such a domain as that discussed above. It is mainly through interest, enthusiasm, and above all a dogged perseverance and incessant readiness to try new things that we can make progress towards the goal.

Preprint, published in *Am. J. Phys.* **41** (1973) 309,
608 and *Phys. Teacher* **11** (1973) 82.

<u>The Committee on International Education in Physics</u>

Michael J. Moravcsik

Institute of Theoretical Science

University of Oregon

Eugene, Oregon 97403 USA

(Chairman, CIEP)

An invited talk given at the annual meeting of the American Physical Society and American Association of Physics Teachers, San Francisco, January 31, 1972.

The Committee on International Education in Physics (CIEP) is one of the many committees of the American Association of Physics Teachers. It is the only organ of the AAPT which deals with international matters, and since the international activities of the American Institute of Physics are, for the moment, defunct, CIEP seems in fact to be the only national entity of physicists concerned with international science activities. In this talk I will briefly outline the structure of this committee and discuss some of its programs.

Since CIEP has no funds to provide for travel expenses of its members, and since its members reside all over the United States, by necessity CIEP had to assume a mode of operation which did not depend on meetings of the Committee has a whole. As a result, CIEP is constituted of a set of subcommittees, each of which consists of one, or at most two, members. The overall direction of these subcommittees is decided by the committee as a whole, through correspondence, but beyond that the subcommittees are independent and report to the chairman of CIEP, who coordinates the various activities and also allocates from CIEP's small budget whatever is needed by the subcommittees.

Beside flexibility, this system has other very beneficial characteristics.
First, it allows for a maximum degree of outlet for individual initiative, so
that CIEP members can engage in whatever projects their special interest and
competence dictates. This is essential, since voluntary, financially almost
unsupported, and "extracurricular" committees like CIEP can be productive only
through a great amount of energy, enthusiasm, and initiative of its members.
If this is lacking, the group quickly becomes a paper committee, and therefore
loses its raison d'être.

Second, the subcommittee system permits CIEP to expand its membership
considerably without causing administrative difficulties. As long as the
chairman can keep track of the various activities, there is no limit on the
number of subcommittees and hence on the number of projects that can be
undertaken.

Third, the subcommittee system permits each member to participate at
his own pace, depending on his work habits, other committements, the difficulty
of the task, and so on. Thus one avoids the prevalent fault of group activities
in which the efficiency of the whole is reduced to the lowest common denominator,
determined by the least enthusiastic or capable member.

Naturally, there are always many more interesting projects that one can
undertake, and so CIEP makes no claim to completeness or even balance. As it
turns out, a majority of the present members of CIEP (of which there are 10)
are mainly interested in problems pertaining to science and science education
in the developing countries, and hence such activities are dominant at the
moment, though not exclusively so.

In the allotted time I could not discuss all the programs CIEP is at

the moment engaged in. I will, therefore, select four of them, which might
be of general interest.

Let me begin with the Physics Interviewing Committee. The purpose of
this program is to generate uniform and reliable information on students
from developing countries who wish to be admitted to and given financial aid
at American graduate schools in physics. It consists of a committee of two
physicists taking a one-month long trip in the Fall to various developing
countries, where they interview personally those students who intend to apply
to American graduate schools in physics. An interview lasts about 45 minutes,
after which the committee immediately completes an evaluation form on the
student. This form is then made available to all those graduate schools the
student is interested in, or all those schools which might be interested in the
student.

The interview covers mainly basic undergraduate physics, and intends
to explore both the background and the capability of the student. It
consists of simple questions and short problems. The interview also
establishes a need, if any, to fill in gaps in preparation (by recommending
the taking of some undergraduate courses in the first year of graduate school),
and also states whether the student is capable of carrying out the duties of a
teaching assistant in his first year in graduate school, something that is of
utmost importance in being able to secure financial aid. Finally, the
evaluation form also comments on the candidate's general personality and on
his command of the English language.

The first interviewing trip took place in the Fall of 1969, covering
South Korea, Hong Kong, Thailand, Singapore, Malaysia, India, and Pakistan.
The two interviewers were Francis Dart of the University of Oregon and myself.

A total of 145 students were interviewed.

In order to test the effectiveness of the new system of selection, no trip was planned for the following Fall, but instead those students who were interviewed during the first trip and did not get admitted to American graduate schools were traced down and their actual performance in graduate school was compared with the evaluation of the interviewers. Only about a quarter of the students interviewed were located in graduate schools, and only a part of those could we obtain detailed information about. Thus the statistics of the survey was not ideal. On the basis of the existing information, however, a virtually perfect correlation was found between actual performances and the evaluations of the interviewers. The only discrepancy was a systematic one, in as much as the interviewers were, on the whole, somewhat more severe with all candidates than their later performance warranted. The difference, however, was small.

On the basis of this favorable indication, a second interviewing trip was organized for the Fall of 1971. It covered South Korea, Hong Kong, Indonesia, Thailand, Singapore, Malaysia, Ceylon, India, Pakistan, and Afghanistan, and the two interviewers were Michael Scadron of the University of Arizona and Andrew deRocco of the University of Maryland. This time 175 students were interviewed. Information on these students is now available to all interested schools.

The financing of these interviewing trips was done through a cooperative effort of several departments of physics in the United States. The 1969 trip was sponsored by four departments, while in 1971 fifteen departments got together. The cost of these trips is about $4,000 - $4,500, which is somewhat less than the cost of the education of one graduate student for one year.

The sponsoring universities are somewhat compensated for their generosity by making available to them, promptly after the completion of the interviewing, the evaluation sheets of <u>all</u> students interviewed. Thus interested departments could directly approach the most promising students. Non-sponsoring departments receive only single individual evaluations, and only on specific request.

It appears from these trips that it is not necessary to cover the same geographical area every year, since in one trip one can catch two generations of students. Thus, the 1972 interviewing trip intends to cover other regions, perhaps Africa or Latin America. Decisions will be made in the near future, and departments of physics all around the country will be contacted to solicit their support. If only 50 departments (of the 180 or so with graduate education in physics) could be interested, the cost per department would drop to about $100, which is a negligible amount even today.

Let me now turn to the second activity of CIEP that I want to mention today. It is called Visitor Registry, and it aims at helping to improve contact between traveling American physicists and local physicists along the itinerary. The program works as follows: A registration card has been distributed to physics institutions all around the world, except in the scientifically advanced countries (i.e. much of Europe, Japan, Australia, and Canada) which have sufficient international ties to make such a program unnecessary. The institutions in question complete the card and return it to CIEP. The card indicates interest in receiving short term physicist visitors (for a few days or so), lists the areas of physics which are of interest, and states whether there are local resources to pay for small detours in itinerary and for local living expenses.

A different type of registration card has been distributed to a very large number of physics departments in the United States. If an American physicist learns that he will soon travel abroad, and if he is interested, he fills out such a card with the date of his intended travel and his itinerary, and sends this information to CIEP.

CIEP then checks the other file for institutions along the proposed itinerary which have expressed an interest in receiving visitors, and sends to these institutions a xerox copy of the traveler's registration card. At this point CIEP's role in the matter ends. If the prospective host institution is interested in the would-be traveler, it contacts him directly.

The need for such a program has grown out of evidence throughout the world that interested institutions have difficulty communicating with would-be visitors, and that travelers have difficulty making personal contacts in areas they have not visited before. The cost of this program is negligible, since it involves only stamps and a very small amount of stationery supplies. Since it has been instituted only recently, its effectiveness is not known yet. At the moment some 50 prospective hosts have registered with CIEP, but so far only a handful of registrations from would-be travelers has been received. More publicity given to this program will undoubtedly enhance its effectiveness.

The third CIEP activity on my list is somewhat similar to the Visitor Registry. It is called Opportunities Registry, and is a clearing house of information concerning longer-range job opportunities for American physicists abroad. I want to emphasize that the idea of organizing this program arose simultaneously in CIEP and in the head of Arnold Strassenburg of SUNY Stony'Brook, and up till now it has been Strassenburg who did all the work. Now that the program is established, CIEP will take over its administration.

The Opportunities Registry is also a bilateral information depository, just like the Visitor Registry. On the one hand, a circular letter was sent to hundreds of individuals and institutions around the world, inquiring about requests for American physicists on a longer term basis (say, 2 months or longer), or perhaps even permanently. On the other hand, a circular has been distributed within the American physics community, soliciting registrations of those physicists who have an interest in assignments or positions abroad. Once information is received from both sides, the program simply coordinates the requests and notifies the two matching parties of each other's existence. This program is also quite inexpensive, since, after its initial establishment, the cost again is only stamps and a small amount of supplies.

This program is also just being established, and hence its effectiveness is not known at the present. Like the Visitor Registry, it grew out of a perceived need in various parts of the world, coupled with a domestic need to enlarge the variety of opportunities available to American physicists.

The last CIEP program I want to discuss today tries to catalyze bilateral arrangements between an American physics institution and a counterpart institution in a developing country. CIEP's role in this is again that of the initial coordinator, as well as a provider of suggestions as to how such a bilateral arrangement can function. There are a number of successful examples of such relationships. Such arrangements can function at a very broad range of intensities, going from a mere skeleton of occasional visits or the acceptance of a few students for advanced education, to an extensive coordination of education, research, and personnel.

CIEP has sent out a large number of feelers to physics institutions in the developing countries to learn about latent interest in such bilateral arrangements. At the same time, another feeler was circulated in the United States to find out about interest at this end. The information was then coordinated. At the moment, three such pairs of institutions have been brought together, and CIEP hopes that the negotiations will result in some de facto bilateral activity. The lack of funds these days should not discourage such experimentation, since a considerable amount can be done without additional resources.

These four examples illustrate that CIEP's general approach to international science problems tends in the direction of supplying expertise and a minimal amount of organization to coordinate already existing resources and opportunities. For an organization with a total yearly budget of around $1,000, this is, of course, by necessity the only option available. It is, however, also a very valuable option, which has been neglected in the past in favor of massive and costly projects operated by large administrative machineries. Many of those programs are also very valuable. At the same time, small-scale projects, taking advantage of the existence of large-scale untapped resources, and involving individual American scientists, must also be pursued, and CIEP is ideally suited to do so.

In conclusion, let me appeal to all of you to participate in any of these programs if you can contribute. None of the activities of CIEP could possibly succeed without the cooperation of the physics community at large. At the same time, with such a cooperation, results will be achieved which will benefit the American physics community as well as our scientist colleagues all around the world.

A physicist suggests new methods of scientific cooperation between advanced and developing countries.

The Research Institute and Scientific Aid

Michael J. Moravcsik

Institute of Theoretical Science
University of Oregon

I. Introduction

☐ My aim in this article is to suggest a number of ways a scientific research institute in an advanced country could promote science and technology in the developing countries without disrupting its own research activities significantly, either in terms of the scientific program or of finances.

The topic is not chosen abstractly. On the contrary, it grew out of an actual conversation I recently had with the newly appointed director of a research institute in the United States, in which we speculated on ways his institute could broaden its activity beyond strict research so as to have a more direct effect on social problems of our times. Scientific development of the Third World is not only one of the, if not *the*, most pressing of such problems, but is also well geared toward the utilization of the experience, expertise, and interest a research institute accumulates in the course of its normal activities.

The suggestions put forward in this article are applicable to research institutes in any of the natural sciences, "pure" or applied. Close association with a university is not a prerequisite, since none of the suggestions are directly connected with educational activities on an undergraduate or graduate level.

The claim is not made that all the suggestions are "original." Some of them specifically overlap with suggestions made in earlier articles of mine,[1] and undoubtedly quite a few of them have also been advocated by others. The purpose is simply to collect some practical ideas which are applicable in the context outlined above.

It is assumed in this article that scientific research is an essential activity in a developing country. This is not a universally accepted assumption, but since I have discussed it previously on a number of occasions[2], I

will not dwell on it here again.

Since scientific aid is a very down-to-earth subject, in which an idea is considerably less than half of the story and the implementation is what counts in the long run, one should at the very outset say a few words about how a research institute, having in principle decided on engaging in some of the activities pertaining to scientific aid, could go about implementing this desire.

There are basically two ways of proceeding. One is to approach one of the many international or national (governmental or private) organizations which are active in scientific aid to developing countries or which represent science management in the developing countries themselves. This approach has the advantage of being able to utilize the administrative machinery already developed by these organizations. Its drawback is that existing organizations would most likely try to channel the interest of the research institute into one of the already molded programs, and hence a certain amount of flexibility, experimentation, and specific suitability is lost.

The other approach for implementation is to utilize the cooperation of one of the numerous members of the scientific community who have interest and experience in scientific aid to developing countries. In the ideal situation there would be one or several such scientists already on the staff of the research institute. Even if this is not the case, however, and outsiders must be used, no difficulty is likely to arise, since undoubtedly many of these scientists would be happy to advise without any complicated or costly formal arrangements. Although they usually have the benefit of their personal acquaintance with local scientists and administrators in the developing countries, they are not likely to have a ready-made administrative machinery to rely on. In the case of most of the suggestions in this

[1] See, for example, my "Some Practical Suggestions for the Improvement of Science in the Developing Countries," *Minerva*, 4, 381 (1966), reprinted in Edward Shils (ed.), *Criteria for Scientific Development: Public Policy and National Goals*, MIT Press, 1968; and my "Education and Research in Scientifically Developing Countries" and "Communication among Scientists and its Implication to Developing Countries," two papers to be published in the Proceedings of the Research and Development Management Seminar of the Scientific and Technical Research Council of Turkey, held in Istanbul, May 1970.
[2] Among them are "Technical Assistance and Fundamental Scientific Research in Underdeveloped Countries," *Minerva*, 2, 197 (1964), reprinted in Shils, *op. cit.*; "Fundamental Research in Underdeveloped Countries," *Physics Today*, 17, 1:21 (1964); and "Fundamental Research in Developing Countries," *International Atomic Energy Agency Bulletin*, 6, 2:8 (1964).

A native of Hungary, MICHAEL J. MORAVCSIK received his Ph.D. in theoretical physics from Cornell University. Later he became Head of the Elementary Particle and Nuclear Theory Group at the Lawrence Radiation Laboratory of the University of California. Since 1969 he has been Director of the Institute of Theoretical Science at the University of Oregon.

article, however, such machinery is really not necessary, and the implementation could very well progress with the help of the amount of administrative staff that is readily available in a research institute.

The suggestions will be divided into four groups: personnel exchange, research, communication, and the stimulation of local efforts. Under each of these headings, a number of independent ideas will be discussed. Thus the content of this article is not an integrated program on an all-or-nothing basis. On the contrary, each suggestion can be considered by itself, and evaluated in the specific context of a given research institute. If all research institutes in the advanced countries took up only one of the dozen ideas discussed below, a very substantial contribution would be made to the scientific development of the Third World.

To conclude the introduction, let me also hasten to emphasize that by no means do I imply that research institutes have done nothing along these lines in the past. On the contrary, a number of them have already developed interest in this area, and some of the suggestions have in fact been already implemented here and there. What is now needed is a realization on a much larger scale that such activities are essential, and that they are also relatively easy and inexpensive to implement.

II. Personnel Exchange

1. Staff Members Visiting in the Developing Countries

Scientific isolation is one of the most serious problems in the developing countries as far as the advancement of science and technology is concerned.[3] One of the very effective ways to break this isolation is through extended visits by scientists from the advanced countries. A visit even for a half a year can have an imprint, but a year is a more effective period.

Research institutes generally do not have a sabbatical leave system as many universities do. It is nevertheless very likely that an institute would not be seriously affected in its research, if, say, one out of 20 of its staff members were at any given time temporarily abroad, working in a developing country. This would be equivalent to a "sabbatical" every 20 years, a considerably more conservative arrangement than what

[3] Salam, Abdus, "The Isolation of the Scientist in Developing Countries," *Minerva*, 4 (1966).

universities have.

The financing of these visitors would not necessarily fall entirely on the shoulders of the institute. There are a number of organizations with programs to help such visits, and in some cases even the host institution in the developing country itself can contribute to the expenses.

It is easier to arrange such visits by theoretical scientists, but experimental scientists could also be effective if their visits are carefully prepared in advance.

One of the lasting effects of such visits is the establishment of personal connections between scientists in the developing and advanced countries, and these connections can play an important role in many other programs that might be established subsequently. In fact, such connections are also pertinent to a number of the other suggestions made in this article. Ideally, such connections can develop into a permanent and reciprocal relationship between two institutions, one in an advanced and one in a developing country. The relationship need not at all be lopsided. The institution in the developing country could well contribute substantially in terms of personnel, locale, or facilities, depending on the particular research done. Astronomical observations in the Southern hemisphere, biology of tropical areas, geology of interesting regions, meteorology outside the moderate belt, medical research in unusual environments, are only a few examples.

2. Staff Members Visiting From Developing Countries

Visits in the reverse direction are equally valuable. There are many "brain drain" cases which could be remedied if some assurance could be given to the productive scientist from a developing country that, say, every fourth year he could spend an extended period of 6-10 months abroad at an active research institute in his field.

Very few developing countries have the resources, particularly in terms of foreign exchange, to underwrite such a visiting program for their staff members. (For an exception, however, see the University of Malaya.) The scientists from the developing countries therefore have to rely on financial sources from the advanced countries to finance such visits. There are some organizations devoted almost entirely to offering a locale and financial help for such visits. The International Centre for Theoretical Physics in Trieste is an example. The combined opportunities for such visits are, however, only a small fraction of what could be utilized.

Most research institutes in advanced countries have temporary positions for visitors. I would like to suggest that such institutes should seriously seek out candidates from developing countries for some of these positions. I do not necessarily advocate a quota system of a specific number of positions being laid aside for such candidates, although if some private contributor wanted to make a donation to the institute for this specific purpose, such specific positions could be established. In general, however, no special treatment is needed to accommodate candidates from developing countries on a postdoctoral level. They are often well trained, enthusiastic, and in addition have more modest financial requirements than their counterparts from advanced countries, because their dislocation expenses are lower. If, as an example, each research institute in an advanced country had one visitor out of ten from a developing country, a momentous impact would be made on the scientific life of these countries.

3. Dual Appointments

A version of personnel exchange that has already proven very successful in some cases is the dual appointment. Under such an arrangement, a scientist of some accomplishment receives a joint appointment at an institution in a developing country and at one in an advanced country. The scientist then shares his time, more or less equally, between the two institutions, in something like two-year sections.

This arrangement combines the advantages of various exchange schemes, and often permits an institution in a developing country to retain the services, at least on a part-time basis, of very distinguished scientists. The expenses connected with such a scheme, assuming that the positions in question already exist, are relatively negligible, since they involve only the transportation of the scientist and his family every two years, which might amount to an annual commitment of no more than $2,000.

There are difficulties connected with such an itinerant

life and the scientist filling such a position must be carefully selected. But successful examples are known and more experimentation along these lines would certainly be welcome.

4. The Group-Repatriation Scheme

Among the many obstacles in the path of scientific activity in a developing country, one of the very serious ones is the difficulty in establishing a "critical mass" of scientists working in one place and in related disciplines so that they form a viable and stable research entity. The following scheme might be helpful to advance the formation of such a critical mass.

Four young scientists, say, all originally from the same country and with strongly overlapping research interests, are given temporary positions in the same research institute in an advanced country. After a year or two, they develop a rapport so that they can help each other in research activities. At that point arrangements are made with an institution in a developing country to transplant this group as a team into that institution. Contact between members of this group and other staff members of the research institute in the advanced country are maintained through correspondence, occasional visits, and other means, so that the feeling of isolation is minimized. Perhaps even help in terms of some research equipment might be feasible in some cases. Once the group has firmly established itself in the new locale and its research activities have taken root, the scale of the cooperative effort might be decreased and instead another group might be "launched."

III. Research

5. Research Brotherhood Arrangements

As suggested by the foregoing discussion, it might be very helpful to establish close cooperation or even coordination in research activities between certain scientists in an advanced country and in a developing country. This can be done on an institution-by-institution basis, or on a group-by-group basis, or even on an individual-to-individual basis. Under such an arrangement, research topics of common interest would be chosen by the two parties and the work at the two locales would be coordinated, perhaps by dividing the planned investigations between the two groups. Results would be exchanged, equipment shared if necessary, and publications could be joint between the two groups.

6. Supply of Scientific Equipment

There might be many instances when a research institute in an advanced country could be of considerable help to a similar institution in a developing country in terms of assistance with the acquisition and maintenance of research equipment. This need not necessarily involve large sums of money. One must keep in mind that the obstacle in the path of having functional scientific equipment in institutions in the developing countries is not always simply lack of money. Sometimes the money is present but cannot be spent in the form of foreign exchange. Often the foreign exchange is ultimately available but the bureaucratic difficulties of securing it produce unreasonably long delays. Sometimes funds are available but limited in such a way that they cannot be easily spent on replacement parts or repair. Again, on other occasions, funds are in fact available and the parts needed are very inexpensive, but they are unavailable close-by and their shipping involves lengthy administrative arrangements. It also happens that what is needed is a technician, familiar with the equipment, to diagnose the cause of a misfunctioning.

In several of the situations outlined above, a research institute in an advanced country could be of invaluable help by sending spare parts or surplus equipment items. Clearly, its contribution would not be large in financial terms, but rather would assure a source of supply of relatively minor items, providing the right piece at the right time with a minimum of complications. It is not at all a pure figment of imagination to contemplate situations when the prompt shipment of a part costing $1 could save the research group in the developing country several months of forced idleness and frustration. It would therefore be important if such a group could in fact rely on a counterpart institute in an advanced country for help when the occasion arose.

7. Joint Research Contracts

Most research institutes in the advanced countries do not finance themselves independently but carry research contracts with outside grant-giving organizations. Once some relationship has been established by staff members of such an institute and the staff members of an institution in a developing country, serious consideration should be given to applying jointly for a research contract, so that part of the funds obtained could be spent by the group in the developing country.

Such arrangements would undoubtedly make it easier for the group in the developing country to obtain research funds, because their credentials would be strengthened by their association with the group in the research institute in the advanced country. And yet the gain would not be one-sided, because the presence of the group from the developing country would open up new sources for research funds, and because research generally costs less to perform in a developing country.

IV. Communications

8. Distribution of Information

Another major aspect of the scientific isolation that has been mentioned already several times is the poor channels of communicating scientific results in the developing countries. Science is par excellence a cooperative undertaking, and advances are closely based on previous discoveries. Thus a constant knowledge of new ideas and recent progress is indispensible to productive scientific work.

Communication in the sciences is through journals, through informal reports and preprints, and through personal contact. Journals are expensive to subscribe to and are very slow in reaching scientists in the developing countries. Reports and preprints are often free of charge, but their distribution is not well organized and heavily favors those in the active centers of research in the advanced countries. Personal contact is also difficult for scientists in the developing countries because relatively few visitors reach them and they have little opportunity to travel.

A research institute in an advanced country could therefore perform a very valuable service if it kept up-to-date one or several of its counterparts in some developing countries on developments within its field. This could be done by forwarding copies of interesting reports received, by writing brief summaries of oral information received, by requesting other institutions in the advanced countries to put their counterparts on the mailing lists for preprints, and possibly even by donating a subscription to one or two major journals. The latter need not be pure charity, since often funds are available in local currency which the counterpart institution could lay aside in reciprocation for use during visits by staff members from the advanced country or for other cooperative ventures.

9. Organization of Lecture Tours

Scientists in advanced countries do more and more traveling and their trips to parts of the world where the developing countries are situated are becoming more frequent. Therefore, what is badly needed is to co-ordinate these opportunities with the willingness on the part of institutions in the developing countries to host such visitors for a day or two and even to contribute the expenses of small detours that need to be made in the original itinerary. It would thus be very useful to have research institutes in various fields take on the responsibility of serving as clearing houses for scientists who intend to travel and for institutions willing to receive visitors. This activity would involve no expense at all on the part of a coordinating institute except for a negligible amount of clerical time.

A more advanced version of this suggestion would be to undertake the organization of special lecture tours by noted scientists from the advanced countries, the expenses of the tours being shared by many institutions in the developing countries, so that the cost per institution per lecturer is not high. For example, a month-long tour, shared among ten institutions, might not be unfeasible, and the lecturer could thus spend about two days at each institution, allowing something slightly more than just superficial pleasantries in the exchange. Again the role of the research institute would be only organizational, and the contribution would be in terms of scientific connections, judgement, and a minimal amount of clerical work.

V. Stimulating Local Efforts

No amount of effort in scientific aid will bear fruit unless it aims at establishing local, indigenous scientific activity which eventually becomes strong enough so that outside help can be reduced to the level of normal scientific exchange. Although the previous suggestions

have this aim in mind, there are some specific programs directed toward such an encouragement of local efforts which could be adopted by research institutes in the advanced countries.

10. Prizes

I would like to suggest that research institutes in advanced countries should establish prizes, to be given out annually, for research ("pure" and applied) performed in the various developing countries in the various fields. For example, one could offer a $1,000 prize for the best piece of applied physics research performed in Pakistan in a given year. Papers describing the research would be submitted to an international jury of respected professionals in physics who would make the award. A prize of such relatively small size would be an attractive stimulant, since it represents 2-5 months salary for a local scientist. The prizes could be established in those fields which need special encouragement in the country involved. Similar prizes could also be awarded to technological achievements.

These prizes would have a number of secondary beneficial effects. They would help in pinpointing meritorious scientists who could then participate in any of the other programs outlined in this article. They would also give local prestige to deserving scientists, thus enabling them to exert more leverage in science policy decisions. The importance of conspicuous prestige is much greater in developing countries than in the advanced countries and hence these prizes would have a great influence.

11. Organizing Regional Seminars

It is peculiar but often true that the relationship between a developing country and a far-away advanced country is much stronger than the relationship of the developing country and one of its developing neighbors. In general, regional cooperation among developing countries is more the exception than the rule. Political, racial, and linguistic differences, as well as prestige and face saving, are among the factors contributing to the lack of cooperation. In addition, historical inertia often drives countries toward the former colonizers or toward the countries where the local scientists are trained.

At the same time, regional cooperation would be an invaluable help for speedier science development.

Examples in Latin America have shown how effective such cooperation can be. Exchange of personnel, co-ordination of research, regional advanced seminars and conferences, and sharing of research equipment in the form of loans or even regional laboratories are among the ways such cooperation can be implemented.

I would like to suggest that a prestigeous research institute in an advanced country could do much in catalyzing such regional cooperation without straining its own resources. For example, the organization of a regional conference in a given field, to be financed by the local participating countries, could be done more easily by such an institute than by any of the partici-pating countries. Proposals for, and advice in, the implementation of regional (as opposed to national) research organizations could also be done better from the outside. The institute in the advanced country could also be instrumental in bringing together under its own roof (through one of the previously discussed pro-grams) scientists from neighboring countries so that they could continue to collaborate after their return to their own countries.

VI. Conclusion

Finally, should an institute become involved in one or several of these activities, and accumulate experi-ence and enthusiasm in them, the institute could also become a center for indirect stimulation of science development in the Third World by using its own con-nections in science and industry to gain further converts to the "activist" point of view. Prestigeous institutes often have some leverage with industry, for example, and could therefore persuade the latter to engage in programs particularly suited to enlarging industrial capacity and know-how. The amount of work to be done is wellnigh infinite, and there is room for parallel work by very many organizations. Furthermore, since there is no well-explored and logical way to proceed in science development, new ideas, many different approaches, and "competing" points of view are sorely needed. Thus it is essential that the interest and com-mitment now felt by relatively few should spread to a much larger group and should be almost universal. Since hardly anything is more impressive than demon-strated success and experience, it should be the func-tion of research institutes, prestigious in science and experienced in scientific aid, to "spread the gospel" and generate more interest and involvement in these problems. □ □ □

Preprint, published as a contribution to the
Symposium on Opportunities in Geomagnetism
and Aeronomy for Developing Countries, University
of Washington, Seattle, 22 August 1977.

SCIENCE AND SCIENTIFIC HIGHER EDUCATION IN DEVELOPING COUNTRIES
OR
SEVEN TASKS FOR GEOSCIENTISTS

Michael J. Moravcsik
Institute of Theoretical Science
University of Oregon
Eugene, Oregon 97403 USA

(Contribution to the Symposium on Opportunities in Geomagnetism and Aeronomy for Developing Countries, at the Joint Assembly of the International Association of Geomagnetism and Aeronomy and of the International Association of Meteorology and Atmospheric Physics, University of Washington, Seattle, August 22,1977)

The building of science in the developing countries is a fascinating, complex, and extensive subject, and so for a 30 minute talk I have many options open to me. I could, for example, assume that many of you have had little acquaintence with this problem area, and thus try to give you a general overview of the whole field. Or, I could select one important aspect of it and attempt a better documented analysis of this narrower topic.

I have, however, decided on neither of these, and I want to tell you why. Since the late 1950's, and throughout the 60's and 70's, a considerable amount has been said and written about science development. Hundreds of articles, scores of conference proceedings, and a few books contain material pertaining to the subject. To be sure, much of   this deals primarily with the development of technology, and science is often mentioned only in passing or indirectly. But even only the material pertaining to science itself is substantial. We cannot say by any means that all problems have, in principle, been solved, and no more differences of opinion remain. It is true, however, that there is a large body of analysis, accompanied by a variety of proposals for action, on which generally all people knowledgeable in this area will agree. If somebody wishes to do something, he will not be impeded by the lack of well defined things to do.

In contrast to all this oral and paper activity, woefully little has actually been done in the past 25 years to assist the developing countries in building their science. This is not the time to go into details of why this is so. For such a discussion, as well as for general background, I would like to refer you to my book on science development[1], and to the references therein, as well as to some of my recent articles, including the one which appeared in the July 1975 issue of Foreign Affairs[2]. Whatever the reasons, the fact is with us that very little has been done, and that international and national organizations, foundations, as well as the worldwide scientific community share the blame for this inaction.

I decided, therefore, that in the few minutes at my disposal today, I will focus on action. In particular,I will describe seven tasks, seven channels of action for you, the geoscientific community, to adopt as action projects in science development for the next year or two. As you will see, most of these projects require negligible financial means, and are completely within the capacity of your scientific community to realize. Yet these projects affect some of the most crucial ingredients of science development, and hence can make very important contributions to this overall aim.

In particular, the seven tasks pertain primarily to science education and to communication among scientists, since the lack of appropriately educated scientific manpower and scientific isolation are perhaps the gravest obstacles to science development.

Task 1. Improve the selection of students from developing countries to be educated for higher scientific degrees in universities of scientifically advanced countries.
Ideally, the best education is the indigenous one. In developing countries, however, at the beginning local education for advanced degrees is simply not available, and hence scientists have to be educated abroad.

The application for admission and financial aid received by graduate schools in the United States or elsewhere from a student in a developing country is typically unevaluable, because it contains transcripts from institutions whose terminology and standards are unknown, and letters of recommendation of a rather formal nature from unknown professors. Furthermore, results of standard graduate admission examinations are either unavailable or unreliable in that context. The result of such a lack of information is either the rejection of the application, or the admission of undeserving students, or the favoring of countries and institutions from which previous experience with students has been favorable.

The solution lies in an oral interview system[3]. It is not a mere suggestion: Such a system has been in operation in physics for eight years, and has proven to be quite successful. Let me tell you how it works.

In the Fall, in October or November, two American physicists together go on a one-month long interviewing trip, to a set of Asian countries ranging from Korea to Iran, talking to students who are about to apply to graduate schools in one of the scientifically advanced countries. Each student is given about 45 minutes, during which he is asked to solve simple problems of undergraduate physics. After the interview a simple, one-page evaluation is written, assessing the student's ability to speak English, his ability in undergraduate physics, and whether he would be able to serve as a teaching assistant during the first year of his studies. The copy of the evaluation is then made available to every school the student is applying to, or every school interested in the student.

The cost of one such interviewing trip is about $6,000, which is collected from voluntary contributions of individual physics departments in the United States and elsewhere. These supporting departments have quick and full access to all evaluation forms and hence can use them also for recruiting purposes. The non-supporting departments get only individual evaluation forms, and only on request.

The program has no overhead and no capital investment.

So Task Number 1. is for the geoscientific community to organize a similar program in the geosciences. I would be happy to assist initially as a free organizational adviser.

Task 2. To equip advanced students from developing countries with an education more suitable for working in their home countries.

In science graduate schools in the scientifically well developed countries, if we compare local graduate students with those who come from developing countries, one crucial difference stands out[4]. The local students, upon receiving the advanced degree, are likely to be employed in an already established organization which is, at least initially, managed by more experienced scientists, so that the young scientist can fully devote himself to research or other directly scientific activities. In contrast, the student from the developing country is likely to be confronted, immediately upon returning to his home country, with the double task of doing science and creating the conditions under which science can be done. For this second task he is not prepared, since we do not teach about such things in our graduate schools.

There are at least two ways to improve this state of affairs. The less expensive, but altogether more time consuming way is for every graduate department to arrange for its own foreign graduate students some exposure to these auxiliary elements of education, using a minimal amount of involvement by some local faculty members. Perhaps the more efficient alternative is to arrange a three-week summer seminar for such foreign graduate students from many institutions together. Such a seminar would deal with a variety of topics, from very mundane things like shop techniques and instrument repair, through library practices and departmental management, to general items like national funding of science or evaluation of scientific output. While such a brief exposure would not make experts in these fields, it would create some awareness, stimulate further self-study, and thus would soften the shock of making a transition from an environment in which the means for doing science are served on a silver platter to a locale where they have to be acquired by skill and hard work.

The cost of such a summer seminar for 50 or so students would be around $40,000. Though a perfectly simple and straightforward idea, such a summer seminar is likely to prove too novel for the conservative tastes of governmental or international agencies or of private foundations, and so it would have to be initially funded otherwise in order to establish its respectibility. But the amount needed is not at all large: If each member of the geoscientific community contributed a yearly sum equivalent to a dinner for two in a reasonable restaurant, the seminar could be easily organized.

Task 3. Form bilateral links.

Perhaps the most effective single measure to break the isolation of scientists in developing countries is to establish bilateral links between small groups of them on the one hand, and analogous small groups in scientifically more active countries, on the other. The two groups should have scientific interests in common, and some previous personal acquaintence of some of their members is also advisable.

The functioning of such a link is quite flexible, and can include a variety of elements: Education of graduate students, a channel for ordering small spare parts for equipment out of order, collaborative research, supply of written scientific information, arrangements for sabbatical exchange leaves, assistance with representation at scientific conferences are only a few of the possible activities. Beyond these specific elements there is the psychological factor of becoming directly tied to the worldwide scientific community.

It is important that such bilateral links be between two small groups of scientists and not between two institutions of larger size[5]. Once the link becomes as large as a university-to=university arrangement, it is likely to turn into a formal relationship between university bureaucrats, and the element of personal acquaintence and responsibility will be weakened. In addition, on a large scale such arrangements become official business of high visibility, with all the concommitant red tape and constraint. Such links must be organized and maintained by scientists themselves.

Since the activities in such a link are flexible, it is difficult to estimate the financial resources needed for it beyond what is available within the normal means groups of scientists have access to. Probably the costliest type of activity is exchange of personnel, but with some skill funds can be procured for such a purpose from regional agencies like OAS, from PL 480 type excess foreign currency reserves, or from programs like the National Science Foundations's SEED. The remaining activities involve only amounts of the size of petty cash, and some involve only thoughtfulness and good will. Remember also that if only one out of every ten groups in the scientifically active countries established such a link with one counterpart group in a developing country, every existing group of the latter type would be taken care of.

Task 4. Distribute surplus back issues of scientific journals[6].

New institutions in developing countries have great difficulties in acquiring for their libraries back volumes of scientific journals because they are partly unavailable and also very expensive in the total amount and in terms of the always scarce hard currency.

At the same time, there is a substantial amount of back issues of such scientific journals on the shelves of scientists in the scientifically active countries. Somebody who has been a working researcher for 15-20 years is generally running out of shelf space in his office, and although having those old journals immediately on hand is a convenience in rare instances, it is certainly far from being a necessity. As a result, a number of such scientists have been seeking recently ways of disposing of such back issues of journals, and instances are known of people actually discarding such journals altogether.

There is clearly a supply and a demand, and so all one needs to do is to bring the two together. This has been done in the area of physics by a surplus journal distribution program[7]. This program identifies institutions in the developing countries and their needs in terms of journals, and also pinpoints potential donors of journals in the worldwide scientific community. The potential donors are then asked to mail journals directly to the potential recipient institutions and, if the shipping charges cannot be covered from either end, the program also reimburses the donor for the postage.

The cost of this program consists entirely of such postage. One volunteer physicist operates the program for the whole world in part of his spare time. A yearly sum of $1,000 goes a long way toward paying for the shipping charges.

A similar program already exists in the geosciences, operated by AGID (Association of Geoscientists for International Development) which some of you might, and all of you should, know about. Information about AGID can be obtained from Anthony Berger, Department of Geology, Memorial University, St. John's, New Foundland, Canada. The AGID program (as well as the physics program to some ex.ent) has suffered from a lack of publicity among the scientific community. So tell about it to your colleagues, and when you return home, look at your book shelves to see if you could not become a donor.

Task 5. Arrange for equitable distribution of journals.

Subscriptions to scientific journals by libraries in developing countries are becoming increasingly impossible as the rates, demanded in hard currency, increase precipitously, and in addition the libraries are increasingly asked to subsidize the already dwindling number of individual subscribers. It is disadvantageous enough to receive journals months late because of the slow surface delivery over huge distances. The prohibitive subscription rates greatly add to this disadvantage.It should also be remembered that journals are even more important in developing countries than elsewhere since other channels of scientific communication (preprints, visitors, meetings, telephone) are very weak in those countries.

The total need for scientific journals in the developing countries is not large. Since 92% of all scientists in the world work in scientifically advanced countries, and many of those have personal subscriptions to journals, it is safe to assume that 5% of each journal's circulation would be needed to saturate the scientific libraries in developing countries. Since some journals get some of their income from page charges, we arrive at the conclusion that if the subscription rates for scientific journals for subscribers in scientifically advanced countries were raised by 10%, libraries in the developing countries could be provided with free subscriptions to these journals. Such a subsidy would not be unprecedented, since, as mentioned above, libraries already subsidize individual subscribers. If it is thought that such entirely gratis arrangements are inadvisable (and there are good reasons for thinking so), subscribers in developing countries could be asked to deposit, in return for the journal, a modest amount in local currency into a local bank account. Funds thus accumulated could be used by the publisher of the journal in a variety of ways, or be given to some worthy local cause.

Many scientific journals are published by scientific societies, and hence the initiative in this matter is largely in the hands of the scientific community itself. In other words, it is up to you, geoscientists, to take action in your own field.

Task 6. Set up a visitors' registry.

Personal communication among scientists is one of the most important channels of interaction and scientists in developing countries have access to it much less than others. In this area again there is both some supply and a considerable amount of demand, and what is needed is information exchange and coordination. This has been the function of the Visitors' Registry, which has been operating to a modest extent among physicists for some time[8]. It works as follows

One physicist during a fraction of his spare time operates the whole program, which consists of two filing cabinets. One contains cards for physics departments or laboratories in developing countries which are interested in receiving a visitor for a day or two. The card indicates the fields of interest of the potential host institution and whether local funds are available for subsistence costs for a day or two, or perhaps also to defray the cost of a small detour in the visitor's itinerary. The other filing cabinet receives cards from physicists from scientifically active countries who, in the next 3-4 months, will travel in or over some developing area. This card indicates the planned itinerary and the fields of activity of the would-be traveller. All the registry does is to send xerox copies of these cards of the second type to addresses on the cards of the first type which lie along the itinerary. If the institutions which receive such a notification wish to invite the would-be traveller, they will contact him directly.

The cost of this program depends on the volume, but after an initial investment of perhaps $200, the yearly running costs are $100-$200. This amount does not include paid advertisements in scientific journals to publicize the program. Such publicity might help in the full-utilization of this system, and in fact the physics program has suffered from a lack of such publicity. Yet, it has been responsible for a fair number of successful visits.

Similar registries would also be useful for scientists from developing countries visiting scientifically active countries, and for longer term visitors as well as for short term ones. Furthermore, a similar registry would also be helpful for scientific positions available in developing countries for which people from scientifically active countries are sought.

Task 7. Induce industries concerned with the geosciences to contribute to science development

A significant fraction of the industry in many developing countries consists of subsidiaries of companies in scientifically leading countries. Thus there are a number of ways such companies could (but presently generally do not) contribute to the building of science in countries where they have subsidiaries.

Perhaps the most important way would be through the sponsoring of some research in the subsidiaries themselves instead of doing it all in the mother company abroad. The arguments for the latter are increased efficiency, better facilities, better qualified staff. The arguments for the former are, however, also numerous: Greater ease of performing research applicable to the local market, lower cost, and, last but not least, a favorable public image in the country of the subsidiary.

A second area of contribution would be the summer employment in the mother company of student from the country of the subsidiary who are receiving advanced science education at a university in the country of the mother company. Such an employment opportunity would provide the students with a much needed introduction into applied research. From the company's point of view, such students, upon returning to their home countries, might contribute to the subsidiaries.

These are but two examples of the benefits of the participation of international companies in science development. In general, there is a need for raising the awareness of these companies to the wants of science development in order for such participation to come about. That part of the geoscientific community which is in contact with these companies could play an important role in this. To do so, these would-be spokesmen for science development would have to become themselves knowledgeable, articulate, and convincing about this field. Reading, combined with the acquisition of personal experience in developing countries, is the best way to accomplish this.

This completes the brief outline of the seven tasks. You will notice that all of them have some elements in common. They utilize the latent potential within the scientific community and hence are particularly well suited for scientists to engage in. They need only minimal financial resources, and are simple to administer, with practically no initial capital investment. Therefore they are well suited to being organized and operated by even a small group of interested scientists.

It is customary to end a talk like this on some euphoric note, expressing confidence about the changes a symposium like this will bring about. Unfortunately I have witnessed too many such meetings before, and hence I would consider it unrealistic to engage in such fantasies. Instead, let me venture the guess that after a possibly entertaining and perhaps even stimulating morning of talks and discussion, all of you will return to business as usual, and none of the suggestions in this talk or any other ones will be converted to reality. Thus scientists in the developing countries will have to continue to struggle by themselves to build science in their countries.

References

1   Michael J. Moravcsik, Science Development — The Building of Science in Less Developed Countries, PASITAM, Indiana University, Bloomington, Indiana, First Edition (1975), Second Edition (1976). This book is available only through direct order from PASITAM.

2   Michael J. Moravcsik and John M. Ziman, Paradesia and Dominatia: Science and the Developing World, Foreign Affairs 53, 699 (1975).

3   For a description of this system, and for some of the experience gathered through it, see Michael J. Moravcsik, The Physics Interviewing Project, International Educational and Cultural EXCHANGE, Summer 1972, p.16, and F.Dart, M.J.Moravcsik, A. deRocco, and M.Scadron, Observations on an Obstacle Course, International Educational and Cultural EXCHANGE 11, No.2, 29 (1975).

4   For a discussion of this as well as of other aspects of the education of "foreign students", see Michael J. Moravcsik, Foreign Students in the Natural Sciences: A Growing Challange, International Educational and Cultural EXCHANGE 9, No.1, 45 (1973).

5   For a skillful analysis and documentation of the effectiveness of links between small groups, see Henry R. Glyde, Institutional Links in Science and Technology: The Case of the United Kingdom and Thailand, Technical Report 55/5, Bangkok, Applied Scientific Research Corporation of Thailand, Bangkhen, Bangkok, Thailand (1972). For an abbreviated version of this report, see International Development Review 15, 7 (1973), also reprinted in Ekistics (Reviews on Problems and Science of Human Settlements) 36, 440 (1973).

6   For an analysis and documentation of this issue, see Michael J. Moravcsik, The Distribution of Surplus Back Issues of Scientific Journals, report commisioned by the scientific information division of UNESCO, Paris (1977).(Copies available from the author.)

7   Information on the surplus back issues of journals being distributed in physics can be obtained from Dr. Luciano Fonda, International Centre of Theoretical Physics, Miramare 34100, Trieste, Italy.

8   For a description of the visitors' registry project in physics, as well as some other programs, see Michael J. Moravcsik, The Committee on International Education in Physics, Amer. Journ. of Physics 41, 309 and 608 (1973) and Physics Teacher 11, 82 (1973).

C. Human Resources

Reprinted by permission of *International Educational and Cultural Exchange* 9 (1973) 45.

*A thorny problem for American graduate schools
is how to evaluate the credentials of prospective
foreign students—how properly to assess their potential for success. Here is an innovative scheme
that has been tried and appears to be effective.*

The Physics Interviewing Project

by Michael J. Moravcsik

Shortage of adequately trained manpower is one of the most
important obstacles in the development of the emerging
countries. In fact, many believe that, as far as the development
of science is concerned, it is *the* most important obstacle.
One of the most valuable forms of help, therefore, that the
more developed countries can offer is assistance in training
personnel.

It should be emphasized, of course, that no program of
training away from the homeland can replace indigenous
education. Hence any such program should be considered only
a stopgap measure that must not interfere with the evolution
of the local educational institutions.

At the moment, however, there are many countries in the
world where the existing institutions of higher learning cannot accommodate all those deserving students who wish to
receive advance education. Such students therefore have been
coming in large numbers to the more advanced countries,
notably the United States. Among the American physics graduate schools, there are hardly any which have no students
from developing countries.

How To Select Foreign Students

The presence of these students in graduate schools creates
responsibilities and problems. One of the most thorny problems has been the proper selection of foreign students for
admission and financial aid. The responsibility is twofold.
On the one hand, education is becoming increasingly more
expensive in the United States, and the expenses are borne
increasingly by the taxpayer. It is not unrealistic to attach
a price tag of $5,000 per year to the education and financial

aid of a physics graduate student. It is therefore a serious political responsibility to assure that the educational facilities are utilized by the best available talent.

On the other side of the coin, it is also crucial for the developing countries that their personnel with the highest potential receive the much needed education. Furthermore, from the point of view of the student, erroneous selection might have far-reaching consequences. The rejection of a talented and promising student could well mean the waste of his talent and promise. The admission of an inappropriate student, on the other hand, might mean the waste of his time and financial resources, coupled with the humiliation of his having to return to his homeland without the much coveted degree. Such a situation is not likely to win friends for us either, since often the failure is rationalized in the form of an aversion to the United States.

In spite of the crucial importance of the selection process, the screening, in the case of physics, of foreign applicants to American graduate schools of physics has been somewhat haphazard, at least compared to the screening an American applicant in physics has to undergo. In a not atypical situation, the information available about a foreign applicant is as follows: Transcript of records from a university whose grading system may be unfamiliar to us; a list of completed courses, whose contents and method of teaching might not be known to us; and letters of recommendation from professors we do not know. In addition, most schools now require the results of some centrally administered test concerning the competence of the candidate in the use of the English language. Finally, many graduate schools now require the candidate to take the graduate record examination (GRE).

It is not even necessary to discuss the vagaries involved in the use of transcripts, lists of courses, and letters of recommendation—they are so obvious. It is, however, important to discuss the implications of using the GRE.

Value of GRE in Evaluating Foreign Students

The GRE was designed to test the ability to function, in an American graduate school, of *American* students coming from a middle class environment and from standard American colleges. There is some difference of opinion as to whether it fulfills that task. Even granting, though, that it does, it appears to have very little relevance to the testing of *foreign* students. It relies on certain terminology used in American schools, and is "culture-bound" in other respects also. Most importantly, being a mechanical, written test, it allows no room for feedback, and hence for an in-depth assessment of the capabilities of the candidates.

17

Those who favor the GRE in this context like to retort that since the potential graduate student has to function in an environment where the norms assumed by the GRE prevail, it is appropriate to test him by these norms. I believe that this is a gross oversimplification of the nature of the adjustment a foreign student has to make when entering an American graduate school. This adjustment is likely to be drastic in any case, and hence what is needed is not to test whether the student has already adjusted to the American system before he even sets foot in the United States, but whether he has the basic background and the inherent potential to make this adjustment after he has arrived. It is my feeling that the GRE does not test these. This is not the fault of the GRE, but it is rather a consequence of its misuse. Just as the GRE is of doubtful usefulness for testing students from minority groups in the United States, it is also of questionable value in testing foreign students.

Naturally there have been efforts by physics departments to get additional assurances about the foreign students to be admitted. For example, departments sometimes have faculty members who are familiar with one or another institution in a developing country, and hence tend to draw their foreign students from that institution. Such chance arrangements nevertheless are hardly a substitute for a more systematic approach to the problem.

Interviews for Physics Students

It was with these considerations in mind that the Physics Interviewing Project was suggested a number of years ago.[1] Since then, it has undergone some minor revisions, but its basic outline has remained the same.

The Physics Interviewing Project consists of a committee of physicists from U.S. universities undertaking a trip in a number of developing countries, interviewing student candidates who wish to study physics in American graduate schools. The committee writes an evaluation of each interview, which then is available to any American physics department interested in the student, or any such department the student is interested in. The committee does not handle directly any matters pertaining to admission, scholarships, assistantships, and the like. Its only function is to provide information about the students.

The suggestion was tossed about for a number of years as a search for a sponsor was under way. Financially, the proj-

---

[1] M. J. Moravcsik, "Some Practical Suggestions for the Improvement of Science in Developing Countries," *Minerva*, IV, 381 (1966). (Reprinted in Edward Shils (ed.), *Criteria for Scientific Development: Public Policy and National Goals*, Massachusetts Institute of Technology Press, 1968.)

ect is a very flexible and modest one. It requires no overhead and no capital investment, and hence can be discontinued at any time if it is found not to yield the desired results. Its budget, at least in the form finally tried, is a very modest one, on the order of what it costs to educate one graduate student for one year. In spite of these advantages and the large potential benefits, only moral and no financial support was obtained from the various governmental and private organizations that were approached. Finally it became evident that in order to *prove* the feasibility of the idea, the participation of physics departments themselves had to be enlisted. Indeed, such a participation finally became a reality when in the spring of 1969 four physics departments, those of the Universities of California at Los Angeles, Michigan, Oregon, and Pittsburgh, joined to sponsor an experimental Physics Interviewing Project.

The trip, in October and November 1969, lasted some 5 weeks, and some 145 students were interviewed in Seoul, Hong Kong, Bangkok, Singapore, Kuala Lumpur, Madras, Bombay, Delhi, Calcutta, Banaras (Varanasi), Dacca, Lahore, and Karachi. Students from universities in other localities traveled to these centers to meet the committee for the interviews. The committee consisted of Francis Dart of the University of Oregon and myself. Each student was interviewed for about 40-50 minutes, after which we immediately wrote down our evaluation.

One of the important elements of such an interview is the method of selecting the students to be interviewed. In this respect, an ambitious, though somewhat illusory goal might be to try to find the best students in each country. A more realistic goal (which we actually aimed at) is to try to line up good candidates, as judged by a variety of local contacts, and then evaluate them through the interviews. In lining up the candidates, we relied on personal professional contacts, U.S. governmental and private organizations, as well as on the indigenous educational and governmental agencies. This multipronged approach helped to avoid the blocking of a candidate by any one channel. We are indeed grateful to all those who helped us in preparing for the interviews.

Conducting the Interview

In the interview we tried to assess both the candidate's basic background in physics and his ability to think and to use familiar physics in solving problems never before encountered. Once we saw how things were going in connection with a particular question, we interrupted and went on to the next question. Thus it was possible to cover a fair amount of terri-

19

tory in 45 minutes. It was our feeling that in that amount of time we acquired a good basis on which to evaluate the potential of the candidate. Incidentally, we could also assess the candidate's knowledge of English, although this was not really one of our aims, since that can be done by one of a number of established organizations. Similarly, we paid no attention to university credentials and letters of recommendation, knowing that they would be surveyed by graduate schools to which the candidate applies.

At the beginning of this article I mentioned my belief in the importance of not interfering with the evolution of indigenous institutions of higher learning. We tried to live up to this belief by emphasizing to candidates that getting a higher degree in the United States is by no means the only road to salvation, and that studying instead in one of the topmost local institutions might in fact be an easier, less costly, and more effective way to begin a scientific career Such advice was particularly appropriate in instances whel advanced students, perhaps only 2 years from a local Ph.D., indicated an intention to trade their present status for the uncertain possibility of obtaining a similar degree in the United States in 5 or 6 years.

I have mentioned that in general we paid no attention to the candidate's formal qualifications. In one instance, however, the candidate volunteered his GRE score during the interview. In this case, at least, there was a large discrepancy between the GRE score and our evaluation: we thought that the candidate was considerably more promising than the GRE score would lead one to believe.

Need for Refresher Courses

In a number of instances we came to the conclusion that the candidate had a good chance of succeeding provided that in his first year in graduate school he take, in part, undergraduate courses for refreshment. Such a suggestion (not likely to be obtainable through written examinations) might be helpful to the student's adviser in deciding the student's program for the first year.

We also stated in each evaluation whether we thought, in view of all elements of the interview, that the candidate would be able to discharge the duties of a teaching assistant during his first year in graduate school. This piece of information, also unattainable through written tests, might be quite valuable to graduate schools, at least judging from past experience in this respect.

Our interviews were somewhat disconcerting to the students. Not only were we obviously more interested in the method and thinking that led to an answer than in the "right" answer itself, but we also concentrated during a considerable fraction of the interview on "elementary" physics which some candidates felt as somewhat of an insult. To quite a few students (and sometimes even to some members of the local staff), the interview was also a useful eye opener, to see how physics is approached in American graduate schools. This was one of the beneficial byproducts of our trip.

There were other byproducts also, not anticipated when the project was formulated. During our trip we had a unique opportunity to compare many different physics educational systems *in situ* and abstract from this an idea of some of the main obstacles a student would have to face when entering an American graduate school.

Remedy for the Information Gap

We have also acquired some useful information on the deficiencies of the information foreign students have about American graduate schools (in physics). Much of the written material available on American universities and on physics departments is naturally aimed at American students and hence takes certain things for granted—which, however, cannot be taken for granted when informing foreign students. It is also important that the information aimed at prospective students be available at the right time and at the right place. To remedy this information gap, we wrote, on our return, a brochure on American graduate education in physics, aimed at prospective foreign students. It is now available from the American Association of Physics Teachers, and has been received very well in many countries.

To check the effectiveness of this interviewing scheme, the Committee on International Education in Physics (CIEP) of the American Association of Physics Teachers in the winter of 1970 undertook a study of the performance of those students interviewed during the first trip who subsequently were admitted to an American graduate school in physics. Only about a quarter of those interviewed were in this category, and detailed information could be gathered about only a part of that quarter. Thus the overall statistics of the survey were not outstandingly good. On the basis of this limited sampling, however, a virtually perfect correlation was found between the evaluation of the interviewers and the actual performance of the student in graduate school. The only discrepancy was a very slight systematic one, the evaluations being somewhat more severe on the students than their actual performance.

Second Trip—to 10 Countries

Encouraged by this confirmation of the efficiency of the new approach, CIEP decided to organize a second interviewing trip for the fall of 1971. This time 15 physics departments got together to sponsor the trip, which cost altogether a little over $4,000. The two interviewers this time were Michael Scadron of the University of Arizona and Andrew deRocco of the University of Maryland. A total of 175 students were interviewed in 10 countries: South Korea, Hong Kong, Thailand, Indonesia, Singapore, Malaysia, Ceylon, India, Pakistan, and Afghanistan. The general format was similar to that of the first trip.

After each trip, the information was handled as follows. Those departments that sponsored the trip received, immediately after the trip, a complete set of all interviewing sheets. A month later copies of individual evaluations were also released to nonsponsoring universities either at their request (in case they received an application from one of the interviewed students) or at the request of the student. Thus the information was made generally available, with the sponsoring departments obtaining a bonus by having direct access to the promising students.

On the basis of the experience so far accumulated, it would appear that it is sufficient to organize an interviewing trip to a given geographical area once every 2 years, since a given trip can catch two consecutive generations of students. Thus it becomes possible to cover, through yearly trips, a larger variety of countries. A trip to a different continent for the fall of 1972 is under active consideration.

One of the often asked questions is whether the Physics Interviewing Project would result in more or fewer students being admitted from those countries visited than there are now. The answer is that CIEP does not know and, from the point of view of the project, does not care either. The purpose has been simply to improve the chances that those students who do come to the United States for education will indeed attain that goal, and with the fewest irrelevant impediments. The indications are that the program has been able to make a significant contribution in this direction. ∎

• MICHAEL J. MORAVCSIK *is professor of physics and Director of the Institute of Theoretical Science at the University of Oregon. Born in Budapest, he had his early schooling there and attended the University of Budapest. He came to the United States in 1948 and earned an A.B. at Harvard and a Ph.D. at Cornell, becoming a U.S. citizen in the meantime. After some years in research he became professor of physics at the University of Oregon and research associate with the Institute of Theoretical Science, of which he became Director in 1969. He has been visiting professor and lecturer in a number of institutions here and abroad. In the fall of 1971 Dr. Moravcsik was appointed a member of the Advisory Committee on East Asia of the Committee on International Exchange of Persons.*

Reprinted by permission of *International Educational and Cultural Exchange* 8 (1972) 16.

A science professor suggests specific remedies for some of the difficulties faced by foreign students—particularly those from underdeveloped countries—who seek an education in the U.S. so as to prepare themselves for careers as scientists back home.

Foreign Students in the Natural Sciences: A Growing Challenge

by Michael J. Moravcsik

THE EDUCATION foreign students receive in the United States is one of the most significant components of our international assistance. Approximately 150,000 foreign students study at American universities and colleges—and about 110,000 of them come from less developed countries. Counting, on the average, $3,000 as the annual cost of educating one such student, above and beyond whatever tuition or fees he may pay, the total expenditure comes to about $450 million, a respectable sum. But more important than the purely financial aspects, the commodity received, namely good education, is rare in the home countries of many of these students and hence is a particularly valuable contribution to their development.

At the same time, foreign students in return contribute substantially to the American educational system. At a time when interest in the natural sciences is decreasing in the United States compared to a decade ago, and hence the quality of American graduate students in the sciences is less than optimal, well-selected foreign graduate students can supply the much needed excellence without which an educational and research institution becomes a dull and unproductive place.

But is this education the foreign students receive in the United States really so good? Judged by the standards and needs of American students and American society, it might very well be good. But is it so from the point of view of students with different backgrounds and requirements who are heading for a society vastly different from ours? There is clearly no single simple answer to this question.

• MICHAEL J. MORAVCSIK *is professor of physics and Director of the Institute of Theoretical Science at the University of Oregon. Born in Budapest, he had his early schooling there and attended the University of Budapest. He came to the United States in 1948 and earned an A.B. at Harvard and a Ph.D. at Cornell, becoming a U.S. citizen in the meantime. After some years in research he became professor of physics at the University of Oregon and research associate with the Institute of Theoretical Science, of which he became Director in 1969. He has been visiting professor and lecturer in a number of institutions here and abroad. In the fall of 1971 Dr. Moravcsik was appointed a member of the Advisory Committee on East Asia of the Committee on International Exchange of Persons.*

In this article I want to consider one particular segment of the foreign student population—graduate students in the natural sciences—and to discuss in what way their education in the United States is deficient and how one could improve it.

Since many of the deficiencies and hence the remedies are interrelated, perhaps the best way to structure our discussion is in terms of a chronological survey of the student's education.

Information Gap

We should start with the student in his home country, contemplating the possibility of getting a graduate education in the United States in one of the natural sciences. First he needs information about schools, degrees, requirements, standards, examinations, teaching methods, methods of preparation. Some such information is sometimes available to him. School catalogs are on the shelves of the nearest U.S. Information Service Library, and so are books and brochures published by individuals and organizations describing various aspects of the American educational system. He may talk to staff members of the local American Consulate or the U.S. Information Service. Is there then need for further information?

There can definitely be such need, and for two main reasons. First the information available might not be specific enough. If he wants to be a chemist, he should know what books are used in chemistry graduate schools, whether examinations are factual, essay-type, or problem-solving, and so forth. These questions are generally not answered in most printed publications, and only by coincidence would the local American contingent include a chemist or somebody sufficiently knowledgeable about chemistry to answer such questions.

Second, whatever literature might be available on such questions will have been written for American students and hence with their background in mind. The terminology used might not mean very much to a foreign student, and many essential but elementary points might be omitted altogether since every American student would know them anyway.

I first became aware of the magnitude of this information gap during the first trip of the Physics Interviewing Project. My colleague, Francis Dart, and I quickly decided to begin filling this gap by writing a simple and short pamphlet [1] for prospective foreign graduate students in physics. After a preliminary edition of 500 was depleted almost immediately, a larger batch of 5,000 published by the Association of American Physics Teachers was also disposed of almost completely in a matter of months. The brochure is inexpensive and light enough for individual distribution. Similar material should be available in other disciplines also.

Application Fee

After our graduate student has sufficient information, and has the application forms, he is ready to apply to a number of graduate schools. Here he runs into a potentially serious obstacle: the $10 application fee required by most graduate schools and nonrefundable if the applicant is not admitted. The justification for this fee is not clear, but in any case it places an insufferable burden on foreign graduate students, particularly those from the less developed countries. Not only is the $10 (multiplied by the number of schools applied to) a huge sum by local standards, but more importantly, it must be paid in hard currency. In many countries the sending abroad of hard currency is almost impossible, and at the least it involves much red tape.

Several specific steps could remedy this problem. Universities should not require students from less developed countries to pay the fee when applying. Such fees should be assessed later and then only if the applicant is accepted. The payment should be deferred until the student has received his first pay check in the U.S. as a research or teaching assistant (the most common form of financial support for a foreign graduate student in the natural sciences). If such a deferred payment proves difficult to negotiate with the sometimes stuffy university admission offices, the departments might set up a small revolving loan fund of $100–$200 for this purpose.

---

[1] F. E. Dart and M. J. Moravcsik, "The Physics Graduate Student in the United States, A Guide for Prospective Foreign Students," American Institute of Physics, 1971.

Evaluation

Upon receipt of the application it is then up to the appropriate university department here to evaluate it. Now further problems may arise. The information the student has sent is often difficult to evaluate: unusual transcripts, unfamiliar grading systems, an unassessable list of courses, an ungaugeable set of letters of recommendation, and the scores of written examinations—like the GRE—that are of dubious applicability to the evaluation of such students.

I outlined this problem in greater detail in an article describing the Physics Interviewing Project,[2] which is a possible answer to this problem. For this project, sponsored by the Committee on International Education in Physics of the American Association of Physics Teachers, a committee of physicists personally interviews potential applicants in their home countries and reports the findings to any graduate school relevant for that student. Three such trips have taken place to date, between 1969 and 1972, and information on their effectiveness indicates the very high degree of reliability of the information generated through the program.

Such evaluation programs could be set up in other disciplines. The evaluation must be done by professionals in those disciplines, and therefore these programs are best generated by—or at least conducted in close cooperation with—groups or organizations of such professionals. The cost of such programs is low enough not to represent an insurmountable obstacle.

Transportation Cost

Let us assume that our student has successfully passed this hurdle also and has obtained admission and some financial aid at an American university. He then faces the next hurdle: the cost of transportation. For the same reasons already mentioned in connection with the application fee, this is also a considerable problem, and greatly magnified, since one-way transportation can amount to as much as $700. One partial remedy for this problem—admittedly not very likely to be adopted for a number of reasons—is the institution of a stand-by fare on international flights for a certain class of passengers. Or the problem can be solved like that of the application fee: the department would issue a plane ticket to the student, with the cost to be repaid by the stu-

[2] M. J. Moravcsik, "The Physics Interviewing Project," *Exchange*, Summer 1972, pp. 16–22.

dent in equal monthly installments during his first year in graduate school.

There are also some sources of outright grants for travel expenses, such as the Fulbright program, but these are not very extensive and hence can cover only a few of the deserving candidates.

But let us continue to follow our student who by now has arrived at the U.S. university. At this point, he runs into numerous problems—climate, food, social customs, personal contacts, language—common to all foreign students in the United States. I will deal here only with those aspects of education peculiar to science graduate students. It is well to keep in mind that these general problems might often burden the student in various visible and invisible ways, and faculty members and particularly advisors must therefore be aware of them.

This brings us to the problem of advisors, which is of central importance in the overall educational scheme of foreign graduate students. Advising practices vary from university to university and from department to department even for American students. Their thoroughness is, however, somewhat less crucial for American students who are in their customary environment, are more familiar with the customs and expectations of a graduate school anyway, and who, in most cases, can be counted on to consult with the advisor or a faculty member if they have serious problems.

All this is very different for a foreign graduate student. He may have come from an educational system where hierarchy is greatly emphasized, where contact between student and professor is usually formal and sparse, and where personal problems are kept separate from "official" academic contacts. It thus takes an extra amount of ingenuity, initiative, and understanding on the part of the adviser to do his job properly.

With the shrinking world in terms of travel, most science departments at American universities have by now several faculty members with personal experience in the less developed countries. They should be involved in the advising of foreign graduate students.

Science Education in the Third World

At the initial stages of advising, two major points should be kept in mind. These pertain to certain dominant characteristics of science education in many less developed countries in the world, as compared to such education in the

United States. The first of these is the tendency to approach science in terms of rote learning, memorization, and formalism, rather than in terms of understanding, usage, and functionalism. As a result, an incoming foreign graduate student might appear, on paper, much more advanced than his American counterpart, but when it comes to demonstrating his knowledge in terms of problem-solving, the student might discover gaps in his background even on a rather elementary level.

The second characteristic likely to be encountered is premature specialization. The foreign student may have terminated broad studies in his area of science at the level of an American B.S. and then proceeded to do "research" with his professor. In doing so, he not only divorced himself from contact with a broad area of science, but also built up unwarranted illusions as to his status in science. Clearly, a B.S.-level student most of the time cannot function in genuine research at a level much higher than a laboratory assistant, technician, or a program for a computer, and he has a long way to go before becoming a research scientist.

The result of these characteristics in the background of many foreign students is that while they need to go slowly and strengthen their grip on areas of science even on a relatively elementary level, they will be reluctant to do so because they believe they have already passed the more elementary stages. They will, therefore, want to go ahead full speed and plunge into more advanced material.

It is the not altogether pleasant, but absolutely necessary duty of the adviser to resist this tendency and persuade the student to go easy, at least in the first year, until he sees how his knowledge and preparation matches that generally required in American graduate schools. Well-selected foreign graduate students perform well in American graduate schools, and often rise to considerable distinction among their fellow students. But almost without exception, their initial adjustment period is a delicate one, and what they are advised to do in the first year is crucial.

Advisers Take the Initiative

As already mentioned, most foreign students at first will relate to the faculty members with excess reverence and formality. It is therefore often up to the faculty adviser to initiate meetings with the student. It is also the adviser's duty to probe into the student's circumstances and discover possible trouble spots early enough. Periodic checks with the student's professors are advisable. The student himself might not always be able to judge whether he is doing well or not, since the teaching methods, examination structure, and evaluation standards might be drastically different from what he is used to.

In this connection it is important that the adviser explain clearly to the incoming student what is expected of him during the various stages of his education in the graduate school. Many unpleasant surprises can be avoided if the expectations on both sides are clarified at the very beginning.

It may be helpful here to list a few general and basic respects in which the future scientific career of the foreign student will probably differ from that of his American counterpart.

Scientists Here and Abroad

Perhaps the most striking difference is that in a less developed country, scientists—even at the beginning of their careers—may well be called upon to practice science *and at the same time* also create the conditions under which science can be practiced. In contrast, a young American scientist will most likely enter a scientific environment already set up for the practice of science, and only considerably later in his career, if at all, will he be asked to participate in the management or the further development of the scientific institutions he is part of.

Second, the student from the less developed country will return to a locale in which the number of scientists is likely to be very small compared to those in the scientifically advanced countries, and hence, specialization along lines predetermined by the student himself might not always be possible. Such a student then needs a much broader education than his American counterpart. This is in direct conflict with his tendency to specialize too early. It is to a large extent the duty of his adviser to re-orient the student's thinking in this respect, and assure the breadth of his education.

Third, the student from the less developed country is likely to prefer theoretical work as compared to experimental activities, and academic work rather than applied science. These two tendencies—which, by the way, are not unknown in advanced countries though generally they do not appear so strongly—are interrelated, having their origin in certain social attitudes toward manual and practical work as opposed to mental efforts oriented in more abstract directions. Partly as a result of this bias, the actual need in less developed countries will, however, be much greater (though not exclusive) for scientists willing to do things with their hands, and able to interact with practical problems in addition to being able to contribute to abstract studies. The student's adviser can play an important role here by assuring sufficient exposure to experimental and practical influences.

Choosing a Specialty

There is also the related question of choosing the field of specialization. As I remarked in previous articles,[3] I believe the best way to influence a student to select a certain specialty is to send him to a department where that specialty is prominent. Most starting graduate students, foreign or American, have only very vague ideas about the selection of a specialty—and are therefore still impressionable in this respect. The adviser still has an opportunity to channel the student into one or another area.

There are specific steps that can be taken to improve the student's skills in manual work, his contact with applied science, and his familiarity with science organizational matters. He should be strongly advised to take all the laboratory courses offered. Second, foreign students should be advised to find temporary summer employment with a research laboratory or with private industry. The latter is not too easy to arrange, for a number of reasons. The industry might feel, perhaps justly, that a brief summer's contact with an unoriented student will not be of much benefit to the company. Industry in the scientifically advanced countries, however, needs to strengthen its contribution to the scientific development of the rest of the world, both for reasons of self-interest and also as an altruistic service function. Among various programs in this direction, one of the simplest and most effective would be to employ some foreign students during the summer vacations. Some of these students might later be able to participate in their home countries in local research activities of subsidiaries or affiliates of the companies where they had worked here. The summer employment program would thus be justified even from a short-term practical point of view. The organization and initiation of such summer opportunities is a joint task for industry and the university department where the student is being educated.

---

[3] E.g., M. J. Moravcsik, "Aspects of Science Development," in *Management of Research and Development* (Proceedings of the seminar organized by the Scientific and Technical Research Council of Turkey, Istanbul, 1970), OECD, Paris, 1972, pp. 189–240, particularly pp. 214–15.

might not always be able to judge whether he is doing well or not, since the teaching methods, examination structure, and evaluation standards might be drastically different from what he is used to.

In this connection it is important that the adviser explain clearly to the incoming student what is expected of him during the various stages of his education in the graduate school. Many unpleasant surprises can be avoided if the expectations on both sides are clarified at the very beginning.

It may be helpful here to list a few general and basic respects in which the future scientific career of the foreign student will probably differ from that of his American counterpart.

Scientists Here and Abroad

Perhaps the most striking difference is that in a less developed country, scientists—even at the beginning of their careers—may well be called upon to practice science *and at the same time* also create the conditions under which science can be practiced. In contrast, a young American scientist will most likely enter a scientific environment already set up for the practice of science, and only considerably later in his career, if at all, will he be asked to participate in the management or the further development of the scientific institutions he is part of.

Second, the student from the less developed country will return to a locale in which the number of scientists is likely to be very small compared to those in the scientifically advanced countries, and hence, specialization along lines predetermined by the student himself might not always be possible. Such a student then needs a much broader education than his American counterpart. This is in direct conflict with his tendency to specialize too early. It is to a large extent the duty of his adviser to re-orient the student's thinking in this respect, and assure the breadth of his education.

Third, the student from the less developed country is likely to prefer theoretical work as compared to experimental activities, and academic work rather than applied science. These two tendencies—which, by the way, are not unknown in advanced countries though generally they do not appear so strongly—are interrelated, having their origin in certain social attitudes toward manual and practical work as opposed to mental efforts oriented in more abstract directions. Partly as a result of this bias, the actual need in less developed countries will, however, be much greater (though not exclusive) for scientists willing to do things with their hands, and able to interact with practical problems in addition to being able to contribute to abstract studies. The student's adviser can play an important role here by assuring sufficient exposure to experimental and practical influences.

Choosing a Specialty

There is also the related question of choosing the field of specialization. As I remarked in previous articles,[3] I believe the best way to influence a student to select a certain specialty is to send him to a department where that specialty is prominent. Most starting graduate students, foreign or American, have only very vague ideas about the selection of a specialty—and are therefore still impressionable in this respect. The adviser still has an opportunity to channel the student into one or another area.

There are specific steps that can be taken to improve the student's skills in manual work, his contact with applied science, and his familiarity with science organizational matters. He should be strongly advised to take all the laboratory courses offered. Second, foreign students should be advised to find temporary summer employment with a research laboratory or with private industry. The latter is not too easy to arrange, for a number of reasons. The industry might feel, perhaps justly, that a brief summer's contact with an unoriented student will not be of much benefit to the company. Industry in the scientifically advanced countries, however, needs to strengthen its contribution to the scientific development of the rest of the world, both for reasons of self-interest and also as an altruistic service function. Among various programs in this direction, one of the simplest and most effective would be to employ some foreign students during the summer vacations. Some of these students might later be able to participate in their home countries in local research activities of subsidiaries or affiliates of the companies where they had worked here. The summer employment program would thus be justified even from a short-term practical point of view. The organization and initiation of such summer opportunities is a joint task for industry and the university department where the student is being educated.

---

[3] E.g., M. J. Moravcsik, "Aspects of Science Development," in *Management of Research and Development* (Proceedings of the seminar organized by the Scientific and Technical Research Council of Turkey, Istanbul, 1970), OECD, Paris, 1972, pp. 189–240, particularly pp. 214–15.

Emphasis on Problem-Solving

Students can also be influenced through a strong emphasis on problem-solving in the courses they take. Such an emphasis is generally present in the American graduate science education system, and so special care should be taken to apply it with unusual intensity to the education of foreign students. It is in this specific area that the student is likely to be most deficient when he arrives, and therefore again, his adviser can play an important role in helping the student to acquire strength here.

Another important area for educating foreign students pertains to the general workings of science, of scientific communities and institutions, to the managing of science departments, laboratories, shops, libraries, and to general science policy. These are areas of immediate utility for foreign students as they return to their home countries, and yet they are usually not even touched upon in American graduate schools.

Though many of these auxiliary educational activities can very well be arranged by individual departments using their own internal manpower resources, this is not done at the present time. For that reason, and also to experiment with a somewhat more centralized approach to these problems, the Committee on International Education in Physics is planning to organize a 3-week summer seminar for science graduate students from less developed countries. Some 50 students would gather to interact with eight lecturers selected for their familiarity with these practical and organizational aspects of the natural sciences. It is hoped that the first such session can take place in the summer of 1973, though sufficient funds have not yet been obtained.

Will He Return Home?

We can now proceed to the next stage in our student's career. Having completed his education and received his degree, he is ready to return home. There are, however, a number of circumstances working against his going. These can be reduced mainly to two primary causes.

The first of these is that during his rather long studies the student tends to lose touch with his home country. He is unaware of the opportunities available there, and as he gains knowledge and experience in his technical scientific studies, he has no chance to take another look at the home environment in the light of that new experience. He is uncertain as to his welcome in the scientific institutions of his home country, and is unsure of his ability to contribute and function in that environment.

The remedy to this problem should really be a two-pronged approach on the part of the student's professors in the United States, on the one hand, and of his former professors or potential future employers in his home country, on the other. The latter could make a conscious effort to keep in touch with him by sending him reports, writing him letters, and generally informing him of developments. I know of at least one instance in a South American country where this approach has been effectively employed. It would also be helpful if the student could visit his home country, say, halfway through his graduate education and work for a summer in one of the institutions where he might go after he receives his degree. This summer employment effort could be a cooperative undertaking between the home country and the American institution, with the latter, for example, providing some financial assistance in terms of support of the student as a research assistant during that period.

Overcoming Fear of Isolation

Many students hesitate to return home after their studies because of the fear of isolation. This is one of the main handicaps [4] in the practice of science in the less developed countries, since communication is so crucial in the natural sciences. There are a number of ways to reduce this isolation.

First, the American professor must realize that his responsibilities do not end when the foreign student has received his degree, as they often do in the case of American students who are "taken over" by the supervisor of their first postdegree position. For the foreign student, therefore, the professor must make an active effort to keep in touch with him, collaborate with him in research, perhaps send him material not available in the student's country. The professor's assurance of his personal interest in the student before the latter's return home might go a long way toward giving the student the fortitude to face his new situation.

Bilateral Exchanges

On a somewhat larger scale, bilateral arrangements between science departments in the United States and counterpart departments in less developed countries are another effective way to lessen isolation. Such a relationship can be established on a wide range—from casual contacts with no

---

[4] Abdus Salam, "The Isolation of the Scientist in Developing Countries," Minerva 4, 461 (1966).

financial resources involved, all the way to huge programs involving large sums and many people.

On a smaller scale, these arrangements, easily initiated and managed by individual science departments, can involve the exchange of graduate students, visits by faculty through sabbatical leaves or through research positions, coordination and cooperation in research, interaction in educational innovations, help with material, and the channeling of scientific communication media. An example for such an effective but very low cost component of such a bilateral relationship is the small-parts service, under which the American participant department can supply the counterpart department with small components of experimental equipment (costing a few dollars each), which are unavailable there and would be very laborious and time-consuming to acquire.

Another way to break down isolation is to include institutions in less developed countries in travel itineraries. Scientists from advanced countries travel often to conferences, summer seminars, and visiting appointments. Much of the time they could include in their itinerary stops at institutions in less developed countries. Many such institutions are interested in hosting such short-term visitors for a day or two, and such an encounter usually results in an enrichment and education of both sides involved. The Committee on International Education in Physics is now maintaining a registry which coordinates physicists about to travel and institutions that have expressed willingness to host such travelers.

The brain drain can of course be reduced by strengthening the enforcement of immigration laws requiring students to return to their home countries after the completion of their education. Besides such formal remedies, however, there is also another approach that might be tried. At the very beginning of the student's educational process, when he is about to come to the United States for graduate education, his future adviser from the department he is about to join could write him a personal letter, explaining the concern of the United States, and the department in particular, with the future of the student's home country, pointing out the large cost of the student's graduate education which is borne by the American taxpayer, and also emphasizing the importance of trained manpower in the future of the student's home country. In view of these considerations, the letter would ask the student to give his personal assurances that after the education is completed, the student would return to his country. Though such assurance would have no legal or official binding force and would only be a person-to-person understanding, it might in the long run be effective in calling the student's attention to these considerations, and also in placing a moral obligation on the student to take them seriously into account.

I have tried in this article to outline some simple and much-needed steps to improve the education which students from developing countries receive in the graduate science departments in the United States. I hope that this list, which is far from complete, may serve to stimulate the thinking of other scientists in this direction so they may come up with further proposals. Most of those I have listed are within the reach of personal action by scientists in American universities, thus demonstrating that one can go a long way by a judicious coordination of existing resources and opportunities.

Directing Attention to the Problem

It is fashionable nowadays to adhere to the conspiracy view of the world, according to which problems are due to the dominance of small but powerful evil forces which suppress the constructive efforts of the good and moral majority. I have always found this world view pure fantasy, concluding instead that, besides our incompetence at any given time vis-a-vis some of the big challenges, the main reason for the existence of problems is the lack of attention paid to them by people with appropriate backgrounds. Thus the shortcomings in the education of foreign science students in the United States are the result mainly of the lack of attention to and understanding of their special needs.

It might be appropriate to appoint a small committee of faculty and students in each American science department to consider these problems and try to work on solutions. Such a committee has, in fact, been appointed in the physics department of the University of Texas in Austin, and it was an inquiry from that group that furnished the stimulus for this article. Similar initiatives in other departments all over the country could make a significant contribution toward substantial improvement in the education of foreign students in the United States. ■

454

Reprinted by permission of *International Educational and Cultural Exchange* 11 (1975) 29.

Observations on an Obstacle Course

Francis E. Dart, Andrew G. De Rocco, Michael J. Moravcsik,
and Michael D. Scadron

*How to clear the obstacles from the path to U.S. graduate education for the
deserving Asian physics student.*

A physics student with a B.Sc. first class honors degree from Sri Lanka (formerly Ceylon) recently was denied admission to graduate study by one of America's leading universities simply because he lacked an M.Sc. degree. The admissions office of this U.S. institution demands an M.Sc. degree from all Southeast Asian applicants. Ironically, Sri Lanka offers no such degree, but it awards only one or two B.Sc. first class honors degrees each year. This particular student would perform admirably at any U.S. graduate school, but due to this misunderstanding of the differences between degrees granted by various Asian countries, he may never have the opportunity.[1]

This is but one example of the numerous difficulties encountered by Asian students trying to obtain an advanced degree in the United States. Additional problems of finance, educational background, cultural differences, choice of an appropriate and relevant course of study, and finally readjustment to their own society confront them. In short, acquiring education in a foreign country for the purpose of being a creative member of the scientific community in one's own country is a difficult process beset with possible mismatches at several stages. It requires understanding and a broad horizon on all sides.

This article is an account of experience gained from firsthand contact with Asian physics departments and students. Much, though not all, of it was acquired by the authors while traveling in connection with the Physics Interviewing Project in 1969, 1971, and 1973. This project, sponsored by the American Association of Physics Teachers' Committee on International Education in Physics, serves as an intermediary between Asian students interested in obtaining advanced physics degrees and many graduate physics departments in the United States. During these trips, the authors visited the Republic of Korea, Hong Kong, Indonesia, Thailand, Malaysia, Singapore, Bangladesh, India, Nepal, Pakistan, Sri Lanka, and Afghanistan. Some 500 students were interviewed in all, and specific comments about the individual students were relayed to roughly 20 U.S. physics departments who sponsored the project, as well as to any department requested by the students.

Differences Between Asian and American Students

Comparing the background of Asian physics students with their American counterparts at the end of college education or its equivalent, we find two major differences: First, many Asian students are used to rote learning, to memorization, to treating science as a fixed set of laws or rules, the knowledge of which is thought to make a scientist. As a result, the Asian student from a good institution may often have more "theoretical" knowledge of science than his American counterpart, and at the same time be seriously (or sometimes even catastrophically) deficient in solving problems or in applying physics to practical situations.

The second characteristic of Asian students is their early specialization compared with Americans. As a result, their backgrounds as well as their outlook and aspirations are rather inflexible and myopic. A corollary to this early specialization is an extremely heavy course load. In some Asian countries, the students take as many as 5 science courses at a time, requiring up to 40 class hours a week.

In addition to these two major differences, there are also some other distinctions in terms of attitudes and

Each author is a professor of physics at his institution—Francis E. Dart, University of Oregon; Andrew G. De Rocco, University of Maryland; Michael J. Moravcsik, University of Oregon; Michael D. Scadron, University of Arizona. Each was or is currently a member of the American Association of Physics Teachers' Committee on International Education in Physics and has participated in one or more of the interviewing trips described in this article.

[1] As this article goes to press, we have learned from this student that, in part because he was interviewed by the Physics Interviewing Project described here, he has now received several offers, one of which he has accepted.

emphasis. The Asian student, for cultural and social reasons, is likely to be greatly predisposed toward theoretical rather than experimental physics. His own scientific community, however, is likely to have a surplus of theorists and a shortage of experimentalists.

Asian physics students interested in coming to the United States often know very little about American higher education in physics. Brief reflection indicates that this could hardly be otherwise. Because of the rather formal relationship between Asian students and their professors, the students rarely benefit from the experience of those professors who have been educated in the United States. Thus, the student is reduced to getting information by word-of-mouth from older students who may now be studying in the United States (sometimes a sporadic and unreliable source), or from the written information available in nearby libraries. This information, however, consists mainly of university catalogs, general works on education, and other such publications, all aimed primarily at American students. For this reason, such discussions never mention the "obvious," namely those aspects of education in the United States which every American would automatically know but an Asian student would not. Hence, while he can obtain much information on American education, it may be confusing or misleading because it is unrelated to his own experience.

In trying to fill this gap, two of us (Dr. Dart and Dr. Moravcsik) wrote a brief brochure about American physics graduate education aimed at potential graduate students still in their home countries. The brochure[2], published by the American Institute of Physics, was produced in a format sufficiently inexpensive to be suitable for individual distribution to students. Now in its second printing, the brochure's success, as measured by demand, seems to illustrate a thirst in

[2] *The Physics Graduate Student in the United States*, Francis E. Dart and Michael J. Moravcsik, American Institute of Physics, 1971.

many countries for realistic and relevant information on this subject.

Ways To Help

How else can we help clarify the Asian students' understanding of advanced education in the United States before they come here?

First, American physics departments should take a positive, perhaps even aggressive initiative in communicating information to many deserving students in Asian countries and in encouraging them to apply. Social customs and rules of interpersonal interaction are often much more subdued in some of these countries, and it might be unreasonable to expect the student to take the initiative in asking for information, or application materials, even when a comparable American student would need no encouragement. Furthermore, considering the large number of applications from Asian countries to many American physics departments, it might be worthwhile for these departments to prepare a mimeographed sheet, aimed specifically at such foreign applicants outlining the relevant information. Since an increasing number of American physics departments now have at least one member with personal experience in an Asian country, such a special brochure could be prepared with considerable understanding.

Another bureaucratic problem is the vicious triangle between the student, the department of physics to which he is applying, and the university admissions office. Admissions offices can be a headache even for American students. For foreign students, as suggested in the introductory paragraph, they are often an incomprehensible maze of form letters, rules, and requirements difficult or impossible to satisfy. Before anything happens, the students are asked to send an application fee, astronomical in terms of local earning power, and often impossible to send in hard currency without a long bout with the local government. (In our opinion, the American university, perhaps at the insistence of its Physics Department, could and certainly should defer pay-

ment of the application fee for students from the less developed countries, at least until the student reaches the American campus where he will begin to have some dollar resources.) Letters of reference sent to physics departments sometimes go unrecorded in the admissions office and the latter often mistakenly send correspondence by surface mail causing the students to miss deadlines.

Much of this could be avoided if the corresponding physics department took an active interest in the matter, intervening to eliminate these incongruities and bypassing those requirements which are absurd for foreign students. Often, however, the physics departments are not even aware of the applicant and his tribulations because his incomplete application—delayed by red tape in the admissions office—has not been forwarded to the physics department.

Special Committee Needed

How to remedy this situation is up to the individual physics departments. In any case, however, it is very useful to have a small but permanent departmental committee of faculty members with personal experience in distant countries. This committee would deal specifically with matters pertaining to foreign students both before and after they arrive. For instance, even obtaining a U.S. social security card can be a problem for newly arriving foreign students. Furthermore, all too often, physics departments feel no obligation to follow the students once they have obtained a higher degree. Perhaps this committee could help to maintain contact with them for some time afterward.

We are also concerned about requirements for written tests required by some universities as a prerequisite for admission. The two most common ones are the TOEFL (Test of English as a Second Language) and the GRE (Graduate Record Examination). Offered infrequently and only in a few geographical locations, neither of these tests is easily accessible to an Asian student. Test fees are huge measured by local earning power, and

> **"... indigenous education of scientific personnel is vastly superior to sending students abroad for education, and hence a strenuous effort should be made to encourage the budding top quality scientific institutions in the country itself."**

also have to be paid in hard currency. Considering that these tests are of necessity given in the applicant's own country where at least part of the cost of administering them will involve local currency, we see little justification for this requirement and we urge that the fee be made payable in local currency. As it is now, some countries have to limit the number of tests for which they give permits.

Neither do the tests have much meaning in the Asian context. Students unfamiliar with them often are seriously handicapped at the outset, and not for lack of knowledge.[1] In some countries, some universities offer drill courses for these tests, thus negating their purpose.

It would be an exaggeration to claim that these two tests are completely useless. However, we have serious doubts about their suitability and fairness for the Asian scene, and instead urge, wherever possible, a personal interview by knowledgeable American physicists. We consider the results of such an interview vastly superior to those obtained from standardized written tests.

So much in the way of general remarks. Let us now turn to several specific observations about the various countries we have visited.

Most important is the lag between the perceptions of American higher educational institutions about the reputation of Asian schools and the

[1] Two of us were shaken by the experience of conversing for an hour in Delhi with a student who communicated easily about sophisticated subjects, only to learn at the end that he had flunked the TOEFL test a few months before, after having received secondary and university education in English for many years.

realities. (Such a lag is understandable, and it also exists in reverse: the reputation of American educational institutions among scientists and students in the less-developed countries also lags far behind the actual realities.) This lag gives rise to a "rich-gets-richer-poor-gets-poorer" effect. Those countries, like India, Taiwan, or Hong Kong, which have supplied some good students to American graduate schools in the past, enjoy continued attention, and students from these countries find it fairly easy to secure admission and financial aid in American graduate schools. On the other hand, countries which have some very outstanding students, but only recently have begun to develop their scientific resources, have a very difficult time attracting attention. In this latter category we would place Indonesia, Bangladesh, and Sri Lanka. Because of a favorable foreign trade balance, foreign education is more accessible to students from Malaysia and Singapore, which have made considerable progress in recent years. It should be noted that students from Nepal are now coming into their own and should be ready to study abroad for advanced physics degrees beginning this year. Afghanistan will be concentrating on its undergraduate program for the near future.

Indigenous Education Better

Coupled with the phenomenon just mentioned, is the great need for well-educated people in countries which just recently began to make strenuous efforts to create their own scientific communities. The number of doctorate-level physicists in Indonesia, for example, a country of over 100 mil-

lion people, is at most 50, and this scarcity hampers both the education of scientists and technical personnel, and the application of science to the practical problems of planning, decision-making, and implementation required for national development. By contrast, India, a country five times larger, has thousands of physicists and a formally well-developed system of scientific institutions. To be sure, a large percentage of Indian scientists are mis-educated and do not creatively contribute to science; hence India remains in great need of top-quality manpower. Yet, its manpower needs are quite different from a country just beginning its development process.

This raises another point. Virtually every observer of the development process in the emerging countries would agree that indigenous education of scientific personnel is vastly superior to sending students abroad for education, and hence a strenuous effort should be made to encourage the budding top-quality scientific institutions in the country itself. If, therefore, there are institutions in a given country which offer good quality advanced education in a given field of physics and have room for a sufficient number of good students, one should be very reluctant to encourage those students to get their education abroad. A much greater service would be to grant them, after they have obtained their Ph.D. in their home countries, an opportunity for a 2-year postdoctoral position in a scientifically developed country. Unfortunately, such opportunities are very scarce indeed.

It should be stressed, however, that many Asian countries do not yet have a Ph.D. program in physics. This includes Indonesia, Sri Lanka, Nepal, Afghanistan, Bangladesh, Thailand, Malaysia, Singapore, and Hong Kong. Furthermore, most of these countries are unlikely to develop Ph.D. programs in the near future. Pakistan is just beginning such a program, while India now has a number of Ph.D.-granting institutions. Most of these countries have reasonably good M.Sc. programs.

Another aspect of this problem is the so-called "brain drain." While opinions on the detrimental effects of the brain drain differ, we feel that it is damaging to the development of the emerging countries, and that American scientists have some responsibility at least not to encourage this exodus of scientific personnel. We must recognize the marked differences in attitudes among various countries and their students on whether to return home upon completing a foreign degree. Students from countries such as Indonesia, Sri Lanka, Nepal, Thailand, Malaysia, and Singapore can, on the whole, be counted on to return. On the other hand, students from India, Pakistan, and Taiwan are more likely to stay abroad. Hong Kong is a special case, since its size hinders the full utilization of its manpower potential. Finally, we are pleased to report that many well-qualified Western-educated Ph.D.'s in physics have recently returned to important positions in Bangladesh.

Quality of Students

We now offer comments on the quality of students from different countries. In the first place, individual differences between students from the same country are much larger than any systematic differences between countries. Contrary to popular belief, no one Asian country has a monopoly on outstanding students.[4] This is another reason why it is risky to rely simply on the country of origin when judging a student's qualifications.

Second, the student's university record is not always a reliable indicator of his ability. For one thing, different countries have different standards for the same degree. For example, even within the Indian subcontinent the quality of a B.Sc. and M.Sc. degree varies considerably be-

---

[4] We base this statement upon our 500 interviews between 1969–73. The top 30 students interviewed were almost evenly distributed over the student populations of Asia, including students from Hong Kong, Malaysia, Singapore, Indonesia, Sri Lanka, most parts of India, as well as from Bangladesh and Pakistan.

tween India, Pakistan, Bangladesh, and Sri Lanka. Recall that although Sri Lanka offers no M.Sc. degree in physics, its best B.Sc. students are fully capable of entering an American graduate school. Since the educational system is highly formalized and rigid in many of these countries, the grades, given on the basis of examinations mainly testing memory, give little indication of the student's preparedness or promise as a potential graduate student at an American university.

"One must use a [evaluation] system which treats individuals as such and has a high reliability of evaluating their potential."

For this reason, we distrust the use of formal records even in situations where sustained experience with various specific institutions has accumulated some statistical information on their "meaning." While the funds for and the desire to admit foreign students to graduate schools in the United States are now decreasing, the demand from the less-developed countries for such education is increasing. Therefore, statistical methods of evaluating applications are no longer satisfactory. One must use a system which treats individuals as such and has a high reliability of evaluating their potential. We believe that the personal approach of the Physics Interviewing Project, or similar programs, constitutes at the moment the only feasible way to accomplish that goal.

Another result of the Physics Interviewing Project was our evaluation of the student's ability to communicate in English. Most Asian countries are now teaching science with English-language textbooks, but this does not imply that all of these students have a working knowledge of English. While generalizations are risky on this matter, it can be said that students

from Indonesia, Thailand, and Korea tend to have problems speaking English. Students from the Indian subcontinent usually have a passable command of the language, but this might change in the future because of nationalistic pressure to study in the indigenous tongue. Interestingly enough, the trend in Indonesia is towards more English-language study. Malaysia, Singapore, Hong Kong and, to some extent, Pakistan, regularly provide students with a working knowledge of English. It would be safe to say, however, that 2 months in the United States prior to the beginning of classes would help most Asian students to improve their mastery of spoken English.

In Summary

In summary, advanced education in physics is of great importance to less-developed countries in Asia, and the demand for such education at American institutions by students from those countries can be expected to increase. At the same time, American physics graduate schools are in need of high quality students because of the momentarily decreased interest in physics within the United States and the corresponding decline in the quality of American applicants to such schools.

Therefore, it is in everybody's interest to offer graduate education in physics at American schools to the top students from less-developed countries. This goal can be achieved successfully only if (a) appropriate information about the American schools is widely distributed to students in less-developed countries, and relevant information about the students is transmitted to the American schools; (b) the selection process, the education itself, and the followup in the form of sustained personal contact between the ex-student and his former institution are all carried out with a realistic understanding of the factors involved in these processes; and (c) these students are encouraged to return home after completing their graduate education. □

AMERICAN UNIVERSITIES AND THE SCIENCE EDUCATION OF THIRD WORLD STUDENTS

Michael J. Moravcsik
Institute of Theoretical Science
University of Oregon
Eugene, Oregon 97403

Quantity, quality, and appropriateness

In purely quantitative terms, American universities and colleges can be proud of their record of educating science students from the developing countries. Approximately 100,000 such students receive education in science and technology in our institutions of higher learning. Placing a price tag of, say, $8,000 a year on such education (which is probably somewhat conservative),we can conclude that close to a billion dollars are spent every year by the United States in educating such students. It should be noted that this figure appears nowhere in the statistics of international assistance, since the lion's share of this amount originates in scholarships, assistantships, and other locally administered sources. Yet, this amount exceeds the total of all other international assistance in science and technology that is offered by the United States.

Although it can be debated whether the education of 100,000 students satisfies the need of the Third World at the maximal rate that could be usefully absorbed by the developing countries, or not, my assumption in this discussion will be that it comes close to doing so, or that at least any quantitative deficiencies that remain are insignificant compared to what improvements could be made by making small changes in the nature of the education given to such Third World students. Hence, my discussion will be aimed exclusively at such changes.

This is not to say that I think the education given to Third World students is deficient in quality. On the contrary, this education is the same as that given to American students, and hence, measured by the indicators of quality used in the American context, the education is quite good. But there are no abstract indicators of quality. Quality is always referred to some set of aims, objectives, and functions, and hence we have to measure the 'goodness' of education given to Third World students in the context of the aims, objectives, functions, and circumstances of science in the Third World, which is the arena these students will be working in.

A different science?

Again to avoid misunderstanding, let me stress that I do not think that there is a different science for the Third World from that practiced in the advanced countries. To be more precise, I think that as far as content, methodology, standards, and substance are concerned, there is only one science which is neither Western nor Eastern, neither Northern nor Southern. There are, however, considerable differences in the present state of scientific infrastructure of various countries, in the type of culture of which science is expected to become an ingredient, in the present state of evolution of indigenous scientific activities, in the social conventions that operate in the local scientific communities, etc. In other words, the one science is embedded in many different contexts around the world, and correspondingly the optimal preparation of scientists has to reflect this variety.

Curricular changes?

Consequently, my discussion will not deal with curricular issues in the sense of my proposing a curriculum for Third World students which is significantly different from that of the American students. For example, it has been proposed that Third World students be given the opportunity to do their Ph.D. thesis research in their home countries, under the supervision of their American thesis advisors in conjunction with a local scientist, so that the thesis could be on a problem 'relevant' to the home country's needs.

I do not have much sympathy with this proposal. A thesis topic is merely an example on which the student can demonstrate his ability to do research, and exactly what that topic is should not matter much. In fact, the student should be emphatically told this so he does not nurture false illusions that by choosing this topic he chose a life-long specialty to work in. It is also important that the student acquaint himself with state-of-the-art equipment, library, and other auxiliary tools so that he has personal aquaintence with standards of scientific research which, ultimately, should be the target to strive toward in his country also. The merit of this thesis-at-home scheme is somewhat greater in areas of science which are geography-specific, but in general, I do not consider this topic of high priority in this discussion. There are a number of other aspects of education which could be improved, and at a cost far below that needed for implementing the thesis-at-home system.

Other reforms

What I want to do then is to list some of these other reforms that are needed to enhance the science education of the Third World student in American universities. As you will see, most of these are extremely inexpensive or, in some cases, even free. The reason why they have not been implemented is not because we live in hard financial times. What is required is a perceptive eye to spot the deficiencies in the present system, and an interested and sympathetic science faculty at American universities to deviate from the course of inertia and to implement these reforms. If the present discussion brings us closer to such action, it will have been well worth it.

To adopt some systematic train of thought, I will follow chronologically the Third World student's course, from the time he prepares for his education abroad, to the time when he returns with a degree in his pocket, or even a bit beyond that.

Student selection

Let us therefore start with the problem of student selection. From the viewpoint of those in charge of student admissions at American universities, the application material submitted by Third World students (let us say, for admission to the Graduate School and for financial aid) is close to unevaluable. It consists of a transcript from a university not known to the admitters, a transcript which is often not even in English and has only a strange English translation, containing course titles but no course content, and grades in an unknown grading system. This is then accompanied by letters of recommendation from unknown professors, written in the context of cultural conventions often radically different from that used in the United States, and thus stating, for instance, that the student comes from a good family, is diligent, and has a high moral character -- information which hardly serves as a basis for a sound judgement as to whether the student will make it in an American graduate school or not.

This material may be accompanied by GRE scores which, however, have dubious validity in the context of the Third World. For one thing, taking the GRE examination is very expensive by local standards, and requires often very scarce hard currency. The test is given only in a few places in the country and only a few times a year, thus involving significant sacrifices in money, time, and effort on the part of the students. The format of the exam is one that the student is unused to. If he is rich and lucky, he will have taken one of the GRE-drilling courses available in some countries. In my view the GRE is not an equitable test even of students within the United States who come from less advantaged schools or environments, let alone of students from the Third World. One clearly needs if not a substitute then at least a supplemental method of assessing these student applicants.

The Physics Interviewing Project

Such a system has been devised, tested, and has been in use in physics for some 17 years, and in chemistry for some 4 years. Unfortunately it has not been taken over yet by the other scientific disciplines. The program, which is called Physics Interviewing Project (and Chemistry Interviewing Project) sends two interviewers from American universities on a 5-6-week trip through Asian countries, where they individually interview students who intend to apply to American graduate schools in physics (or chemistry, respectively). Each interview takes 45 minutes, during which simple problems are asked to be solved to check the student's facility in problem solving in undergraduate physics (or chemistry). After each interview 15 minutes are spent to write a one-page report on the student's background and potential in physics (or chemistry), including an assessment of his English and an overall recommendation as to whether the student could be employed as a teaching assistant in the first year (which is the most common way of offering financial assistance to a graduate student).

A copy of the report is then sent to any university the student is applying to. In addition, a complete set of all reports is available, right after the trip, to those departments in American universities which by voluntary contributions financially supported that year's trip. These departments then can use this material also for recruiting purposes. The share of each department depends on the number of departments volunteering that year and on the total expenses. In 1985 the share was $600 per department, that is, less than 10% of what it costs to support one student for one year. Since the program has no bureaucracy, no central office, no overhead, no fixed expenses, it is extremely economical. The expenses consist of the two airplane tickets, the room and board during the trip (a large part of which is assumed by the institutions where the interviewing takes place), and a minimal amount for postage, xeroxing, etc.

The results of this interviewing system are particularly rewarding in the case of countries where science is just developing, from which not many students have come to the United States previously, and hence where other methods of evaluation are particularly weak. Of the countries covered in Asia by the present interviewing program, Nepal, Sri Lanka, and Indonesia can be particularly stressed as being in this category.

Where to apply?

The second step in the chain of events I want to dwell on is the choice of departments to which Third World students would apply. The selection on their part is done in very strange ways. The list typically includes a few longstanding and famous institutions, and next to them a peculiar assortment of much lesser known institutions, chosen because some friend or relative happened to go there previously. The selection is clearly done by criteria which are not based on factual considerations.

There exists a small brochure, printed almost 20 years ago, but still available today, which at least lists the names and addresses of departments in the United States which offer a Ph.D. in physics, and also discusses briefly the main features of American graduate physics education. The content of the brochure, nevertheless, is of little help in choosing departments to apply to. Whereas it would be difficult to go too far in providing such students with tips on which departments are 'good' and which are less so, one could at least make the students aware of some of the criteria to be taken into account when choosing schools. The chemists also have a brochure, much more extensive than that for physics, but even that could be improved in this respect.

But let us proceed, and assume that the student has been admitted and has received some financial help, and he has arrived at his future department in the United States. The next problem is advising.

Advising

In most American universities the professional (as distinct from logistic) advising of even the American students is a disgrace. Particularly undergraduate students can, if they choose to, roam around on campus for a full 4 years without having any meaningful interaction with a faculty advisor. In the case of graduate students the same is true for the first 2-4 years, that is, until the student has passed his big exam which qualifies him to choose a thesis advisor and begin research. Unfortunately, most of the mistakes, false steps, mishaps, or even tragedies occur prior to that point in the graduate student's career, that is, during the period when advising, if it exists at all, is extremely lax.

A Third World student is likely to have come from an academic situation in which informal interaction with faculty members is rare. Therefore it is unlikely that the student will initiate interaction with his nominal advisor. The advisor, on the other hand, usually assumes that there is no need for him to initiate a dialogue, because if the student has problems, the student will speak up -- a quite unjustified assumption vis-a-vis Third World students, not only because of the above mentioned reason, but because the student is often quite unaware of his problems until it is almost or actually too late for remedies. What are the main problems?

Deficiencies in background

There are two conspicuous characteristics of science education in the Third World which cut across regional, national, or cultural boundaries and appear to be omnipresent. They are reliance on memorization and rote-learning, and premature specialization. Thus a Third World student, even coming from one of the best undergraduate programs of his home country, tends to have had, formally, advanced courses (often ahead of his American fellow students) while at the same time his ability to solve problems even in freshman physics courses may be minimal. His natural tendency, therefore, will be to take very advanced courses in his first and second years. Since, strangely enough, problem solving is less stressed in advanced physics courses than in elementary physics courses in the American educational system, the student may manage to get by in these advanced courses. The gaping hole in his ability to solve problems therefore becomes evident only when he takes his big exam (variously called qualifying or comprehensive examination),

which tests such problem solving on a fairly elementary level.
The discovery of this gap is both shocking to him, and is
difficult to remedy by then.

More on advising

Good advising immediately after he arrives to enter graduate
school could avoid such mistakes. The knowledgeable advisor would
ignore the seemingly 'advanced' incoming record of the student
and give him an initial schedule of undergraduate courses (say,
on junior-senior level) for a term, a semester, or even for a
year. Such 'remedial' work (which need not be called that)
never can hurt, even if some of it turns out to be unnecessary
for some of the students. Alternately, the incoming student could
be given immediately an examination (e.g. via the type of
interview used by the Physics Interviewing Project) to
ascertain the need for such a refresher curriculum.

In order to inform the science faculty in American
universities about this and other aspects of advising Third World
students, and about other aspects of the education of Third
World students, the American Association for the Advancement of
Science a few years ago produced a brochure and distributed it
in some 10,000 copies to such faculty. It was a good start, but
much more awareness building needs to be done if lasting changes
are to be brought about.

Other aspects of advising: Battling isolation

What are the other aspects of advising? To illustrate one
more, let me mention a great opportunity almost always missed by
Third World students during their education abroad. One of the
most serious shortcomings of the environment of a scientist
working in a developing country is isolation. Whether we talk
about the written modes of communication such as books, journals,
reprints, preprints, conference proceedings, or about oral modes
of communication, such as personal visits, the use of the
telephone, the attendance of conferences, sabbatical leaves,
etc., the scientist in a developing country is at a great
disadvantage compared to his colleague in a scientifically well
developed country. Thus he is severely isolated.

One way to lessen this handicapping isolation is to have
private communication, through letters and perhaps otherwise
also, with particular scientists in the advanced countries that
the scientist in the developing country is personally acquainted
with. The best time to begin to evolve such personal
acquaintances is when the scientist from the Third World is
being educated abroad, since at that time he can, with relatively
little effort and resources, visit many florishing scientific
centers and get to know people.

This is an aspect of the 'education' of the scientist that he
at that time is completely unaware of and hence does not
practice.

It is the responsibility, therefore, of the advisor to make him aware of this potential, to encourage him to get to know scientists visiting at his university, and also to spend the vacation periods visiting other scientific centers. Taking advantage of this opportunity costs very little if any money. What is needed is an advisor who is aware of the opportunity and who can patiently but insistently explain it to the Third World student.

But let us leave now the question of advising and let us turn to another important way in which the science education of a Third World student at an American university can be made more effective.

A difference in context

To appreciate this next opportunity, let me stress a very important difference between the situation of a fresh American Ph.D. in physics and a Third World student who has just received his Ph.D. from an American university. The former, when he takes up his first scientific position, is almost always asked only to do good science. He is not asked to also work on the creation and maintenance of the institutions and services he is embedded in, since those institutions and services already exist. The scientific infrastructure in the United States is already in place, and although it might undergo further evolution, that development is relatively slow from one year to the next. Whatever is needed to be done in that respect is usually relegated to older scientists who perform the managarial and science policy functions, and thus the young scientist can entirely devote himself to doing science.

The situation is exactly the opposite for the young scientist from the Third World who has just received his Ph.D. from an American university and has just returned to his home country to take up a job. Almost certainly he will be called upon, simultaneously, to perform two functions:To do good science, and to help in creating the circumstances under which good science can be done. The scientific infrastructure in the developing countries is either not yet in existence, or is fragmentary, in the state of constant turmoil. Hence the young scientist often has to create his own infrastructure before he can seriously settle down to do good science.

Unfortunately, this young scientist is completely unprepared by his American education and diploma to undertake such a science-building task. In American graduate schools of science we do not teach about the context of science, about science policy, about the methodology, sociology, economics, history, or philosophy of science. We avoid these topics since we presume that the young American scientist need not know about them at the beginning of his career, and that by the time he does need to know something, he will have had the opportunity to pick it up by some process of osmosis through exposure.

I think this presumption is questionable even in the case of the American student. It has, in any case, a very negative effect on the Third World student educated in an American graduate school, since that student would very much need some background in these contextual aspects of science, but he does not get it.

Providing education in context

It would, however, be quite possible to provide such a background, or at least the beginnings of it, for the Third World student. There are at least two ways of doing it. The first way is more costly but involves fewer people. It is the organization of 2-3-week long summer seminars specifically for Third World students now being educated in science graduate schools in the United States. In such a summer program the above mentioned auxiliary and contextual problems of science are discussed, particularly in the framework of the developing countries. These topics range from very mundane things like how a library works, how machineshops operate, an elementary, course in glass blowing or in electronic construction and repair, to more abstract topics like how scientists communicate, how one gives out money for science, or how the link between science and technology can be strengthened. This way of accomplishing the aim of bringing these contextual subjects to the Third World student has in fact been tried once, through a summer program held in Nova Scotia. While that experiment appeared to be quite successful, the method needs further exploration. This way is not without expense either, since funds are needed to bring the students together, cover their room and board, compensate the lecturers, etc.

The other way of accomplishing the same aim is a very decentralized one which therefore requires widespread cooperation. In every science department of every American university there are some faculty members who have some personal experience in these contextual aspects of science, simply because they have themselves been involved in them. Such faculty members could, in their own institutions, and through variuous formal or informal opportunities, provide such a background for the Third World students. In this case the resources needed are minimal, and such a program can be instituted at each university department quite independently of what any other department is doing. Hence anybody can start such a program in his own department quickly and inexpensively. As far as I know, however, practically nobody has done so.

Combatting the brain drain

I want to stress that the main aim of such an auxiliary element in the education of Third World students is not to make great experts out of them in matters of the science of science and science policy. Indeed, this could not be accomplished in a short, 2-3 week summer program or in occasional discussions and seminars in a university. The most important aim is to create the awareness in the Third World student's mind that, 1) the infrastructure in his home country for science and technology does not exist, and this is to be expected; 2) such an infrastructure can be slowly evolved and every scientist in such a country needs to participate in that evolution; 3) the absence of such an infrastructrure is not a good reason for giving up scientific research, for joining the ranks of inactive civil servants, or for emigrating to an advanced country. The young scientist must be convinecd that the situation is not hopeless, but that skill and hard work can create opportunities in a developing country for good scientific research. I believe that the damaging effect of the brain drain could be considerably blunted by creating such an awareness and by providing the young scientist with at least the elements of the skills needed to create a scientific infrastructure where there is none.

Summer job in applied research

This special angle of science education could also be strengthened by giving the Third World science student an opportunity to acquire personal experience in a more applied and more industrial scientific research atmosphere by providing him with a summer job in such an applied and more indiustrial research laboratory. While scientific research is in general quite weak in the developing countries, applied scientific research is particularly weak as measured in terms of output, even though the lion's share of the input resources placed into science in those countries is earmarked for applied science. The reasons for this could not be discussed adequately in the course of our present brief survey which in any case has a different topic. So let us just take that as a fact, and see to it that it is remedied in part through such summer job opportunities for Third World students while they are being educated in the United States.

Summer job at home

A similarly useful, though more expensive program would take such Third World students, in the midst of their graduate science education in the United States, back to their home countries for the summer vacation, to some position connected with science, so they could gain some practical experience in doing science at home, could make acquaintance with some of the scientific personnel in the home country, and at the same time could have a renewed opportunity, after their return to their education in the United States, to make 'midcourse corrections' in that education to harmonize it better with the experience just gained in science at home.

Returning home

To conclude this very sketchy survey of some of the opportunities, let me turn to the time when our Third World student returns to his home country to take up his first scientific position. An American student in such a situation can rely on advice, guidance, help, and cooperation from his new bosses at his new institution. Not so for our Third World student. As we have seen, his infrastructure does not exist, he does not have peers or elders experienced in doing science whom he could rely on for advice, guidance, help and cooperation. On the contrary, he might very well face a quite hostile environment, in which society does not understand the activity of doing science, in which the similarly uncomprehending bureaucracy is quite unhelpful, or even obstructive. It is this time when this young scientist would need moral and de facto help from somewhere, for example, from his former institution in the United States where he received his degree. Even if such a help copnsists only of correspondance with his former thesis advisor, or of a somewhat easier way for obtaining much needed small spare parts for equipment, or of a better channel for receiving reprints, the impact can be considerable, since such a help also provides encouragement, a visible sign that the young scientist is indeed a member of the worldwide scientific community. Such a help costs practically no money, and only very small amounts of time. The key here again is being aware of the need.

Epilogue

As I promised, my list of ways in which the science education in the United States of a Third World student could be made more effective without spending serious sums of money is simple, extensive, and within the reach of any American university. It is indeed sad to contemplate how little of this simple but effective agenda has ever been experimented with or implemented. This is even more puzzling if one considers that the American Association of Physics Teachers has had, for many years, a committee for international educational programs, and that the American Physical Society, a few years ago, finally also created a group within itself that is supposed to be concerned with physics in the developing countries. The American Chemical Society has an even more developed machinery for international activities, since, in addition to a committe from the membership for this purpose, it also has a full time administrator for international activities. Then there is the Association of Geoscientists for International Development (AGID) formed specifically for cooperation with the developing countries. Yet, much of even as simple and elementary an agenda as the one enumerated here has remained unimplemented.

The agenda was first articulated many years ago, so the delay in tending to it is not due to the unavailability of guidance of what to do. The same holds for the many American university science departments. We can only hope that occasions like the present one, when such an agenda can be reiterated, further publicized, and propagated, will eventually bring this problem area across the threshold beyond which a sufficient number of people are involved with it so as to precipitate the action necessary to realize the agenda. For this reason I want to express my appreciation to the organizers of this particular conference for the opportunity to discuss the agenda.

POSTGRADUATE EDUCATION FOR DEVELOPING COUNTRIES

Michael J. Moravcsik
Institute of Theoretical Science,
University of Oregon,
Eugene, Oregon 97403 U.S.A.

I. INTRODUCTION

In this paper I want to give a brief overview of some of the motivations for postgraduate education in developing countries, followed by an analysis of some of the main problems of such education as it exists at the present and then by some suggestions for programs that may improve the present situation. The discussion, by necessity, will be rather brief. I would like to refer those who are interested in more details to some of the previous articles of mine [1-23] on this subject.

II. THE SOCIAL MOTIVATIONS FOR SCIENCE

This is a subject on which there is extensive literature and therefore, I want to summarize here only the very basic outline of such a social motivation. It can be listed under three headings. [14] [18]

A. *Science as a basis of technology*

This is probably the motivation for science which is most often thought of or at least most often talked about. In the twentieth century science has become the basis of most technological developments and therefore the capability in indigenous scientific activities is a prerequisite for a country to have its own technological independence.[23]

Since there is a time interval between the development of a certain scientific field and its technological applications, science in any country must keep ahead of technological development and extend into pioneer areas of scientific research which only eventually will become the basis for new technology.

B. Science as an Important Human Aspiration

All civilizations, all countries, all groups of people, all communities, all scientists in human history have thrived on certain individual and collective human aspirations. These aspirations are sometimes material but much more often non-material in nature. In historical examples one encounters such non-material aspirations connected with religious ideas, formation of great empires, excellence in arts and learning, geographical or space exploration, etc.

For society to have such aspirations is very essential for its psychological well-being and for its dynamic force of growth and development. In the twentieth century scientific exploration has become one of the very important human aspirations and the greatness of a country is measured, in part, in terms of its contributions to the sciences. This is reflected in the number of specific activities, such as the watching of the national scores of Nobel prizes, the sending of national delegations to scientific conferences and so on.

It is especially important for developing countries to be aware of this aspect of science[21] because the state of being a developing country is a psychologically difficult one, particularly from point of view of aspirations. One of the hallmarks of a developing country is that its state of development and its activities are primarily determined by forces outside that country. This is true even if, strictly speaking, the country is no longer in a colonial state. As long as in economic, cultural, scientific, technological, artistic and other aspects, the leading edge of development is outside the country, the country will remain in a state in which decisions about its activities are, in fact, made abroad. This contributes to a very severe feeling of frustration within the country, which in turn creates a low morale with respect to development and can also contribute to different international political problems. Science is one of the areas in which even a small country can make important contributions without the investment of huge material resources and therefore science can contribute importantly to the realization of some of the aspirations of that country.

C. Science Has an Impact on the World-View of Man

While science is not a culture in itself, it can be a very important ingredient in any culture. Science provides a certain world-view and provides a certain attitude which can have a considerable influence on the overall world-view of Man. Science has in fact had such an important influence on the Western civilization for the last three hundred years. In the developing world, science as a cultural force has appeared only recently and has not had the opportunity to make a sufficiently significant impact.[24]

Many of science's attitudes and beliefs are very similar to the attitudes and beliefs that are required for development in general.[10] Some of these overlapping elements are the belief that knowledge is open and that through an energetic pursuit of further knowledge one can learn new things about the world and can apply that to the betterment of one's state. Change is a very important element in our scientific thinking just as it is in the concept of development. There are many similar overlaps between science and development and therefore science has a very important conceptual effect on the state of developing of a country.

III. A GENERAL ANALYSIS OF THE REQUIREMENTS AND THE REALITIES

A. *Requirements*

In developing countries science education, particularly on a postgraduate level, has a number of very important requirements.[3]

a) Broad Education: The education of a scientist has to be very broad in a developing country for a number of reasons.[22]

First, if the person's education is very narrow, it is unlikely that the person can find a job opportunity which fits his education exactly. This is so because the scientific community in a developing country is very small, the job opportunities are rather limited and therefore, it is not likely that a perfect match can be made.

Second, the scientific requirements in developing countries change very much with time and therefore, a person educated in a certain area at a given time cannot hope that throughout his career his particular narrow expertise will be the one that is in demand in his country.

Finally, most scientists in developing countries are needed for several reasons simultaneously. It is likely therefore that an active and productive scientist in a developing country will have to be concerned with several subdisciplines, several types of activities, several problem areas simultaneously. As a result, it is extremely important to educate scientists as broadly as possible.

b) Problem Solving Orientation: That science is a method for asking interesting questions which so far have not been answered and then following a course which solves the problems described in the question is a very important element of scientific activities. In developing countries, in particular, this is an element that must be communicated to people in the society as a whole because this is one of the assumptions and beliefs of science that could be most useful for non-scientists in daily activities pertaining to development of any sort. It is therefore very important that scientists in developing countries be highly skilled in the solution of problems of various sorts.

c) Strong Personal Motivation: In developing countries, the context and tradition of science have not been established, and hence it is a difficult task to be a scientist there. Quite often, science and scientists do not have social prestige, are not understood by society as a whole, and science and scientific activities are mistreated by government bureacrats who themselves have not had any background in science. Furthermore, science and scientists there are very much isolated and therefore the scientist has to rely on his own resources. For all these reasons, it is very important that scientists in developing countries have a strong personal (inner) motivation to pursue science,[11] because the external motivations and reinforcements simply may not be there, or in fact, may be negative.[15]

d) The Context of Science: In a developed country, the scientific infrastructure is more or less already in place, the scientific institutions are formed, the population on the whole has had some contact with science and has at least a fleeting understanding of the role of science in society. Therefore, a young scientist can afford, at the beginning of his career, to deal mainly with direct scientific activities in research or teaching. In contrast, in a developing country, the context of science is extremely important because it has not been established yet. Science as a societal undertaking has not grown its roots yet, scientific institutions are not present, the auxiliary services for science such as machine shops, stockrooms, glass-blowing, laboratories, equipment repair, etc. may not exist. Furthermore, the administration and management of science are not established yet or are done by people inexperienced in these activities. As a result, scientists in a developing country must be able to deal with these contextual problems of science at the same time when they are involved in the direct scientific activities also.[13]

e) Personal Ties: One of the most debilitating factors in being engaged in scientific activities in a developing country is isolation.[1] For a number of reasons and in all areas of scientific work, a scientist in a developing country is much more isolated from his colleagues in other parts of the world than it is the case for somebody who lives in a scientifically advanced country. Since science is a highly collective and universal undertaking, this isolation is a very serious handicap in carrying out scientific research in developing countries. Consequently, it is very important for a scientist in a developing country to develop an extensive network of personal ties with scientists in the worldwide scientific community. This is a difficult task which needs to be approached with a considerable amount of detailed knowledge of the communication system in the scientific community as well as with skill and devotion.

B. *Realities*[25]

I will now discuss the actual realities of the present day situation under two headings. The first one will deal with local education, that is, postgraduate education in the developing countries themselves. The second part will deal with education abroad, that is, the postgraduate education of students from the developing countries in countries which are scientifically more advanced.

a) Local Education:

1) Narrow, premature specialization: One of the serious defects of the science education (both pre- and postgraduate) in many developing countries is narrow and premature specialization. Students are asked to select a very narrow speciality very early in their educational process, for instance, at the age of twenty-two, that is, just after they have finished what would correspond to their college education in the United States. From then on, their activities will be confined to this narrow field. Tragically enough, this is often justified by local science policies and decision makers by the desire to see these scientists working in applied research. In actuality, however, working in applied science research requires a very much broader scientific base than work in basic research because problems in the applied fields do not arrange themselves conveniently into narrow disciplinary and sub-disciplinary areas but usually involve knowledge from various disciplines and various areas of the sciences.[1][22]

2) Rote Learning: Another general characteristic of science education in developing countries is the emphasis on rote learning and memorization instead of understanding and problem solving.[24] The examination system is quite often stilted so as to require menial memorization and the lectures and textbooks are also oriented in this direction. All of us are familiar with students arriving from developing countries for graduate education abroad who have on their record courses in advanced subjects but who at the same time are unable to solve problems even on a beginning college level. This is probably the most important area of science education that needs to be improved in developing countries.

3) External Motivation: The social structure and political arena of many developing countries is such that it is a tradition to rely on external motivation coming from parents, from society's ranking of professions, or from governmental exhortations and political pressure. It is generally not the tradition to allow an individual to shape his life according to his own internal likes, interests and preferences.[11]

4) No preparation for Context: Science education for developing countries rarely involves any exposure of the students to matters surrounding science as a human activity and matters pertaining to the context of science in society. History of science, philosophy of science, scientific methodology, science management, science policy and similar subject matters are not part of the curriculum and therefore, the student often places science in an ivory tower with very little relationship to the world around it.

5) Isolation from Scientific Community: For reasons which are often beyond the control of the local educational system, students who are preparing for science in developing countries hardly have any opportunities to meet, to interact with and to develop personal ties with members of the worldwide scientific community. Visitors are few and if they come, they are treated with too much respect and are placed too much on a pedestal so that they do not have an opportunity for direct interaction with young students. The acquisition of written information in terms of journals and other scientific communication tools is also very difficult in developing countries. There is, therefore, a tradition of isolation right from the time when the student gets his postgraduate education.

b) Education Abroad:

1) Tendency to specialize too much: Even though the science education in the scientifically more advanced countries, and particularly in the United States, attempts to give the students a broad education, students from developing countries still retain a tendency to constrain their education and attempt to specialize too early and too narrowly. This tendency is not sufficiently inhibited by the advisors of these students or by the rules and regulations of the graduate education itself.

2) No preparation for context: Since the young scientists from developed countries, after getting their degree, are absorbed in an institution or organization which is already formed, they do not have to know, at the beginning of their career, about the auxiliary aspects of science and the context of science. Therefore, these skills are not taught in the graduate schools in these countries, and since they are not taught to the local students, they are not taught to the students from developing countries either.

3) Insufficient opportunities to develop ties: Graduate students from developing countries studying in an advanced country often retain a certain reticence in initiating interaction with scientists who are senior to them in age and experience. Furthermore, their graduate careers in the graduate schools abroad do not offer them easy opportunities to travel around and meet a large number of scientists. To be sure, usually there are a number of visitors coming to each graduate school to deliver seminar and colloquium talks but opportunities during these talks are not very plentiful for making the personal acquaintance of the speakers. As a result, it is quite possible for a student from a developing country to go through five years of graduate education without having made the personal acquaintance of more than a handful of scientists.

IV. SPECIFIC PROGRAMS FOR IMPROVEMENT

A. *In Local Education*

a) Broadening the scope of graduate education: It is important to convince those few graduate schools which already exist in the developing countries to offer their students as broad an education as possible. This might run into difficulties because of the lack of faculty and because of the feeling that the time span of graduate education should not be expanded excessively. Nevertheless, it is important to inject an element of breadth into graduate education. It might be possible to coordinate several institutions in a given area to offer some broadening postgraduate seminars to their students jointly. It is particularly important to convince the science administrators that creative applied research in the sciences must be based on a broad postgraduate education. A very effective way of doing this would be to ask outstanding applied scientists from the advanced countries to lecture in the developing countries, talking specifically to managers of science, governmental bureaucrats, administrators and people in general who are in charge of science policy and decision making, and convincing them of this particular point. [2] [5] [18]

b) The problem solving approach: It is extremely important to reform science education from their rote learning, memorizing mode into a mode which emphasizes problem solving and understanding rather than memorization. There are a number of tools through which this can be carried out. One very useful one is the open book examination. An examination which allows the student to bring in any written material he wants for the examination, greatly reduces the incentive on the part of the student to memorize for the examination. Another important technique is to inject into the education the solving of open ended problems, that is, problems which do not have exact solutions and which have to be approached in the same spirit as actual, practical problems in scientific research.[12] These problems could be taken from real life and from technological applications pertaining to problems in the country so that it becomes clear to the students how science and scientific understanding are related to practical problem solving.

c) Strengthening personal motivation: As I mentioned earlier, personal motivation does not fare very well in the developing countries. Therefore, one needs a strong re-inforcement of this personal motivation through sympathetic teachers who interact with the student on an individual basis and encourage him to undertake scientific work for his own personal motivation even in the face of adverse external societal influences.

In many developing countries students who end up in the sciences are those who have been rejected from medical schools, who have not been allowed to become engineers or lawyers and who therefore are in science by default rather than by personal choice. One should try to catch, at the very beginning of the educational process, those who do have an internal, personal motivation to become a scientist but whose motivation would later be overwhelmed by external factors and as a result, would end up in other professions. It is important to establish special programs for these interested and talented students who want to go into the sciences. In that way they are exposed to a stimulating atmosphere, can meet other students with a similar motivation, and become reinforced in their desire to become a scientist.[28]

d) The context of science: As I mentioned, exposure to the context of science during science education is of extreme importance. It is therefore mandatory that science curricula in developing countries should include some exposure to the science of science, to science policy, and to the practical skills which are needed in scientific activities. This includes interaction with non-scientists, interaction with engineers and technologists, and interaction with administrators and science policy makers who are not scientists themselves. This is a difficult task because most of the teachers, professors and practical scientists in the developing countries have not been exposed to this additional contextual education in science and therefore, it is difficult to find teachers and people who would be able to handle this particular aspect of science education. Nevertheless, one has to build up slowly a cadre of people who are, in fact, interested in the knowledge about these aspects of science and can inject it into the curriculum.

e) Ties with the scientific community: It is important to make sure that scientific institutions in the developing countries have access to the tools of scientific communication with the rest of the world. We should increase the number of visitors to these places, we should make sure that the channels of written communication reach these institutions, and we should make sure that people from these institutions have an opportunity to participate in scientific conferences. This is a problem that involves the whole scientific communication system on a worldwide scale and is a subject that is a large one even for an entire book let alone for a part of a brief talk like this one. I nevertheless want to emphasize the importance of this aspect in the context of postgraduate scientific education.

B. *Education Abroad*

a) Selection of students: The selection of students from developing countries for graduate education in the scientifically advanced countries is at the present time a somewhat haphazard procedure. For a school in a developed country is is difficult to evaluate the credentials of students who apply from developing countries. In this context I would like to refer to the Physics Interviewing Project[4][7][9][28] which has tried to remedy this problem, which has been in operation for 10 years, and which has produced very useful results. This project could be utilized much more fully by the community of physicists and could also be extended to other disciplines besides physics.

b) Advising of students: There is a need for improving the advising of graduate students from developing countries who are studying at institutions in the developed countries.[6] There are a number of reasons why such a student needs special encouragement and special advising in terms of his professional goals, in terms of his choice of courses, in terms of his curriculum development and in other respects. At the present time this advising is quite imperfect because of reasons which pertain both to the students and to the advising professor. Recently, the American Association for the Advancement of Science developed a short brochure[27] which is now circulated in all the science departments in the United States which discusses some of these problems, is directed at the science faculties in graduate schools in the universities in the United States and elsewhere, and aims at the improvement of this advising process. Similar improvements could be made in other countries and in other ways also.

c) Opportunities to establish personal links: It would be very important for departments of science in universities in the developed countries to pay special attention to encouraging their graduate students from developing countries to establish personal links with as many scientists as possible during their education in that department. Special opportunities should be made available to these students to attend scientific meetings and to spend their summers in other universities or other laboratories making new ties and getting to know additional people.

d) Education for the context of science: Students from developing countries who are receiving graduate education in the sciences in an advanced country should be given, as an auxiliary element in such education, exposure to those elements of the context of science which are of particular interest in the developing countries. This may include some practical skills like glass blowing, electronics, and machine shop work, as well as managerial information on the operation of universities, libraries, laboratories and industrial organizations, including items like how to order from catalogues. Furthermore, the exposure should also include science policy and the science of science. This auxiliary exposure can either be done by knowledgeable members of the department he is educated at, or through summer seminars which gather students from a number of universities.[2] [5] [18]

e) Continuing links after education: After the student receives his Ph.D. degree and returns to his own country, he enters a phase of his career when it is especially important for him to have access to interaction and collaboration with other scientists. It is therefore very important that his former institution in the developed country conscientiously maintain a link with him. Particularly his former advisor should maintain collaborative correspondence with him. This may involve letters, the sending of communication material such as reprints, preprints and general articles, the sending of small spare parts of equipment and the providing of advice in the choice of research subject matter. Further elements may be collaborative research, personal visits by the former advisor to the institution in the developing country, and personal visits, from time to time, by the former student back to his former institution. There are some countries, including Canada, which have mechanisms for organizing such return visits by former students from the Third World who received their education in those countries. This practice, however, is not universal enough. Much could be done to encourage scientists in developing countries if such personal interaction would be made possible, especially in the early stage of their careers.

V. EPILOGUE

As said at the beginning, this talk was a very brief summary of some of the elements important in graduate education that are deficient in developing countries from the point of view of the social motivations for science. I also tried to outline a few specific suggestions of how to improve the situation. In conclusion, I want to emphasize one point. Many of the suggestions for improvement are of the type that can be advanced and implemented entirely by the worldwide scientific community itself.[17] For instance, the deficiencies in the equitability of the worldwide scientific communications system is almost entirely under the control of the worldwide scientific community. I would like to urge everybody, therefore, not to wait for ponderous international organizations to develop complicated programs or for government bureaucrats in the advanced or the developing countries to begin to suggest projects to be implemented through the development plans. If anything comes from those sources, the better. In my opinion,[2] however, the much more likely and much more productive source of relief for those who work in the sciences in the developing countries is the worldwide scientific community itself. I would like, therefore, to urge everybody in the scientific community to consider these problems and to begin implementing, even on a very small scale, measures which will directly benefit their colleagues in the developing countries.

[1]M. J. Moravcsik, *Science Development*, International Development Research Center, Indiana University, Bloomington, Indiana 47401, Second Printing (1976).

[2]M. J. Moravcsik, Some Practical Suggestions for the Improvement of Science in Developing Countries, Minerva *4*, 381 (1966).

[3]M. J. Moravcsik, Education and Research in Scientifically Developing Countries, in *Management of Research and Development*, OECD, Paris (1972).

[4]M. J. Moravcsik, The Physics Interviewing Project, International Educational and Cultural Exchange, Summer, 1972, p. 16.

[5]M. J. Moravcsik, Some Modest Proposals, Minerva 9, 55 (1971).

[6]F. E. Dart and M. J. Moravcsik, The Physics Graduate Student in the United States, (A Guide for Prospective Foreign Students), American Institute of Physics, New York (1971).

[7]M. J. Moravcsik, The Committee on International Education in Physics, Amer. Journal of Physics 41, 309 and 608 (1973) and Phys. Teach. 11, 82 (1973)

[8]M. J. Moravcsik, Foreign Students in the Natural Sciences: A Growing Challenge, International Educational and Cultural EXCHANGE 9, #1, 45 (1973).

[9]F. Dart, M. J. Moravcsik, A. deRocco, and M. Scadron, Observations on an Obstacle Course, International Educational and Cultural EXCHANGE 11, #2, 29 (1975).

[10]M. J. Moravcsik, Scientists and Development, Journal of the Science Soc. of Thailand, 1, 89 (1975).

[11]M. J. Moravcsik, Motivation of Physicists, Physics Today 28:10, 8 (1975).

[12]M. J. Moravcsik, Two Views of Science — As a Student and "Vingt Ans Apres" The Physics Teacher 15:1, 32 (1976).

[13]M. J. Moravcsik, The Missing Dialogue — An Obstacle in Science Development, International Development Review: Focus 18, #3, 20 (1976).

[14]M. J. Moravcsik, The Context of Creative Science, Interciencia 1, 71 (1976), (Reprinted in Everyman's Science (India), 11, #3, 97 (1976).)

[15]M. J. Moravcsik, How to Pick the Right Person (The Choice of Personnel in the Building of Science), Science and Culture (India) 44, 339 (1978).

[16]M. J. Moravcsik, Developing Countries and the Fruits of Science, Leonardo 11, 214 (1978) (Also in Internat. Develop. Review, Focus 20:1, 27 (1978).)

[17]M. J. Moravcsik, Science and Scientific Higher Education in Developing Countries or Seven Tasks for Geoscientists, Proceedings of the Symposium on Opportunities in Geomagnetism and Aeronomy for Developing Countries, at the Joint Assembly of the International Association of Geomagnetism and Aeronomy and of the International Association of Meteorology and Atmospheric Physics, Seattle (1977).

[18]M. J. Moravcsik, Science and the Developing Countries, Position paper requested by the U.S. National Science Foundation. Copies obtainable from Andrew Pettifor, Room 1229, National Science Foundation, Washington 20550.

[19]R. H. B. Exell and M. J. Moravcsik, Third World Needs 'Barefoot' Science, Nature 276, 315 (1978).

[20]M. J. Moravcsik, The Two Ways of Mountaineering, Physics Bulletin 29, 249 (1978).

[21]M. J. Moravcsik, What Motivates the Developing Countries to Do Science and Technology, Theory of Knowledge and Science Policy, Communication and Cognition, Ghent, Belgium, Vol. 2 (to be published in 1980).

[22]M. J. Moravcsik, Linking Science with Technology in Developing Countries, Approtech 2:2, 1 (1979).

[23]M. J. Moravcsik, The Role and Function of a Scientific and Technological Infrastructure in the Context of Development Policy, Invited paper at the symposium on Science and Technology in Development Planning, Mexico City, 1979, Pergamon Press (to be published, 1980).

[24]F. E. Dart and Pradhan, Panna Lal, Crosscultural Teaching of Science, Science 155, 649 (1967).

[25]E. F. Fuenzalida, Relevance and Development Style, Paper presented at the AID/NAFSA Workshop, Washington, D.C. March 1980 (to be published).

[26]E. Callen and M. D. Scadron, The Physics Interviewing Project: A Tour of Interviews in Asia, Science 200, 1018 (1978).

[27]The Third Worlds Graduate Students in American Science Departments, American Association for the Advancement of Science, Washington, D.C. (1979). (Copies available from Denise Weiner, AAAS, 1776 Massachusetts Sve., Washington D.C. 20036).

[28]M. J. Moravcsik, Creativity in Science Education, Science Education (to be published, 1980).

SOME OFF BEAT SUGGESTIONS FOR SOFTWARE TEACHING AIDS

MICHAEL J. MORAVCSIK
Institute of Theoretical Science
University of Oregon
Eugene, Oregon 97403
USA

Abstract

Six suggestions are advanced for the creation of software teaching aids, each resulting in a locally produced book, brochure, or meeting. These six suggestions deal with a) a collection of problems which are interdisciplinary, open ended, and not solvable by mathematical tools in a closed form; b) a collection of examples taken from daily life, together with their physical explanations; c) a set of case studies of applications of science which are interdisciplinary; d) a set of case studies showing the application of science in the activities of local industry; e) a set of case studies showing the application of science in rural life; f) workshops or meetings in which students are asked to solve realistic, practical problems in science policy. The creation of such teaching aids require no foreign currency, and they also help in bringing together the scientific community of a country from the various different sectors of university, governmental research laboratory, and private industry.

Introduction

Most people, when hearing the phrase "teaching aid", visualize some sort of a gadget: movies, film strips, computer displays, televized demonstration experiments, laboratory equipment, and alike. Indeed, such teaching aids can play an important role in science education and time is well spent on discussing how they can be developed, made widely available, and used fruitfully by an average instructor in an average institution. I have little doubt that this conference will receive an ample number of contributions on this subject. Partly because of that, but partly because I feel the need of emphasizing different types of teaching aids also, my talk will not deal with any "hardware" at all. Instead, I want to discuss "software", that is, teaching aids the tangible form of which is very conventional indeed: books, brochures, notes, meetings. To be sure, some of this could be turned into hardware, such as movies, but that is quite incidental to my exposition. Instead, I want to stress what I believe is the novel **content** of·these teaching aids, and I want also to suggest ways in which they could be created in return for·a very minimal out-of-pocket expenditure, practically none of which is in foreign exchange.

The Background

In order to place the discussion in the proper context, I want to list three major shortcomings of science education as practiced in many places in the Third World. In doing so, I want to apologize to you, who are predominantly science teachers in Third World institutions, since I will undoubtedly be covering ground that you are much more familiar with than I am, and so I am running the risk of telling you some things that you very well know already. Yet I feel that in order to motivate and support the suggestions I will make, it is necessary to place on the record the shortcomings which these suggestions aim at remedying, even if they have been discussed before.

The first of these shortcomings is the reliance on memorization rather than on understanding. The instruction of science, together with the examination systems, encourages the students to memorize "facts", formulae, "laws", instead of aiming at the understanding of the concepts and the way of thinking that is at the basis of these facts, formulae, and laws.

There are many factors that contribute to this shortcoming. Perhaps the most basic is an epistemological one. Science, in such instructional systems, is viewed primarily as a closed system of already formed knowledge, rather than a method of asking new questions and of finding partial answers to those questions. This closed model of science is also more comforting to the instructor, who, once "mastering" this closed set of facts, is under no pressure to undergo continuing education, to admit that he is less than omniscient, to experience a situation when a student can ask questions that he cannot answer. The often practiced external examination system also contributes to the strengthening of this view of science as a set of facts, since such an examination system discourages educational experimentation with more creative ways of teaching science.

This is not the place to discuss in detail the catastrophic effects such a model of science has on the scientific manpower development of countries of the Third World. I want to mention only one such effect, namely on the country's ability to perform applied scientific research. Such applied research consists of asking the "right" questions in a given situation, and then trying to find approximate answers to those questions. In the latter phase, the most important element is to judge the choice of approximations so that they are solvable and yet realistic. None of these activities involve the mechanical regurgitation of well known facts formulae, and laws. Indeed, if science were such a closed epistemological system there would be no need for additional research at all.

31

The second shortcoming I want to list here is premature specialization, and the narrowness of science education in general. Students are often directed to specialize their range of studies and their specific activities to a quite narrow field at a time when their grasp of their scientific discipline is still very rudimentary. To be more specific and to use physics as an illustration, students in many Asian countries are considered generally educated in physics when they reach their B.Sc. degree, and already for their M. Sc. they are narrowed down to specialized subjects. In the four year they studied for their B.Sc., even in an academically outstanding institution, they could become acquainted only with the elementary stages of mechanics, electromagnetic theory, atomic physics, nuclear physics, particle physics, plasma physics, statistics physics, optics, acoustics, thermodynamics, solid state physics, and the auxillary subjects of mathematics, electronics, and others. These foundations are insufficient for the student later to be able to turn to any of these areas in his scientific career. As a result, the student is locked into the particular narrow specialty he enters when he works for his M.Sc., a specialty which might depend on the interes of one professor, which may be out of date five years hence, and which, in any case, is of limited importance and utility in the overall spectrum of scientific work.

Again, this is not the place to fully elaborate on the detrimental effect of such a curriculm, but, again, I want to stress the particularly damaging influence on applied scientific research. In so-called "basic" research it is to some extent possible to limit work to one particular scientific discipline, and to one particular field in that discipline, and sometimes even to one particular subspecialty in that field, since the problem is to a large extent defined by the researcher himself. In contrast, in applied scientific research the problem is defined by the intended application, and hence tends to be broadly interdisciplinary and within that encompassing many special fields in each discipline. I consider premature specialization probably the most important contributing factor behind the often very low quality of applied scientific research in many Third World countries.

Finally, the third shortcoming I want to stress in the frame work of this discussion is the absence of what I call contextual science education, that is, education on the context in which science is embedded in the real world. This involves the rationale of doing science in a developing country, the way "science policy" (if there is such a thing) is determined, the way research is supported, the way research is organized, the way research proposals and completed research are evaluated, the way a library is run, the way machine shops are managed, the way research and demonstration equipment is improvized in the local context, the way science can be communicated to the population as a whole, the way the international scientific community operates and on ways isolated scientists in Third World countries can be integrated into this community, etc. The environment in which science exists in countries where science is new and the infrastructure is rudimentary is much more hostile, much more burdesome on the scientist, much more challenging for the researcher than it is the case in the countries with a longstanding scientific trádition. Yet, the science student in Third World countries, upon the completion of his formal education, is thrust into the lions' den without any preparation for how to deal with the situation. It is not surprising that, in response, many young scientists emigrate, or turn to bureaucracy, or loose their enthusiasm and settle down to a busy but only marginally productive scientific career.

These three observations, as well as others, were derived in my interaction with science and science education in many parts on the world during the past 20 years and they are particularly well illustrated during the interviewing trips of the Physics Interviewing Project, which many of you probably know. This program aims at obtaining objective, comparative, and relevant information on students in Third World countries who are applying for admission and financial aid to graduate schools in physics, particularly those in the United States, Canada, and Australia. The program sends two physicist on a five-week trip to educational institutions in Third World countries (usually in Asia), during which they personally and individually interview such students, for about 45 minutes each, and then produce a one-page report which is available to any school the student is applying to, or any school that is interested in the student. The interview deals with simple problem solving in elementary and intermediate level physics, since this is probably the most important factor that determines whether the student will be able to succeed in graduate school.

As I mentioned, illustrations of the above listed three shortcomings can often be found throughout these interviews. But let me also mention to you a different feature of these interviews, one that is most heartening both to the interviewer and to the countries where the interview takes place. Regardless of the state of science education in a country or in a particular educational institution, there is always a chance (and this chance is not very dependent on how good or bad science education is there) to find a most exceptional student, one who has a deep internal motivation to do physics, one who lives in physics, one whose face lights up when an interesting problem is given to him, one who has absorbed physics as a fascinating activity of problem-generation and problem-solving. The encouragement and the further development of such students should be a supreme goal of programs in science education, since it is likely that these exceptional scientists-to-be will be able to make decisive contributions to the building of science in their countries.

Six Proposals

Against the background outlined in the previous sections, I want now to suggest six "software" teaching aids. As mentioned, each of these consists of a book, a set of notes, a brochure, or a meeting. Since these aids are to be produced locally or regionally, and not taken over from a developed country, their production and distribution could be done and carried out quite inexpensively in local currencies. Publishing companies like the World Scientific Publishing Company in Singapore, and perhaps others, already operate and have an international distribution network to deal with the logistics.

Thus the main problem is not in producing the teaching aid once it is prepared, but in creating it. It is my suggestion that ASPEN should appoint, from within its members, a small group (2–3) of volunteers for each of the six proposed teaching aids. It is very likely that in order for these physicists to devote enough time and energy to the creation of these teaching aids, their respective institutions would have to release them from some of their teaching and/or administrative duties over a limited span of time. Although this represents cost, it is not an out-of -pocket cost, in that no new funds need to be found to cover this expense.

The most difficult task for ASPEN will be to find appropirate volunteers for these tasks. If these were in great abundance, the shortcomings which these aids are aimed to remedy would not exist. The participants in these tasks will have to be physicsts and physics teachers who can reach beyond the present educational system and philosophy, and who have the energy, imagination, and courage to be innovative.

A.   Problem Compendium

A remedy to obsessive memorization is the insistence on problem solving rather than on the recitation of formula and lws. There are, however, several obstacles to this practice. For one thing, not any problem solving will do. As we all know, solving a problem has at least two parts: The "setting up" of the problem, that is, the decision concerning the procedure to be used to answer the question implied in the problem, and then the actual carrying through of this procedure. The first part involves a deep understanding of the physics of the problem, while the second uses mathematical manipulating skills. While both are needed for successful problem solving, we want to test mainly the first, namely the understanding of the physics of the problem, since this is by far the more difficult of the two attain. Unfornately a large fraction of the problems found in textbooks and given by instructors do exactly the opposite. In them the "setting up" is trivial, and only the "working out" takes any time at all. In other words, they are problems in which we simply have to "plug into" obvious formulae.

Second, problems found in textbooks and given by instructors tend to be highly non-interdisciplinary. They clearly pertain to only one subdiscipline (mechanics, electricity, atomic physics, etc.), and in fact usually to only one particular aspect of that subdiscipline (e.g. in mechanics, to conservation laws, or rotational motion, etc.) As indicated in the previous section, this is quite far from the way problems arise in real life.

Third, problems found in textbooks and given by instructors virtually always can be solved exactly, in a closed form. As mentioned earlier, this is also completely at variance with research problems in real life, when first an appropriate physical approximation procedure must be found, and then an approximate or numerical mathematical method needs to be applied to get a quantitative answer.

My first suggestion therefore is that a problem compendium be prepared containing, say, 500 problems in physics, together with their solutions, problems that require understanding of physics to be "set up", which are interdisciplinary or at least not definitely relegated to one particular little corner of physics, and which are "open ended", with no exact answer in closed mathematical form.

A good criterion to apply to the selection of these problems is that they should be appropriate for open-book examinations, that is, for examinations in which the student is allowed to use the textbook, his notes, or any other written material. Open examinations themselves would be a very effective measure to counter memorization, but their actual introduction into the examination systems of most institutions in the Third World would be so difficult at the present time that this aim can be expected to materialize on a wide scale only a decade or two from now.

B.   Example Collection.

It is easy for a student to gain the impression, consciously or subconsciously, that physics is something "academic" which has little to do directly with everyday life and the observations and experience one accumulates in the course of the most mundane daily activities. It is therefore important to refer, in the course of teaching, to the physical explanation of such everyday occurrences. Textbooks tend to do this from time to time, but textbooks are often written in the developed countries and used in the Third World either unaltered in the original English or in a direct translation into the local language. Naturally, the examples in those books, even if they are numerous (which is seldom the case) pertain to the observations in the daily life of an inhabitant of a developed country. While physics is universal, the particular phenomena which have physical explanations vary from one environment to another.

I therefore like to suggest that an example collection be prepared by ASPEN, in which hundreds of everyday phenomena encountered by students in Asian countries are explained in physical terms. Particularly useful would be to give such explanations for ways of doing things which are traditional in Asian countries. Some such practices are inefficient and counterproductive from a physical point of view, but many of them are very effective and incorporate empirical experience of many centuries that now can be expressed very easily in terms of modern physics. Such an example collection would therefore also help to bring physics closer to the local cultural traditions.

C.   Nature is Interdisciplinary

Another effective teaching aid would be a small book or brochure which would contain case studies of important applications of physics, taken from historical but recent situations, in which the problem in question was clearly interdisciplinary and where therefore many branches of physics, combined with ingredients of other scientific disciplines had to be combined to solve the problem. Twenty such examples, each treated on not more than five pages, would suffice to make the point. Since applied science is more often practiced in industrial or governmental laboratories than at universities, this task would also serve to bring into science education the nonacademic sector of the local scientific community, and thus bridge a gap that is all too sharp and deep in many countires.

D.   Science And Industry

One of the acute problems in many countries is the lack of a link between science as taught and practiced in the country and the country's industrial activities. It would therefore be very useful to produce a small book or brochure which would present actual case studies of how science is used in practice in the everyday life of local industry. It is not necessary that such a use of science involve the latest pioneer front-line discoveries of science. Much of such an application relies on "old" science, science that is already in the textbooks, and on the basis of that science and of the use of the "scientific method" practical problems are solved. Since industrial acitivities vary from country to country, or at least from region to region, such a collection of case studies should perhaps be prepared regionally. This project, just as the previous one, would promote the cooperation between scientists and technologists from universities, governmental research laboratories, and private industry, a cooperation that surely needs strengthening.

E. Science And Rural Life

My next suggestion is very similar to the previous one on science and industry, except that it pertains to activities in the rural sector. After all, most of the countries within ASEAN are strongly or even predominantly rural, and hence science can have and has had a major impact on rural life. Yet an explicit illustration of this is seldom presented, and students are likely to gain the impression from their studies that science is a lofty and esoteric activity, relevant, perhaps, to sophisticated "high-tech" activities but quite isolated from the life of the millions in the villages. A study, brochure, or book which would present specific and concrete case studies of how science can be used in the everyday activities in the rural areas would therefore make a major impact.

F. Practice In Science Policy

My sixth and last suggestion would not result in a brochure or book, but instead could be realized in the form of a seminar, meeting, conference, or workshop. Students would gather to consider a problem in science policy and to carry out the steps needed to make decisions and to implement them. The implementation would of course not necessarily take place in reality, and yet it is important to carry through the exercise including the plans for implementation, since it is at that stage that practical and realistic problem solving is needed. For example, such gatherings could deal with questions on how to give out money for science projects, how to assess and evaluate proposals for scientific projects or final reports of research already performed, how to assure an adequate supply of scientists for a country, how to organize a mixed science-and-technology research-and-development project in connection with a specific problem, etc. It is crucial that such workshops be led by scientists with substantial practical experience in actual management of research and development, and it is also important that at such meetings the questioning and the critique of past practices or of methods proposed be free and extensive. The meeting may want to record its conclusions in a short report which could then be circulated to other students and to other meetings.

Epilogue

As mentioned earlier, the six suggestions just discussed have certain distinct advantages. They require no foreign exchange, no formal cooperative arrangements with international or national agencies, and they directly involve the local scientific community by calling on their skill and energy. Each of them can be organized easily, since it requires no new funds. On the other hand, exactly because these suggestions rely on the volunteering of outstanding members of the local scientific communities, they are dependent on the scarcest commodity, namely highly trained scientific manpower. In a way it is easy for me, an outsider, to make these suggestions, when the burden of the actual implementation lies entirely on the local scientists.

And yet, I feel that any of these suggestions represent invaluable "teaching aids" in the real sense of the word, since they are likely to affect not only the form but also the spirit of science teaching. Furthermore, the creation of such teaching aids would also contribute to the self confidence and the feeling of accomplishment of the local scientific communities, far beyond what any kind of gadgetry could do. Through any of these proposals "live science" would be communicated to the future generation of the country's scientists which is, it seems to me, the ultimate aim of science education.

HOW TO PICK THE RIGHT PERSON

(The Choice of Personnel in the Building of Science)

Michael J. Moravcsik*

For the selection of personnel to manage science, three criteria are offered, involving personal serenity, a proper tension between doing and managing science, and articulateness. It is also urged that scientific managers should be chosen from the scientific community itself. The article concludes with a plea for proper consideration of exceptional individuals.

I. Introduction

SCIENCE is a quality undertaking. There are many human activities where quality, while desirable, is not altogether crucial. A bad carpenter will make furniture which is perhaps ugly and which will perhaps fall apart in a relatively short time, but his product is furniture, nevertheless, and is functional, though perhaps in a more limited way than one might have hoped for. The same can be said about farmers, lawyers, tradesmen, or even doctors.

The same, however, does not hold in science. To practise or support bad science is completely meaningless. The purpose of science is to provide new knowledge about nature which contributes to our understanding of natural phenomena and which might be used for man's control and modification of adverse natural phenomena. Bad science, which deals with questions we are not interested in, or supplies information which has no impact on the overall structure of our understanding or on our ability to control nature, has no justification whatever, even if it is "new" in the sense of nobody having undertaken it previously.[1]

It is therefore of utmost importance in any scientific community to constantly guard against deterioration of standards,

and against the dilution of scientific manpower by incompetence, lack of motivation, or laziness. However unpleasant this may sound in an age which strives to be egalitarian (at least rhetorically), science is a fundamentally elitist endeavor. It can be pursued only by a relatively small fraction of human beings, and even among them, relatively few will contribute the lion's share of progress.[2]

In as much as development is usually conceived in terms of quantitative indicators there is an eternal tension and conflict in science development between quality and quantity.[3] National development plans as well as organizational reports and political discussions contain exclusively quantitative goals, targets, and achievements, with the result that in some countries serious situations have arisen due to the unmanageability and productivity of a large but at best mediocre body of scientific manpower, created in the hasty atmosphere of chasing formalistic and quantitative targets.

Such as sub-mediocre majority of the scientific community affects the entire scientific atmosphere and tradition in the country. Activities will be based on obsolete knowledge and narrow horizons determined by the desire to stay within comfortable boundaries rather than charting new territories. Partly as a result, scientific priorities

*Institute of Theoretical Science, University of Oregon, Eugene, Oregon.

are established by catchwords and cliches rather than by an expert assessment of the alternatives. The "relevance" of an activity is thus gauged by whether it can be linked in a facile way with "in" words like technology, agriculture, or import substitution, rather than by whether it really contributes to the cultural, spiritual, material, and intellectual development of the country.

The elitist character of science can also be disturbing on an individual level. To be sure, in the social and political sense, science is strongly anti-elitist, in fact almost ideally democratic. Financial assets, race, religion, political ideology, nationality, cultural background, social background are all subjugated to the sole criterion of being a good scientist, and to assess this, science offers quite reliable criteria. In practice neither the criteria nor the lack of depedence on the above mentioned other factors works perfectly, but still, science perhaps approaches the ideal of ranking by merit more closely than any other human activity. Yet the fact remains that, even though initially selected on purely meritorious grounds, some who enter science will succeed splendidly, while others will find themselves toward the bottom of the ladder or even out. Since becoming a scientist implies an arduous and lengthy investment in time and effort, and since most people crave for some security and dislike being outdone by others (perhaps younger than themselves), this risky element in being a scientist might be disturbing psychologically.

In fact, this desire for seniority, job security, and absolute predictability is the motivation for the civil service systems, now rampant in every country of the world. Such a system assures us that if we are not flagrantly incompetent or dishonest, then we can count on a position forever, can anticipate regular promotions depending only on the length of service, and we are also protected against the threat of some young and talented upstart getting ahead of us.

Needless ot say, the application of civil service mentality to scientific manpower is a disaster[4]. Its philosophy is exactly antithetical to the requirements of science as I discussed at the beginning of this article. To be sure, one can hear arguments contending that by placing science and scientists within the civil service system a certain permanence and stability is achieved which otherwise might be questionable in a country where science is just being born and exists only on a tiny and inconspicuous scale. While this might be valid to some extent, such a short term benefit is heavily outweighed by the long term detrimental effects and by the quality of science that is likely to be evolved via the civil service channel.

Let me now turn around, however, and consider certain manpower problems from a human point of view. In actuality there are, in any scientific community, people who were selected into it even though they should not have been, and who have served to the best of their ability for a long time, even though such a best of their ability was not very good. Such a situation is particularly likely to arise in a scientific community which is undergoing fast growth and development. For example, the first, pioneer generation often has an inferior opportunity to acquire a good education, and thereafter inferior opportunities to practise science and to expand its talent and competence. When such scientists, later in their life, and after they have failed to turn into truly productive persons, are confronted with members of a younger generation, who received superior education, who have better research facilities, and, in general, a better developed infrastructure to live in, a sharp conflict of generations arises. By criteria of instantaneous merit only, the old guard should be eliminated. Yet one feels that humanly this would be unjust as well as psychologically undesirable, since it might create too much of a feeling of insecurity in the scientific community.

There is, nevertheless, a solution of this clash between two conflicting considerations. I believe that it is possible to separate financial and positional job security on the one hand, and influence and decision making power on the other, and thus I advocate granting financial and positional job security to mediocre or inferior scientific personnel who have been in the system for a long time and who cannot be redirected into other professions any longer. At the same time, I urge that the responsibilities for decision making in research, scientific education, organization and policy be handled purely on a merit basis, independently of past chronology, age, social influence and the rest. This is not the place to explore the details of mechanisms which can accomplish this, but I have no doubt that it can be done.

The best medicine being the preventive one, however, the focus must be primarily on preventing, as much as it is possible, inappropriate people from entering science and rising through its hierarchy. To put it it in a more positive way, we must concentrate on selecting the right people initially and later continue to choose the right people for promotion. How to do this is the subjust of this article. In Section II I will endeavor to offer some criteria for choosing scientists, followed in Section III by criteria to select scientific managers. The article will conclude with some general remarks.

II. Criteria for Scientists

There are five dimensions along which the assessment of a scientist is of great importance.

1. *Motivation*

I consider this the most important criterion, and at the same time also one of the most difficult ones to assess. The motivation of people to do science is always a mixed one[5], but according to my experience with creative scientists, the indispensible and perhaps even primary factor must be genuine personal interest in and fascination with natural phenomena. Such a drive will have notning to do with the social or technological relevance of science, and will not be concerned with external recognition either. It is a purely internal motivation, fueled by the person's own intellectual and emotional make-up.

I want to emphasize that I do not advocate a disregard of the other motivations for doing science, most of which are external. They can play an important additional role, and they also contribute greatly to society's motivation for supporting science. All I am claiming is that a person whose incentive to do science is exclusively or primarily in such externalities is unlikely to develop into a creative scientist. While I cannot prove this contention, I have never come across a counterexample to it.

How can we test or ascertain such an internal motivation in a person ? According to my experience, the best way is through a personal encounter with the person. Internal motivation will make the person appear possessed, he will live in his subject, and his demeanor will quickly reveal this. He will seize upon any little scientific puzzle or problem that he is given. When asked a question that was not "covered" in his previous educational experience, he will not be outraged, he will not protest, but will immediately turn to trying to figure out how to answer it.

As it is evident from the above description, to test such a person's motivation througn the usual examinations given in the schools or universities is absurd. In fact, given the method of science education practised in many parts of the world, it is not unlikely that the person who is highly motivated in the sense suggested here will have been by no means distinguished in terms of the formal criteria of scores, ranks and the like. This points up the great importance of talent search through personal interviews, a method that proved quite successful in at least one scientific context[6].

2. *Scientific competence*

For a student or a scientist at the very beginning of his career, competence can again be best tested through personal interviews given by other scientists who have well established their reputation by the criteria to be discussed below. For scientist who have had time to chalk up some accomplishments, however, there are somewhat more systematic ways to make the evaluation. In particular, it can be done through the parallel methods of assessing his written contributions and of subjecting him to peer judgment.

I elaborated on this[7], in some detail in a previous article, and hence here I will only give a bare synopsis. The measure of written communications (whether in "basic" or "applied" science) can be done in a quite quantitative fashion by counting articles and reports by the scientist, and by looking at the number of citations and the citation patterns that his articles received. While such a procedure is not unobjectionable, it provides a quite reliable indicator of the person's contribution to science.

Such quantitative and impersonal measures should be supplemented by direct evaluation by scientists of recognized reputation in the appropriate fields of science. Contact with the international scientific community is very important in this respect, since local communities might very well be too small or too entangled in personal animosities to yield a reliable assessment.

For the details of such an evaluation of scientific competence and promise, I urge the reader to turn to Reference 7.

3. Flexibility

A crucial quality for a scientist who wants to be successful in the infrastructure of a developing country is flexibility and versatility. In a small community one cannot expect to find very productive niches for highly specialized scientists, and in fact the multiple demands on scientists to perform personal research, to collaborate with other scientific and technological personnel, to participate in the creation of a scientific infrastructure, to take an initiative in scientific and technological education, etc. demands breadth.

It is a tragedy that this commodity of flexibility is in such a short supply in most developing countries. The blame must be laid primarily to the educational system which, for well-intentioned but completely fallacious reasons, aims for a narrow education with premature specialization[8], claiming that for applied research such education is the most appropriate. In reality, applied scientific research[9] requires a much broader education than basic research, mainly because the problems in applied research are, at least partly, externally defined and hence are often interdisciplinary.

How to test for flexibility and versatility ? Again the best way is through personal interview, during which the interviewee can be confronted with scientific problems outside his present speciality. His reaction to such a challenge and his ability to handle it will go far toward revealing his capability and inclination to be productive in a variety of areas and to be able to switch from one area to another.

4. Self-reliance

This is another crucial factor in a nascent infrastructure. Potential collaborators might not be available, technician help might be absent, equipment might not exist or might not be in working order, and even the problem definition might be missing. Under such circumstances, many have just thrown up their hands, declared the impossibility of doing science in such an environment, and then either emigrated or settled down to "doing science the civil service way', in other words, not at all. It is common in developing countries to encounter expensive equipment, donated by international agencies, rusting idly[10] on the shelves or in the basements because of the helplessness and lack of self-sufficiency of the scientists who were supposed to utilize it. In such an atmosphere, mutual charges and recriminations fly back and forth between the scientists and governmental agencies, and besides the waste of resources and manpower, science itself acquires a bad name in the minds of some politicians.

How can we test for self-reliance ? A person ranking high by this criterion will improvise his own equipment, fix his own instruments, formulate his own research problems and even those of others, will follow the scientific literature, will have a broad circle of personal acquaintances in the world-wide scientific community, and will radiate a certain self-assurance, quite distinct from bombastic arrogance. His theme will be to give rather than to receive, and he will create his own infrastructure rather than being absorbed in writing memos to the government about the lack there of.

Another way of stating this requirement is to say that the scientist in a developing country must be a self-propelled doer. He must have the ability, courage, energy, motivation, and skill to *do* things on his own, not only talk. For example, in establishing a fruitful dialogue with the technological community in his country, he must take the initiative, he must find ways to get across ideas, he must be persuasive to assure a functional bridge between scientific results and technological problems.

492

5. *Scientific Leadership*

In a young community, there are always many more awaiting to be led than seasoned scientists ready to lead. While this deficiency is expected to decrease automatically as time goes on and the young communities mature, at the moment it is a serious obstacle in the path of scientific success.

It is not easy to tell the leadership potential of a scientist who has not had an opportunity to exercise such a potential yet. One can say, however, that such a potential leader should have a good overview of a segment of science larger than his momentary speciality, he should have plenty of ideas for future exploration, he should show good judgment in scientific matters where an objective assessment is not yet possible, and of course he should have the human characteristics which can enable him to stimulate, inspire, and hold together group of collaborators.

III. Criteria for Scientific Managers

In this section I will enumerate some criteria for choosing scientific managers in the broad sense of the word, that is, people who are called upon to participate in the policy making, organization, decision making, management, and administration of scientific activities. Before going into these criteria, however, I want to state what I consider a general principle with respect to such science managers : I firmly believe that with extremely few exceptions, all such managers, at all but the most routine levels, should be chosen from the ranks of scientists themselves. In other words, I do not believe that on can "train" a scientific manager without giving him a thorough scientific education *and* the opportunity of practising science for at least a few years.

I consider the validity of this principle an empirical fact, proven by scores of case studies in dozens of countries. As a result, I take a very dim view of the practice of taking a B. Sc. degree holder in one of the sciences, sending him abroad for a year or two of "training" in scientific management, and then placing him in a responsible and non-routine position within the science management structure.

There are at least two reasons for my skepticism. First, I really do not believe that anybody who has not practised science can acquire the personal knowledge and intuition about the working of science that is needed in scientific management. Our understanding of the process of doing science is in itself not a science yet[11], and hence empiricism plays a dominant role in science management. Those who have not "tried the thing" personally are too apt to plunge into meaningless formalities and counterproductive manipulations.

Secondly, scientists, rightly or wrongly, consider as 'bureaucrats" all non-scientists who sneak into the "scientific system", will look down upon them and will refuse to cooperate with them in spirit. This might be regrettable, but it appears to be a fact, and since scientific management involves the multidimensional human elements needed to create productive collaboration among many individuals, appearance and attitudes, whether 'justified" or not, play a crucial role.

In the following discussion, therefore, I will automatically imply that I am talking about scientific managers who are or have been scientists themselves. In that context then, I now list the three criteria I consider important.

1. *Serenity vis-a-vis past scientific accomplishments*

A successful scientific manager must have a personal satisfaction with his own scientific career. This does not mean that he needs to be a Nobel prize winner. In fact, serenity with regard to a person's own achievements and standing in the scientific community has very little relationship to that person's "absolute" or external standing in the community. There are outstanding scientists who are disturbed, unsure of themselves, unreasonably competitive about priority, and obviously lack serenity, while some others, much less accomplished on an absolute scale, appear content and satisfied with their past achievements and seem to live a balanced and basically happy and relaxed life.

A discontented manager will quarrel with his subordinates, will display envy and anger at younger scientists he suspects are more talented and more promising than he was at their age, and will constantly cling to externalities to prove to himself his worth and value. These qualities are so conspicuous that it is seldom a problem to pinpoint such a person even on the basis of only fleeting contact with him.

2. *Tension between doing science and managing it*

In a good scientific manager there must be a healthy tension between the desire to do science on the one hand, and the interest and ability to manage science. If the former overwhelms the latter, and the person considers management just a chore (even if a necessary chore) which distracts from his doing science himself, there is trouble. On the other hand, a person who too easily drops personal involvement in science and turns altogether to management might reveal a certain personal disillusionment with science which might prevent him from advocating scientific causes with undaunted enthusiasm. Furthermore, if he is too ready to leave science, he will also lose touch with progress in science, and in a few years he will then be managing something he knows very little about.

To maintain the right amount of tension and the proper mix between being a scientist and a science manager at the same time, it is advisable to make scientific managerial positions temporary, perhaps even part time, and rotate them among the appropriate candidates from the scientific community. Even while somebody holds a managerial position, the structure of his duties should be arranged so that he has some time to continue to keep up with happenings in his field of science, or perhaps even continue his research to some extent. An ideal science manager should always be in the position of being able to give a seminar talk to scientists in his speciality about some subject of interest to working scientists. Only when considerably older (say, in his 60's) should a science manager be allowed to divorce himself substantially from ongoing science since by then retirement is sufficiently close that there will not be enough time for him to get out of step to a disastrous extent.

3. *Articulateness*

Especially in the context of science in developing countries, which is in a state of constant crisis[12], a scientific manager must be a highly articulate person who can serve as a spokesman for science vis-a-vis a multitude of different elements in society. He will be called upon to deal successfully with the scientists in his organization, with competing organizations, with government bureaucrats, with politicians, with representatives of international agencies, and with a variety of critics of science. In handling such a task, the manager must have a broad and well thought-out position on the role of science in a developing country, on the problems of science building in such countries, on the practical means of organizing such science, and on many other aspects of science, and he must have the ability to communicate these ideas and convictions in different forms and languages appropriate to the audience he is facing. He must, in doing so, radiate assurance and energy, rather than pleading defensively for casual consideration of his cause. He must have a high morale which can be transmitted to his listeners, a confidence in the future, a belief that the task on hand can in fact be carried out.[13]

To prepare himself for such a role, the manager should have the opportunity and inclination to think and read about the broader questions pertaining to science. Such thinking and reading will congeal his own personal experience and supplement it with that of others. His reading should cover a very broad area, including pertinent items in the philosophy, history, and methodology of science, and in the "science of science". Formalistic documents should be avoided (there is a surfeit of them on the market), and instead personal accounts of successful science managers should be consulted.

Whether a person is capable and amenable to pursue such a course of self-education can generally be predicted on the strength of personal acquaintance with him. Intellectual breadth, reading habits, multidimensional curiosity, and the ability to enter and absorb new fields and areas are qualities that cannot be acquired instantaneously, and hence they are either present at an earlier stage of the scientist's career, or they are not, in which case the chances of it suddenly appearing are not great.

IV. Conclusion

Having just offered some criteria for choosing the "right person", let me quickly add that even in the presence of such criteria creating scientific manpower is a chancy process, and the rate of success should, *a priori*, be considered relatively small. In view of this, I want to return to the theme of the introduction of this article and plead for special consideration for exceptional persons. If a country can somehow produce an exceptional scientist, and retain him in the country, he should be given special attention so that he can maintain his creativity and productivity. There are, of course, various levels of· exeptionality, including one which is affirmed only by the person himself, and so one must have strong evidence along the lines indicated in this article that the person is indeed a scientist of exceptional promise, creativity, and ability. Once this is ascertained, however, support of such a person will be rewarded hundredfold, through his personal work, through his ability to energize others around him, and through his contribution to establishing high standards in the indigenous scientific community. In this sense the existence of the superelite will insure a democratic distribution of the benefits of this existence.*

---

*I am indebed to Dr. R. O. B. Wijesekera of Shr₁ Lanka for discussions which stimulated the writing of this article. and for his comments on a draft version.

Notes and References

1 I discussed "good" science and "bad" science in M. J. Moravcsik, The Context of Creative Science, *Interciencia* 1, 71, 1976.

2 The quantitative demonstration of the fact that a small fraction of the scientific community contributes a large fraction of the literature has become an active area of the "science of science". For a few key references, see Derek, J. de Solla Price, *Little Science, Big Science* (New York : Columbia University Press, 1963) ; Derek. J. de Solla Price, "The Scientific Foundations of Science Policy ; *Nature.* 206 (17 April 1965) ; "The Structures of Publication in Science and Technology" in William H. Gruber and Donald R. Marquis (eds.), *Factors in the Transfer of Technology* (Cambridge, Mass. ; M. I. T. Press, 1969), 91-104 ; Jonathan R. Cole, "Patterns of"

Intellectual Influence in Scientific Research,' *Sociology of Education,* 43, 377-403. (Fall 1970 ; Joel Yellin, "A Model for Research Problem Allocation among Members of a Scientific Community," *Journal of Mathematical Sociology,* 2, 1-36, 1972.

3 For an unusually witty, amusing, and yet penetrating analysis of quality vs. quantity, see Jorge Sabato, Quantity versus Quality in Scientific Research (1) : The Special Case of Developing Countries, *Impact* 20, 183, 1970.

4 A perceptive description of this civil service mentality in science is also contained in Reference 3.

5 For an example, see M. J. Moravcsik, Motivation of Physicists. *Physics Today,* 28, 10, 8, 1975.

6 The example in question is the Physics Interviewing Project, see M. J. Moravcsik, The Physics Interviewing Project, International Educational and Cultural Exchange, Summer 1972, p. 16, and F. Dart, M. J. Moravcsik, A. de Rocco, and M. Scadron, Observations on an Obstacle Course, International Educational and Cultural Exchange, 11, 2, 29, 1975.

7 For such a detailed discussion of evaluation methods, see Reference 1. For a brief review of quantification methods for the output of science, see also M. J. Moravcsik, A Progress Report on the Quantification of Science, *Journal of Scientific and Industrial Research* (India).

8 I discussed premature specialization (together with another curse, rote learning,) in the second chapter of M. J. Moravcsik, Science Development, Pasitum, Indiana University, Bloomington, Indiana, Second Printing 1976.

9 For a more thorough discussion of the status of applied scientific research, see M. J. Moravcsik, Applied Scientific Research and the Developing Countries, (to be published).

10 See for example the discussion in A. B. Zahlan, Science in the Arab Middle East, *Minerva,* 8, 1, 8, 1970.

11 To illustrate this point, I might mention that the professional society for people interested in "science of science" has just been formed, and so far had only one general meeting. The record of this meeting, Proceedings of the First international Conference on Social Studies of Science, Cornell University, November 1976, and the references given in the contributions to this conference, are a representative cross section of the many intriguing directions such studies progress in and at the same time they also demonstrate

the very rudimentary state of this field from the point of view of a scientific manager interested in functional mechanisms to improve the quality of science policy decisions.

[12] For an emphasis on this "crisis" state of science building in developing countries, see Reference 3 and also Jorge Sabato, Atomic Energy in Argentina, *Estudios Internacionales* 2, 3, 1968.

[13] The lack of such qualities was analyzed in detail in M. J. Moravcsik, The Missing Dialogue—An Obstacle in Science Development, International Development Review : *Focns*, 18 : 3, 20, 1975.

ASEAN JOURNAL ON SCIENCE AND TECHNOLOGY
FOR DEVELOPMENT, 3 (1) (1986) 1 – 15

THE MOBILIZATION AND UTILIZATION OF SCIENTIFIC AND TECHNOLOGICAL HUMAN RESOURCES IN DEVELOPING COUNTRIES

Michael J. Moravscik
Institute of Theoretical Science
University of Oregon
Eugene, Oregon 97403
USA

I. INTRODUCTION

Human resources (henceforth abbreviated as "manpower"[1]) in the development and the functioning of science and technology in developing countries is the most crucial ingredient, and forms, most often, the bottleneck in evolution. This is so because it takes an extended period of time (decades) to establish an indigenous scientific and technological manpower of sufficient quality and quantity, and because it is a much more difficult and uncertain undertaking to do so than to acquire sufficient financial resources. For this reason problems related to the generation and utilization of scientific and technological manpower are at the very centre of attention when it comes to managing science and technology in developing countries.

In such discussions usually more emphasis is placed on the creation of manpower (e.g. through educational institutions), and relatively less attention to how to mobilize and utilize manpower already in existence. It is, therefore, very appropriate to devote a discussion to this latter topic.

There are two immediately evident hallmarks of the situation regarding scientific and technological manpower in developing countries, both related to the very small size of such manpower. First, because of the scarcity of good scientists and technologists, there is an extraordinarily urgent need not to waste this precious manpower and provide for it the best possible circumstances for work. Second, because of the many tasks to be attended to and the few people who are prepared to handle them, there is enormous pressure on scientists and technologists in developing countries to get involved in a large number of different tasks and activities simultaneously. These two factors often work against each other: The pressure to get involved in many activities results in the person spreading himself so thin that none of the functions are performed satisfactorily, resulting in effective under-utilization and also in a lowering of morale. More about this later.

A second consequence of science and technology being in a nascent state in developing countries is that when it comes to creating the circumstances under which manpower is well utilized, there is a shortage of appropriate manpower to manage the situation. It is generally agreed on that a necessary (though not sufficient) condition for becoming a good science manager is some past personal experience with and involvement in scientific and technological activities on the research and development level. In a country in which there are only a few scientists and technologists who have had the opportunity to do research and development work themselves, such creative science managers may be difficult to find. In their absence, however, the evolution of science and technology will be slow, which in turn will delay the emergency of good science managers. This is one of the many vicious circles[2] that occur in the development of science and technology in developing countries. They are dissolved only gradually, by successive iterations.

These general introductory remarks allow us now to discuss some of the more specific issues in connection with the mobilization and utilization of scientific and technological manpower in the developing countries. The outline of the discussion will be as follows. In section II I list some of the problems in finding indicators through which we can assess whether in fact manpower is utilized well or not. Then, in section III, a review of the aims of science and technology is given, in terms of which utilization is judged, and some questions of quality and of entry into the science and technology profession are covered. In section IV the reader will find an elaboration of the point that there are often too many conflicting demands on the scientists. Section V is devoted to supporting services which can enhance the efficiency of utilization. In a similar vein, section VI contains some elements of a conducive environment for scientific and technological work. Section VII turns to the very crucial question of how one can ensure the use of scientific and technological research and development results in the productive sector of the economy as well as in the other realms of the country's life. This is very often a notable bottleneck in the effective functioning of science and technology in a developing country. Section VIII dicusses interaction with abroad, and particularly the utilization of scientists and technologists with an origin in the country but living abroad. Then, in section IX, the discussion turns to what formal manpower planning and control can and cannot do with regard to the mobilization and utilization of scientific and technological manpower. All this leads up to the "bottom line", section X, which contains suggestions, recommendations, and proposals for steps that can enhance the mobilization and utilization of such manpower. Emphasis is placed on simple but effective measures that can be implemented in a decentralized way, even by individual science managers and communities without massive financial and organizational resources, since such measures are most likely to be actually incorporated into practice in the near future.

II. THE ASSESSMENT OF THE DEGREE OF UTILIZATION

I will call an imperfect utilization of scientific and technological manpower "waste", even though this term is, in everyday parlance, ambiguous and sometimes used in tendencious ways. In discussing the assessment of the waste in manpower, it is useful to differentiate between two types.

The first kind of waste is independent of opinions, values, or policies. It would generally be agreed on as being waste. I recall, for example, a quite high level scientist in a developing country who was forced to spend about an hour every day signing expenditure forms submitted by others and covering any sum, no matter how tiny. Since he had hundreds of these to sign, he had of course no way to ascertain the worth or even genuine nature of these forms, so that his signature served no purpose at all. Such a situation is a waste of scientific manpower by anybody's judgement — and yet it can occur.

In the above example the waste is quite avoidable. In other situations, the waste, although universally agreed on as being waste, is unavoidable because of the lack of an appropriate scientific and technological infrastructure in the country at that particular stage of development. For example, using a well trained scientist as an electronic technician to solder, hour after hour, circuits while the work could be done just as well (or better) by a much less educated technician (who is, however, not available), or when the whole circuit could be bought premanufactured (if hard currency were available but is not), is certainly a waste by anybody's definition, and yet may be unavoidable in that country at that particular time.

The second type of waste is not so universally tagged as such but may depend on what opinion, value system, policy frame, and set of aims one has to judge by. Let me, as examples, list a few board issues of science. development which definitely influence whether we call a certain deployment of manpower a waste or not.

The first dichotomy is that of short term aims **versus** long term objectives. Developing countries face many material problems that are known to be solved by science and technology and these short term objectives often play a prominent role. On the other hand, without paying attention to longer term problems also, the crises of short term problems will persist and continue forever: As soon as one short term problem is solved, the next looms ahead without the country having had any preparations to attack it. Rice production in Thailand was a most important concern in, say, 1960, but some effort dedicated to developing expertise in the then very esoteric field of molecular biology would have resulted, in 1985, in Thailand being able to participate fully in the biotechnology revolution. Indeed, the short term **versus** long term dichotomy is often parallel with the dichotomy of immediate mission-focused research and development **versus** infrastructure building. Those not sensitive to the second but very intent on the first would consider scientific manpower involved in the strengthening of the infrastructure a waste. Conversely, those who believe that the primary aim of scientific activity in a developing country in the initial stages should be long term preparation for the future would consider scientists devoted to immediate scientific-technological problems a waste.

Let us now assume for a moment that such differences of opinions have been settled, and there is some consensus on what type of activities scientists should be involved in. Even then, we are faced with the formidable problem of finding indicators of output for those activities by which we can tell if the utilization of manpower was good or not.

This is not the place to deal with the general problem of output indicators of science and technology other than mention some of the conceptual and practical issues that arise. One has to decide whether to try to gauge activity, productivity, or progress. Then, one has to delimit the interacting system of science an technology that one wishes to assess, an find indicators for the various links. A discussion of this problem area was given relatively recently in a different context[3].

One possible way to gauge certain aspects of the utilization of scientific manpower is to determine the productivity distribution of scientists in the country in terms of scientific publications. For such a distribution in scientifically advanced countries the so-called Lotka law[4] holds. There is no accurate empirical information on what the corresponding situation is in developing countries, though some fragmentary studies have been carried out[5]. Since presumably any differences between such distributions between advanced and developing countries is due to differences in the environment in which science is practiced, such studies would provide very valuable information on the mobilization and utilization of scientific manpower in the developing countries. Similar indicator distributions could also be established for other functions of science.

Finally, it is important to note that practicing science is a very low-efficiency human undertaking in general, and hence in judging the utilization of manpower by some absolute standards, one must be realistic. This is particularly important when one is engaged in long-term manpower planning, as some countries like to be. Studies in the science of science in the last 20 years have given us much concrete information on this low efficiency of the practice of science. Of the many students entering educational institutions in the natural sciences, only a small fraction becomes productive in science at all. Of all those that publish scientific research, an overwhelming fraction does so only once a twice in a lifetime. Scientists with, say, a hundred publications in a lifetime are relatively speaking very rare. All this holds for any scientific community, even for those which would be considered very well utilized. Since the distribution of the productivity of scientists is so skew, and since a relatively small percentage of all scientists and technological manpower, to place great emphasis on the full utilization of those very few exceptionally able, talented, and productive scientists the country may have. If they become discouraged or emigrate, the seeds of creative scientific activities may go with them.

III. INITIAL STEPS

Under this heading I would like to discuss three factors which are essential in full utilization of scientific and technological manpower.

First, it is important to keep in mind the broad variety of aims of science and technology in a country. These can be summarized under three headings[6]

1. Science and technology are the bases of the material development of a country
2. Science and technology are paramount aspirations of mankind in the 20th century and hence contribute to national consciousness, national pride, national morale
3. Science and technology constitute a major influence on Man's view of the world and of his role in it.

It is difficult to separate these three aims of science and technology in that they must be nurtured simultaneously in order for any of them to florish. The kind of contribution a scientist or a technologist can make to these three objectives are usually very different. For example, the third objective involves carrying science as a way of thinking (as a world-view) to the whole population of the country, perhaps through public communication media. Efforts devoted to that are of a very different nature than those involved in doing scientific research in connection with a specific technological problem. It is important, right from the start, to spell out explicitly what expectations a group of scientists, a scientific organization, or a country as a whole has with regard to these different aims of science and what the relative weights are in assessing the performance of a given scientist. Such systems should be geared to the particular talents particular scientists may have in different directions. The pluralism of talents within a scientific community can be best utilized by offering a pluralism of opportunities and careers. Such a flexible system is very seldom encountered in developing countries, where the aims of science and the criteria for contributions tend to be defined in a much more formal and mechanistic way. This is, for example, reflected in documents stating criteria for promotion of science faculty members in universities.

The second point I want to dwell on in connection with the utilization of scientific manpower is the importance of attracting talent into the scientific manpool. If we look at the problem of utilization at its most fundamental level, we can say that this utilization in most developing countries is atrociously low. Of the undoubtedly large number of potentially talented people in the population who would make excellent scientists or technologists, only a tiny fraction is in fact attracted to entering the profession. Here we are talking about career choices, and the fact that a scientist must begin his commitment to science at an early age, while still in school. Since being a scientist is not a traditional and common occupation for a person in a developing country, and since the social fabric of many developing countries gives parents and other older people a large or even predominant say in the career choices of young people, science as a profession has, right from the start, a tremendous handicap in attracting people. Medicine, law, business, and even engineering have a large advantage in this respect over science. This must be compensated for by an enhanced, individual effort to search out and encourage young students to come to science. Although this is done in a few countries, most of the efforts I see involve only impersonal, finance incentives: scholarships or other support of a material kind. This is not enough. The young student, who "secretly" loves science, must be given an opportunity to develop a personal relationship with a scientist from which the student will gather inspiration and strength to decide in favor of science in spite of parental or social pressure.

The third point I want to make in this section is the perennial issue of quality **versus** quantity. The classic writing on this subject, to my mind, remains the most perceptive, educational, and entertaining article by the late Jorge Sabato[7], who described what happens to science and technology in a country in which policy stresses quantity and is oblivious of quality. This is relevant to our discussion in that in the utilization of manpower activities must be weighted by the quality of the personnel considered. The crucial element is to identify the high quality part of the manpower pool and to utilize them fully. Whether the low-quality tail of the productivity distribution in the scientific and technological community is well utilized or not is, in some sense, unimportant, since that tail will not contribute much, no matter what.

IV. MULTIDIMENSIONAL DEMANDS

In this section I want to elaborate on the illustrate the point made earlier that the demands on a scientist or technologist in a developing country are very numerous, sometimes mutually conflicting, and altogether too large and numerous in relation to the small scientific and technological community in the country. To illustrate this point, I want to discuss four different conflicts which arise in the demands and which complicate the full utilization of manpower.

The first one is the already mentioned conflict between infrastructure building **versus** solving immediate problems. In practice, the first of these is usually neglected, and yet some activity in it is a prerequisite for doing the other. For research on any problem, laboratory facilities must be procured, auxiliary personnel must be attracted and trained, equipment and supply must be made available, communication facilities with other scientists must be built up, contact with potential users of the research results must be made, etc. None of these elements can be counted on as being automatically present, or having been established by predecessors. For this reason the efficiency of research and development activities in developing countries will most certainly be automatically lower than the efficiency of the same activity, other things being equal, performed in a country where the scientific infrastructure is well developed.

The second example of conflicting demands is between the scientist as a provider of knowledge through his personal research and the scientist as a provider of information by being a funnel of information from the worldwide scientific community. In most frames of science policy one encounters the tacit assumption that a scientist contributes to the solution of local problems through the results of his personal research. Thus the expectations of science management agencies will focus exclusively on that aspect of the work of the scientist. In reality, the most important contribution of a scientist to the solution of local problems is through his ability to funnel worldwide scientific information to the technologists and others directly involved in problem solving. This facet of the scientist's utility is, however, very rarely tapped. Thus, in effect, most scientists in the developing countries are extremely poorly utilized, since this most important aspect of their potential contribution is left completely unexploited.

A third example for the conflicting demands arises from the mis-utilization of personnel in the scientist-engineer-technician chain. Due to a scarcity of technicians, engineers in developing countries are often forced to do the work of a technician. This then leaves some of their work as engineers undone, and this gap is then filled by scientists who are impelled to do work that would be more properly that of an engineer. The result is that everybody involved performs a task that he is not ideally suited for, and as a result has a decreased morale and a lessened enthusiasm for the work. Thus everybody is under-utilized or mis-utilized.

Finally, as the fourth example, let me turn to the conflict between doing science and doing science adminitstration. This is a particularly complex issue with no clear "rights" and "wrongs". As said earlier, it is imperative that managers of science have some personal experience in scientific research and some successes in such activities. This will enable them to exercise better judgement, to have self-respect and self-confidence when dealing with active research scientists, and to have a positive image in the eyes of the scientist they deal with. Thus science managers should rise from the ranks of the scientists of the country.

On the other hand, good scientists are needed to pursue scientific research, and hence they must not be diverted from research into other channels, because thay may mean the death of science in the country. At the same time, the rapidly mushrooming science management bureaucracies around the world have a voracious appetite for consuming worthy scientists and losing them forever for science. In principle, it is possible for a scientist to continue research part time and tend to management functions part time also, but in practice the examples for this are very few. Going from science to adminis- tration is a one-way street. It takes great energy, determination, and some resources to resuscitate somebody's capability for doing scientific research once he dropped out for even just a few years. The path of least resistance, strengthened by the reward system, strongly dictates leaving science forever.

V. SUPPORTING SERVICES

The utilization of scientists and technologists is greatly influenced by the level of supporting services they are given access to. Thus this aspect of the problem deserves special discussion.

The most important element in the supporting services is communication. It has been said many times and in many places that since science is a more universal, more objective, more collective, and more cumulative human undertaking than anything else, communi- cation among scientists is an absolute necessity for the productive pursuit of scientific research. The same can be said also for technologists, even though their **means** of com- municating might be somewhat different from that in science.

Problems of communication exist both within a country and internationally. The improvement of the latter is a task for which the scientific communities in all countries need to cooperate. The improvement of communication among scientists within the country is, however, a task that is within the powers of each country separately.

The first task is to realize and to create awareness about the degree to which scientists in developing countries is isolated even from their own fellow scientists in their own country. Scientific visits and lectures by scientists from other institutions are rare, even though domestic transportation is usually inexpensive, and even though often several institutions are located in the same city or nearby. Collaborative research between two scientists, each located in a different institution but in the same country, is made difficult by bureaucratic obstacles. Scientific societies are often weak and not very active, and in any case do not provide a good forum for scientific exchange. Very little of this is a matter of money. It is more a matter of attitude. Yet, in terms of utilization, the sum could be much more than the aggregate of the individual parts, and therefore we can say that the utilization of scientists is weak as long as they work just by themselves, in isolation.

As an example of an aid to such communication, let me mention the telephone. Studies have shown that in the scientifically advanced countries the telephone is the most important single mode of communication among scientists, even if one includes journals, computerized data bases, scientific meetings, and all the rest. Thus the often poor and rudimentary telephone system that exists in developing countries, plus the exorbitant cost of international calls compared to local financial resources, place a serious handicap on the communication tools a scientist in a developing country has access to. This one single improvement of making the local telephone system functional and making international calls financially accessible would improve the utilization of scientific manpower in developing countries enormously. The effect would not only be material but also psychological, since the feeling of isolation is a very debilitating one of an active scientist.

The second large group of supporting services pertain to equipment and repair. I want to dwell here particularly on the problems of repair and spare parts. I have encountered many cases of researchers in developing countries, immobilized for months in want of a simple spare part costing $5, or in want of a technician who can repair a piece of equipment. This is waste of scientific manpower on a massive scale. To remedy the situation involves action from all sides: better spare part and service facilities by companies selling scientific equipment to developing countries, more functional programs by bi-national, regional, and international scientific assistance agencies, as well as purely local measures. A national stockroom of spare parts would not cost more *in toto* than the present system of piecemeal purchase of parts but would cut down on the delay time. Roving mechanics, shared among institutions or laboratories, would help combining the resources. Foresight in training technicians in upcoming technological areas instead of waiting until it is too late would also help. Many spare parts can be fabricated or manufactured locally. Somehow the ingenuity exhibited in most developing countries in fixing old imported autombiles in most successful ways needs to be transferred also to the more esoteric but not necessarily more complicated field of scientific instruments.

VI. CONDUCIVE ENVIRONMENT

In this section I want to concentrate not on the material environment (which was discussed in several of the preceding sections) but on the psychological environment in which the scientist or technologist in a developing country finds himself. In particular, I want to discuss two aspects of this psychological environment: bureaucracy and politics.

Pursuing science in a developing country is a demanding task in any case, as it is evident from the discussion in the previous sections. It is therefore essential for a scientist or technologist to feel that he receives as much support from his own country institution, or community as he could reasonably expect. Even if the motivation of a scientist to do science is to a large extent internal, external factors such as communal support play an important role. The direct contact a scientist makes with his community, his institution, his country is usually through some bureaucracy. It is therefore very essential for a scientist to have the feeling that the bureaucrat he has to deal with is "with him" and not "against him". In practice, this means that the bureaucrat needs to find ways of responding to the request of the scientist and not to enumerate rules which prohibit him from taking the request seriously. Many of my colleagues in developing countries have told me instances, both pro and con, of dealing with bureaucrats which support this general picture. This availability of bureaucratic support is truly not a matter of finances or resources, but a matter of understanding, attitude, respect, and purposefulness. Even if in the cooperation of the scientist and the bureaucrat in finding a solution to the problem the scientist ends up without a solution in that particular case, the nature of the interaction will encourage or discourage the scientist in his future endeavors.

As to politics, for an optimal utilization of scientific manpower politics should be kept apart from science. A vivid example of the consequences of not doing so may be illustrated in various Latin American countries. In that region of the world social moves among intellectuals tend toward considering being apolitical as immoral. As a result, university scientists and sometimes even those in the laboratory are or feel compelled to get deeply involved in partisan politics, with the result that when the government changes (which often happens), half the scientific community goes into exile, to return later when the other half leaves. Under such circumstances it is very difficult to maintain productive scientific activities and the utilization of the scientific or technological sinks to a very low level of efficiency. Whether such political activities by scientists are more or less important than their contribution to science may be a matter of opinion, but that their involvement in politics is deeply detrimental to science is a demonstrable fact.

In the above example the choice of becoming involved in politics was made primarily by the scientists themselves. In other situations we might encounter governments which require scientists in the country to be actively political, of course on the government's side. Under such circumstances also, scientific work suffers, not only because time, energy, and resources are drained away from science, but also because political criteria replace scientific ones in the choice of personnel, in the decisions regarding scientific activities, and sometimes even with regard to the content of science that is permitted. Under such circumstances, the utilization of scientific manpower becomes close to zero.

A particularly pernicious consequence of a prolonged state of chaos in that regard is that the next generation of scientists, who would have received their education, training, apprenticeship, and start in career, is then also wiped out. Thus the lack of utilization of manpower in such a case extends over a considerably longer period of time than the duration of the chaos. Some very vivid recent examples are quite obvious and need not be mentioned by name.

VII. UTILIZATION OF RESULTS

Because science is a collective, universal, objective, and cumulative activity, and because the aims of science pertain to the links of science with other domains of human activity, the results of scientific research, just as the results of technological development work, need to be communicated to its users. Without that, the research or development will remain useless, and hence the utilization of the manpower that created it will be zero.

Many aspects of such a transfer of research results could be discussed, but I want to limit this discussion to only one of these, namely the communication of research and development results to potential users in the economic sector of the country. I chose this to discuss because it is perhaps the weakest of the links in most developing countries and hence contributes most to the poor utilization of scientific and technological manpower.

Since I focus on the utilization of the results of a given effort in research and development activity by specific users in the same country and shortly after the research and development work was completed, I am considering an instance of **applied** scientific research and the subsequent technological development work. By applied I mean that the primary motivation for carrying out that particular scientific research was in the hope that the results could be used soon thereafter in a particular technological development activity. This development work, in turn, is assumed to be taken up with the intention of the results being used soon thereafter by a particular segment of potential users among the economic forces in the country.

The most common error that is often committed is to plan such research and development activity "from top down". I have been asked repeatedly in various developing countries, by heads of national research councils and similar organs, to specify what areas the country (or, specifically, certain institutions or laboratories of the country) should perform research in. The underlying assumption is that once a good area is chosen and good research and development are performed, the results wlll somehow be welcome and utilized by the agricultural or industrial entrepreneurs, operators, and workers.

Unfortunately in reality this almost never happens. As to anything, there are exceptions to this statement also, but it is certainly true that even in the case of the exceptions very arduous work needed to be done after all research and development was completed to convince the agricultural or manufacturing sectors to get interested in these results and utilize them. Neither is such a top-down approach to the choice of applied research topics practiced in the scientifically developed countries. The model for doing it nevertheless must have emerged from the mind of somebody inclined toward logical thinking but having very little practical experience in science and technology management.

Instead, in order to be able to utilize the results of research and development effectively, the potential users must be involved from the very start in the choices of research areas. The potential users must feel that they have a stake in the outcome of the activity since they were given a voice in defining the activity. In reality, it might very well happen that the potential user (says, an industrial entrepreneur) knows very little about the area of science and technology in which the applied research and technological development work will take place, and may not have a very clear conception of how that area may pertain to his own manufacturing activities. Such a vagueness in these initial stages does not matter. What does matter is the opportunity for cooperation, for communication, for joint undertaking. It might even happen that the research and development will be aimed at something which does not relate to any of the already existing manufacturing activities in the country, but aims realistically at setting up new industries. Even then, initial contact with the potential entrepreneurs and the drawing them into the venture from the start is essential.

Time and time again, extensive resources in developing countries were poured into perhaps in principle applicable research and perhaps in principle usable development work which, nevertheless, remained as reports in filing cabinets or as prototypes in museums, and failed to contribute to the economic development of the country. Sometimes this was due to the low quality of the work performed. But it seems to many observers that much more often the situation coud have been saved or slightly redirected if the potential users had been drawn into the activities from the start. This "trick" of planning applied research and technological development work does not cost any additional money. It rests simply on the recognition of the nature of the connection between science, technology, and the productive sector, and on the recognition of certain character traits of human nature. And yet, the "trick" may easily turn the un-utilization of manpower into a very effective utilization.

VIII. INTERACTION WITH ABROAD

Since science is highly international, and technology is in many ways equally so, the international dimension of science management issues is a very strong and extensive one. In the present context, however, I want to discuss only two specific problem areas directly relevant to the utilization of scientific and technological manpower. They are a) the brain drain, and b) the reattraction of scientific and technological manpower originally from the country but now living abroad.

The brain drain, that is, the emigration of highly trained manpower, is a problem that has been much discussed. Conferences have been devoted to it and volumes have been written. Statistical data pertaining to it indicate that while numerically, in terms of the percentage of manpower, brain drain in most developing countries is not a dominant effect, there is no statistical information either to prove or to disprove the feeling many observers have that since the drain captures, preferentially the exceptionally outstanding individuals in the scientific and technological community, the effect is much more severe than the percentage figures imply.

From the point of view of measures to lessen the effects of the brain drain, I want to stress here two pivotal elements.

First, coercive measures do not work. For example, some countries ask their students to sign a bond before going abroad for advanced education. The bond specifies that should the students brain drain away, they are obliged to pay a certain sum to the home government for compensation. Such a measure is ineffective partly because the sum state in the bond is too low and so its payment is not a very major task for somebody with a good position in an advanced country. But more importantly, the measure is counterproductive because it places the obligation of the student toward his home country on a purely commercial basis. This and the signing of the bond creates deep resentments and the signer will therefore be more inclined to brain drain away. I know of no actual situation in which the bonding resulted in anything else but the return of those weak elements in the scientific and technological community who were unable to find employment abroad anyway. This is of course not to say that all good people brain drain away. On the contrary, there are many reasons why nationals will return to a home career in science and technology even if they are of high quality and could easily find employment abroad. These reasons, however, do not include being bonded.

Second, and, so to speak, presenting the other side of the coin, there are many reasons why a developing country, even with salaries low on an international scale, can successfully compete with any other country for their own scientific and technological personnel. Indeed, the **absolute** level of salaries is not an important factor at all. Salaries need to be good in comparison to other comparable professionals in the country in question. Beyond that, however, factors other than financial play a much more important role. Good scientists and technologists feel an internal urge to fulfill themselves in their work, to utilize their talents and potentials, and they want to have sympathetic support from their environment to achieve this goal. They want to feel that, within the realistic limitations that exist in developing countries, they are given all the opportunities to do their work optimally. In other words, their aim of optimal efforts and the country's aim of full utilization completely overlap. The discussion in the previous sections contained many areas in which such support is necessary, and I also stressed the importance of the psychological relationship between bureaucrats and scientists in the determination of the level of morale within the scientific and technological community. These are the directions in which efforts to prevent the brain drain should primarily proceed.

The second issue I want to discuss in this section is the utilization of scientific and technological manpower originally from the country but now living abroad. There may have been many reasons why these individuals left originally: some political, some due to an apparent oversupply of manpower at home, some due to the then rudimentary state of science and technology in the country. Even if these individuals have lived abroad for a number of years, however, it is a mistake to give up on them. There are shining examples of countries which managed to reattract their nationals from abroad even after many years to occupy important positions in the scientific and technological community at home. There are several reasons working in the direction of success in such an endeavour. When the children reach school age, the very real decision needs to be made as to whether they will grow up in the language and tradition of the home country or whether they will separate from their parents' tradition and become a product of the different culture of the host country. Also, many prefer to be in a leading position in a relatively small arena of science and technology, where they have much leeway and support to shape things in their own image, rather than being a solid but rank-and-file member of a huge scientific or technological community in an advanced country. Political systems may have changed since the scientist's original departure from his home country, thus eliminating certain negative factors.

To encourage such a return of scientific and technological manpower and the consequent fuller utilization of scientific and technological manpower, it is very helpful to keep steady contact with such scientists and technologists during their years abroad. This can be (and in some instances has been) done, for example, through the cultural or scientific attaches of the developing countries stationed in the advanced countries, or through contact initiated by home institutions and laboratories. In some cases joint research proposals were constructed between scientists home and scientists from the country living abroad, or the latter were invited to lecture at workshops or seminars or attend conferences organized in the home country. The resources for such activities need not even come entirely from the home country. For example, programs of the U.S.

National Science Foundation have repeatedly helped in the support of such programs, and many recipients of the Fulbright grants were also former nationals of the country to which they went under the grant. This dimension is of great essence in the full utilization of scientific and technological manpower, and yet, very little attention has been paid to it by the developing countries.

IX. FORMAL MANPOWER PLANNING AND CONTROL

Most developing countries have some centralized science policy organs, and most of these are engaged, among other activities, in manpower planning and control. In this last section before I turn to the specific suggestions and recommendations, I want to discuss what role such agencies can play in the above discussed improvements in the mobilization and utilization of scientific and technological manpower.

It is evident from the above discussion that improvements in the utilization of scientific and technological manpower in the developing countries will not come from one magic move which will suddenly bring about great results. Instead, the picture that has emerged is that the need is for a large array of detailed measures in almost all realms of science management, and that the implementation of such an array of measures must be done in a very decentralized and personalized manner. Thus, in order to bring about progress, the need is to influence the thinking, attitude, and activity of scores or hundreds of people connected with science, technology, and their management.

The fact that the need is for a decentralized, grass root type of action does not mean that the centralized governmental science agencies can play no role. On the contrary, they can have a very positive influence on such a development. The task, however, is not an easy one. The mere preparation of manpower plans and policy proposals constitutes only a small fraction of what needs to be done. The rest hinges on the indirect, decentralized, sensitive and non-formalistic implementation of what needs to be done. Unfortunately this is exactly the type of activity in which large centralized organizations have proven themselves the weakest. This is illustrated, for example, by juxtaposing the many thousands of recommendations, resolutions, plans, and proposals created in the last 40 years of national science agencies, international agencies and conferences, regional bodies, and other similar organs with the tiny fraction of that which has in fact been implemented and appears as an organic part of the science and technology in the developing countries. For this reason it is essential that such policy bodies place much less emphasis in the future on planning, and much more emphasis on effect methods to implement and to experiment with such implementation.

Since, however, planning is a much practiced activity in such agencies, one pertinent remark on planning might also be useful. There are two extremes of manpower planning. One is no planning at all. This may result in having no manpower available at all in a particular field when the need arises, no provisions whatever made for providing supporting services for scientists and technologists, etc. In other words, a completely laissez-faire approach to the building of a brand new type of activity in a country is not likely to produce good results.

The other extreme is rigid and long-term centralized planning, in which number of people needed are projected years ahead in great detail, in which the assignment of each individual of the scientific and technological community is fixed and allocated far ahead, and in which scientific and technological tasks are specified in great detail for many years to come. Although such a system may appear to utilize manpower well by actually placing every individual in some position, in actual fact it is extremely wasteful of manpower as well as scientific and technological opportunities. What areas of research and development will become important and urgent cannot be foretold very well far ahead, except in very general outline[8]. The performance and potential of an individual is also unknown ahead of time. As said earlier, in order to have a sufficient number of good scientists and technologists available in a specified time into the future one must educate a much larger number and be reconciled with the seeming inefficiency of the process. Mobility among institutions and even by changing fields is one of the great morale booster of scientists and technologists, and this must be preserved in any system claiming to utilize manpower well. This is, however, a wellnigh impossible task within a rigid scheme of planning.

The skill and wisdom of a centralized science management agency is therefore tested by the requirement that it finds the right middle road between these two extremes. To what extent we can plan science and technology remains[9] an open question, and creative research and experimentation with regard to it should be part of the duty of a science management agency.

These remarks conclude the brief survey of various aspects of the mobilization and utilization of scientific and technological manpower in developing countries. Throughout the discussion various measures suggested themselves as helpful in improving such a utilization. The last section, therefore, will collect, point by point, some of these specific steps.

X. SUGGESTIONS, RECOMMENDATIONS, PROPOSALS

As mentioned earlier, all recommendations made in this section are of the type that can be implemented by a single developing country within its own confines and without massive organization or financial resources. There are also measures to improve the utilization of scientific and technological manpower in the developing countries which require the cooperation of the advanced countries and may require large scale organization and resources. These may be just as important as those listed below, but their implementation is much more difficult and hence unlikely to occur in the near future. In contrast, many of the improvements listed below can be implemented little by little, in a decentralized way, often without additional resources. They are therefore most suitable for making a start.

1.   It is essential to have an openly articulated, unambiguous statement of the aims and directions of science and technology in a country, based on the consensus of all those involved in the science and technology process. In such a statement the various dimensions of science and technology and the conflicting dichotomies discussed in earlier sections should be included. This lays the foundations for specifying and assessing the degree to which scientific and technological manpower is utilized.

2.   The system described in (1) should be flexible so as to allow individual scientist to find and pursue those activities within it in which they are best, for which they have the greatest motivation. Morale depends to a high extent on being allowed to do what you like to do within the broad array of possibilities and aims of the system as a whole.

3.   Particularly in connection with such statements pertaining to **applied** scientific research and technological development work, it is crucial that the potential users of the results of this work be included from the very start in the process of defining these aims and priorities.

4.   Since new manpower development needs a lead time of about a decade, all plans for activities in the near future must be based on the already available manpower. Such plans, therefore, should start by asking the questions: Whom do we have to do the work?

5.   In allocating, mobilizing, utilizing manpower and in assessing the degree of utilization, emphasis must be placed on quality. This requires having a substantive and functional assessment system of scientific and technological performance. The formulation and articulation of such an assessment system should have high priority in any country. Much discord can be avoided if scientists and technologists know the criteria that will be used in assessing their performance.

6.   The assessment system mentioned in (5) implies having functional indicators of scientific performance. In this connection more research and experimentation is needed to find such indicators. For example, it would be interesting to investigate whether the productivity distribution among scientists, as measured in the number of publications over a lifetime, is different for communities in developing countries from what it is in the scientifically well developed countries, and if so, whether the differences can be correlated with factors in the utilization and mobilization of manpower.

7.   Special and personalized efforts need to be made to attract to the professions of scientist and engineer talented and motivated young students, beyond just offering scholarships and other financial inducements to such students.

8.   Scientists represent the funnel through which scientific information from around the world enters and get processed in the country. This capability of scientists should be better recognized and utilized, since most of the utilization of science in the country has to rely on science produced outside the country.

9.   Special and personalized efforts need to be made to reattract scientists and technologists originally from the country but now living abroad. Even if such personnel continues to maintain permanent residence abroad, it can be very helpful in various cooperative activities with scientists in the home country.

10.   The communication network among scientists and technologists within a given country need to be strengthened through mutual visits, active scientific societies, exchange arrangements, and other methods.

11.   Facilities for the repair of scientific equipment in the country and for spare parts for such equipment need to be arranged so as to eliminate long delays. Roving technicians, lending stockrooms, and other mechanisms may be ways of accomplishing this.

12.   The interaction of the scientific and technological community with the bureaucratic structure of science management organizations need to be made smooth and one that transmits sympathy and encouragement to the scientists and engineers.

13.   The management of science and technology need to be strictly separated from politics so as to ensure continuity in the deployment of scientific and technological manpower and so as to prevent the entrance of political criteria in the utilization and assessment of such manpower.

The Mobilization and Utilization of Scientific and Technological   15
Human Resources in Developing Countries.

References and footnotes

1. There is no concise English term for human resources other than manpower. The word "man" is used here in the generic sense (corresponding, for example, to the German "Mensch") and hence includes human beings of both sexes.

2. For an elaboration of this phenomenon, see for example M. J. Moravcsik, "Science Development − A Network of Vicious Circles", APPROTECH 5, 70 (1982).

3. The reference here is to the project organized by the United Nations Centre for Science and Technology for Development (UNCSTD). United Nations, New York, and the study document for that project, Michael J. Moravcsik, "An Assessment Scheme for Science and Technology for Comprehensive Development", UNCSTD, GRAZ/P/3 (1984), copies of which are available from Dr. M. Anandakrishnan at UNCSTD, Room 1040, 1 UN Plaza, New York 10017.

4. The original reference for this is A. J. Lotka, "The Frequency Distribution of Scientific Productivity", J. Washington Acad. Sci. 16, 317 (1926, No. 12). The law, however, became widely known among scientometricians through the writings of the late Derek de Solla Price, for example his "A General Theory of Bibliometric and Other Culmulative Advantage Processes", J. Amer. Soc. Inform. Sci. 27, 292 (1976, No. 5).

5. For example see P. G. Blasco, "La Produccion Cientifica Espanola de 1965 a 1970. Un Estudio Comparado.", Revist. Mex. de Soc. 37, No. 1 (1975); T. Saravic, "Evaluation and Potential Use of the Data Bank at the Brazilian Institute of Bibliography and Documentation (IBBD)". UNESCO report 3055/RMO.RD/DBA, Paris, 1974; W. O. Aiyepeku, "The Productivity of Geographical Authors − A Case Study from Nigeria", of Document. 32, 105 (1976, No. 2).

6. For a more extensive discussion of the pluralistic aims of science, see for example Michael J. Moravcsik, *How to Grow Science*, Universe Books, N.Y., 1980, Chapter 2.

7. Jorge Sabato, "Quantity versus Quality in Scientific Research (1): The Special Case of Developing Countries", Impact 20, 183 (1970). With the recent death of Sabato the developing countries lost one of their most brilliant, incisive, articulate, and active personalities in science development. I would like to dedicate this analysis to his memory.

8. For an excellent recent study on the difficulties of forecasting in science, see John Irvine and Ben Martin, *Foresight in Science − Picking the Winners*, Francis Pinter, London, 1984.

9. For an analysis of some of the considerations regarding planning in science, see Michael J. Moravcsik, "Can We Plan Science − Semantics and Pitfalls", Bull. Sci. Techn. and Soc. 4, 361 (1984).

D. Science and Technology

Reprinted by permission of *Approtech* 2 (1979) 1.

Linking Science With Technology in Developing Countries

by Michael J. Moravcsik

In the developoment literature, as well as in political speeches and national development plans, the seemingly inseparable twin-phase of "science-and-technology" appears profusely. Indeed, the impression is given that these two activities are strongly and smoothly interwoven. In reality, this is not at all so. For one thing, the function of science in a developing country goes far beyond just being a sibling to technology. Furthermore, the link between science and technology in developing countries is not at all smooth. On the contrary, the weakness of the link and the many obstacles that separate science from technology in these countries are some of the most disconcerting elements in the complex problem of development. This article will make an attempt to articulate reasons why the link is so weak, in the hope that such an analysis will facilitate eventual remedy of the situation.

In order to describe the situation fully, it is advisable to sketch briefly the history of the link between science and technology during the last 300 years or so.

Until about 150 years ago, technology and science were hardly linked at all. Science at that time was rather undeveloped, and was preoccupied with the study of phenomena that were directly observable through our senses of hearing, seeing, touching, smelling, etc. These phenomena were characterized by scales commensurable with those of our own body: lengths between, say, 1 mm to 1 km, time periods between, say, 1/1000 second and 100 years, temperatures between, say, -40°C and 300°C, etc.

At the same time, technology also concentrated on the utilization of phenomena that could be directly observed by human beings. Mechanical devices, simple machines, non-chemical weapons can serve as illustrations. Such technology developed through trial-and error, through tinkering with everyday experience, and hence it did not need to rely on any systematic knowledge of natural phenomena

Michael J. Moravcsik is on the faculty of the University of Oregon and an associate of the Institute of Theoretical Science at that University.

which, in any case, science at that time was ill-prepared to supply. Thus at that time, technology was almost completely empirical.

The situation began to change about 150 years ago. By that time the scientific phenomena directly perceptible to our senses were fairly well explained, and science went on to explore other domains of phenomena, which could not be directly observed by us, and the scales of which were different from those "natural" to our body. Phenomena of heat, electricity, and later atomic and nuclear reactions fall into this category.

Parallel to this, technology, having also fairly well exhausted the utilization of direct everyday experience, turned to the use of phenomena more distant to direct human experience. The utilization of electricity and electromagnetic waves in particular are good examples. In doing so, the old trial-and-error method had to be deemphasized in favor of an approach based on scientific knowledge. There is no way of successfully tinkering with, say, the potential ingredients of a transistor radio. A merely random trial-and-error method of assembling such a device from raw components would remain unsuccessful even after a million years of hard work. Thus technology became science-based.

I do not mean to imply that empirical manipulations were thereby altogether excluded from technology. On the contrary, they have remained very important in the final phases of invention when the general approach has been well delineated by scientific knowledge, but when it would be too tedious (or practically impossible) to resolve the remaining details on the basis of scientific calculations. Nevertheless, without the guiding hand of scientific knowledge in suggesting the overall outlines of the technological process, trial-and-error nowadays would not get far.

We can thus speak of two kinds of technology, which I will call empirical and science-based. Nowadays most of the significant technological innovations are of the second kind. Yet, in a developing country, because of the relatively recent origin of scientific and technological evolution, both kinds of technology may play an important role.

The main problems, however, are with science-based technology. As to empirical technology, this has existed in most countries for some time, although it might be more flourishing in some countries than in others. It may also be different from country to country, depending on geographical, economic, and other factors. On the whole, however, one can rely on the ability of local people everywhere to make simple inventions of an empirical kind to help them in their local problems.

This point was completely missed, for example, in a little book which was popular a decade or two ago, called **The Ugly American**. Intended as a critique of international assistance efforts, and written by two people who seemingly never set foot in a developing country, it contained a story in which the American hero visits a remote village in a backward country and teaches the local village women to use a broom with a long handle instead of the traditional short-handled broom. The implication of the story was that what developing countries needed was Americans teaching them the simplest uses of empirical technology which, the story implied, they were too dumb to invent themselves.

Absolute nonsense! There is no reason why empirical technology would not be available to any group of people around the world, and in fact it is. The problem of technological development is, instead, in the area of science-based technology, to which I will now turn.

The influence of science in science-based technology can be two-fold. Some of it is the use of well-established science, that is, scientific knowledge that is already systematically in standard textbooks and university curricula, and which is part of the background of any doctoral-level scientist. The second kind of influence of science on science-based technology is through frontline science, through the scientific discoveries just made, which are hardly even available in journals, and which are shared, at that time, mainly by those scientists who carry on research in that or neighboring fields.

These two types of influences are, most of the time, intermixed and hard to distinguish. For one thing, science races on at a fast rate, and scientific knowledge which at a given time might appear highly esoteric and advanced turn, within 5-8 years, into standard textbook knowledge with broad technological applications already based on it. The history of transistors or or nuclear fission and of artificial isotopes illustrates this point vividly.

With this background, let us now try to pinpoint the primary defects in the scientific and technological environment of the developing countries which prevent the effective interfacing of science with technology.

The first group of such defects occurs in the education of scientists and technologists. I would like to name three crucial ones.

First, the education of scientists and engineers in a large number of developing countries is plagued by the practice of rote learning. Teaching and examinations rely heavily on memorization, and the student that is deemed to be most successful is the one who can regurgitate the professor's lectures quickly and without fault. The solving of problems is neglected, and the understanding of the material in an operational sense is ignored. Knowledge is conceived as a closed entity, already available, that needs to be absorbed by diligent committing to memory. That science is the art of asking interesting new questions and attempting to find novel answers is not revealed or practiced in such an educational system, and engineering is made out to be a matter of referring to standard tables when encountering standard problems, rather than the skill of innovation and invention.

Is it a wonder if students raised in such an atmosphere fail to become productive scientists or ingenious engineers? Is it a surprise if, when encountering novel problems and unusual challenges in the course of the development of their country, these scientists and engineers fail and falter, while desperately searching in their beloved textbooks for the answer to these new problems?

The second main defect in the education of scientists and engineers in many developing countries is premature specialization. Students of both science and engineering receive an extremely narrow education, and are asked to specialize at the end of four years of college level experience. For example, by the time a student on the Indian subcontinent receives his M.Sc. degree (which is roughly equivalent to a bachelor's degree at a strong college in the United States), he will be committed not only to a particular subfield of his discipline (e.g. to solid state physics) but even within that, he will be asked to engage in "research" in a specialty. As a result, students acquire only a minimal degree of flexibility and their ability to fit into positions that happen to be available is very meager.

Particularly damaging is this narrow specialization for activities in applied research, since there the research problems are determined to a greater extent by circumstances external to the researcher, and hence the problems tend to require broader, in fact often interdisciplinary expertise. In contrast, in basic research the problems can be posed, to a greater extent, by the researchers themselves and hence can be fitted more easily into the existing background of the person in question. It is ironic that the premature specialization and the narrow education in developing countries are promulgated by the argument that they are required by the needs of applied research. This statement is based on a complete misunderstanding of the nature of applied scientific research, as it will be evident from the later parts of our discussion.

Premature specialization also has another catastrophic consequence in the inability of the scientists and engineers thus educated to keep up with new developments in science and technology and hence continue to be functionally employable throughout their careers. Scientific knowledge doubles every ten years, and technological developments occur at least that rapidly. A specialty acquired at the time of obtaining a degree is likely to grow out of date and archaic a few years thereafter, unless the person has the ability to grow with the field and acquire new knowledge. In order to do so, one needs a broad basic education on which to build this continuing professional development. In the absence of such a basis, scientists and engineers professionally die a mere decade after leaving school and become a burden on society.

The third deficiency affecting science and technology in developing countries is motivational. It happens quite often in those countries that the choice of career for a student is determined, to a large extent, by factors external to the student. Internal motivation, the personal drive toward and love for a certain profession plays a secondary role compared to the will of parents, the attitude of society, the weight of tradition, or the edict of the commissar. Such a conservative and externally driven environment is particularly hard on professions which are relatively new in the country, because they have no image yet and hence do not have a place in the perceived hierarchy of careers. In a large number of countries the most gifted students choose medicine as their first preference, followed perhaps by law, and only then comes engineering. Science is even lower than that, and often serves as a last resort. Since both science and technology are activities in which the exceptional person contributes an immense amount more than the run-of-the-mill employee, the loss of many highly creative and internally motivated people to other careers has a very detrimental effect on the quality of science and technology in the developing countries.

Having discussed three main deficiencies in the educational system of developing countries, let me turn to defects in other areas pertaining to the link between science and technology. One of these is the perceived role of applied scientific research.

Applied scientific research is an activity which leads to new knowledge about natural phenomena, just as basic scientific research is. The difference between the two is that in applied scientific research the **motivation** is in the eventual use of that knowledge in technological work, while in basic science the emphasis is on the broadening of knowledge **per se**. But both activities result in new knowledge.

In contrast, technological development work aims at producing a new gadget, a prototype, a process, a procedure for doing something. It is aimed at producing something tangible, while applied scientific research produces intangible though very important and sometimes very specific knowledge.

The difference between these two activities is often completely ignored in developing countries. In fact, many institutions bearing the name of "Applied Research Institute for so-and-so" perform no applied research at all, only simple technological trial-and-error manipulations. As a result, the country attempts to do what, as I tried to explain at the beginning of our discussion, cannot be done, namely to make progress in science-based technology purely by the trial-and-error method. The result, not surprisingly, is dismal.

Another shortcoming of the science-and-technology chain in

developing countries is in miscasting the role of a scientist and of an engineer. In the sciences, for example, in the case of an average developing country, well over 99% of the new scientific research is performed outside the country, in other parts of the worldwide scientific community. Thus, the results of the personal research work of a scientist in a developing country is of secondary importance, from the point of view of his contribution to the technology of his country. Much more important from that point of view is his ability to serve as a source of scientific knowledge, both with respect to established science and with respect to frontline science. This is a crucial role, since the country must import virtually all science it uses, and in this importation process a key element is the country's ability to receive, assess, evaluate, select, and then utilize the imported scientific knowledge. Practically all scientific knowledge is freely available to anybody who can understand and use it. The bottleneck, therefore, is not in the availability at the source, but in arrangements for the country actually importing and absorbing such knowledge.

It follows from this that the personal research of a scientist in a developing country should be regarded primarily not from the point of view of its immediate utility to a particular technological project of the country, but instead as a tool whereby the scientist can maintain his expertise, his enthusiasm, his international connections, and his scientific communication network so he can continue to serve as a recipient and transmittor of scientific knowledge from abroad to the country in question. Practices in developing countries are faulty on both counts: scientists are frequently assessed according to the specific "relevance" of their personal research projects, and at the same time the provisions are very poor for extracting from the scientist the general expertise he has in his field.

In technology the situation is similar though probably not quite as extreme. But there, also, when it comes to the selection, assessment, importation, and purchase of patents, to the decision about licensing one or another foreign company to operate within the country, and to the choice between alternative means of achieving a certain technological task, general know-how and up-to-date expertise are much more important than the particular development work the technologist is engaged in.

Another problem area in developing countries is the contact between scientists and technologists. There is no tradition of research in industrial companies, and instead applied research (or what is called that) is carried out in huge governmental institutes which are generally isolated both from university scientists and from industrial engineers. Partly as a result, problems for research and development in those institutions are often selected on theoretical grounds and only on the basis of the interest of the people already at the institute. The resulting research or development frequently ends up being deposited only in progress reports which in turn vanish in the bowels of huge filing cabinets and seldom make contact with the productive sector.

The situation is somewhat better in agriculture simply because there is a somewhat longer tradition in agricultural research and development. But even there, problems often arise when the result of agricultural research and development has to be transmitted to the farmer in the field.

I want to return for a moment to the problem of patents. There is, nowadays, much talk about restructuring the international patent system so as to allow greater or easier access to patents by the developing countries. I support such moves, and believe that equitable arrangements can be worked out. At the same time, I do not believe that this issue is nearly as important as it is made out to be. I consider it primarily a manifestation of the fact that science and technology have a significance to the developing countries which far transcends practical utility, and this symbolic, aspirational role of science and technology is at stake in the patent controversy. Having access to patents is an external sign of being an equal member of the worldwide community of "modern" nations, and hence the issue is pressed.

The point is that as far as practical utility is concerned, free access to patents would make little difference to most developing countries. Many of them would be unable to utilize patents even if they were freely available, since the infrastructure necessary for such utilization, in terms of scientists, technologists, managers, skilled labor, organizational structure, capital, marketing, transportation, etc. does not exist. In fact, there is an enormous amount of technological information freely available to anybody who can use it, but much of it remains unused for the reasons mentioned. Clearly the countries which will benefit most from an egalitarianization of patent practices are the most advanced of the developing countries, which do have some hope of being able to compete with the industrially well-developed countries. The large majority of the developing countries, and in fact the least developed and poorest ones, are not likely to gain much. As I mentioned before, this is not to be construed as an opposition to the restructuring of patent policies. It is simply a warning that one must not expect a significant change in the position of developing countries as a result of such restructuring. The key is the country's creative and inventive infrastructure, which can produce science and technology on its own.

This also brings us to the subject of the so-called "appropriate technology". It is an example of what might be called one-dimensional thinking, that is, an attempt to remedy a complex, multidimensional situation with one, simple, and one-dimensional suggestion. It claims, in its extreme form, that if only the "appropriate" kind of technology were pursued in the developing countries, the disparity between the developed and developing countries would be greatly decreased.

It is difficult to discuss this area in a concrete way since what is "appropriate" and what is not is seldom specified unambiguously. There are, however, some general remarks that can be made.

1. What is the "appropriate" technology for a country at a given time must be determined from within the country and by the scientists and technologists in the country. An organization in a developed country which claims to be doing "appropriate technology" for the developing countries does not seem to make much sense. There is an enormous amount of "appropriate" technology freely available to developing countries if they wish to, and have the ability to select, adapt, and use it.

2. It follows from the above as well as the rest of this discussion that the most effective way to advance "appropriate" technology in a country is to strengthen its scientific and technological infrastructures so that the evaluation, selection, creation, adaptation, and production of such technology can materialize as fast and as efficiently as possible.

3. In determining what is "appropriate" technology, a very complex array of considerations must be taken into account, including some which are connected with values, psychological factors, emotions, and other non-material elements. (It is for this reason that the decisions must be made locally.) To mention a few of these considerations, such technology should utilize existing manpower and yet provide for a continuous upgrading of the capabilities of such manpower; it should be feasible in terms of knowledge, know-how, and expertise available locally, and yet be sufficiently challenging to stimulate a growth in knowledge, know-how and expertise; it should contribute to the long-term growth of the country so as to avoid recurring crises; it should contribute to the material well-being of the country but at the same time should cater to the country's non-material aspiration of being a respected member of the world community of practitioners of technology. It is very likely that for almost every country, "appropriate" technology will mean something different, which is another reason why such technology cannot be evolved from the outside in a centralized fashion.

Summarizing our discussion, we might point at some specific actions that might advance the cause of linking science and technology in the developing countries. I want to mention five of these among the large number that could be listed.

First, our assistance to developing countries should include a considerable amount of infrastructure building rather than providing help only for specific projects of short term importance. Such projects are valuable and should not necessarily be abandoned altogether, but they must be balanced by some of the funds (perhaps 20-30%) going into an enhancement of the countries' capability in science and technology, without having in mind any particular project.

Second, we have a great responsibility toward those students from the developing countries who study science and technology at the universities of the developing countries (there are some 100,000 of them in the United States alone). We must insist on providing them with an education which is broad, functional, has a problem-solving spirit, and includes elements of the context of science and technology that will allow such students to simultaneously undertake doing science and technology in their countries and creating the environment in which such science and technology can be practiced.

Third, we should continue to maintain contact with such students after they have returned to their countries and encourage them in their activities by providing them with information, equipment and spare parts, and with recurrent opportunities to visit scientific and technological institutions abroad in order to interact with the worldwide community of scientists and engineers. In this way they will also be encouraged to evolve the links and contacts of their local science with the local technology, using as precedents those many countries around the world where such a contact is traditional and flourishing.

Fourth, provisions should be made for an extensive set of cooperative research projects in science and technology, the partners of which are a group of researchers or developers in an advanced country, and a counterpart group in a developing country. Experience has shown that the smaller the two groups are, the more effective such a bilateral relationship will be. It would be of particular importance to select small interdisciplinary applied research groups for such linkage. The governmental or national research laboratories of the developed countries have many such programs and groups. At the present, it is very cumbersome if not impossible for an American national laboratory to undertake the formation of such a bilateral link, since international collaboration, and, in particular, cooperation with developing countries with the aim of assisting the infrastructure of such countries is not among the "missions" of such national laboratories.

Finally, the governments of the developed countries should offer inducements (taxwise or tied to other transactions) for multinational companies to establish research laboratories in those developing countries in which they have subsidiaries. Such laboratories, which should perform both applied scientific research and technological development, would continue to have a commercial, profit-making incentive and hence would set a pattern for successful research and development activities as measured by down-to-earth criteria pertaining to the user himself.

The weakness of the link between science and technology in developing countries is just one of the symptoms of the overall weakness of science and technology in those countries. That, in turn, is mainly due to the very recent vintage of such activites in those countries. Even though the situation is still disconcerting, considerable progress has been achieved, at least in some developing countries, in the last 20-30 years. With appropriate international cooperation, such progress will continue and accelerate.

Bull. Sci. Tech. Soc., Vol. 2, pp. 135-140, 1982. Printed in the USA.
0270-4676/82/020135-06$03.00/0 Pergamon Press, Ltd.

CREATING AN EFFECTIVE
APPLIED SCIENTIFIC RESEARCH PROGRAM
IN A DEVELOPING COUNTRY

William J. Pardee
Rockwell International
1049 Camino Dos Rios
Thousand Oaks, California 91360 USA

Michael J. Moravcsik
Institute of Theoretical Science
University of Oregon
Eugene, Oregon 97403 USA

ABSTRACT

A systematic approach to the development of an applied scientific research program to meet a developing country's future technological needs is briefly described. The essential features common to all applied science programs are discussed, and approaches to the special problems of a developing country suggested.

INTRODUCTION

Although 80-90% of the research and development funds in a developing country are intended for applied scientific research and technological development, the actual results of these activities are often disappointing. We suggest that this reflects a lack of understanding of the nature of applied scientific research and of the steps needed to establish a productive applied science program. In this note we[*] outline a plan to determine needs, prioritize these needs, assess resources, and initiate and maintain an applied scientific research program. We then discuss implementation of this program in the presence of some of the special problems of developing countries.

A GENERAL APPROACH TO MORE EFFECTIVE APPLIED SCIENCE

The meaning we ascribe to "applied science" can be explained in terms of the distinction between two words: "Knowledge" and "Understanding."

[*]*The recommendations made here are the opinions of the authors, one of whom is an applied scientist from a high-technology environment, and the other an academic scientist with many years experience with scientific research in developing countries.*

"Knowledge" here refers to quantified experience with natural phenomena, while "understanding" refers to quantitative, predictive (falsifiable) conceptual relations among those observations.

Knowledge about phenomena that appear relevant to a given technological task can be obtained by an engineering testing program. Such purely empirical approaches have their use. Tables of mechanical properties are valuable to the design engineer, easy (though sometimes time-consuming and expensive) to construct, and not reliably obtained except by empirical means. When entering new territory, however, it is usually difficult in the absence of scientific understanding to guess which parameters influence the property of interest. When, for example, an alloy designer needs to improve a particular property like wear resistance, he doesn't know whether to change the hardness or toughness or texture of the alloy's thermodynamic stability under strain, or some combination of these. The purely empirical approach becomes prohibitively slow, expensive, and ineffective as the number of relevant parameters increases, and extrapolation from it is always unstable. Moreover, a laboratory engineering test is always idealized, and structure designers must incorporate a large, wasteful "ignorance factor" in the absence of deeper understanding. The "Edisonian" approach is often inappropriately applied when, for example, a developing country wants to create domestic substitutes for imported materials.

Understanding of the mechanism of certain phenomena is, in contrast, the objective of applied scientific research; it provides not only graphs of dependent _versus_ independent variables, but also a falsifiable, quantitative, explanation of the relationships between parameters. This distinction between the mere gathering of information and the development of scientific understanding of potentially applicable phenomena appears to be missing from much of the activity that is labeled "applied scientific research."

With this definition of applied science as background, we can describe the four steps to a technologically productive applied scientific research program.

The first step is to assess which areas of technology will be important to the country some 10-15 years hence. These areas may include "low technology" sectors like agriculture, or "high technology" sectors like some modern industrial processes, or anything in between. This assessment would presumably be based primarily on economics, social considerations, environmental needs, defense requirements, and the fulfillment of some non-material aspirations of the country. Current technology consumers such as the military, medical, agricultural, and energy industries can be asked what they see as their future needs. The assessment should also seek to identify new technologies which should be created. The evaluation in this first step should prioritize the areas into A (essential), B (important), and C (desirable). The basis would naturally be a mixture of scientific and technological judgments and extra-scientific (i.e., political, social, cultural, etc.) values. Though scientists, economists, businessmen and others can assess the implications of various choices, this decision about what the country wants must ultimately be a political choice, in the best sense.

To make the results of this first step useful, the study should be completed in a relatively short time (say 12-18 months); the number of areas studied should be arbitrarily limited to a small, manageable number. A developing country or an economic sector of a developing country may not possess enough research potential to cover all of its needs, and it is certainly reasonable to sacrifice comprehensiveness for effectiveness.

The second step is to identify strengths and weaknesses in the already existing technology related to the technological areas prioritized A and B

in Step 1. This identification will result in a long list of specific research problems. This list then should be examined for unifying themes
(e.g., "corrosion," "entomological ecology," etc.), and problems should be
combined where possible. Combinations can, of course, themselves be combined ("chemistry"), but a convenient level of abstraction is one where a
single well trained scientist could do effective research on any problem
in the group.

The objective of this second step is to select those technical problems where the level of performance is limited primarily by a lack of scientific understanding of the phenomena. We are going to recommend an
applied science policy which does not fund a project unless success of
that project can be related clearly and realistically to economic benefits whose value substantially exceeds the inflation adjusted research
costs. That is, be as optimistic as possible about scientific success,
and examine the benefits. Accordingly, the technological problems on that
list should be classified a if increased understanding is absolutely necessary for a particular desired performance level, b if such an understanding is not absolutely necessary but would offer considerable advantages over existing approaches, and c if approaches based on already existing understanding are adequate.

It is desirable, at this point, to discard problems where the probability of success is less than say 15%, but not to otherwise consider
probability of success.

The third step is to determine the human, material, and institutional resources available to address the Aa and Ba problems (in this order).
We discard Ab while retaining Ba and even Ca because of our view that it
is wasteful to devote applied science resources where even complete success will be of minor importance. The human resources thus assessed
should include both the available scientific talent and the quality of
its technical support. The study of institution resources should recognize that in many cases no existing institution (government laboratory,
university, or industrial laboratory) will be suited for the task at hand
without modification. It may be judged desirable, in those cases, to create a new laboratory, or a new branch of an existing one.

Finally, the fourth step is implementation. Implementation is, however, so dependent on the nature of the particular technology, resources,
national and cultural setting, etc., that only a few general suggestions
can be offered. One of these is the warning that without special efforts
"applied scientific research" tends after a while to cease to be at least
one of the two: "applied" or "scientific." The activity may transmute
either into a purely internally motivated "basic" scientific research project, or into a program that produces hardware or knowledge, but not deeper understanding. To avoid this "derailment" requires specific organizational measures. To maintain the scientific character of the work, one
must hire good scientists, encourage these scientists to publish in the
world scientific literature, and provide (through seminars, sabbaticals,
and similar devices) for continuing communication between these scientists
and the basic scientific research community. On the other hand, the applied orientation can be preserved through periodic reviews by representatives of the technological areas that will eventually become the cus-
tomers of the applied research in question. Those reviews can provide a
nucleus for the essential and difficult technology transfer process.

The results of the applied scientific research must be transferred
to a development project, and from there to a production facility. Because of its vulnerability, the transfer should be planned from the beginning, recognizing that the transfer of knowledge and understanding requires good communication at the technical level, probably including exchange of personnel. Close communication between those in applied scientific research and those in technological development may contribute to a

stronger motivation by the applied scientists. Even scientists with a very "basic" orientation and motivation are delighted to see their work have a practical consequence. This attitude should therefore be encouraged, perhaps by making the connection between research and technological success shorter and more visible.

Even in the best circumstances, however, the transfer process is impeded by factors that make many scientists reluctant to complete a mature project and begin a new one. Laudable scientific qualities such as perseverence, thoroughness, and a passion for excellence can combine with a human reluctance to relinquish the position of expert for that of novice to undermine the applied scientist's commitment to his nominal technological objective. Indeed, he may expect that explicit communication of the project's results will result in the termination of support. Specific answers must be sensitive to cultural values, but, at the very least, management must ensure that completion of successful technology transfer should result in attractive new opportunities for the involved scientists.

PRACTICAL PROBLEMS IN IMPLEMENTING THE FOUR STEPS

The first major problem is: Who will perform the various tasks and functions in these four steps? The countries frequently lack an adequate number of broadly educated scientific and technological personnel, and seemingly similar tasks are commonly performed by civil servants without experience in science or technology, or by quickly flown in "international experts" ignorant of the country and sometimes without even appropriate technical qualifications. Next, if manpower competent for these tasks is somehow available, what should be the composition of the groups involved in these activities? What should be the distribution in their backgrounds? Should teams be formed? If so, what kind? Furthermore, step 2 may reveal the requisite scientific understanding may already exist, but not in the country in question.

In some areas the technological application envisioned for, say, 10 years in the future may not even exist at present. How, then, does one provide a user orientation to the research?

Most nations have appreciable psychological barriers between industrial, government, and academic laboratories, but these seem particularly forbidding in some developing countries. Planning of the sort we are discussing is commonly done entirely by local economists, and they cannot be expected to manage all the nuances which contribute to an effective program. Before commenting on these questions, two general remarks are appropriate.

First, most developing countries will be unable to complete the four steps without some collaboration with and assistance from foreign scientists and technologists. In this post-colonial era, developing nations are sensitive to suggestions of foreign dominance. A number of measures can be taken to protect the national interest of the developing country while still obtaining the benefits of collaboration. The foreign scientists should be selected from as many different nations as practical, and the selection process should avoid the heavily political national and international organizations. The personal contacts of a country's own best basic research scientists are probably a good starting point. Moreover, such foreign contributions need not be entirely from developed countries; regional cooperation can provide the involved nations the benefits of specialization, permit each nation to give as well as receive, and increase the sensitivity of the process to local cultural values. This use of foreign consultants should be viewed by the host country as a creative use of one of many resources. Even the scientifically well-developed countries cannot always handle their own needs entirely with their own citizens, and it is common for them to seek foreign scientists for particular tasks, including major policy advisory roles. No country in the history of mankind has ever been able to evolve its own scientific and technological capacity without considerable reliance on the cooperation with "predecessor" countries, even if the "apprentice" eventually overtook and dwarfed the "teacher."

Second, it is very important to devise, ahead of time, a method for evaluation and assessment of the success of each step. The basis of this evaluation must be the effect of such research on technological and production activities. The assessment of step 2 and the user reviews provide a starting point for these evaluations.

Let us now review the four steps in view of the questions asked earlier.

The needs assessment panel of step 1 might be a group of 8-10 people. This group should contain people with a mix of backgrounds and origins. Some representatives of potential local users of technology should be in this group, even if, at present, the connection between this potential user and the technologies in question appears somewhat remote. The group should also include outstanding members of the local scientific and technological community, chosen for their general productivity and broad vision rather than for eminence along a very specialized line. The group also needs one or two local economists and spokesmen for various broad groupings of society. Finally, at least one scientist and one technologist should be added from abroad, representing worldwide experience in the fields and serving as a channel to obtain specific expert advice from abroad if appropriate.

The needs assessment may seem a very difficult task. Nevertheless, some needs in applied research can be foreseen, and most countries will have some people with sufficiently extensive experience, background, and vision to pinpoint such needs. Moreover, the purpose of this panel is to provide a rational basis for the political process to answer the question "What do we want?"

The second step is strictly scientific and technological in nature, and hence should be performed by people from those fields. It is in this step that collaboration with the worldwide scientific and technological community is of greatest value. A group of 10-15 people, perhaps half of them from abroad, and working in smaller subgroups dealing with each priority area, and including both "basic" and "applied" scientists, might be an appropriate mechanism. In both steps 1 and 2, good local scientists are invaluable. This is one of the many reasons why it is important for most developing countries to possess at least a modest number of knowledgeable people in most major areas of science and technology. The only way known to accomplish this is to offer to these people adequate facilities for personal research without paying much attention to whether the results of that research fit in with the latest national development plan. Such discretionary research, on a very modest scale, must be considered an overhead needed to provide the country with at least a modicum of "receptors" of scientific information generated elsewhere.

In selecting the members of these groups who come from abroad, it is also valuable to have local scientists and technologists who are familiar enough with the field to know who are the good people in it. The fact that somebody is recommended or even sponsored by some international organization is by no means a guarantee that he will be helpful in the task or even competent in it. Peer evaluation is still the best way to identify scientists with the appropriate blend of depth and judgment.

The third step (resource evaluation) is one for which adequate local expertise in science, technology, economics and management is likely to be available. A broad view of resources should be taken. There are, for example, notable precedents of developing countries bringing home previously brain-drained scientists with a challenging, well-supported research program. Multinational corporations doing business in the country are one institutional resource. An oil company, for example, might find it advantageous to establish an applied chemistry laboratory in an oil-producing nation, using indigenous personnel augmented by short term (1-3 year) participation of scientists from the company's other laboratories. This is a very effective way to transfer existing science and science management skills to the developing country, and could promote goodwill. Caution must be exercised in the establishment of new laboratories; if adequate manpower is unavailable, the effect may be detrimental to the overall national research program. If a new laboratory is to be established, it may be helpful to vendor it to a strong existing institution with demonstrated strength in science and science management.

The fourth step requires thoughtful administration and patience. If the previous steps are effectively carried out, adequate human and other resources will be available, but even so, it might take several years before tangible results emerge from the project. During that period, periodic visits for peer assessment of ongoing work might be very helpful, using, if necessary, appropriately selected members of the world-wide applied scientific community.

CONCLUSION

It should be evident that the prescription offered in this article can be applied on a broad variety of scales. It applies to the needs of one particular organization in one particular field as well as to the overall needs of an entire country. In practice it would be a major and probably unwieldy task for a developing country to use the prescription in a centralized fashion for the whole national science and technology policy. It might be preferable to select one or several important areas in which the ingredients needed for making use of the prescription are available, and proceed there. Progress is almost always achieved in small steps and in localized areas where leadership and manpower are available. The method we suggest could be applied by a particular industry (e.g., electric power) or a single company, as well as, and perhaps more easily, than by an entire nation.

The prescription we offer is a demanding one, one that requires considerable human resources in a developing country, not so much in quantity but in quality. As a result, in some countries it may take time before the type of scenario we envisage can occur. "Never" is too big a word to use in human history in which "miracles" do occur if enough time is allowed, and countries which are developing today can rocket into leadership within a few generations.

We are indebted to Jean-Jacques Salomon for some perceptive comments on a draft of this article and to Peter Cannon for a stimulating discussion of the issues.

E. Measuring Science

Preprint, published in the *Proceedings of the First Pan American Workshop on Quantitative Methods in Science Policy and Technology Forecasting,* 1983, Vol. II, p. 30.

PRACTICAL PROBLEMS AND STEPS OF IMPLEMENTATION IN THE APPLICATION

OF SCIENTOMETRICS IN THE THIRD WORLD

Michael J. Moravcsik
Institute of Theoretical Science
University of Oregon

I. Introduction

The purpose of this discussion is to present a frank and functional discussion of the status of scientometrics in developing countries and the deficiencies that now exist, together with specific suggestions on how these deficiencies can be remedied.

In doing so I cannot help stressing that such discussions alone amount to nothing. In the area of science policy of the Third World, there has been a mountain of proceedings of conferences, reports, plans, articles, resolutions, and recommendations. Almost all of these, whether good or bad, realistic or naive, formalistic or operational, broad or specific, massive or small scale, have been gathering dust in the filing cabinets or shelves of libraries, national and international agencies, and other depositories of documents. In fact, there is a large degree of duplication in this material, and workshops keep being organized, resulting in the same resolutions as those passed by their predecessors. There is something in human nature that believes that once a fancy resolution has been passed, about half the work has been completed. The real figure is, instead, about 1 percent.

I would therefore like to urge this workshop not to pass any more resolutions at all. Instead, each participant of this workshop should focus on one of the many specific suggestions discussed in this workshop (including in my discussion), and then personally work on implementing it. Among the action needed there are some which require broad cooperation and some new resources, but there are also some which can be put in place on an individual level and require practically no new funds at all. Through the sum of many such small, incremental steps, implemented individually by many people, significant and functional progress can be made in the improvement of the context of science and technology in the developing countries.

My plan for the discussion will be as follows. Section II will discuss reasons why the infrastructure for scientometrics is, at the present time, practically non-existent in the developing countries, and what needs to be done to create such an infrastructure. In Section III I will outline the

adaptive research and follow-up action needed to generate scientometric
material in the developing countries. Finally, in Section IV, the problem of
how to incorporate scientometrics into the management of science and technology
in the developing countries is discussed.

II. Creating an infrastructure

At the present time, scientometrics is practically unknown in the developing
countries. Scientometric articles by authors from the developing countries are
very rare, the international professional society of scientometricians (together
with other practicioners of the science of science), namely the 4S (Society
for the Social Studies of Science), has practically no members from the
developing countries. Furthermore, if somebody in the developing countries
wanted to educate himself in scientometrics, it would be most difficult for
him to do so. There are no library resources, no curricula at universities,
no older colleagues who could guide such studies, and no institutional
incentives to conduct such studies. To be sure, science and technology themselves
are predominantly practiced in the "advanced" countries, to the extent of
more than 90%. But even compared to this degree of skewness, the lopsidedness
of scientometrics is even greater, perhaps 99% vs. 1% or even worse.[1]

Let us see what the reasons may be for this gross neglect of scientometrics
in the developing countries. Among the undoubtedly many reasons, I want to
point to two.

a) Scientometrics is a new field.

Scientometrics, a quantitative study of science and technology as human
activities, is a quite new field. It would be difficult to set an exact
time for its birth, but it is safe to say that 20 years ago, it was
practically non-existent. This of course does not mean that no statistics had
been related earlier at all to scientific and technological activities, but
it does mean that scientometrics, as a field of study and as a serious and
functional tool of for the management of science and technology did not
exist before.

Since scientometrics is so young, it is unlikely that the developing
countries would have caught up with it by now. Most unfortunately, developing
countries continue to be followers rather than leaders in scientific and
scholarly undertakings, and it takes followers a decade or two to pick up
the latest trends, ideas, and achievements. Such an adaptation involves the
education of appropriate manpower, the formation of institutional frameworks,
the opportunities for research and practice, the supply of auxiliary services,
and hence the time delay is considerable. It is, as said above, most
unfortunate that the developing countries continue to be such followers.
In my view,[2] the chief blame for this perpetual retardation of the developing
countries must be placed on the simpleminded, myopic, and mistaken interpretation
of "relevance" by the policy makers in the developing countries and by their
international and binational collaborators and advisers. What is "relevant"
in their eyes is what is already obviously utilitarian in the short term
sense, and hence what is already "old stuff" in science and technology,
well as in the related areas of scholarship. The reasons given publicly

for this retardation are, of course, different ones, and always external ones, since it is much more comforting to believe that one's failures are due to the conspiracy of external forces. Thus the reasons given are the lack of raw materials, the lack of capital, the oppression of multinational corporations, etc. But, as I explained recently, [2] in the explosive recent development of the biotechnology industry, there was no need for raw materials or of substantial capital investment, and there was no competition by multinational corporations. Neither are any of these elements present in the development of scientometrics.

b) The tradition of formalistic science policy.

Science in its modern form was mainly imported to the developing countries. [3] This does not mean that there was no science at all in these countries in their previous history. On the contrary, Arab, Indian, Chinese and other cultures contributed much to the ancient development of science, and indeed, some 400 years ago there was little difference in the state of science between Europe and some of these other civilizations. The scientific revolution, however, as we know it today, was an overwhelmingly European phenomenon, and in fact most of what we know today in science was discovered and formulated by European scientists. Our total scientific knowledge 400 years ago was negligibly small compared to the present one.

Thus science was imported to the developing countries, partly during the colonial period, and much more intensively, since independence. As it often happens with foreign imports, it was much easier to assimilate the form than the content. As an example, the British university system was imported into most of the former British colonies, with all its external trimmings, robes, titles, and terminology, but in extremely few of these countries was the content, spirit, and conceptual base of the British university system transferred also. The importation was thus purely formalistic and not functional.

Science also suffered from this fate, and particularly the management of science. Since many developing countries began, in the earnest, to import science only after the Second World War, that is, after the beginning of the era of "Big Science" in Europe, the United States, and Japan, they had before them, as examples, often quite elaborate structures of governmental organizations for the planning, management, and performance of science. Vast offices, thousands of people, spendid science plans, an overlay of organizations, title administrators, progress reports, committees, policy documents and many other parephernelia dazzled the eye. It was not difficult to imitate these externalities, especially because both binational and international agencies greatly encouraged the establishment of such a machinery, as a visible sign of progress.

At the same time, the people who filled these structures in the developing countries had had little opportunity to learn and practice science, and in fact most often came from a purely bureaucratic and civil servant background, the education for which included little if any science. Even the few scientists who participated in these structures had modest scientific credentials, not necessarily because of lack of innate talent for science but because when they were educated and spent their young years in trying to practice science, the circumstances in the country were too rudimentary to allow a substantial development of their talents and the significant acquisition of experience with what science really is

Today the situation in many developing countries is much better, and quite a few of the scientific leaders and managers are now well educated and have had the opportunity to acquire significant experience in scientific research. Unfortunately, however, by now the scientific bureaucracy and its traditions have been firmly entrenched in these countries, and represent a horrendous burden on them. A tradition has grown up of formalistic science management, with research councils, committees, ministries, agencies, and hierarchies, a maze in which a productive scientist intent on accomplishing something in scientific research can operate only with great difficulty.

In particular, such a formalistic science management is inimical to the spread of scientometrics. The formalistic management likes set rules, regardless of how inappropriate they may be to the case at hand, while scientometrics is a wide open subject in which whatever methods have so far evolved must be used with caution and with the full knowledge of its limitations. Formalistic science management pays most of its attention to input indicators, while the most interesting and operationally fruitful part of scientometrics is its treatment of output indicators. Formalistic management often glides over the conceptual ambiguities and intricacies of the subject in order to attain its simple rules, while in scientometrics we increasingly find that the analysis of the basic concepts and the related problems of semantics are perhaps the most central and important questions to resolve, in the absence of which results, rules, and statements have little meaning and therefore either no operational impact or the wrong one. There are many other dimensions also in which formalistic science management and scientometrics are at odds with each other.

I have just outlined two main causes (scientometrics being a new field, there has been not enough time yet for the developing countries to adopt it, and scientometrics being impeded by the tradition of formalistic science management in the developing countries) why scientometric infrastructure is generally missing in the developing countries. What can we do to encourage the evolution of such an infrastructure?

I would like to suggest five ways to do so, but before going into these details, a general remark might be helpful. In the absence of any infrastructure, the first steps are always difficult and, by necessity, involve help from and collaboration with people and organizations abroad. In this respect. I would recommend making contact with the already mentioned Society of the Social Studies of Science (4S). Members of the society or collective action on the part of the society may be helpful in any of the five directions in which my suggestions reach. Advice in educational curricula, assistance in building up library holdings, cooperation in getting access to reports and preprints, collaboration in the arrangement of workshops like the present one, and suggestions of how to inject scientometrics into day-to-day science policy and management are all areas in which 4S has the expertise and could easily act. Although in the past 4S has not been involved in such collaborative activities with people outside its membership, I am quite certain that it would respond. This year's president of 4S is Arnold Thackray, History of Science, University of Pennsylvania, Philadelphia 19104.

Now let me list the five ways.

1. Curricula at universities

Scientometrics should be incorporated into the education of future scientists, technologists, and science managers. This may be achieved, at first, through short-term visiting lecturers, and later through university personnel who, during their education abroad, purposefully included scientometric material in their own curricula.[4] It is particularly important for scientists and technologists in the Third World to be aware of and knowledgeable about the context of science and technology and not only about the disciplinary details of these fields, and in such a contextual education scientometrics should have an appropriate part.

2. Local libraries should evolve small holdings in scientometrics.

At the outset, a very modest amount of material will suffice. Three relatively inexpensive journals (Research Policy, Social Studies of Science, and Scientometrics) would constitute a good start in the area of journals, and a set of appropriately selected 30 books would form a sound basis in the category of books. In choosing the books, cooperation from 4S would be very useful.

3. Access to preprints and reprints.

Although preprints (i.e. prepublication copies of articles) play only a moderately important role in scientometrics, they, together with reprints, allow a person who has only meager access to journals to have access to a larger part of the current literature. The distribution of preprints and reprints is usually done in a rather haphazard manner, with each author having his own mailing list. By requesting prolific authors to include addresses in the developing countries in their mailing lists, a stronger link can be achieved with the scientometric community. In becoming aware of what has recently been published in the field, the current bibliography of the science of science, published by A. Rahman at the CSIR in New Delhi, India is particularly useful.[5]

4. Meetings and workshops.

There are many meetings in the science of science each year, almost all held in the advanced countries. The annual meeting of the 4S is, probably, the most comprehensive such meeting in the field. Sending one person from a country or even from a region to each such meeting would ensure that events at such meetings would at least indirectly spread throughout the region, and the presence of such a regional representative would also make 4S members aware of the growing scientometric community in the developing countries.

In addition, workshops like the present one are valuable for broader coverage of the discipline. Especially meetings of small size like the present one are important because of the opportunity for informal discussions and debates.

5. The education of science managers.

Although this goal overlaps in part with some of the four mentioned earlier, I listed it separately since it requires special logistic preparations. Once somebody is in a programmatic position, and has ongoing, daily administrative responsibilities, the greatest problem is finding time and energy, over and above the daily responsibilities, to further learn, to reflect, to synthesize piecemeal experience, to look at the broader issues. As Henry Kissinger remarked,

when somebody is placed in a prominent position, he must exist on resources
accumulated prior to being placed in that postion, since in the postion he may
learn something about how to do things, but not about what things to do.

For example, drawing away busy administrators from their work for a meeting
like this one is most difficult, and can be done only rarely. Thus some much
more ordinary opportunities and incentives are needed to assure a steady
contact between science administrators and the ongoing events in scientometrics.
The reading of books or articles, placed right into the hand of the science
manager, lunch meetings at which a brief discussion of some scientometric
development takes place, outsiders invited to visit an agency, for 2-3 weeks,
for the expressed purpose of providing a channel to scientometric developments
around the world, might be such mechanisms. "Sabbatical years" for science
managers, during which they can immerse themselves fully in the latest
developments in the field would also help. Such a program would show definitely
beneficial effects on the daily work of the science managers, by improving
the management of science for which they are responsible and making it more
realistic in terms of the actual requirements of a productive scientific
and technical community.

III. Adaptive research in the science of science for the developing countries.

One of the exciting and intellectually challanging aspects of the
application of scientometrics in the developing countries is that such an
adaptation cannot be done mechanically and automatically but the issues,
results, methods, and procedures must be reexamined and possibly modified
for the particular situation. Although there is only one science,[6] and
although technology also has a very large degree of universality, the
practicioners of science and technology, their context, society, and infrastructure varies from country to country, and hence scientometrics as applied
to science and technology in a given country or institution must be adapted
to do its function. In this section, therefore, I want to mention some areas
in which such adaptive research is needed. Since we cannot cover everything
in a short time, I will concentrate on the creation of functional output
indicators for science and technology in the developing countries. I chose
this topic also because in practice the output indicators (especially when they
also deal with quality and not only quantity) have suffered more than
anything else because of neglect, misunderstanding, and misuse.

Science and technology have three broad functions. First, they contribute to the material betterment of life in a country. Technology does
that directly, and science indirectly, through technology. This is a
conspicuous aspect of science and technology which need not be elaborated
on here.

Second, participating in the evolution of science and in the increase
of Man's technological capacity is an aspiration in our century that all
countries want to share. It is one manifestation of throwing off dependence,
of being on an equal footing with the other countries around the globe. It
provides one goal (though of course not the only one) that gives meaning to
the efforts to improve our material well being. There are numerous statements
by leaders of developing countries attesting to the importance of this
aspirational motivation for science and technology.

Finally and thirdly, science and technology have a profound effect on Man's view of the world and of his own role in it. This aspect of the scientific revolution as a cultural force has been very evident in Europe since there this effect has been working on subsequent generations for some 400 years. Almost all definition of "development" implies attitudes and perceptions that strongly overlap with the beliefs and methods of science. This cultural impact of science and technology has also been recognized by many leaders of the developing world.

In terms of this tripartite goal of science and technology, let us discuss the status of output indicators. We might begin with the second and third functions of science and technology, namely the aspirational and the outlook-forming ones, since, unfortunately, there is little to be said about those. Whether in the advanced countries or in the developing ones, quantitative indicators for measuring the effect of science and technology along those two functions are practically non-existent. This is most regrettab especially in the context of the developing countries since having no indicator means a temptation to ignore those functions altogether. Yet in the overall context of development, those two functions are at least as important as the effect science and technology have on the material growth of the country. Therefore, formulating output indicators for measuring these two types of impacts of science should be considered important research problems in scientometrics.

In the remainder of the discussion, therefore, I will focus on output measures for the scientific research and technological development work, which are directly related to the function of improving the material standards of the country. It would be tempting to measure simply the final output in terms of the material production of the country, but this would be foolhardy, since there is a long chain of various transfers, actions, middlemen , and economic factors between science and technology at one end, and material production at the other. Thus unsatisfactory production might be caused by a large number of other factors beside poor science or technology, and, on the other hand, good production in certain areas of goods can come about even if the scientific and technological basis is weak. Thus we should use more immediate output measures for science and technology.

Let us start with science. Scientometric output indicators for science have so far been concentrating on measures derived from the communication system of science, since it is believed that any progress scored in scientific research must be communicated to the scientific community in order to ensure its constructive role in the further evolution of science and technology. This is not the place to present a critique[7] of these indicators in general, but instead we will accept their existence and focus on their adaptation to developing countries.

The exclusive tool for scientometric output measures relating to the communication network within the scientific community is the journal article. All the measures, such as scientific authors, scientific publications, citations to other works in science all use journal articles. It is thus assumed that all scientific output worth measuring manifests itself in such articles.

In the context of developing countries, however, one can think of possible modes of communicating the results of scientific research which are different from journal articles. For example, one might think that a very directly applicable but quite specialized piece of research might be communicated to the technologist through reports only. I personally believe that if we consider also local scientific journals, very little scientific research that is valuable will avoid making the journals. This is, however, only an opinion. It would be very instructive and interesting to have a few case studies in which it would be determined how much and what kind of scientific work in the developing countries does not make a trace in journals.

Let us, however, for the moment constrain our output measures to the use of material published in journals. The next question is "What journals?". Scientometric work can be performed most effectively by computers, and hence the material in journals (i.e. titles, authors, citations, etc) needs to be computerized in order to be of practical use. The most prevalently used computerized information system on material in journals is that of the Institute of Scientific Information (ISI) in Philadelphia, an organization that publishes the Science Citation Index, among various other compilations.

ISI does not survey all scientific journals around the world. There are several times ten thousand such journals, the precise number depending on definitions and on whose count one wants to use. Of those, ISI surveys some 3,000. This sounds like an overly small sample, but there are two factors which may alter that judgement. First, these 3,000 journals represent about 80% or more of the total amount of scientific journal literature, since the journals which are included are published frequently and are thick. Secondly, and perhaps more controversially, ISI uses a "quality indicator" to choose which journals to survey, the indicator being related to how often articles from the journal in question are cited in the scientific literature. In that sense, ISI claims to survey not only most of the scientific literature but also the most important part of it.

ISI's compilation is not the only one, but the few others that are available have, broadly speaking, the same general characteristics as ISI's.

It is a matter of heated controversy whether ISI's method of surveying the scientific literature is or is not prejudicial to the developing countries. Those who claim it is say that many of the small journals left out of ISI's set are in fact journals published in the developing countries, often dealing with applied research, and often written in local languages. Furthermore, they claim, judging the quality of a journal by the degree to which it is cited is a circular argument, since in order to be cited journals need to be included in international compilations, but in order to be included in such compilations, it is required that they be cited.

It would be most important to resolve this controversy to almost everybody's satisfaction, since suspicions about the ISI roster results in the developing countries not using it. On the other hand, nothing more acceptable exists in its place, and hence nothing is being used. I have suggested holding a three-day workshop with perhaps 30 participants, selected from both the developing countries and from the counterparts in the advanced countries, with the specific mandate of arriving at a modification of the presently available scientometric compilations which would be acceptable to all sides.

Since, however, such a meeting is not likely to result in a modified compilation that would be appropriate for all possible uses of science management in the developing countries, I would also like to suggest that the developing countries undertake, regionally, the establishment of their own scientometric publication compilations. Beside the local utility, this undertaking would also serve as a most educational project in practical scientometrics. A number of conceptually and practically important questions would have to be faced, like: What is a scientific publication? What is a scientific journal? What contitutes "double publishing"? Is citation a good weighting factor to account for the quality of articles? How can one deal with mechanical problems like the misspelling of names, authors with identical names, etc.? Are the institutional affiliations given on articles a correct indicator of where the work was done? Does the existence of a compilation in itself have an effect on publication patterns, and if so, in what way? Should reports be included in the compilation, and if so, what types of reports?

Let me now turn to technological indicators. The one most commonly used as a direct indicator of technolgical activity is patent statistics.[8] In the context of advanced countries this is a reasonably good measure, because most technological inventions are patented, and because patentatibility itself provides at least a minimum level of qualitative indicator also. The measure is far from perfect, since some inventions are kept as trade secrets rather than being patented, and because patentatibility is a very weak indicator of quality.

In the developing countries, however, patenting is generally not a good indicator. The system of patenting may not exist at all in the country, or may be rather rudimentary, and if only patenting in other, advanced countries is counted, most of the inventions made in developing countries would be missed.

It is clear, therefore, that there is a need for developing new indicators to measure technological activity in developing countries. In doing so, it is important to keep in mind some of the present shortcomings[9] of such activities. In most developing countries such technological development work is done mainly in governmental laboratories, with a Civil servant mentality and with no competition. Developments projects very rarely emerge from a cooperation with the potential user and hence from a consideration of what the realistic practical needs are. Upon the completion of the development work, the result is generally displayed in the form of a report which, after a perfunctory and cursary inspection by people not well versed in technology, is deposited in a filing cabinet. Mechanisms for transmitting such results to potential users are only fragmentary. Thus, we see, at the moment there is practically no assessing of such technological development work at all. Thus, in a sense, almost any assessment system would improve on the present situation. Research and experimentation in this area would be most important.

To conclude this section on what is needed to utilize scientometric tools in the science and technology management of developing countries, I want to discuss the importance of a peer review system for assessment both in science and in technology. Peer review is often not considered a scientometric tool, because in some forms it is not as numerically quantitative as publication measures. Imaginative versions of peer review, however, can be made meaningfully quantitative and numerical.[10] In any case, peer review is so important that it must be built into the assessment system of any system of science and technology.

Peer review involves soliciting the opinion of competent scientists and technologists in order to assess the standing of an individual, or institution, a research project, or even a whole country. In its basic foundations it is not too different from bibliometric tools, because citation ratings also reflect peer judgement. Yet peer review can give a more detailed, multidimensional, and functional assessment than the single numbers produced by bibliometric indicators.

Peer review is rarely used in developing countries. There are several reasons for this which are worth enumerating.

The scientific and technological community of a given country is usually much too small to form a sufficiently large pool from which peer reviewers could be selected. Furthermore, in such a small community, almost everybody knows (or knows of) everybody else, and personal dislikes and obligations develop easily. Hence it is not feasible to use the local scientific and technological community for peer review.

Thus the developing country, like any other country in the world, would have to turn to the worldwide scientific and technological community for peer review. To do so is a daily activity in the scientifically advanced countries and is done automatically, without any negative feelings about it.

In a developing country, however, this is not done, partly for reasons which I consider extraneous and irrelevant, but partly for reasons which are legitimate.
There are two extraneous reasons. One is a feeling of national pride which prevents going abroad for peer review, as if such review were a reflection on the image of the country. Since science is international, and since scientific peer review is independent of politics, using the international peer review system has no implications for status or image. On the contrary, participating in the international peer review system is a sign that the country is an active member of the worldwide circle of science and technology.

The other reason which is extraneous though much more difficult to remedy is that in order to use the international peer reveiw system, one has to know in each field who the experts are. To know that, the developing country needs to have solid expertise in all fields, which is often not the case. Thus it is difficult to "break into" the international peer review system at the beginning, when the scientific and technological manpower and infrastructure are just being established. As in many instances, here also we have a vicious circle: To achieve expertise, one needs international peer review, and to have access to such peer review, one needs expertise.

There is much, in this respect, that the professional societies in the well developed countries could do to assist their colleagues in the developing countries. Unfortunately little is actually done. It is up to the science managers and scientists in the developing countries themselves to initiate such a collaboration for the identification of appropriate peer reviewers.

The one reason for the reluctance to turn to international peer review by developing countries that I find legitimate is a concern that active scientists working in developed countries will not be familiar with the infrastructural problems in science and technology that prevail in the developing countries and hence would offer unrealistic evaluation and advice. This is indeed something to be concerned with. In fact, I myself have seen instances when "experts" sent to developing countries have proven themselves of little use because of their inability to understand the circumstances under which science and technology are practiced in the developing countries. On the other hand, this problem is certainly not insurmountable, since there are enough scientists and technologists in the advance countries who are both prominent in their respective specialties and have had sufficient personal experience in the developing countries to be able to set their assessment and advice in the proper context.

My conclusion therefore is that there are no insurmountable obstacles in the path of the widespread use of peer review in the developing countries, and therefore I would like to urge most strongly all science managers in these countries to utilize peer reveiw as widely as possible: In connection with research to be supported, with research already performed, with the selection and promotion of personnel, with national development plans, etc. There is hardly any other single step to be taken in the management of science and technology in the developing countries which would bring so much benefit for the amount of effort expended than the usage of the peer review system.

IV. Incorporating scientometrics into the management of science and technology.

We have been discussing scientometric tools for the assessment of the output of science and technology. In this last section I want to relate these tools to the actual management of science and technology.

I want to start with a very strong appeal for the actual use of evaluation and assessment as part of science and technology policy. It always struck me as most incongruous that exactly those countries which have the most modest resources and which therefore would have to weigh most carefully every allocation of such resources are the ones which do the least amount of evaluation and assessment of their scientific and technological activities.

Part of the reason (or perhaps the consequence) of this lack of assessment is the method used to distribute funds for science and technology. There are basically two types of such systems, although combinations and variations of these also exist. One is what I called elsewhere[11] the "dribbling-down egaltarian system" (DDES), in which the total sum of centrally appropriated money is shared in an egalitarian way through a succession of organizational hierarchies among all organizations that have scientific or technological labels. Differences between funding for various organizations are caused not by any scientific or technological merit but only by political machinations. The second system is the "reaching down merit system"(RDMS), in which individual scientists and technologists submit their own research proposals, and get funded directly on the basis of the merit of their proposals.

The overwhelming fraction of all funds for science and technology in the developing countries is given out on the basis of DDES system. This is so because not only is the DDES system easier to administer, easier to fit into the bureaucratic image of a pyramid—like organizational structure, but it can be done automatically, without the need of ever doing any evaluation or assessment at all. Although the DDES system has some advantages (for example, it is helpful to institutions or groups of individuals just about to start out on a scientific or technological career), the overall net effect of the system on the country's scientific and technological activities is strongly negative, since the lack of assessment and evaluation fails to keep efforts on a creative and productive course.

I do not advocate a pure RDMS system either. The proper balance is somewhere in between, perhaps equally sharing resources between the two systems. I therefore strongly urge science and technology managers in the developing countries to give greater role to the RDMS way of distributing resources, and to switch over to the system which is half-and-half between the two systems instead of being entirely DDES. In order to be able to do this, however, the country has to establish a good evaluation and assessment system, using the methods and tools we discussed in the earlier sections.

A second appeal I want to make to science and technology managers in the developing countries is somewhat related to the first. It is to place as much emphasis on output indicators in the planning and managing of science and technology as they place on input indicators. At the moment, virtually all science policy documents, whether national or originating in international agencies, use almost entirely only input indicators. We are told of the amount of money to be spent or already spent, the number of scientists and engineers to be created or already in place, the number of buildings to be erected, etc. Input numbers like these have, of course, their role, but they describe only the quantitative input into science and technology. The implicit assumption that the output will somehow be proportional to the input is quite unwarranted, especially if neither the quality of the input nor the quality of the output are taken into account. The shelves of agencies around the world are filled with science and technology plans, created in terms of quantitative input indicators, plans which have turned out to have only the slightest resemblence to the actual output of the scientific and technological communities of the countries. Since, however, output is not measured, such discrepancies are seldom even noticed, and hence when time comes along for drawing up the next science and technology plan, the same academic exercise is performed with equally irrelevant results.

It would have been impossible to cover in one discussion adequately all the implications of scientometrics for the practical management of science and technology in a developing country. What I tried to do is to use a limited area, namely that of output measures for science and technology, to illustrate the type of problems that arise in the application of scientometrics in the developing countries. The problems pervade all aspects: Education, research, and management. I also tried to indicate that although the problems are numerous, the can all be resolved eventually. This is not going to occur in a day, a year, or even a decade. If however, each person in   science, in technology, or in the management of these two selects a modest but realistic goal of implementation of just one or two of the elements we have been talking about, visible progress will in fact be achieved well within the lifetimes of most of those present here today.

References

1. To the best of my knowledge there is no statistics on the fraction of scientometrics that resides in the developing countries. A somewhat related datum is the percentage of historians of science in the developing countries which is about 4%, see Derek de Solla Price, "Who's Who in the History of Science: A Survey of a Profession." Technology and Society 5:2, 52 (1969).

2. I discussed this recently in "Generating Innovative Capabilities in Science and Technology in Developing Nations", lecture at the Woodrow Wilson International Center for Scholars, Washington, D.C., December 1982 (to be published).

3. A classic account of this was given in George Basalla, "The Spread of Western Science", Science 156, 611 (1967).

4. To include the discussion of contextual problems in the science education of students from the developing countries studying in the universities of the United States has been the goal of many efforts in the past. For an early suggestion in this area see M.J. Moravcsik, "Some Modest Proposals", Minerva 9, 55 (1971). It is now virtually certain that, due to the vision and organizational skill of J.W. McGowan of the Physics Department at the University of Western Ontario in Canada, the first summer seminar with such a goal will take place in the summer of 1983.

5. "Current Literature on Science of Science", Centre for the Study of Science, Technology, and Development, CSIR, Rafi Marg, New Delhi 110 001 India.

6. See for example M.J. Moravcsik, "Do Less Developed Countries have a Special Science of Their Own?", Interciencia 3:1, 8 (1978) and an extended debate on this subject in Interciencia 6:3, 167 (1981).

7. For one of the most articulate and vocal critics of quantitative measures, see David Edge, "Quantitative Measures of Communication in Science: A Critical Review", History of Science 17, 102 (1979).

8. For a discussion of patent statistics as well as of many other aspects and problems of science indicators, see the duoannual publications of U.S. National Science Foundation entitled Science Indicators. The last volume published was that for 1980, and the 1982 volume is now in the process of being published.

9. For a good summary and bibliography see Diana Crane, Technological Innovation in Developing Countries: A Review of the Literature, Research Policy 6, 374 (1977).

10. For a particularly imaginative and successful peer review system, see Cornelis LePair, "Decisionmaking on Grant Applications in a Small Country", presented to the Conference on the Evaluation in Science and Technology - Theory and Practice, Dubrovnik, July 1980, and published in the proceedings of that conference in Scientia Yugoslavica 6:1-4, 137 (1980).

11. M.J. Moravcsik, How to Grow Science, Universe Books, New York, 1980. Chapter 11.

Additional Readings

A. BOOKS

Alatas, S.H., Intellectuals in Developing Societies, London: Frank Cass (1977).

Eisemon, Thomas, The Science Profession in the Third World, New York; Praeger (1982).

Moravcsik, Michael J., Science Development--The Building of Science in Less Developed Countries, Bloomington, Indi.; PASITAM (1976, second printing).

Morgan, Robert P., Science and Technology for Development, New York; Pergamon (1979).

Price, Derek deSolla, Little Science, Big Science, New York; Columbia Univ. Press (1963).

Rahman, A., et al, Science and Technology in India, New Delhi; Indian Council for Cultural Relations (1973).

Shils, Edward, The Intellectual Between Tradition and Modernity: The Indian Situation, The Hague; Mouton (1961).

Srinivasan, Mangalam (Ed.), Technology Assessment and Development, New York; Praeger (1982).

Zahlan, A.B., Science and Science Policy in the Arab World, New York; St. Martin's (1980).

B. ARTICLES

Blickenstaff, J., and Moravcsik, M.J., "Scientific Output in the Third World", Scientometrics 4, 135 (1982).

Djerassi, Carl, "A Modest Proposal for Increased North-South Interaction among Scientists" The Bulletin of the Atomic Scientists 32, 56 (Febr. 1977).

Frame, J. Davidson, et al, "The Distribution of World Science", Soc. Stud. of Sci. 7, 501 (1977).

Frame, J. Davidson, "The International Distribution of Biomedical Publications" Federation Proceedings 36, 1790 (1977).

Frame, J. Davidson, "National Economic Resources and the Production of Research in Lesser Developed Countries", Social Stud. of Sci. 9, 233 (1979).

Frame, J. Davidson, "Measuring Scientific Activity in Lesser Developed Countries" Scientometrics 2, 133, (1980).

Freites, Y. and Roche, M., "Opinions of an Academic Group in Venezuela on the Planning of Science and Technology", IVIC, Caracas (unpublished), (1982).

Glyde, Henry, "Institutional Links in Science and Technology: The United Kingdom and Thailand", International Development Review Focus, 15:7-11, (1974).

Herrera, Amilcar, "Social Determinants of Science Policy in Latin America: Explicit Science Policy and Implicit Science Plicy." Journ. of Devel. Stud. 9, 19 (1972).

Inhaber, Herbert, "Distribution of World Science", Geoforum 6, 45 (1974).

Lomnitz, Larissa, "Hierarchy and Peripherality: The Organization of a Mexican Research Institute", Minerva 17, 527 (1979).

Nayudamma, Y., "Decentralized Management of R & D in a Developing Country" Minerva 11, 516 (1973).

Price, Derek deSolla, "Measuring the Size of Science", Proceedings of the Israel Academy of Science and Humanities 4, 98 (1969).

Pyenson, Lewis, "The Incomplete Transmission of a European Image: Physics at Greater Buenos Aires and Montreal, 1890-1920.", Proc of the Amer. Philosoph. Soc. 122:2, 92 (1978).

Roche, Marcel, "Early History of Science in Spanish America", Science 194, 806 (1976).

Sabato, Jorge, "Quantity versus Quality in Scientific Research (1): The Special Case of Developing Countries", Impact 20, 183 (1970).

Salam, Abdus, "The Isolation of the Scientist in Developing Countries", Minerva 4, 461 (1966).

Vlachy, Jan, "Publication Output of World Physics", Czech. J. Phys. B29, 475 (1979).

Wijesekera, R.O.B., "Scientific Research in a Small Developing Nation--Sri Lanka" Scientific World 1, 6(1976).

Scientometrics, Vol. 7, Nos 3–6 (1985) 165–176

APPLIED SCIENTOMETRICS: AN ASSESSMENT METHODOLOGY FOR DEVELOPING COUNTRIES

M. J. MORAVCSIK

*Institute of Theoretical Science, University of Oregon
Eugene, Oregon 97403 (USA)*

(Received May 8, 1984)

A United Nations sponsored project is described to formulate a practicable method for assessing the impact of science and technology in the developing countries and to propose further research to improve the development of such indicators. After a discussion of the importance of the project, the aims of science and technology are summarized, followed by the elements that need to be considered in such an assessment procedure, and the structure of the relationships among these elements. The first step in the assessment process is to make a map of the part of the system to be assessed. The types of indicators that can be used are then listed, and it is suggested that the status of these indicators is weak, especially with respect to their applicability to developing countries. It is proposed that a small number of specific pilot projects be undertaken to test the general ideas contained in the discussion and to experiment with novel kinds of indicators.

Introduction

This article caters to two dimensions of *Derek Price*'s interests. *Derek*, besides being a most influential person in the conceptual foundations of scientometrics, was also interested and active in the utilization of the results of scientometrics in the management of science and technology. He was asked to testify before committees of the U. S. Congress, and also interacted with the executive entities in the US government. He also had contact with international organizations pertaining to science. More details on these involvements emerge in connection with other contributions in this memorial volume.

The second aspect of *Derek*'s interest which will be commemorated in this article is the evolution of science in the developing countries. *Derek* had close contacts with scholars and scientists in various countries, including India, and showed much interest in my work pertaining to developing countries, including a public appraisal of my book on science development published in 1976.

On both these counts, I feel that *Derek* would be happy to see, among the contributions commemorating him, an article on practical science policy in the context of the developing countries. Coincidentally, the time is just ripe for making such a contribution. At the request of the Advisory Committee on Science and Technology for Development of the United Nations (ACSTD), the Centre for Science and Technology for Development of the United Nations, headquartered in New York, recently undertook a project with the dual aim of formulating a practical system of assessment for science and technology in developing countries and of formulating a research program for the coming years aimed at making improvements on systems of assessment in this context. This project began by a background paper created by a group of about a dozen knowledgeable and interested people from around the world. The paper was written by the coordinator of this group, and then was circulated, in three iteration, among the group for critique, additions, deletions, or modifications. The resulting paper[1] was then submitted to a one week long workshop in Graz, Austria, at which some 15—20 participants, mostly different from the original group, would discuss the paper and formulate final recommendations to be forwarded to ACSTD for specific action.

At the time this account is being written, the background paper has been completed, but the Graz workshop has not yet taken place. The background paper, for which I served as coordinator, is some 70 pages in length and is being circulated in such a report format. Its conclusions, however, may be of interest to a broader segment of the scientometric community. My contribution to the *Price* memorial issue of *Scientometrics,* therefore, will be a summary of this background paper.

The importance of the project

Assessment of scientific and technological activities is inherently essential for such activities and yet is often neglected. It is inherently essential because quality in science has a very skewed distribution. We know from *Lotka*'s law[2], for example, that the lifetime contributions of individual scientists have such a strongly skewed distribution, a subject the *Derek* was instensely interested in and which he interpreted in terms of a purely statistical model[3] dependent only on parameters internal to science itself. Unfortunatedly we have almost no evidence on whether the *same* skewness of distribution appears also in scientific communities in the developing countries.[4] Yet we do know that a strong skewness does occur there also. A similar disparity between input and output can also be observed with respect to scientific institutions. For example, while in most developing countries the lion's share of support for research goes into large governmental institutes, the output of these institutions is often disappointing compared to the output of much smaller and much less generously supported university research.

At the same time, science and technology are in our times key elements in development, whatever definition of development one subscribes to. Hence the functioning of science and technology and of their links with other aspects of human activity is a pivotal factor in the evolution of a country.

And yet, assessment of science and technology is very much neglected in the developing countries. The egalitrian dribbledown institutional funding schemes[5] often used there do not require assessment of projects either prior to funding or after it. The division between working scientists and technologist, on the one hand, and civil service personnel handling "science policy" but having little if any personal background in science or technology, on the other, makes such assessment even more problematical. The logistic means for such an assessment (bibliometric data bases, peer review teams, data on the details of industrial and other production, studies on the status of science in the minds of the population as a whole) are also often missing. Thus the curious paradox arises that those countries which are least affluent and hence presumably can least afford the resources invested into the development of science and technology pay the least attention to ascertaining that this investment actually produces benefits. For this reason, the project I am describing is timely and challenging from the points of view of both basic scientometrics and practical management of science.

The goals of science and technology

Assessment is the establishment of the extent to which a given segment of science and technology functions favorably in the context of a given set of goals and objectives. We therefore have to start with listing the goals and objectives of science and technology.

There are a very large number of these, since science and technology have impacts on almost all human activity, practical, abstract, material, social, intellectual, esthetic, or spiritual, and in each of these domains the impact can work toward a variaty of goals.

Yet this complex array of goals, objectives, and impacts can be brought under three primary headings. They are

(a) Science and technology are closely linked with material (economic) development.

(b) Science and technology are prime human aspirations in the 20th century.

(c) Science and technology have very important impacts on Man's view of the world and of his place, potential and role in this world.

In any particular assessment, therefore, one of the first tasks is to specify the goal or objective (one or several of them) from the point of view of which the assessment is to be made. It is one thing to assess Burmese biology from the point of view of cancer research and quite another to do the same from the point of view of rice production, or of the health practices of the population as a whole.

The elements to be considered

The assessment in any particular situation deals with a number of elements of a situation and of the links among them. What the elements are will depend on the particular scope and purpose of the assessment in question. For example, if we want to make an assessment of the contribution of Burmese biology to rice production in the country, the elements will include, among many other things, biology education in schools, the management of applied research institutes, the attitude of farmers toward science, logistics of distribution fertilizers and pesticides, etc. The various large scale elements that might enter in such an assessment are science, technology, production, the utilization of production, individual and national aspirations and attitudes, the prevalent world view, and others. On a more detailed level we deal with education, manpower, communication among scientists and technologists and with people outside science or technology, management, international connections, etc. It is often useful to divide the elements between that is local and what is international, since mechanisms may very well differ in these respects. In drawing up these elements other distinctions also come to the fore, such as that between "basic" research and "applied" research, between theoretical and experimental research, between scientific research and technological development work, etc. A clear delineation of the way these concepts are used is necessary in order to produce an assessment which is unambiguous and functional.

The structure of the relationship among elements

As mentioned earlier, the assessment involves elements together with the links among them, and in fact the investigation of these links is often the most important part of the assessment procedure. If, for example, we are interested in the impact of science on the solution of practical hydrological problems in a country, a crucial factor in such a picture may be the extent to which scientists with relevant capability interact with employees of the governmental water board. Thus the structure of the linkages connecting the elements is a primary focus of attention.

In virtually all cases the connections between the elements will be arranged in a multidimensional way.[6] Each element will be linked with a number of other elements, often both ways in a feed-back fashion, and almost anything will be linked, however indirectly, to almost anything else. This almost trivially sounding statement has two important implications.

First, we must avoid, even as an approximation, to think of the system to be assessed as a one-dimensional linear chain-like structure, in which an event has only one cause and in which cause and effect can be neatly separated. Although very prevalent in analyses of situations, such a one-dimensional substitution for the real situation is completely unrealistic. In actuality all we can say is whether something is a necessary condition for something else or not. It will practically never be a sufficient condition.

A consequence of this is that simple correlations will tend to be very weak and misleading indicators. A correlation may be found even when no unique causal relationship exists because the two events have a common necessary condition behind them, while correlations may not be found convicingly because although an event may be a condition for another one, the latter has other precursers also which all must occur in order for the event to take place. As an example, scientific efforts on rice research may be excellent in a country and yet rice production may falter because social and economic incentives for agricultural production may be absent. In such a multidimensional work we must go beyond simple correlations and try to understand the true causal relationship among the elements of the system we are considering.

The second consequence of multidimensionality is that an important initial task for a given assessment is to delimit the number of elements and links we will take into account. If we had a one-dimensional situation with a linear chain, this problem would hardly arise. In reality, however, we must make, at least as an initial approximation perhaps to be modified later, a guess as to which elements in the picture need to be included in the assessment.

An example may illuminate this problem.

The main agricultural crop of a country is being infested by a destructive pest. Efforts and resources are being invested into scientific research, technological development work, and industrial production to create a new pesticide specifically effective against this rare pest. Yet the problem is not alleviated. An assessment is initiated to investigate the situation. It includes an evaluation of the scientific community, the technological manpower, the production facilities, the distribution outlets, and the fields. It is established that the scientists were effective in creating a pesticide proven potent in laboratory tests, the technologists did well in working out the production technology of the new substance, the factories produced the material with sufficient quality control, the distribution outlets were reasonably successful in selling the pesticide, and the farmers did spray their fields. Clearly, something was left out of the set of elements and links that should have been included in the assessment, since every segment of the assessment was positive and yet the problem remained. Had the initial delineation of the network to be assessed been done with more perception, it would have been discovered that the farmers could receive no financing for their purchases of the new pesticide, hence could afford to buy only insufficient amounts which therefore were used in an overdiluted solution thus making the effort ineffectual.

In this example the solution of the puzzle is not very difficult, in that one can relatively easily pinpoint the missing elements and links to be included in the assessment. In other instances, however, this may be quite difficult, especially if the missing elements and links are intangible factors like morale, motivation, personal or collective discoragement, personal animosities between two key figures in the network, etc. The situation is made more complex by the fact that the "right" outcome of the assessment is not always known: Often it is not a matter of finding the causes of a specific and conspicuous failure somewhere in the network, but simply a desire to determine the unknown efficiency of a system. In that case there may be no signals indicating that important factors were left out.

The first step: The making of the map

In view of the foregoing, the first concrete step in our assessment procedure is to prepare, in view of the announced objective of the assessment and in possession of the elements and links to be considered, a map of the elements and the linkages among them. Such a geometrical analogue of the network to be investigated is helpful in visualizing and concretizing the tasks to be tended to. Circles denote the elements and lines connecting the circles depict the links.

Maps like this can be prepared on various scales and with various amounts of detail. Let us again take our example of the infestation of the crops discussed in the previous section. On the most "macro" level science would be denoted by one circle, technology, by another, production by a third, distribution by a forth, financing by a fifth, the farming community by a sixth, etc. On a more "micro" level the circle of science may be subdivided and narrowed down into a few smaller circles indicating specific biological institutions or individual biologists, or they may indicate the elements of scientific activity such as education, manpower, communication, etc. Similarly the other large circles in the macromap would also be subdivided and narrowed down. In doing so we refine and increase the number of links also.

In this process it may often be useful to follow the procedure of a televison repairman who, when faced with a faulty set, would first perform a "macroinvestigation", trying to determine in which of the 6 or 7 large sections of the set the fault resides. This would correspond to our performing an assessment using a macro map. Once the faulty section is located, it may be investigated in greater detail, corresponding to our using a more micro map for the second step of the assessment. In reality, the TV repairman often does not bother continuing this procedure of nested and increasingly more detailed investigations, but simply replaces a whole faulty section (in which 99% of the parts are perfectly healthy). In science policy this cannot be done, and hence the sequence of increasingly more micro assessments need to be continued until the faulty elements or links in the system are located.

Types of indicators

Now that we have constructed our map with elements and links, we need to find indicators to measure the working of the elements and the linkages. Since there is a multitude of aspects of the various elements and linkages depending on the objectives of the assessment, it would be both impossible and constraining to try to list all the various indicators that may prove useful in various situations. It is, however, useful to enumerate *types* of indicators so as to create a conceptual framework in which the creation of indicators for a particular purpose is facilitated.

Indicators can be classed according to whether they measure *input* or *output*. This is useful because in general it is much easier to find indicators for input than for output. For example, the output of science is an intangible commodity (namely scientific knowledge), while the input to it are manpower, finances, buildings, equipment, communication facilities, and other rather tangible elements. The ultimate aim of an assessment is, however, the gauging of output for the purposes specified in the assessment.

Indicators can be *quantitative* or *qualitative*, with some gradations in between. There are situations when exact numerical assignments are not possible or meaningful but where a comparative ranking can be made, which, after all, is also a quantitative activity. Whether the science of science is quantifiable is hotly debated even in the academic circles of the study of science, something that readers of *Scientometrics* need not be reminded of.

Indicators can measure[7] *activity, productivity,* or *progress.* Activity is simply the expenditure of energy without necessarily resulting in movement in the "right" direction. Productivity is the extent to which such activity is oriented in the desired direction. Finally, progress is measured by the fractional extent to which the final goal is attained.

Indicators may also measure[8] *quality, importance,* or *impact.* Quality denotes excellence in terms of internal, not necessarily goal-oriented criteria. Importance denoted the potential influence on the aim in question, whereas impact measures the actual results attained in connection with the goal in question. These two trichotomies have contributed to the conceptual sharpening of the procedures in the case of real and very specific assessment efforts.

Indicators can also be divided into *functional* and *instrumental* ones. The latter evaluate the *tools* used in the scientific or technological work aimed at a given purpose, while functional indicators measure the actual *functions* performed by such work. For example, improving the self-reliance of the population may be a function of science, while evaluating a system of roving mechanics to repair equipment could better be considered an instrumental indicator. The distinction is not always clear and yet often helpful.

Finally and perhaps most importantly, indicators can be classed according to whether they are *"data-based"* or *"perceptual".* The former type is sometimes called, quite misleadingly, objective indicator, and includes, for example, bibliometric measure, patent counts, production statistics, counts of literate persons, etc. The class of "perceptual" indicators implies a personal evaluation, by inspection, on the part of knowledgeable investigators of the particular situation to be assessed. This kind of assessment is often referred to as "peer review."

Because of the multidimensional nature of the structure of elements and links, in almost all situations that we want to assess, we have to work with not a single indicator but an array of such indicators, each representing one element or one link, and often in fact several indicators will be needed to characterize the state of a given element or a given link. It is conceptually impossible and hence practically erroneous and misleading to compose, out of this set of indicators, one single measure to describe the situation. Thus the outcome of all assessments should be a complex composite of statements and results, reflecting information on all the relevant aspects of the situation. For this reason alone, not even speaking of others to be discussed, it is not possible to devise an assessment procedure which is automated and simplified in such a way that a lower level administrator, simply by following a set of recipes, could carry out an assessment. In as much as one aim of the project I am reporting on was to formulate a realistic assessment system that could immediately be applied to problems of science and technology in the developing countries, this above statement is important since it definitely circumscribes the nature of such a system even before further details are discussed: The system will be complex and will have to be administered not mechanically but creatively by a group of knowledgeable and imaginative people.

The status of indicators

In view of the above requirements and the typology of indicators, what is the present situation with regard to the availability of such indicators in the context of the developing countries?

Perhaps the best developed type of indicators is the scientometric one. As the readers of *Scientometrics* well know, such indicators have been used widely both for assessment and for research on the science of science. There are, however, some problems with using such indicators in the context of developing countries. For one thing, there is some question and controversy about the extent to which the computerized data base by the Institute of Scientific Information, which is the basis of almost all scientometric work, includes work performed in the developing countries. That data base includes only a small fraction of the scientific *journals* around the world (less than 10%), though at the same time it does include a large percentage (around three-quarters) of all scientific *articles* published. It is claimed, however, and probably correctly, that among the journals and articles omitted there is a very high fraction published in the developing countries. On the other hand, the selection of which journals to scan for the data base is made on the basis of an objective criterion, which involves mainly the citation rating of the particular journal. In a rejoinder, it is claimed that citation rating of journals and their being listed in the data base form a vicious circle into which it is difficult to break in from the outside. The resolution of this controversy involves a number of very interesting issues, for example the determination of quality of a scientific article by means other than citations. In any case, a workshop is now in progress, which, by mid-1985, will have studied this general problem area and then will make and implement some recommendations for its resolution.

Bibliometric indicators, however, have other possible deficiencies also when used in the context of developing countries. They measure parameters largely internal to the scientific community, and do not necessarily reflect on links with technology, with aspirations, or with the forming of the world view. In particular, they do not include modes of written communications that are different from books and

journal articles, such as reports, workshop notes, etc., which may be much used
in transmitting scientific material to people outside the sciences.

Perceptual indicators, that is peer reviewing, are also on shaky grounds in the
context of developing countries. Scientific communities in individual countries
are mostly too small to allow internal peer reviewing. On the other hand, developing
countries are often reluctant to turn to external peer reviewing, because those
in charge of science policy are not acquainted sufficiently with the worldwide
scientific community to know whom to ask, because going outside the country
is thought to reflect on the national sovereignty, because there is fear that the
request for outside help will be rebuffed, and because financial resources may
not be available to organize such external peer reviews. In some cases external
review teams arranged through international organizations turned out to be both
ignorant of and insensitive to the local conditions under which scientific work
must be performed in the developing countries.

To look at the problem area from a different point of view, it can be stated
that in general input indicators pertaining to science and technology are much more
readily available in the developing countries than output indicators. Statistical
information on financial resources to be spent, on trained manpower, on the
buildings built, on the equipment purchased or received, etc. can be had easily
from national development plans or from the filing cabinets of science management agencies. To be sure, such data will state only quantitative information, and
will say little if anything about quality. One will know the number of scientists
(subject to vagaries of definitions and other uncertainties), but not the quality of
their background, expertise, or output. One will know the number of pieces of
equipment, but not whether they are inoperative or not due to a lack of a technician,
the lack of spare parts, or the lack of a competent or interested user. It is in these
areas of quality where peer review is invaluable and indispensible.

To look at the other end of the linkages, economic statistics may very well be
available on the various areas of production in the country. This is likely to
indicate, again, only quantity and not quality. Furthermore, as said earlier, simple
correlations between these economic output statistics and the scientific input
statistics would give next to no information at all between the linkages between
science and production. Here also, perceptional methods looking in detail at the
causal linkages are necessary.

The situation is even less in control when we examine the impact of science
on national or individual aspirations or on national and individual attitudes
through an influence on the world view. Whereas in the context of, for example,
the United States we now begin to have some indicators pertaining to this aspect
of science[9], I know of no analogous information about developing countries.

In summary, it is fair to say that considering the complex system of elements and interconnections in the picture of science, technology and its areas of impact in the developing countries, and considering the sophisticated procedure that appears to be necessary to undertake a meaningful assessment, the tools that are currently available are woefully rudimentary.

A blueprint for the future

The conclusions thus far can be summarized as follows: Making an assessment of the effects of science and technology in some particular context is a very sophisticated task the details of which will depend very much on the specific circumstances of the particular assessment which, therefore, will have to be specified carefully and clearly. Thus, in general, one can only describe the broad outlines of the procedure to be followed. The actual methodology for the assessment will be different in each case. Correspondingly the repertory of tools that can be used in such assessments is virtually limitless and in many cases needs to be improvised after the situation is specified. Some tools are readily available, but even those may need to be adapted and developed for the purposes of an effective use in the context of developing countries.

The strategy for the future follows quite naturally from these conclusions. The primary task is to try out the general procedure described above on a number of concrete pilot studies. Indeed, the recommendation of the report is that a small number of such pilot projects be initiated as soon as possible, independently of each other, each dealing with a different type of specific situation in which some effects of science and technology need to be assessed. It would be very advisable to select situations which pertain to quite different segments of the network into which science and technology are embedded. One such situation may deal with the influence of science and technology on people's attitudes and their world view, while another may be directed toward the effect of science on industrial or agricultural production. One situation may involve a very "micro" segment, another may be broad and general. There should also be a geographic distribution in the locale of these projects.

Each such project should be placed in the hands of a small group of independent researchers in the form of a grant from an international, national, or private organization, the grant being awarded by competition after the general outlines of the assessment project to be carried out are specified. It is important that large agencies not be directly involved in the performance of pilot projects of actual assessment, and that each such grant should also contain a "theoretical" research component which would encourage the generation of new assessment methods and the improvement of older ones. These innovations could then be immediately utilized in the operational part of the same grant.

At a time when both international and national organizations are straining under financial constraints, real or imagined, substantive or only bureaucratic, it is also important to state that the funds needed for pursuing the type of research and implementation program are very modest indeed. As mentioned, 3–5 pilot projects during the next few years would be sufficient for a start. Counting, say, $200.000 for the *total* cost of *each* of these projects, the *annual* expenditure needed for all projects together would be the same order of magnitude. With cooperation among international, national, and private organizations, such a sum should represent no serious obstacle. If this article in any way increases the probability of such a set of programs being actually implemented, my aim in writing it will certainly have been attained, and, I feel, *Derek Price,* the par excellence combination of the theorist and empiricist in scientometrics, would have been equally satisfied.

References

1. "An Assessment Scheme for Science and Technology for Comprehensive Development", Working Paper GRAZ/P. 3 for the Panel on Indicators of Measurement of Impact of Science and Technology on Socio-economic Development Objectives, Graz, Austria, May 2–7, 1984. Copies available from the Centre for Science and Technology for Development, Room 1040, 1 UN Plaza, New York 10017, USA.
2. A. J. LOTKA, *Journal of the Washington Academy of Science* 16 (1926) No. 12, 317.
3. D. deS. PRICE, Journal of the American Society for Information Science 27, 292 (176)
4. See e.g. P. G. BLASCO, La produccion cientifica espanola de1965 a 1970. Un estudio comparado. *Revista Mexicana de Sociologia* 37 (1975) No. 1; W. O. AIYEPEKU, The productivity of geographical authors – A case study from Nigeria, *Journal of Documentation* 32 (1976) No. 2 105.; T. SAREVIC, Evaluation and potential use of the data bank at the Brazilian Institute of Bibliography and Documentation IBBD, UNESCO report 3055/RMO. RD/DBA, Paris, 1974.
5. M. J. MORAVCSIK, *How to Grow Science,* Universe Books, New York 1980, Chapter 11.
6. M. J. MORAVCSIK, *Scientometrics,* 6 (1984) 75.
7. M. J. MORAVCSIK, *Research Policy,* 2 (1973) 256.
8. B. R. MARTIN, J. IRVINE, Assessing Basic Research: Some Partial Indicators of Scientific Progress, Research Policy (in press)
9. See, for example, *Science Indicators 1982.* National Science Foundation, Washington, 1983, pp. 143–162.